Evolutionary Biology – New Perspectives on Its Development

Volume 3

Evolutionary biology has been a remarkably dynamic area since its foundation. Its true complexity, however, has been concealed in the last 50 years under an assumed opposition between the "Extended Evolutionary Synthesis" and an "Alternative to the Evolutionary Synthesis". This multidisciplinary book series aims to move beyond the notion that the development of evolutionary biology is structured around a lasting tension between a Darwinian tradition and a non-Darwinian tradition, once dominated by categories like Darwinian Revolution, Eclipse of Darwinism, Evolutionary Synthesis, and Post-Synthetic Developments.

The monographs and edited volumes of the series propose an alternative to this traditional outlook with the explicit aim of fostering new thinking habits about evolutionary biology, a multifaceted area composed of changing and interacting research entities and explanatory levels. Contributions by biologists and historians/philosophers are welcomed. Topics covered in the series span from (among many other possibilities):

- An Overview of Neutralist Theories in Evolutionary Biology
- Developmental Biology: From Reductionism to Holism and Back
- Selection Theories Beyond Hard and Soft Inheritance
- Divergent, Parallel, and Reticulate Evolution: Competing or Complementary Research Programs?
- The Rise of Molecular Biology: Between Darwinian and Non-Darwinian
- Biologizing Paleontology: A Tradition with Deep Historical Roots
- The Darwinian Revolution and the Eclipse of Darwinism: Blurring the Historiographical Lines
- Darwinism, Lamarckism, Orthogenesis: Can We Really Define Them by Their Hard Explanatory Cores?
- The Evolutionary Synthesis: A Fabricated Concept?
- The Opposition to the Evolutionary Synthesis: Criticizing a Phantom?
- A Reversed Perspective: Approaching Charles Darwin from the Pre-1859 Period
- The Long Development of the Multilevel Paradigm in Evolutionary Biology
- Self-Organization: A Research Tradition from Morphology to Cosmology
- Human Evolution: Sociobiological or Sociocultural?

More information about this series at http://www.springer.com/series/16175

Richard G. Delisle
Editor

Natural Selection

Revisiting its Explanatory Role in Evolutionary Biology

Springer

Editor
Richard G. Delisle
School of Liberal Education
and Department of Philosophy
University of Lethbridge
Lethbridge, Alberta, Canada

ISSN 2524-7751 ISSN 2524-776X (electronic)
Evolutionary Biology – New Perspectives on Its Development
ISBN 978-3-030-65535-8 ISBN 978-3-030-65536-5 (eBook)
https://doi.org/10.1007/978-3-030-65536-5

Cover photo by GLady is licensed under CC0
https://pixabay.com/photos/mosaic-fish-tile-art-ceramic-200864/

This Springer imprint is published by the registered company Springer Nature Switzerland AG.
The registered company address is: Gewerbestrasse 11, 6330 Cham, Switzerland

Contents

Chapter 1
Introduction: In Search of a New Paradigm for the Development of Evolutionary Biology

Richard G. Delisle

Abstract This Introduction raises a number of interpretative difficulties facing the standard view of the development of evolutionary biology. It challenges the central tenet of this view, the claim that the field has largely been organized around a fundamental divide, comprising, on the one hand, theories focusing on a strong selective approach and, on the other, theories embracing a weak selective one. It is argued that the main historiographic labels—Darwinism, Darwinian Revolution, Eclipse of Darwinism, Modern Synthesis, Extended Evolutionary Synthesis, Non-Darwinian Synthesis—are increasingly less clearly supported by historical, epistemological, theoretical, and empirical analyses. The co-optation of historians and philosophers under the rhetorical discourses of a limited number of influential evolutionists has, apparently, played a key role in the persistence of a static and uncritical historiography. This Introduction calls for a new and more consistent paradigm that would make sense of the overall development of evolutionary biology, one based on a realignment of the alliance between all partners pursuing research in this area.

Keywords Darwinism · Darwinian revolution · Eclipse of Darwinism · Modern synthesis · Extended evolutionary synthesis · Non-Darwinian synthesis · Selective theories · Rhetorical arguments · Co-optation

For several decades now, the field of evolutionary biology has been envisioned as organized around a profound and fundamental divide: theories relying on strong selective factors and those appealing to weak ones only. Of course, it is also believed that the first ones are in keeping with the true Darwinian spirit, unlike the other theories. On closer analyses—empirical, theoretical, epistemological, and

---

R. G. Delisle (✉)
School of Liberal Education, University of Lethbridge, Lethbridge, AB, Canada

Department of Philosophy, University of Lethbridge, Lethbridge, AB, Canada
e-mail: richard.delisle@uleth.ca

R. G. Delisle (ed.), *Natural Selection*, Evolutionary Biology – New Perspectives on Its Development 3, https://doi.org/10.1007/978-3-030-65536-5_1

historical—this divide proves to be largely illusory and is collapsing at a rapid pace, opening up an era devoted to the search of a new paradigm for the understanding of the development of evolutionary biology.[1] Theories favouring the agenda of strong selective factors are unquestionably legitimate (Alcock 2001, 2017; Dawkins 1976, 1996; Dennett 1995; Mayr 1983; Wilson 1975; Wray et al. 2014) but are both mainly the product of fairly recent developments and part of a minority view. The widely held assumption that the adaptationist programme (or "pan-selectionism") finds its intellectual roots in Charles Darwin's *Origin of Species* is not supported by recent analyses (Delisle 2014, 2017a, 2019).

As shown by several contributions in this volume,[2] "Darwinism" as a historiographic category is rapidly approaching a definition crisis.[3] The concept of natural selection does not clearly mark out a distinction between *most* evolutionists from the nineteenth century onwards, even for so-called Darwinians, no matter how acrimonious debates might be today among evolutionists. For instance, Richard Delisle (Chap. 4) holds that Darwin could not have put natural selection in the driver's seat of evolution as a core explanatory element, for the simple reason that Darwin severely restricted its action under a pre-established overall divergent view of life, reducing it to the explanatory role of an auxiliary hypothesis. Carlos Ochoa (Chap. 10) shows how George Gaylord Simpson reintroduced the notion of orthogenesis by the back door—although he officially opposed it in the name of a contingent evolution driven by natural selection—through the concept of parallelism. Georgy Levit and Uwe Hossfeld (Chap. 9) establish strong ties between Theodor Ziehen and Bernhard Rensch, both working under a monistic and law-like universe reducing, in the case of Rensch, natural selection to an important yet subordinate role (see also Levit et al. 2008). Mark Adams (Chap. 8) analyses how a score of scholars declined to commit themselves to a basic assumption presumably at the heart of the Modern Synthesis—the microevolution/macroevolution equation—depriving natural selection of its creativity for macroevolutionary events. They avoided making this assumption in various ways: (1) some scholars denied this equation (DeVries, Johannsen, Filipchenko, Goldschmidt, Guyénot, and J.S. Huxley); (2) others failed to take the question seriously (Fisher and Wright); (3) while the others recognized the lack of evidence for it (Severtsov and Rensch). In fact, scholars who supported such an equation often did it on the basis of a programmatic principle only, using various rhetorical strategies to hide their reservations (Haldane, Simpson, and Dobzhansky). Jitse van der Meer (Chap. 11) throws

---

[1]The word "paradigm" in this chapter is used in its ordinary and non-technical sense.

[2]Of course, the opinions expressed here are only mine. The readers are urged to review all the chapters contained in this volume and judge for themselves. It would be impossible for me to do justice here to the many stimulating insights each author brings to it.

[3]I have argued elsewhere that the time has perhaps come to dispose of the label "Darwinism" altogether (Delisle 2017a: 157). It seems to me that this move is warranted, if only because at our current state of understanding, such a label conflates major issues rather than revealing them. For a similar conclusion, but based on a different argument, see also the comment of Mark Adams (Chap. 8, footnote 14).

more light on this state of affairs as it appears in Dobzhansky, finely analysing the various explanatory levels and mechanisms he jostled with, thus depriving natural selection of its sole and dominant role in evolution. Given these considerations, it is less and less clear what the Modern Synthesis really is.

If the views of Charles Darwin and of a number of so-called proponents of the Modern Synthesis are not quite as presented in the traditional historiography (see also Cain 2009a; Delisle 2008, 2009a, b), then what is "Darwinism" all about? The question raised by Lorenzo Baravalle (Chap. 15) in this volume seems to me most relevant: "Darwinism Without Selection?". The intellectual space seems quite limited (Delisle 2017b), if non-existent, between, on the one hand, "Darwinians" who would deny the centrality of natural selection in evolutionary explanations and, on the other, "non-Darwinians" who use it in company with a number of other evolutionary mechanisms, as in the case of Henry Fairfield Osborn's efforts to build a synthetic biology (Ceccarelli, Chap. 7). Have we been far astray in our understanding of the development of evolutionary biology? "Co-opting" historians of biology in the service of evolutionary biologists seems to have been part of the problem, in addition to the fact that scientists have actively engaged in self-promotional arguments (Adams, Chap. 8 and Schwartz, Chap. 12).[4]

1.1 The Domino Effect

The professionalization of the fields of history and philosophy of biology arose at the time when the "Modern Synthesis" was widely thought to have achieved a unification of evolutionary disciplines. Ernst Mayr (1980) and Stephen Jay Gould (1980a), for instance, characterized this synthesis as organized around two main explanatory components: (1) gradual evolution is explained by small genetic changes (variations) oriented by natural selection, a process leading to adaptation; (2) evolutionary trends and speciational events are macroevolutionary phenomena explainable solely in terms of the extension of processes and mechanisms occurring at the preceding microevolutionary level. On this view, natural selection holds a central explanatory role in evolutionary theory. Strangely enough, historians and philosophers uncritically bought into this self-serving narrative (Adams, Chap. 8), perhaps for reasons more sociological than scientific, under the implicit desire for a wider recognition of their newly acquired status among scientific disciplines.[5] Thus was the historiographic category "Modern Synthesis" canonized, externally and artificially reinforced in the context of post-World War II ideological and religious pressures

[4]For years I was myself entirely co-opted under labels, such as "Darwinian Revolution" and "Modern Synthesis". See also a somewhat similar admission by Mark Adams (Chap. 8).

[5]The co-optation of historians and philosophers by evolutionary biologists promoting the "Modern Synthesis" constitutes an essential research topic for future "critical" historians and philosophers.

in favour of a pro-science agenda in the United States (Adams, Chap. 8; Smocovitis 1999).

It would appear that, from this unassailable citadel, the other labels created to organize the development of evolutionary biology were either *interpolated* or *extrapolated.* The Modern Synthesis needed historical roots for both prestige and credibility, so Charles Darwin was placed in command of a "Darwinian Revolution". Clearly, the argument went, natural selection was the core of Darwin's theory. From there, it was easy enough to imagine a label for recalcitrant scholars who lived in a no-man's land somewhere in between, called the "Eclipse of Darwinism". The logic was pushed to its ultimate conclusion through the creation of a category for post-synthetic developments, defined more or less in opposition to the one incarnated by the Modern Synthesis. Depending on whether or not one sees himself or herself as departing radically from this older label, one might belong either to the "Extended Evolutionary Synthesis" or the "Non-Darwinian Synthesis".

It is only in hindsight that one can tell how fragile this historiographic edifice was, having only the Modern Synthesis as a foundation. The issue being raised here is not merely whether or not the category "Modern Synthesis" should be kept intact, expanded, or replaced. The question might be more gripping, even existential: does history show us there ever really was a Modern Synthesis, as traditionally defined? Whatever happened between 1930 and 1960 in evolutionary biology, and once one goes beyond the rhetorical discourses, what is left at the conceptual level doesn't appear quite as originally advertised. It remains to be investigated how to fill the void created by a reconceptualization of that time period. And this can only be achieved at the expense of rethinking the entire development of evolutionary biology. Keeping in mind that all the labels listed above originated from the starting point of the "Modern Synthesis", one sees why removing such a central piece instigates in its wake a redefinition of others. Indeed, what would an "Extended Evolutionary Synthesis" look like if there had been no Modern Synthesis to extend in the first place? If Darwin's image is the product of looking through the distorting lens of the Modern Synthesis, then what were Darwin's real achievements? Finally, if the "Eclipse of Darwinism" was invented merely to fill the void between the "Darwinian Revolution" and the "Modern Synthesis", what was such an interstitial moment really about?

It may be tempting to argue that it does not matter whether or not the "Modern Synthesis" had (or has) a historical reality, provided that the perception was (is) that it did (does) exist and that scholars behave(d) accordingly. Fair enough, as long as the issue is the analysis of the impact of rhetorical argument on science. Parenthetically, Ernst Mayr (1993: 32) had this to say of the Modern Synthesis in retrospect: "Historians (perhaps even Mayr and Provine) have overemphasized the unity achieved by the synthesis". It does matter, however, whether or not the Modern Synthesis did exist for both historians and evolutionists currently thinking about the field. For the latter, there would be little point asking for greater diversity and tolerance of views if a part of that pluralism already existed in the past. One must be careful not to make a straw man of the "Modern Synthesis" and thus engage in a self-promotional narrative stressing the so-called novelties of recent views. In my understanding, pluralism has prevailed throughout the history of the field to a degree sufficient to allow us to give credit to scholars of the past.

1.2 A New Alliance

As strange as it seems, the future of evolutionary biology may in part belong to a re-discovery of its past. The richness of the area since its inception—both in scholars and ideas—is plainly overwhelming and largely understudied (e.g. Amundson 2005; Esposito 2013, 2017; Gissis and Jablonska 2011; Laubichler and Maienschein 2007; Levit and Hossfeld 2017). The inspiration sought by contemporary biologists in pursuit of greater explanatory pluralism can be accelerated, I would argue, by the rediscovery of older ideas capable of reformulation. Just as no one can legitimately claim that solutions to all current and future problems are to be found in past proposals, neither can it be denied that re-engaging with them may lead to some fruitful inspirations.

There is something admirable about Ernst Mayr's (1982) and Stephen Jay Gould's (2002) way of doing science: they realized the necessity of also engaging with the past of their discipline. Of course, this comes with the inevitable risk of self-serving strategies. This is exactly what has happened. For instance, Mayr's (1982: 45–47; 1991: 40–42) attempt to read the development of evolutionary biology through the lens of "population thinking" is markedly tendentious (Greene 1992, 1999; see also Sepkoski 2019), as is Gould's (1980b) notion of a so-called constriction of the Modern Synthesis (Adams, Chap. 8 and van der Meer, Chap. 11; Cain 2009b; Sepkoski 2019). And things only get worse when historians and philosophers become co-opted by such arguments. Maurizio Esposito (Chap. 2) provides us with a long-overdue reflection on the conflicting and complementary agendas of the two camps involved (scientists and historians/philosophers).

Whereas historians and philosophers have been slow at questioning the traditional historiography, practicing biologists had to go through it alone when rereading the development of evolutionary biology, busy as they were undermining the traditional narrative and uncovering a concealed pluralism. Jeffrey Schwartz (1999; Chap. 12) does this for evolutionary biology in general. Ian Tattersall (1995; Chap. 14) is engaged in a similar quest with respect to the field of human evolution. Some of these studies take up the specific angle of reinterpreting the contribution of past scholars, as is done by Carlos Ochoa (Chaps. 6 and 10) with respect to William Bateson and George Gaylord Simpson (see also Ochoa 2017). For his part, Andrei Granovitch (Chap. 13) requires more intellectual space to account for other evolutionary processes. Perhaps the issue of a mythical Darwin created by recent scholars should be raised, a question that impinges on how and what should be taught about natural selection where wider audiences are concerned (Watts, Chap. 16). Indeed, natural selection's epistemological role in science was not equally agreed upon even among Darwin's contemporaries (Bellon, Chap. 3).

Just as Maurizio Esposito (Chap. 2) rightly stresses that scientists and historians-philosophers do not have identical agendas, it is hoped that *a new alliance* will be forged that will generate an overall narrative that is, at least, *consistent* throughout. Avoiding the co-optation problem raised above, a new motto might be "no submission, no contradiction". We should not be naive about the difficulty of the task, but the aim of working our way toward a coherent and unified narrative by "moving across the aisle" seems to me a desirable goal.

1.3 A New Paradigm: Seeking New Historiographic Labels

More and more studies show that the development of evolutionary biology was never organized around a central and dominant narrative called Darwinism, as currently defined. Considering that natural selection was not a discriminatory factor between most scholars, it seems difficult to believe that the labels created from within the traditional historiography can justifiably remain unmodified. It may even be necessary to eventually replace them entirely:

The Darwinian revolution: Far from being a scholar who subscribed to modern evolutionism, Charles Darwin shared many uniformitarian commitments (steady-statism, to name but one) with contemporaries, such as Charles Lyell, Thomas H. Huxley, and Richard Owen, making him a less-than-ideal candidate for a torchbearer (Delisle 2019; Chap. 4). In addition, Darwin's eclecticism with respect to evolutionary mechanisms matches those encountered among Darwin's contemporaries and immediate successors (Levit and Hossfeld 2011; Chap. 5).

The eclipse of Darwinism: The eclecticism just mentioned spills over into later decades, with natural selection now being taken seriously by most evolutionists but only as one factor among others (Ochoa, Chap. 6 and Ceccarelli, Chap. 7; Delisle 2017a; Largent 2009; Levit and Hossfeld 2006).

The modern synthesis: Serious doubts have been raised about both the internal/conceptual coherence of such a synthesis and its external/contextual isolation from the rest of the evolutionary field, the label also serving strong sociological ends (Adams, Chap. 8; Ceccarelli, Chap. 7; Cain 2009b; Delisle 2008, 2009a, b, 2011; Levit et al. 2008; Smocovitis 1999; Sepkoski 2019).

***Extended evolutionary synthesis* or *non-Darwinian synthesis*:** In light of the profound reassessment of the pre-1960 period currently underway, it is unclear to me what content should be attributed to these recent labels. It seems that only the piecing together of the emerging views concerning the pre- and post-1960 periods will allow us to give them any meaning.

The "mechanism-centered" approach exploited thus far as a way of organizing the development of evolutionary biology has proven far too restrictive to capture its essential features. This approach is also closely tied to a view that consists in reducing the complexity of the scientific enterprise to mere "theories" (see Chap. 3 by Bellon, who finds supports for his view in the analyses of Jane Maienschein

(1991) and Philip Pauly (2000)). It seems that we will have to cast our empirical, theoretical, epistemological, and ontological net wider. The next step in our collective inquiry, I suggest, lies in finding a new paradigm to make sense of that complexity.

Acknowledgements I thank James Tierney (Yale University) for assistance in improving the quality of my English.

References

Alcock J (2001) The triumph of sociobiology. Oxford University Press, Oxford
Alcock J (2017) The behavioral sciences and sociobiology: a Darwinian approach. In: Delisle RG (ed) The Darwinian tradition in context: research programs in evolutionary biology. Springer, Cham, pp 37–59
Amundson R (2005) The changing role of the embryo in evolutionary thought. Cambridge University Press, Cambridge
Cain J (2009a) Rethinking the synthesis period in evolutionary studies. J Hist Biol 42:621–648
Cain J (2009b) Ritual patricide: why Stephen Jay Gould Assassinated George Gaylord Simpson. In: Sepkoski D, Ruse M (eds) The Paleobiological revolution. University of Chicago Press, Chicago, IL, pp 346–363
Dawkins R (1976) The selfish gene. Oxford University Press, Oxford
Dawkins R (1996) Climbing mount improbable. WW Norton, New York
Delisle RG (2008) Expanding the framework of the holism/reductionism debate in neo-Darwinism: the case of Theodosius Dobzhansky and Bernhard Rensch. Hist Philos Life Sci 30:207–226
Delisle RG (2009a) Les philosophies du néo-darwinisme. Presses Universitaires de France, Paris
Delisle RG (2009b) The uncertain foundation of neo-Darwinism: metaphysical and epistemological pluralism in the evolutionary synthesis. Stud Hist Philos Biol Biomed Sci 40:119–132
Delisle RG (2011) What was really synthesized during the evolutionary synthesis? A historiographic proposal. Stud Hist Philos Biol Biomed Sci 42:50–59
Delisle RG (2014) Evolution in a fully constituted world: Charles Darwin's debts towards a static world in the Origin of Species (1859). Endeavour 38:197–210
Delisle RG (2017a) From Charles Darwin to the evolutionary synthesis: weak and diffused connections only. In: Delisle RG (ed) The Darwinian tradition in context: research programs in evolutionary biology. Springer, Cham, pp 133–168
Delisle RG (2017b) Introduction: Darwinism or a kaleidoscope of research programs and ideas? In: Delisle RG (ed) The Darwinian tradition in context: research programs in evolutionary biology. Springer, Cham, pp 1–8
Delisle RG (2019) Charles Darwin's incomplete revolution: the Origin of Species and the static worldview. Springer, Cham
Dennett D (1995) Darwin's dangerous idea. Simon & Schuster, New York
Esposito M (2013) Romantic biology, 1890-1945. Pickering & Chatto, London
Esposito M (2017) The organismal synthesis: holistic science and developmental evolution in the English-speaking world, 1915–1954. In: Delisle RG (ed) The Darwinian tradition in context: research programs in evolutionary biology. Springer, Cham, pp 219–242
Gissis SB, Jablonska E (eds) (2011) Transformations of Lamarckism: from subtle fluids to molecular biology. MIT Press, Cambridge, MA
Gould SJ (1980a) Is a new and general theory of evolution emerging? Paleobiology 6:119–130

Gould SJ (1980b) GG Simpson, paleontology, and the modern synthesis. In: Mayr E, Provine WB (eds) The evolutionary synthesis: perspectives on the unification of biology. Harvard University Press, Cambridge, MA, pp 153–172
Gould SJ (2002) The structure of evolutionary theory. Belknap, Cambridge, MA
Greene JC (1992) From Aristotle to Darwin: reflections on Ernst Mayr's interpretation in the growth of biological thought. J Hist Biol 27:257–284
Greene JC (1999) Debating Darwin: adventures of a scholar. Regina Books, Claremont, CA
Largent M (2009) The so-called eclipse of Darwinism. Trans Am Philos Soc (New Ser) 99(1):3–21
Laubichler MD, Maienschein J (eds) (2007) From embryology to evo-devo: a history of developmental evolution. MIT Press, Cambridge, MA
Levit GS, Hossfeld U (2006) The forgotten "Old-Darwinian" synthesis: the evolutionary theory of Ludwig H. Plate (1862–1937). NTM 14:9–25
Levit GS, Hossfeld U (2011) Darwin without borders? Looking at "generalised Darwinism" through the prism of the "hourglass model". Theor Biosci 130:299–312
Levit GS, Hossfeld U (2017) Major research traditions in 20th century evolutionary biology: the relations of Germany's Darwinism with them. In: Delisle RG (ed) The Darwinian tradition in context: research programs in evolutionary biology. Springer, Cham, pp 169–194
Levit GS, Simunek M, Hossfeld U (2008) Psychoontogeny and psychophylogeny: Bernhard Rensch's (1900–1990) selectionist turn through the prism of panpsychistic identism. Theor Biosci 127:297–322
Maienschein J (1991) Transforming traditions in American biology, 1880-1915. Johns Hopkins University Press, Baltimore, MD
Mayr E (1980) Prologue: some thoughts on the history of the evolutionary synthesis. In: Mayr E, Provine WB (eds) The evolutionary synthesis: perspectives on the unification of biology. Harvard University Press, Cambridge, MA, pp 1–48
Mayr E (1982) The growth of biological thought. Belknap, Cambridge, MA
Mayr E (1983) How to carry out the adaptationist program? Am Nat 121:324–334
Mayr E (1991) One long argument: Charles Darwin and the genesis of modern evolutionary thought. Harvard University Press, Cambridge, MA
Mayr E (1993) What was the evolutionary synthesis? Trends Ecol Evol 8:31–34
Ochoa C (2017) El eclipse del antidarwinismo: la historia detrás de la síntesis moderna. In: Centro de Estudios Filosóficos. Políticos y Sociales Vicente Lombardo Toledano, México
Pauly PJ (2000) Biologists and the promise of American life: from Meriwether Lewis to Alfred Kinsey. Princeton University Press, Princeton, NJ
Schwartz JH (1999) Sudden origins: fossils, genes, and the emergence of species. Wiley, New York
Sepkoski D (2019) The unfinished synthesis?: paleontology and evolutionary biology in the 20th century. J Hist Biol 52:687–703
Smocovitis VB (1999) The 1959 Darwin Centennial celebration in America. Osiris 14:274–323
Tattersall I (1995) The fossil trail: how we know what we think we know about human evolution. Oxford University Press, Oxford
Wilson EO (1975) Sociobiology: the new synthesis. Harvard University Press, Cambridge, MA
Wray G, Hoekstra H, Futuyma D, Lenski R, Mackay T, Schluter D, Strassmann J (2014) Does evolutionary theory need a rethink? No, all is well. Nature 514:161–164

Part I
Crossing Perspectives About Evolution: Historians Versus Biologists

Chapter 2
Cathedrals, Corals and Mycelia: Three Analogies for the History of Evolutionary Biology

Maurizio Esposito

Abstract In *The Structure of Evolutionary Theory*, Stephen Jay Gould argued that the history and structure of the evolutionary theory could be represented as a fossilized coral constituted by a central branch representing the mechanism of natural selection. In analysing Gould's analogy—and some of the assumptions informing the traditional historiography of evolutionary biology—I defend three related points. First, while hierarchical analogies might be adequate for biologists outlining the current structure of evolutionary biology, they are not fit for historians reconstructing the history of evolutionary biology. Second, the historical reconstructions of evolutionary biology have been shaped by different approaches, purposes and questions. On that account, I distinguish between scientist and humanist historiographies of evolutionary biology, whereby the concept of "natural selection" obtains diverse connotations. Third, beyond Gould's static historiographic analogies, I propose an alternative, more flexible, analogy for the humanist historiography of evolutionary biology: the *mycelium*. In fact, contrary to hierarchical analogies employed by Gould—which emphasize the search for foundations and essences—the metaphorical mycelium inspires a different kind of historical understanding.

Keywords Darwinism · Evolutionary biology · Historiography of science · Natural selection · Philosophy of history

2.1 Introduction

At the beginning of his monumental monograph published in 2002, Stephen Jay Gould recalled the correspondence between Darwin and the Scottish surgeon and paleontologist Hugh Falconer. The two naturalists discussed about the alleged structure of the theory of evolution. Falconer observed that the theory had to be

M. Esposito (✉)
Department of History and Philosophy of Sciences, University of Lisbon, Lisbon, Portugal

R. G. Delisle (ed.), *Natural Selection*, Evolutionary Biology – New Perspectives on Its Development 3, https://doi.org/10.1007/978-3-030-65536-5_2

conceived as the Cathedral of Milan: a solid structure with very firm foundations rooted in the principle of a common ancestor and the mechanism of natural selection. Darwin was less inclined to consider the structure of his theory as being as solid as the robust bricks of a Gothic dome. He suggested instead that the structure of his theory was best understood as a broad framework that could undergo major revisions without questioning its general outlines, i.e. descent with modification and adaptation. While for Falconer the metaphorical "founding stones" constrained and oriented all future theoretical developments, for Darwin the identity of the entire structure was condensed in its "style" enduring significant historical changes (like the Gothic style for Milan's cathedral). As Gould explained: "Consider the different predictions that flow from the disparate analogies chosen by each author for the Duomo of Milan—Falconer's 'foundation' vs. Darwin's 'framework.' After all, a foundation is an invisible system of support, sunk into the ground, and intended as protection against sinking or toppling of the overlying public structure. A framework, on the other hand, defines the basic form and outline of the public structure itself" (Gould 2002, 2). And again: "Darwin's version remains Gothic, and basically unchanged beyond the visual equivalent of lip service. Falconer's version retains the Gothic base as a positive constraint and director, but then branches out into novel forms that mesh with the base but convert the growing structure into a new entity, largely defined by the outlines of its history" (Gould 2002, 6).

From Falconer and Darwin's correspondence, Gould drew a very interesting conclusion. There are two ways of considering the structure and history of evolutionary theory: a Falconerian foundational view and a Darwinian non-foundational view. If we prefer the former, then we need to pinpoint the few essential elements that sustain the entire theoretical structure. If, on the contrary, we opt for the latter, then our task is to figure out in what such a general framework really consists. In both cases we need to decide which concepts are central, peripheral, fundamental, subordinate, typical, or atypical. While Gould preferred the Falconerian approach, he believed that the structure of evolutionary theory was a blend of both models and could be represented as a fossil coral with three branches: basic, revising and subsidiary ones. Gould thought that the evolutionary theory had foundations (a basic branch) and a framework (revising and subsidiary boughs). There was a Darwinian logic at the heart of evolutionary theory and this logic had permeated evolutionary biology since the beginning. Whether we agree or disagree with Gould, his analysis touches very important points that anyone interested in the history of evolutionary biology, and the place of natural selection within it, should address.

In this chapter, in revising Gould's analogies and historiography, I defend three related points. First, while architectural (and hierarchical) analogies might be adequate for biologists outlining the modern structure of evolutionary biology, these analogies are not fit for historians reconstructing the history of evolutionary biology.[1] The reasons for shaping and justifying the structural content of a theory are

---

[1]Although I am aware that, for many contemporary biologists, Gould's hierarchical analogies would be equally inappropriate for describing the present structure of evolutionary biology. Yet, this does not change the substance of my argument.

different from the reasons leading historians for reconstructing the history of evolutionary biology. We can certainly endorse the idea that any contemporary synthesis of evolutionary biology should have at its centre the concept of natural selection, but this does not imply that the history of evolutionary biology has at its centre such a concept. For the evolutionary biologist, Darwin might stand as the founding father of evolutionary biology: for the historian of biology, Darwin (and his theory) is part of a larger, more complex and convoluted, narrative. In short, we need to distinguish between the synchronic structure of evolutionary theory (contemporary structure) and its diachronic development, and we should not expect an obvious and consistent overlapping between the two.

Second, the historical (diachronic) reconstructions of evolutionary biology have been shaped by different purposes and questions. In that sense, I distinguish between scientist and humanist historiographies of evolutionary biology (SH and HH onwards). The former has frequently used history for underpinning particular scientific options (i.e. Mayr's or Gould's historiographies) and is based on a linear narrative emphasizing great figures, core concepts and foundations. It usually identifies the current structure of evolutionary theory (SET onwards) with its history. The HH, instead, has no explicit interest in supporting specific research traditions and tends to question linear narratives in favour of more twisted ones. Unlike the SH, the HH is more inclined to sever synchronic questions (What is the SET or what should it be?) from diachronic ones (How did the SET arise and develop?). As a consequence, we should expect that the significance of natural selection within the history of evolutionary biology changes whether we assume a scientist or humanist historiography. While we can learn from both kinds of historiographic approaches, we should be aware that both have their own specificities and agendas.

In the third section of this chapter, I propose an alternative analogy for the humanist-historian of evolutionary biology: the *mycelium*. In fact, beyond the architectonic or hierarchical analogies employed by Gould—which emphasize the search for granitic foundations and fossilized essences—the analogy of the mycelium, with its ramifying *hyphae*, inspires a different kind of order. If we look at the structure of a developing mycelium, we observe that there is a ramifying network of filaments without central branches. The history of evolutionary theory could therefore be compared to a complex and dynamic network of hyphae, where new offshoots emerge and old ones dry out. Accordingly, unlike Gould's view, Darwin's mechanism of natural selection would represent only one conceptual "filament", which thrived within a broader and overarching structure of a dynamic developing "mycelium".

2.2 The Historiography of Evolutionary Biology and the Place of Natural Selection Within It

Gould believed that evolutionary biology was essentially Darwinian and that the reason for such a conclusion was mainly conceptual. His argument for defending this position is as interesting as it is typical for many biologists with historical sensibilities, and therefore, it is an excellent starting point for exploring the relation between the current (synchronic) SET and its historical (diachronic) development.[2] Gould's argument is based on two fundamental and connected premises: first, the evolutionary theory has a hierarchical structure, and second, such a structure is underpinned by a fundamental conceptual core. The premises are supported by Gould's conviction that if we want to preserve some form of intelligibility of the evolutionary theory, then we cannot refrain from identifying a specific structure in concert with some basic elements that resist historical change:

> In order to enter such a discourse about "the structure of evolutionary theory" at all, we must accept the validity, or at least the intellectual coherence and potential definability, of some key postulates and assumptions that are often not spelled out at all, and are, moreover, not always granted this form of intelligibility by philosophers and social critics who do engage such questions explicitly. Most importantly, I must be able to describe a construct like "evolutionary theory" as a genuine "thing"—an entity with discrete boundaries and a definable history—especially if I want to "cash out", as more than a confusingly poetic image, an analogy to the indubitable bricks and mortar of a cathedral. (2002, 6)

Gould assumes that if the evolutionary theory is a "genuine thing", then it has a structure. And if it has a structure, then it also has an order. And if it has an order, then we should define its hierarchical order and, therefore, its basic core. In that sense, Gould's argument is a kind of transcendental argument: the very condition for starting our conversation about the SET and its history is that we are talking about "something"—a genuine thing which, for definition, has an order that can be expressed hierarchically. The argument is quite sound so far and places all of the burden of proof on those who want to deny it. After all, a scientific theory supposes the existence of a given structure, even though the experts might disagree about its "real" structure. Conversely, if we argue that the theory has neither a structure nor an order, how can we talk about the evolutionary theory at all? Yet, Gould does something more than just expressing some general conditions of historical and conceptual intelligibility: from individuating the "transcendental" conditions of the SET, he abruptly applies the same argument to Darwinism itself. He does this through an effective (although weak) syllogism: the evolutionary theory must have a structure and a core. Darwinism must have a structure and a core. The core of evolutionary theory is Darwinism, and the core of Darwinism is natural selection.

The reason why I consider the syllogism both effective and weak is because, while it might sound persuasive, it implicitly assumes (without demonstrating) that the core of the evolutionary theory must be Darwinism, both conceptually and

[2]For a very thoughtful review of Gould's last monograph, see Radick (2012).

historically. The entire argument overlooks the fact that for many historians, this assumption would be far from obvious. Gould insisted that to discuss about the SET, we need to suppose the existence of some kind of order, and to discuss about Darwinism, we need to presuppose the existence of some basic concepts. But nowhere he supported the argument that the history of the SET must be rooted in some fundamental Darwinian concepts. Of course, Gould's solution killed two birds with one stone: we have a SET that is stable enough to be a "discrete" object of study but also flexible enough to be able to integrate new findings and developments. We can have both Falconer and Darwin, although the former wins on the structure and the latter on the content. Yet, what is left unclear is whether Gould's argument referred to the history of the SET or to how the SET should be understood today. If the SET is understood in its diachronic development, then the centrality of Darwin should be demostrated historically, not only assumed. We can all agree that Darwin is a crucial figure in the history of biology. But to recognize the relevance of a historical figure does not automatically imply placing them at the centre of a historical account. To say that Copernicus was a really important astronomer does not imply that the history of astronomy has a Copernican core. Astronomy has a much larger history that cannot be reduced to Copernicanism, and we can assume something similar as regards evolutionary biology. Thus, far from disparaging Darwin's fundamental contributions, I want to emphasize an inclination that is very typical among various important historically minded biologists, i.e. to confuse diachronic (historical) studies with synchronic (scientific) arguments.[3] The confusion gets even worse if we delve deeper into Gould's analysis. In fact, in his both diachronic and synchronic enquiry, Gould conceived the nucleus of Darwinism as being composed of three fundamental concepts: *agency*, *efficacy* and *scope*. By *agency* he means the fact that organisms are the agents of natural selection. By *efficacy*, he intends the fact that the selection is a creative force. And by *scope* he refers to the fact that small changes (adaptations) produce, over very long geological times, all the diversity we can observe today. These three concepts define the nucleus of Darwinism and are represented by Gould as a tripod coral:

> ...the image of a tripod suits my major claim particularly well—for I have argued, just above, that we should define the "essence" of a theory by an absolutely minimal set of truly necessary propositions. No structure, either of human building or of abstract form, captures this principle better than a tripod, based on its absolute minimum of three points for fully stable support in the dimensional world of our physical experience. (Gould 2002, 15)

Gould uses a plate by the artist and paleontologist Agostino Scilla representing a fossilized coral (Fig. 2.1). The central trunk represents the theory of natural selection as instantiated by the aforementioned three concepts. Its roots are well embedded into the ground (experience). If we cut off the main trunks, the *Ks* (killing) sections,

[3]The confusion is related to the following questions: does Darwin stand at the centre of the SET because history shows this to be the case? Or is it because science shows this to be the case? In other words, does the reason for putting Darwin at the centre of the SET depend on diachronic (historical) arguments or synchronic (scientific) evidence?

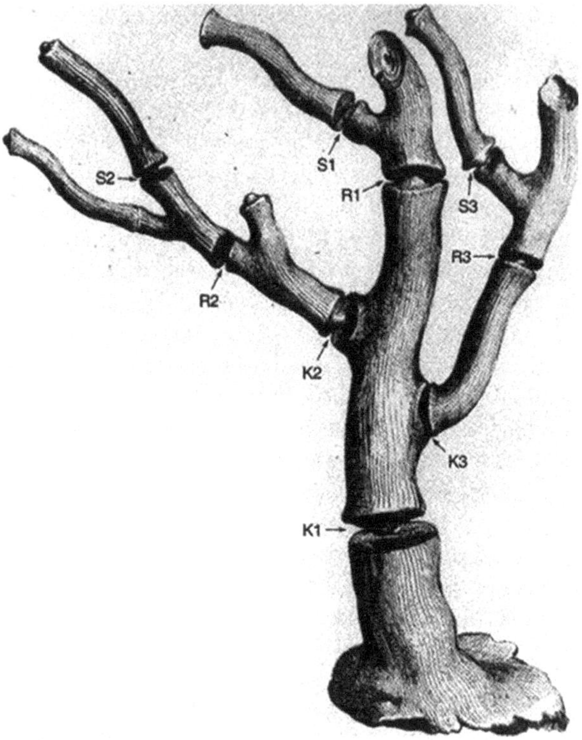

Fig. 2.1 Agostino Scilla's picture of a fossilized coral used by Gould for illustrating the logic of Darwin's theory. In Gould (2002), p. 18

we destroy the whole structure. But if we cut the secondary branches—the *Rs* (revising) and *Ss* (subsidiary) sections—we only obtain a reinforcement of the central structure (similar to the pruning of a tree). In short, in the course of history, the structure suffered periodic alterations that strengthened the basic logic of Darwinism. In this way, Gould provides a modern version of a Falconerian structure stressing both the firmness of the core structure and its more flexible parts situated in its periphery. We might condense Gould's model as follows: the SET has two parts, a basic structure and a superstructure. The former is Darwinian and its pulsing heart is the mechanism of natural selection. The latter is a complex and ramified set of non-Darwinian hypotheses, although they are not in contradiction to Darwinism (and, therefore, natural selection).

But, again, is the coral a diachronic or synchronic representation of the SET? What kind of relation exists between the synchronic structure (what biologists suppose it is or should be today) and the diachronic development (how the structure itself came to be)? Gould seems to imply that there is a profound connection between the two: the historical reconstruction grounds or justifies the contemporary structure. The "fossilized coral", as the current representation of the SET, reveals its long history. The structure we see today is the outcome of many historical adjustments and changes, whereby the central trunk—Darwinism–resisted innumerable strains and shocks. Interestingly, Gould contrasted his view about the subordination of the diachronic reconstruction to the synchronic structure with David Hull's analogical argument assimilating the history of science with biological evolution. In *Science as a Process* (1988), Hull argued that science advances through a selective process akin to the biological process of species formation. Ideas, like organisms, change over time, producing lineages. These lineages split over time, thrive, or become extinct. Like modifying organisms, cognitive systems are open to all sorts of mutations and, as a consequence, have no essence. The cognitive system we call Darwinism would be whatever Darwinians think it is within a given timeframe. The SET would be whatever biologists believe it is in a given epoch. For Hull, if you want to understand a cognitive system like Darwinism, you must look at its lineage, not its ethereal essence. Thus, as Hullian scholars, we should not chase solid cores behind superficial changes; we can only observe how the entire Darwinian cognitive system changes over time. Of course, for Gould this was unacceptable:

> Hull correctly defines theories as historical entities, properly subject to all the principles of narrative explanation—and I shall so treat Darwinian logic and its substantial improvements and changes throughout this book. But theories of range and power also feature inherent "essences," implicit in their logical structure, and operationally definable as minimal sets of propositions so crucial to the basic function of a system that their falsification must undermine the entire structure, and also so necessary as an ensemble of mutual implication that all essential components must work in concert to set the theory's mechanism into smooth operation as a generator and explanation of nature's order. (2002, 11)

We can resume Gould's counterargument as follows: without essences, we have no Darwinism and no SET. The lineage has produced both, and the historian's task is to understand how such essences have resisted historical idiosyncrasies and changes. Therefore, the important questions for Gould cannot be: is Darwinism the main subject in the history of evolutionary biology? Or, is natural selection central in that history? But instead: how did Darwinism persist throughout all historical and theoretical challenges becoming the heart of the SET? And, accordingly, how did natural selection endure as a core concept of evolutionary biology? Both sets of questions are really different because the latter two suppose that the former have been already conclusively answered (at least Gould supposes so). Once again, I am not suggesting that Gould was necessarily wrong and that natural selection cannot be conceived as central in the SET. I am only suggesting that we have a problematic overlapping between synchronic and diachronic perspective of SET.

But is Hull's diachronic view any better? Many arguments have been advanced against it and would not be easy to defend it today (see Bradie 1986; Koertge 1990; Dupré 1990 among others). I only want to stress one thing: the discussion between Gould and Hull hides the contentious belief that the understanding of the contemporary structure of evolutionary biology relies on the reconstruction of its history. Gould might be right in pretending that a Darwinian biologist needs something more than lineage of similar ideas to structure her theory. Hull might be right in stressing the historical and contingent nature of ideas and cognitive systems. I think that, in very general terms, they might be both right provided that Gould and Hull were considering different sorts of questions. The first is essentially asking: how does the SET need to be understood today in the light of its history? The latter is asking: how do cognitive entities, such as scientific theories, evolve? In other words, the former is primarily interested in the synchronic morphology of evolutionary biology, while the latter is more interested in its diachronic becoming. My argument is that the two questions, although they might be related, do induce to give very different, and possibly even contradictory, answers. The reasons for structuring a contemporary theory, and the aims of doing so, are not the same as those for organizing its history. In short, when we analyse a structure like Gould's coral, we are naturally led to individuate central and peripheral elements and therefore to pinpoint what is really essential and what is accessory. But when we approach evolutionary biology from a diachronic view—and therefore ask how it evolved in all its complexity—any hierarchical model will hinder the heterogeneity of the evolving cognitive systems.

I have lingered on Gould's proposal because it represents one of the most cogent examples of what I call the scientist historiography of evolutionary biology (SH). This includes most of the modern architects of synthesis, as well as some of their descendants: the so called extenders of the modern synthesis (Pigliucci and Müller 2010). Providing a better definition of the SH will be the task of the next section, but as a first approximation, I see one basic element that characterizes it: the presupposed coincidence between the effort to concoct a modern structure of evolutionary biology and the attempts to reconstruct its history. Gould assumed that thinking about the structure of evolutionary biology has to go hand in hand with thinking about its historical reconstruction. His great 2002 monograph is indeed a mixture of history, philosophy and contemporary biology, and this was not something new. Ever since he published his first book, *Ontogeny and Phylogeny*, Gould has masterly intermingled scientific contents with fascinating historical narratives.[4] Yet, Gould is not alone in doing this. An important part of the historiography of evolutionary biology has been written by biologists with a historical interest, and Ernst Mayr is probably one of the most significant examples (see Mayr and Provine 1980). Mayr is particularly interesting for the purposes of this chapter because he was one of the few biologist-historians who felt the need to think about how to write history. Mayr

[4]So far as I know, Gould's most insightful reflections on the philosophy of history were devoted to clarify the explanatory use of history in evolutionary biology. In several occasions, he remarked how important history was for understanding evolution (on the relation between evolutionary biology and history in Gould, see Blaser 1999). He declared that evolutionary biology "...embodies one theme even more inclusive than evolution—the nature and meaning of history. History employs evolution to structure biological events in time" (Gould 1985, 15). In the address for the Scientific Research Honor Society Sigma Xi, Gould even argued that Darwin was a "historical methodologist" (Gould 1987). At that time Gould did not wrap evolutionary biology around the mechanism of natural selection, but linked it to a series of epistemic strategies for inferring the past from the present world provided that Evolution was a deeply contingent process which could be represented as "...a bush, not a ladder" (1987, 69). Strikingly, Gould's conception of history as applied to evolutionary biology was much more flexible and "historicist" than his historiography of evolutionary biology. He granted much more contingency to evolution than to the conceptual history of evolutionary biology.

introduced his well-known book *The Growth of Biological Thought* (1982) with a short historiographic section titled "How to write history of biology". There he listed different kinds of possible narratives: lexicographic, chronological, biographical, cultural, sociological and problematic history. Drawing on Robin Collingwood's posthumously published essay, *The Idea of History*, Mayr saw his approach as a strategic blend between chronological and problematic history (see Mayr 1982, 6). The latter was particularly congenial to him because he conceived of the history of science as a history of problems and of the different attempts to solve them.

In the next section, I will contextualize Collingwood's influence on Mayr's historical thought. Yet, what I would like to stress now is a specific aspect of Mayr's historiographic approach: his distinctive awareness that he was writing history from the viewpoint of a scientist. As he warned at the beginning of his 1982 book: "The reader should be aware of the fact that this is not a traditional history of science...Since I am a biologist, I am better qualified to write a history of the problems and concepts of biology than a biographical or sociological history" (1982, 7). The view of the scientist is not the view of the professional historian. "A scientist tends to select for analysis and discussion rather different problems from a historian or sociologist" (1982, 13). Although Mayr was not entirely explicit on the matter, he repeated in different occasions that the importance of the history of science for scientists lies in clarifying contemporary issues. "...most scientific problems are far better understood by studying their history than their logic" (1982, 6). History should show how "erroneous beliefs" are gradually replaced by better beliefs and: "The historical demonstration of the gradual replacement of these prescientific or early scientific beliefs by better scientific theories and concepts greatly assists in explaining the current framework of biological theories" (1982, 19). History has a fundamental scientific role for Mayr: it helps us to clarify contemporary problems. But in order to engage in this task, the scientist-historian needs to begin his/her analysis from a clear understanding of what the structure of biology is: "It is my conviction that one cannot understand the growth of biological thought unless one understands the thought-structure of biology" (1982, 8). In other words, a monograph on the history of biology resembles a biology textbook that is organized historically and, as Mayr himself declared: "Perhaps this is what a problematical history of biology ought to be" (1982, 8). The difference that Mayr recognized between the historiography of biologists and the historiography of professional historians is similar to the distinction I draw between scientist and humanist historiographies, but with some important additions, which will be the object of the next section.[5]

Like Gould and Mayr, many self-taught historians of evolutionary biology have been biologists who see history as a strategic ally. However, as we have seen, while

[5]As Mayr put it: "Two groups of scholars with entirely different viewpoints and backgrounds—historians and scientists—have claimed the history of science as their own. Their respective contributions are somewhat different, dictated by differences in their interests and competence" (Mayr 1982, 13).

it is safe to assume that the contemporary structure of evolutionary biology has something to do with its long history, it is much more problematic to assimilate such history with contemporary evolutionary biology. Back in 1953, the French historian Marc Bloch clearly expressed this issue in his celebrated *Historian's Craft.* In criticizing the myth of "origins" in history, Bloch warned that the past does not necessarily account for the present (and it frequently does not).[6] I think that it would be wise to keep in mind Bloch's insight and apply it to our case: the history of evolutionary biology does not necessarily justify or explain the current SET, while the latter cannot serve as a reliable guide for reconstructing its past (and the place of natural selection within it).[7]

2.3 Scientist and Humanist Historiographies of Evolutionary Biology

Let us be a bit over-schematic and assume, on the basis of previous analysis, that there are at least two ways to interpret the history of evolutionary biology.[8] On the one hand, we have Gould's and Mayr's approaches, which are the typical models of many biologists looking at the history of their discipline. This is what I call the SH and can be applied to a large part of modern synthesis historiography (see Mayr and Provine 1980). Some usual questions driving the SH have been: how did Darwinism (broadly understood) triumph in spite of all of the internal and external challenges? How was the consensus that natural selection is the central mechanism in evolution reached despite intellectual and social resistance? How did the neo-Darwinian agenda win against neo-Lamarckians and other sorts of "cranks"? On the other hand, we have an alternative approach that I call the HH of evolutionary biology. Unlike SH, the HH considers different sort of questions. For instance, were Darwinism and Neo-Darwinism coherent and unitary projects (Delisle 2009a, b, 2017; Cain 2009)? In what did internal and external challenges to those projects really consist (Reif et al. 2000)? And furthermore, why and how were the alternatives to

[6]See Bloch (1992), pp. 26–27.

[7]Bloch's argument is a variant of what Morris Cohen and E. Nagel called "genetic fallacy" in their 1934 book *Logic and Scientific Method.* Simply stated, the fallacy states that the origins or history of x cannot account for the current meaning, use and merits of x. The argument I defend in this chapter makes a much broader use of the genetic fallacy. In my view, the main issue is not that the history of the SET is irrelevant for understanding its current organization. Rather, I am arguing that the reasons for justifying the current SET are different from the reasons that might lead historians to reconstruct the history of evolutionary biology (and the place of natural selection within).

[8]I am aware that I am proposing a very schematic and simplified map of the historiography of evolutionary biology, but, as the Polish philosopher Alfred Korzybski famously declared: "A map is not the territory it represents, but, if correct, it has a similar structure to the territory, which accounts for its usefulness" (*Science and Sanity*, 1933, p. 58). I do think that my general scheme is an interesting approximation of the "territory" of the historiography of evolutionary biology and can be useful for revealing some of the presuppositions that inform different practices of historical craft.

natural selection eventually sidelined (Esposito 2017)? Is the concept of "natural selection" a historically stable and coherent notion (Schwartz 2021)? While scientist-historians frequently see resistance (deviant and "unscientific" ideas) as obstacles to be shunned and then consigned to the dustbin of history (until we get to Gould's solid tripod coral), HH often sees those "resistances" as objects of accurate and focused investigation, which, at the same time, attest to the complexity and idiosyncrasies of the history of evolutionary biology. In the light of this still-vague distinction, we should not be surprised to notice that the historical significance of the concept of natural selection will change depending on whether we uphold SH or HH perspectives. We have different questions producing different historical landscapes in which the concept finds different places.[9]

But before exploring further the continuities and differences between the two historiographies, I need to warn the reader against a possible alluring argument, namely, that SH accounts are inexorably "Whiggish," while new accounts largely avoid such a sin. I think that to consider the SH as simply "presentist" would be inaccurate and unfair toward Gould, Mayr and many of those biologists who have written seminal works on the history of evolutionary biology. Ever since Herbert Butterfield wrote the *Whig Interpretation of History* in 1931, the epithet of "Whig" history or "presentism" has become one of most derogatory for historians. Especially from the 1970s onward, as Nick Jardine (2003) reminds us, writing the history of science in terms of one's present outlook is considered the hallmark of intellectual unsophistication. As Butterfield himself observed back in the 1930s: "It is part and parcel of the Whig interpretation of history that it studies the past with reference to the present...the total result of this method is to impose a certain form upon the whole historical story, and to produce a scheme of general history which is bound to converge beautifully upon the present..." (1965, 12). And "The Whig historian stands on the summit of the 20th century, and organizes his scheme of history from the point of view of his own day..." (1965, 13). Hence, the Whig historian arranges historical materials according to his/her present prejudices, domesticating, at the same time, the past in terms of the more familiar categories of the present. Of course, it would be easy to point the finger at Gould, Mayr and many other historically minded protagonists of the modern and extended synthesis, in this regard. Their narratives seem to mirror present interests and successful ideas. Their history of evolutionary biology might look like an odyssey of Darwinists struggling against anti-Darwinian and/or Lamarckian deviations.

[9]It is important to clarify that the distinction I am proposing here between scientist-historians and humanist-historians is not absolute but relative. The distinction refers to two "ideal types" of historical methodology and understanding which are much messier in reality. For example, excellent historians such as Peter Bowler, Jon Hodge, Janet Browne and few others would not easily fit in those categories. In fact, many historical narratives might be classified as a mixture between those types. Accordingly, the distinction does not describe two communities of real historians, but denotes *very general* historiographic approaches over the contentious relation between science (Biology) and its history.

Yet, it would be inappropriate to accuse Gould, Mayr, Provine and other traditional biologist-historians of Whiggism, because the specific form of presentism that Butterfield argued for does not really apply to them. Butterfield reproached "presentists" for the fact that they assemble historical evidence according to their present viewpoint, as if history was teleological and determined to reach one particular end. For Butterfield, the real perniciousness of Whiggism was that historians read past events as an inevitable progression toward liberal democracy. Although we might find here and there examples that fit with this form of presentism in SH, overall I resist the idea that Gould, Mayr and many other first-rate biologists-historians have committed such crass mistakes. In order to justify my unwillingness to accuse SH of Whiggism, it might be useful to distinguish between two kinds of presentism: an unsophisticated and a sophisticated one.[10] The unsophisticated version would correspond to Butterfield's Whiggism, which is impregnated with teleologism and, therefore, anachronisms (reconstructing the past with present categories). After all, the concept of "Whig history" is a different—more refined—version of Voltaire's philosophy of history, which famously preached the idea that history is always modern. Voltaire believed that history was modern because before modernity, there was nothing interesting to talk about unless it contributed to the formation of the modern world. In that sense, history is interesting insofar as it illuminates how our present "enlightened" ideas came about (Collingwood 1946; Force 2009).

The sophisticated version of presentism was very well outlined by certain idealist philosophers at the beginning of the twentieth century.[11] In fact, one of the foremost philosophers of history in the twentieth century, the Italian philosopher Benedetto Croce, who would later deeply inspire Robin Collingwood (who, in turn, influenced Ernst Mayr and his historiography), observed that history is always present. But his notion of "present" has nothing in common with Voltaire's presentism or Butterfield's "Whiggism". For Croce, Collingwood and his fellow idealists, history was always contemporary for the simple fact that historians are always situated in some present time in relation to their object of inquiry. Even the history of the most remote past is written from the perspective of a particular present, and there is no escape from this conclusion unless we endorse a very simplistic form of positivism and unrealistic objectivity (like Leopold von Ranke's positivist view according to which history "...wants only to show what actually happened"). Croce stressed the obvious point that historians make choices. From the set of virtually infinite historical facts, historians only select those facts that are meaningful to them. And what "meaningful" or "meaningless" depends on the interests of historians in the present.

---

[10]In a paper published in 1979, David Hull argued that some form of presentism is not only justifiable but also necessary. Although his defence of "presentism" was not entirely different from Croce's and other idealist historiographers, he emphasized the fact that historians cannot avoid using some form of present knowledge for reconstructing past knowledge. Historians need to do that for communicating their results to their contemporary fellows (see Hull 1979).

[11]For more recent discussions on the Philosophy of History, see Ankersmit and Kellner (1995), Bentley (1997), Koselleck (2002), among others. Yet, I think that the philosophical acumen reached by Croce, Collingwood or E. H. Carr remains unsurpassed, and many of the problems they presented are far from being resolved today.

Therefore, historical research consists in a constant reconsideration of particular facts, events and figures, so that we have a continuous reinterpretation of the past in the present. Our comprehension of the historical past changes with our presents. This also explains why every century has its diverse histories of Greek philosophy, Portuguese colonialism, or French revolution (and, we might add, diverse Darwins, as well as diverse histories of evolutionary biology). As Croce observed: "...for it is evident that only an interest in the life of the present can move one to investigate past fact. Therefore, this past does not answer to past interest, but to present interest, in so far as it is unified with an interest in the present life (1921, 12)".

In short, with Butterfield, Croce might have agreed that history should not be teleological, but unlike Butterfield, Croce would have maintained that historians cannot write history in abstraction from their present curiosities and issues. In the face of the sheer quantity of historical events, historians can only select and order those events that fit their own contemporary interests. Of course, Croce, as later Collingwood, did not make free concessions to historical relativism. Not all kinds of narratives are good historical accounts. In that sense, Croce distinguished between real history and chronicles (or philosophical history versus philological histories), the former referring to the capacity of the historian to meaningfully orchestrate documents and evidence supporting a coherent and convincing narrative, while the latter points to a collection of unarticulated facts and events which have neither structure nor causal order. Following Croce, Collingwood argued that good history, what he called problematic history (Mayr's idea of history), had to be a concrete and coherent *reenactment* of the past in the mind of the historian.[12]

I will not enter further into the intricacies of the different philosophical conceptions of history of Croce, Collingwood, or Butterfield (for more details, see Ankersmit and Kellner 1995; Bentley 1997; Jardine 2003). This short digression serves only one purpose: to show that "Whig history" itself is not a useful conceptual category for distinguishing SH from HH. For example, Mayr was aware of the pitfalls of what I call "unsophisticated presentism". In the abovementioned *Growth of Biological Thought*, Mayr lamented that "The history of biology is rich in such biased Whig interpretation" (1982, 12). The historian, for Mayr, should avoid at all costs the evaluation of past theories in terms of contemporary ones. Indeed, Mayr's presentism was not that of Butterfield, but that of Collingwood (and Croce). For Mayr, history was always present insofar as histories: "...depend on how the author interpreted the current zeitgeist of biology and his own conceptual framework and background" (1982, 19). And yet: "Subjectivity enters at every stage of history writing, especially when one is seeking explanations and asks why, as is necessary in problematic history. One cannot arrive at explanations without using one's own

[12]As Collingwood explained: "We shall never know how the flowers smelt in the garden of Epicurus, or how Nietzsche felt the wind on his hair as he walked on the mountains; we cannot relive the triumph or Archimedes or the bitterness of Marius; but the evidence of what these men thought is in our hands; and re-creating these thoughts in our minds by interpretation of that evidence we can know, so far as there is any knowledge, that the thoughts we create were theirs" (1946, 296).

personal judgment, and this is inevitably subjective" (1982, 10). Both passages are entirely Collingwoodian. Mayr was aware that histories need to be rewritten every century precisely because the interests of the historians change: "The main reason why histories are in constant need of revision is that at any given time they merely reflect the present state of understanding" (1982, 1).[13]

Now, if we cannot use the unsophisticated form of presentism to distinguish SH from HH, does it make sense to distinguish between two or more types of historiographies? After all, we might conclude that more recent historians are producing better and more accurate narratives, avoiding the shortcomings of the older accounts: we have one historiography of evolutionary biology, and this historiography is about to get the past right. Yet, while this conclusion might sound persuasive, I think that it would dismiss too easily some of the important distinctions that inform the historiography of evolutionary biology (and the historiography of science more generally). In order to characterize more precisely such distinctions - implicitly or explicitly perceived by Mayr and many others before and after[14] it might be helpful to use

[13]Further, in 1990, Mayr even felt the need to defend his 1982's monograph against several accusations of "Whiggism". As he thundered: "...I feel the pejorative label 'whiggishness' has been used increasingly in such an irresponsible and often completely unjustified manner, that one might want to hope that it will disappear altogether from the literature of scientific historiography" (Mayr 1990, 309). On that occasion, Mayr clarified that his historiography was not "presentist" but "developmental" in the sense that the "...interest of the modern scientist-historiographer is the development of ideas, from their origin through all their permutations up to the present day. The reason for this interest is that it is impossible to understand many of the current controversies and prevailing concepts without studying their history" (Mayr 1990, 304). In short, for Mayr, scientists' historical reconstructions serve to clarify conceptual issues in contemporary science. Most scientist's historical reconstructions are purposefully presentist, although certainly not "Whiggish".

[14]So far as I know, after Lucian Febvre (1934), George Stocking was one of the main scholars who explicitly distinguished between science-history (which to him was essentially presentist) and "historicism" in the history of science. For Stocking, while the former approach tended to track down how particular agencies (great men, ideas or "deterministic forces") direct progress in the field, the latter was rather focused on describing the convoluted processes producing meaningful changes in history. Interestingly, Stocking compared the *historicist* approach to a mountain climber ascending the Everest. He climbs it only because "it is there". The historicist in fact: "...demands no more of it that the emotional satisfaction which flows from understanding a manifestation of the changing human self in time" (1965, 213). On the contrary, the science-historian pretends much more than an emotional satisfaction; the past is useful and needs to be employed "...for furthering his professional activities in the ongoing present" (1965, 213). The distinction was later developed in different ways and for different scopes by Debus (1971), Graham (1981), Reingold (1981), Russell (1984), Roger (1984), Harrison (1987) and Brush (1995), among others. While I find such literature illuminating, I think that it relies too much on the notion of "presentism", which is considered as the real difference maker between science-history and professional history of science. I do not think that "presentism" is a very interesting concept for distinguishing different kinds of historiographic approaches for the simple reason that all histories are, in some general sense, presentist (Croce 1921; Collingwood 1946; Hull 1979). In general, after the 1970s, the concern over presentism was often associated with Thomas Kuhn's maxim of the "internal historiography". Kuhn argued that the historian of science should ignore, as much as possible, contemporary science in order to study and understand past science (Kuhn 1968). Ignorance (or epistemic self-restraint) about contemporary science would eventually guarantee a certain degree of historical objectivity

Maurice Mandelbaum's informative taxonomy of historical writing's "modes": i.e. *sequential*, *explanatory* and *interpretative*. The *sequential* mode denotes the typical approach in which the historian first selects a research subject (i.e. neo-Lamarckism, Darwinism and Modern Synthesis) and then "...seeks to trace the strand of events making up that history" (1977, 27). In such account, and beyond the general heading orienting the inquiry, historians do ignore what they would eventually find out. Scholars texture the evidences as they discover them. In the *explanatory* account, instead, the historian already knows what happened and try to explain why it happened as it did. "In such a case—Mandelbaum argued—he [the historian] starts from a fact taken as present and seeks to trace back its causes...the inquiry itself moves back from what is known to have occurred and seeks an explanation of it through tracing its antecedents" (1977, 26). Finally, in the *interpretative* accounts, the historian focuses on broad historical periods (i.e. Renaissance, Medieval or Modern science and Enlightenment) and tries to identify the most salient elements (whether cultural, social or metaphysical) exemplifying such periods. Of course, Mandelbaum was aware that these were not "pure" forms of historical narratives. Historians often use these three "modes" in the same research. Yet, historical works tend to be dominated by one of those forms over the others.

We can add that all those forms of historical inquiry stem from present concerns (sophisticated presentism) but have very different aims: explanatory history pretends to explain how and why some past or contemporary options succeeded (or did not), assuming that we already know what happened (i.e. Gould's or Mayr's approaches). Sequential and interpretative histories organize historical facts according to general headings (i.e. Darwinism) or periods (i.e. Enlightenment) without being directly concerned about the success of some contemporary ideas. Following Mandelbaum's insights, I argue that the SH tends to be *explanatory* in form, while the HH leans towards *sequential* and *interpretative* historical reconstructions. Accordingly, the SH and HH are informed by different presuppositions, which I summarize under the categories of questions and purposes.

Let us start with the latter. Most humanist-historians today are not biologists concerned with reconstructing the history of evolutionary biology from the standpoint of the present interest in contemporary biology. They do not pretend to "explain" why some particular ideas succeeded for present purposes. Their principal purpose is rather focused on assessing how different traditions thrived, coexisted and eventually declined, and that includes the reconsideration of forgotten figures, ideas and institutions. Such diverging purposes are connected with diverse conceptions of what evolutionary biology is. In fact, for the HH, evolutionary biology appears not

---

avoiding, at the same time, presentist temptations. I do not think that today we would need such a kind of restriction for framing a humanist historiography of science. After all, contemporary science has been, and still is, an important and preliminary starting point for writing excellent historical narratives. The problem emerges when the historian (or the scientist) assumes that a present scientific field is a transparent and lean reflection of its past.

as a Gouldian "genuine thing", but as a very complex field of contradictory stances. The development of evolutionary theory does not consist of a linear succession of extraordinary figures and their controversial ideas. Its history is not an itinerary of an "essential entity" crossing the blusterous seas of time. The constitution of evolutionary theory is rather a dynamic process that is shaped by different subjects and communities struggling for authority and visibility. The humanist-historian does not use history to understand why and how some present options succeeded but is rather concerned with comprehending how evolutionary biology grew and changed, including all its idiosyncratic contradictions and dead ends.

The same discussion applies to the concept of natural selection and its place within the history of evolutionary biology. As we have seen, for Gould natural selection represents the stable core of evolutionary biology, both from the historical and conceptual perspective. But is natural selection really a stable concept after all?[15] The answer depends on the aims and modes of historical accounts, whether we want to *explain* its success within the current theory or simply reconstruct its conceptual history. An example taken from Collingwood's the *Idea of History* might help clarify further the connection between "stable concepts" in history and the explanatory historiographic "mode". In the book, the British philosopher distinguished between *substantialist* and *non-substantialist* historiographies. He believed that the ancient—Greco-Roman—historiography was generally *substantialist*, and the Latin historian Livy was one of its main representatives. Collingwood deemed Livy a *substantialist* because in his *Ab Urbe Condita* (History of Rome) he had conceived Rome as an unhistorical object: "Rome is the agent whose action he is describing...Rome is a substance, changeless and eternal. From the beginning of his narrative Rome is ready-made and complete" (1946, 44). In short, Livy conceived of Rome as a *substantial* entity crossing history. The "eternal" city was not a subject of change because what changed was all that surrounded it. And we might add that Livy conceived of Rome in this way because his aim was basically *explanatory* and driven by the following question: How did such a "genuine thing" called Rome succeed, becoming the heart of a great empire and resisting all possible disrupting threats?[16]

I see here an interesting parallel with the SH and its conception of how historical ideas thrive (or vanish). "Natural selection", like Livy's Rome, is conceived as a substance keeping its identity ever since it came out from Darwin's mind. In Falconer or Gould's picture, natural selection is a substance that stands at the core of the Darwinian project, and Darwinism itself is seen as a historical agent persisting despite an ever-changing environment. In short, both "natural selection" and "Darwinism" are conceived as substantial things. They are not changing things but entities that resist historical change. For a non-substantialist historian instead, "natural selection"

[15]For the scientist-historian, the answer is yes, because to explore the history of such an ideal entity is equivalent to revealing how it became the heart of contemporary evolutionary biology.

[16]Of course, Collingwood's interpretation of Livy's historiography has been challenged. Yet, this would not invalidate the argument I am defending.

and "Darwinism" are not *enduring* concepts crossing history or solid "bricks" that can be added to whichever conceptual "edifice" without changing its identity; they are rather complex notions acquiring different meaning in different contexts. While a scientist-historian already "knows"—supposedly—what natural selection is and whence it come so that he/she can *explain* how it succeeded, the humanist-historian tries to recover the changing meaning of it, tracking its historical variations (without assuming a metahistorical and definitive connotation attached to the concept itself).[17]

All of the above can also be applied to other ideas and research traditions. For instance, for the humanist-historian, Darwinism and Neo-Darwinism do not appear as linear and unified research programs (substantial things), but as fragmented and patchy research strategies. The very project of modern synthesis does not look like a coherent agenda (a "genuine thing"), but rather as an assorted and heterogeneous field of distinct, and sometimes opposing, tendencies. For the HH, any "ism" is better understood by the plural "isms". A scientific synthesis is better grasped as an ensemble of syntheses, a research tradition as a conjunction of connected and disconnected traditions. Accordingly, where the SH see continuities, unities and generalities, the HH sees breaks, disarticulations, incoherencies and particularities. In short, the SH aims to simplify *unnecessary* complexity, and the HH wants to complexify *spurious* simplicity. And the reason why this happens is because we have two alternative kinds of purposes: one aiming to use history for *explaining* how contemporary options succeeded and the other one reconstructing the different possible historical *sequences* and *interpretations* of evolutionary biology.

Different purposes are connected to different kinds of questions. We have already seen that Gould did not doubt about whether Darwinism was at the centre of evolutionary biology. He assumed it in his coral analogy insofar as one of the main questions driving his inquiry was: how did Darwinism became the heart of our current SET? The question makes total sense if we want to *explain* how the contemporary structure (whatever we assume it is) came about. Yet, for the HH perspective, such an option would preclude the consideration of other interesting questions, namely; are Darwin and Darwinism really at the centre of the history of evolutionary biology? When, where and for whom has that happened? How many other figures and ideas have been left out and sacrificed on the altar of the SH's tidiness? Does Darwinism really have a core? Is it a name referring to different, and often contradictory, research programs? Is it something biologists manufacture periodically in order to adapt it to new research contexts? And, most of all, do we really need a linear and general narrative? For Gould the answer to this last question would be affirmative because without such an assumption, the intelligibility of his overall project would be jeopardized (we would lack any "genuine thing" to talk about). For the scholar adopting the HH, the answer would be negative: we do not

[17]For example, if we consider the following questions: "How did natural selection succeed?" and "How did the concept of natural selection change in history?", we are presupposing two different historiographic approaches and two different understandings of "natural selection" as object of historical inquiry.

have such a narrative, and we do not need one. While biologists might placed Darwin at the centre of the structure of evolutionary biology, natural selection or whatever they regard as fitting in any given moment, the historian doubts that the actual structure of evolutionary biology (synchronic SET) is a neat reflection of its history. The history of evolutionary biology can be much more idiosyncratic and hectic than any tidy and clean structure might suggest. What I am arguing here is that different narratives are shaped by different modes of historical reconstructions as well as particular interests and questions. There will be scientists' histories (that are mostly explanatory in form) and humanist's histories (which are generally sequential and interpretative) together with their more or less *substantialist* accounts. Whether we prefer a scientist or humanist historiographic approach, we should be aware of the presuppositions informing them. The SH deems history as a tool for clarifying actual science explaining how contemporary options succeeded; the HH considers the tool (history) as an end in itself. But conceiving of the tool as the end in itself does not make it necessarily any more objective. We have different, complementary, stories.

Now we have a further and final task: If we question hierarchical analogies, as many humanist–historians implicitly or explicitly do, then it might be wise to think of alternative, non-hierarchical, analogies that can represent the complexity of the historical process.

2.4 Beyond Corals and Cathedrals: Do We Need a New Analogy for the HH?

We have seen that according to the SH, the synchronic structure of evolutionary biology orients historical research. As a consequence, the present state of knowledge in biology inspires the selection of facts, events and figures that are pertinent for building up a coherent and linear narrative. History itself becomes a powerful tool for clarifying, explaining and strengthening modern scientific options. After all, as Mayr keenly observed, scientists' history of biology looks like a biology's textbook organized historically. Hence, evolutionary biology is a "genuine thing", an agent with a specific and hierarchical structure that has resisted many historical vicissitudes. The idea that the history of evolutionary biology requires the existence of a hierarchical structure composed of *substantial* entities is a condition of possibility of the SH. And a further important condition is that such a structure needs to be understood in terms of analogies that evoke firmness and stability. Falconer's architectonic analogy is one of these: the development of the evolutionary theory (as a *substantial* thing itself) is constrained and shaped by the existence of solid pillars that cannot be removed without destroying the whole edifice. Its history, in that sense, is all that emerges from the conceptual foundations of such a solid structure. Yet, while the SH needs such analogies to satisfy its goals, the HH does not need them and is also sceptical about them. Humanist-historians tend to separate discussions about the contemporary SET from its history. Biologists might well argue that the Darwinian mechanism of natural selection must stand at the centre of any modern SET, but for the humanist-historian, it is not obvious that this argument applies to history itself. From a historical viewpoint, the significance of natural selection has changed constantly in different epochs and places. The very meaning of "natural selection" has changed, and this should not be surprising because, for the humanist-historian, there are no essences, *substantial* things or inflexible foundation stones in history.[18] In other words, for SH, natural selection is a "thing" to be explained and clarified; for the HH, it is a concept to be understood in its historical (and dynamic) dimension.

However, if humanist-historians can dispense with all the hard architectonic analogies, shall they distrust of any historical analogy? No doubt, historians can easily dodge the question, but if they do, they would probably assume an answer. Is the history of the evolutionary theory something linear that can be represented like a dome or a tree? Or is it something very different, more flexible, chaotic and open-ended? Both questions imply that there is, after all, an indefinite "something" we might need to elucidate. Perhaps the evolutionary theory is not a "genuine thing" in the sense advocated by Gould (like the "Rome" of Livy), but is it "something" which can be represented and broadly defined in its historical development? What I am referring to here is Gould's condition of historical intelligibility that I mentioned in the first section, but with a fundamental difference: we do not need to presuppose that, from the historical perspective, evolutionary theory has a persisting core and an enduring hierarchical structure. In order to maintain some minimal conditions of historical intelligibility, we might conceive the evolutionary theory as a "thing" referring to an intricate bush of representations, traditions and communities spreading over more than 300 years of development (or even more). The historical "something" we are thinking about is not a "genuine thing", but a complex conceptual network that can be metaphorically represented as a growing "mycelium".

A mycelium is a part of a fungus that is structured in a complex mass of branching hyphae. Unlike the more famous analogy of the tree, the mycelium does not have a clear hierarchy, but an initial point, a fungal hypha. The structure can be represented as in Fig. 2.2.

We can consider the initial small dot as the beginning of evolutionary biology as a field of study. What the dot represents can be discussed by historians (whether Lamarck, or some previous figure or some old theory of transmutation). From the initial point, several branches take shape. These might represent distinct research lineages, which, in turn, will undergo further divisions and subdivisions (we might think about Darwinism and Neo-Darwinism, as well as Lamarckism and neo-Lamarckism). The analogy even contemplates that distinct lineages can be reunified later on. In the course of time, old branches might die out, leaving the space for new ones. There is no end, in principle: the structure grows indefinitely

[18]Whether or not we conceive natural selection as a creative or non-creative agent and whether or not we attribute to it great or little explanatory powers.

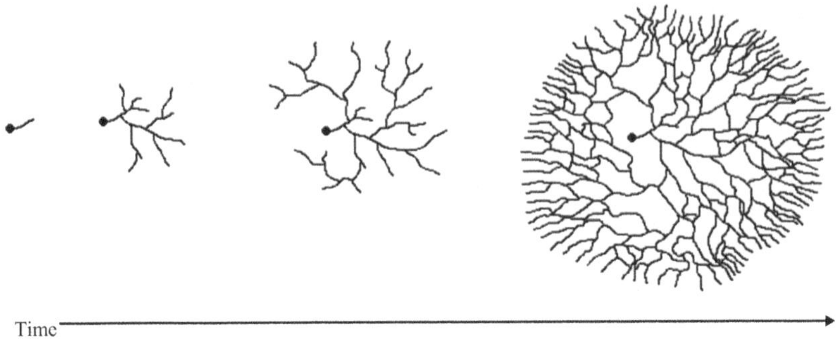

Fig. 2.2 A branching mycelium representing the history of evolutionary biology

while the overall morphology changes constantly (it is well known that mycelium can grow very large). What this picture exludes is historical linearity. The growing field ramifies in different and unpredictable directions.

The historian's task would therefore consist of exploring the different sections and subsections of the growing mycelium. There is not one central structure that might orient the historical investigation. There is not one foundation or essence which remains unchanged. There is not one end toward which the structure moves (i.e. the contemporary structure). There are, instead, many different possible, and even contradictory, ends, so that the historian cannot use the present structure as a model for reconstructing the past. In this metaphoric mycelium, we do not have central and peripheral branches representing what "really" happened. We do not have *substantial* things crossing history. We have an indefinite number of branches from which historians and biologists can extract what they deem more appropriate for building their own narratives. Yet, this does not imply that any narrative would be equally correct. Historical narratives would be the outcome of historical facts sifted through the interests of the historian who is willing to explore one or more sections of the overall structure. In this picture, natural selection would not be conceived of as a "genuine thing" (a central "branch" or "stone") crossing time and space and resisting several theoretical challenges, but rather as a term acquiring different meanings depending on the conceptual network (hyphae) in which it appears. The historian should not expect that the concept of natural selection in Darwin would be the same as that of Ernst Haeckel (Levit and Hossfeld 2021), August Weismann, Henry Osborn (Ceccarelli 2021), William Bateson (Ochoa 2021), Thomas Morgan, Ronald Fisher, J.B.S. Haldane, or Theodosius Dobzhansky (Schwartz 2021; Van der Meer 2021; Adams 2021), or any modern or contemporary understanding of the concept.[19] The meaning and use of the concept are not defined

[19]Even the place that such notion plays in Darwin's *Origins of Species* is far to be obvious (Delisle 2021). Another interesting question is whether Darwinism—in any of its historical and theoretical forms—necessarily includes Natural Selection as its main and fundamental conceptual component (see Baravalle 2021).

once and for all, but change and depend on the dynamic networks in which it appears (and that the historian needs to uncover). For Gould, as we have seen, natural selection is one central brick at the base of the great building of evolutionary biology. For a humanist-historian, natural selection would be a term referring to distinct networks (or sets of hyphae) and therefore different cognitive systems.

Of course, for Gould (and for Mayr and many other scientist-historians), the analogy of the mycelium would complicate the goal of clarifying or justifying contemporary scientific options because their historiographic approach is mainly *explanatory*. The mycelium analogy would also suggest that contemporary biology is only a small subsection of many other possibilities that have been forgotten or discarded (whether for good or bad reasons). From the large historical set of models, concepts and representations, biologists have constantly sifted those contents that fit their interests and conceptual syntheses. And from the large ensemble of research programs that have existed, scientist-historians have filtered a small subsection of those (i.e. those that are interesting as regards clarifying our modern structure). In short, the mycelium's historiographic analogy would reveal the lack of any necessary relation between the contemporary SET and its history. It would also show how the historian's approach for reconstructing the history of evolutionary biology is different from the biologist's use of that history for explaining how contemporary options succeeded.

2.5 Conclusion

In this chapter, I have argued that historical analogies are not neutral. They inspire different sort of approaches and conclusions. Mayr was aware that scientists see history differently from professional historians. The former often conceive history of science as a strategic tool for clarifying the conceptual issues haunting contemporary science. Such an attitude characterizes what I have called the scientist historiography (SH) in opposition to the humanist historiography (HH). I have argued that the specific difference between the SH and the HH cannot be reduced to the presence (or lack) of "presentist" accounts, in the sense conveyed by Butterfield in the 1930s. If we consider a more sophisticated conception of "presentism", all history is, in some general sense, presentist. I have therefore explored the possibility that the difference between the SH and the HH lies in distinct "modes" of historical accounts (whether explanatory, sequential or interpretative) as well as in different purposes and research questions. I have tried to expose the seemingly uncontroversial idea that the current structure of evolutionary biology might be the transparent reflection of its history. The scientist-historian tends to assume this, especially because history can be a useful tool in the service of contemporary science. But for the humanist-historian, this is a highly contentious matter. For the scientist-historian, Darwin (and Darwinism broadly understood) might stand at the centre of evolutionary theory, but for the humanist-historians, this is far from being obvious because the conceptual reasons for structuring evolutionary biology (with natural selection as its basic stone) can be quite different from the historian's motives for writing its history.

I have also argued that we should be wary of reconstructing the history of evolutionary theory using architectural or foundational analogies. Such history might be much more convoluted, tortuous and erratic that any geometrical or foundational analogy might suggest. I therefore proposed to use a mycelium as a historiographic analogy. Unlike architectural and hierarchical analogies (cathedrals or corals), the mycelium analogy suggests that the evolutionary theory has no centres, no substantial cores, no metahistorical essences and no firm foundations. Instead, it expresses the idea of an intricate network of concepts, theories, models and representations in constant change and refinement. This has interesting implications for our historical reconstructions of evolutionary biology and the place of natural selection within it. In fact, from a historiographic perspective, it matters whether we conceive natural selection as a substantial—*ahistorical*—thing or as a more plastic concept acquiring different meanings in diverse theoretical contexts. I have suggested that for the humanist-historian, natural selection does not appear as a *substantial* entity crossing history, but as a relatively plastic concept getting diverse meanings in diverse cognitive spaces. The concept of "natural selection" has itself a history, and such a history would correspond to the different uses and meanings that the concept acquired in disparate theoretical systems (see Schwartz 2021). In other words, the concept would not be conceived as "thing" moving into history, but as a notion shaped *by* history. Marc Bloch beautifully defined historical concepts as "...variable according to time and space; they become coefficients of emotivity leading to further equivocations" (Bloch 1992, 142). When Bloch wrote these words, he had the concept of "capitalism" in mind. Although the term "natural selection" might be much more specific and unambiguous than the word "capitalism", for many humanist-historians, the concept of "natural selection" would appear as an equally problematic (and sometimes emotional) coefficient, and not as a solid keystone keeping the entire vault of evolutionary biology in place.

I think that today we are headed towards an increasingly humanist approach to the historiography of science. This is mainly due to the fact that in the very last decades sociocultural history (broadly conceived) and the history and philosophy of science have gradually meshed together. Nonetheless, the force and influence of traditional "scientist" historiography are still very powerful. Indeed, in my view, the sign showing the persistence of the SH is not the resolute unwillingness to question it, but rather the propensity to amend it as far as possible. For many humanist-oriented historians, the way to deal with the SH is not to challenge its basic elements, but to criticize its scarce inclusiveness; the traditional view would be fine if only it were more comprehensive (and eventually less Anglo-centric). The strategy of what I call the "revisionist agenda" can be resumed as follows: The history of evolutionary biology is much more complicated than the account we inherited from the received view. We need to extend and complement it including more actors, ideas, traditions and countries. While such an agenda (to which I have contributed with enthusiasm) is scholarly important and has helped us in identifying the limits and weaknesses of the old accounts, it has stimulated the proliferation of studies focused on forgotten or marginal figures which explicitly or implicitly avoid to tackle more ambitious and controversial questions.[20] The result of such prudent revisionism is a chaotic historiography which, while multiplying disconnected voices and perspectives, leaves intact the traditional master narratives. Yet, to use a Kuhnian analogy, perhaps the old "historiographic paradigm" cannot be amended with more data and information. Perhaps we might need to reconsider the whole thing. I am persuaded that the only way to do that is to provide a broad philosophical reconsideration about what the history and historiography of evolutionary biology is about (and can eventually be).

References

Adams MB (2021) Little evolution, big evolution rethinking the history of Darwinism, population genetics, and the "synthesis". In: Delisle R (ed) Natural selection: revisiting its explanatory role in evolutionary biology. Springer, Cham, pp 195–230

Ankersmit F, Kellner H (eds) (1995) A new philosophy of history. University of Chicago Press, Chicago

Baravalle L (2021) Darwinism without selection? A lesson from cultural evolutionary theory. In: Delisle R (ed) Natural selection: revisiting its explanatory role in evolutionary biology. Springer, Cham, pp 441–464

Bentley M (ed) (1997) Companion to historiography. Routledge, New York

Blaser K (1999) The history of nature and the nature of history: Stephen Jay Gould on science, philosophy and history. Hist Teach 32(3):411–430

Bloch M (1992) The historian's craft. Manchester University Press, Manchester

Bradie M (1986) Assessing evolutionary epistemology. Biol Philos 1:401–460

Brush S (1995) Scientists as historians. Osiris 2nd Ser 10:214–231

Butterfield H (1965) The Whig interpretation of history. W.W. Norton

Cain J (2009) Rethinking the synthesis period in evolutionary studies. J Hist Biol 42(4):621–648

Ceccarelli D (2021) Recasting natural selection: Osborn and the orthogenetic view of life. In: Delisle R (ed) Natural selection: revisiting its explanatory role in evolutionary biology. Springer, Cham, pp 171–194

Collingwood RG (1946) The idea of history. Clarendon, Oxford

Croce B (1921) Theory and history of historiography. Harrap, London

Cunningham A, Williams P (1993) De-centring the "Big Picture": The Origins of Modern Science and the modern origins of science. Br J Hist Sci 24(4):407–432

Debus A (1971) The relationship of science-history to the history of science. J Chem Educ 48 (12):804–805

Delisle R (2009a) Les Philosophies du néo-darwinisme. PUF, Paris

[20] I largely share Andre Cunningham and Perry Williams' argument that big historical pictures of science are necessary and desirable, although they should be "de-centred" big pictures (see Cunningham and Williams 1993).

Delisle R (2009b) The uncertain foundation of neo-Darwinism: metaphysical and epistemological pluralism in the evolutionary synthesis. Stud Hist Philos Sci Part C Stud Hist Philos Biol Biomed Sci 40(2):119–132

Delisle R (2017) Introduction: Darwinism or a Kaleidoscope of Research Programs and Ideas? In: Delisle R (ed) The Darwinian tradition in context: research programs in evolutionary biology. Springer, Cham, pp 1–8

Delisle R (2021) Natural selection as a mere auxiliary hypothesis in Charles Darwin's *Origin of Species*. In: Delisle R (ed) Natural selection: revisiting its explanatory role in evolutionary biology. Springer, Cham, pp 73–104

Dupré J (1990) Scientific pluralism and the plurality of the sciences: comments of David Hull 'science as a process'. Philos Stud 60:61–76

Esposito M (2017) The organismal synthesis: holistic science and developmental evolution in the English speaking world 1915-1954. In: Delisle R (ed) The Darwinian tradition in context: research programs in evolutionary biology. Springer, Cham, pp 219–242

Febvre L (1934) A propos d'un précis d'histoire des sciences: sciences et techniques, Annales d'histoire économique et sociale, No. 30, pp 606–607

Force P (2009) Voltaire and the necessity of modern history. Mod Intellect Hist 6(3):457–484

Gould SJ (1985) The Flamingo's smile: reflection in natural history. W.W. Norton, New York

Gould SJ (1987) Evolution and the triumph of homology, or why history matters. Am Sci 74 (1):60–69

Gould SJ (2002) The structure of evolutionary theory. Harvard University Press, Cambridge

Graham L (1981) Why can't history dance contemporary ballet? Or Whig history and the evils of contemporary dance. Sci Technol Hum Values 6(34):3–6

Harrison E (1987) Whigs, prigs and historians of science. Nature 329:213–214

Hull D (1979) In defense of presentism. Hist Theory 18(1):1–15

Hull D (1988) Science as a process: an evolutionary account of the social and conceptual development of science. Chicago University Press, Chicago, IL

Jardine N (2003) Whigs and stories: Herbert Butterfield and the historiography of science. Hist Sci 41(2):125–140

Koertge N (1990) The function of credit in Hull's evolutionary model of science. Proc Biennial Meet Philos Sci Assoc 1990:237–244

Koselleck R (2002) The practice of conceptual history: timing history, spacing concepts. California University Press, Stanford

Kuhn T (1968) The history of science. Int Encyclopedia Soc Sci 14:74–83

Levit SG, Hossfeld U (2021) Talking about the Darwinian revolution: natural selection in Ernst Haeckel's legacy. In: Delisle R (ed) Natural selection: revisiting its explanatory role in *evolutionary biology*. Springer, Cham, pp 105–136

Mandelbaum M (1977) The anatomy of historical knowledge. John Hopkins University Press, Baltimore

Mayr E (1982) The growth of biological thought. Belknap, Cambridge

Mayr E (1990) When is historiography Whiggish? J Hist Ideas 51(2):301–309

Mayr E, Provine W (eds) (1980) The evolutionary synthesis: perspectives on the unification of biology. Harvard University Press, Cambridge

Ochoa C (2021) The origins of theoretical developmental genetics: reinterpreting William Bateson's role in the history of evolutionary thought. In: Delisle R (ed) Natural selection: revisiting its explanatory role in evolutionary biology. Springer, Cham, pp 137–170

Pigliucci M, Müller GB (eds) (2010) Evolution: the extended synthesis. MIT Press, Cambridge

Radick G (2012) Review—the exemplary Kuhnian: Gould's structure revisited. Hist Stud Nat Sci 42(2):143–157

Reif W, Junker T, Hossfeld U (2000) The synthetic theory of evolution: general problems and the German contribution to the synthesis. Theory Biosci 119:41–91

Reingold N (1981) Science, scientists and historians of science. Hist Sci xix:274–283

Roger J (1984) Per una Storia Storica delle Scienze. Giornale Critico della Filosofia Italiana, LXIII, fasc. III:285–314

Russell C (1984) Whigs and professionals. Nature 308:777–778

Schwartz JH (2021) What's natural about natural selection? In: Delisle R (ed) Natural selection: revisiting its explanatory role in evolutionary biology. Springer, Cham, pp 329–390

Stocking GW (1965) On the limits of "presentism" and "historicism" in the historiography of the behavioral sciences. J Hist Behav Sci 1(3):211–218

Van der Meer J (2021) The concept of natural selection in Theodosius Dobzhansky. Its development and interpretation. In: Delisle R (ed) Natural selection: revisiting its explanatory role in evolutionary biology. Springer, Cham, pp 291–328

Part II
Different Views of Charles Darwin

Chapter 3
Guiding a Train of Discoveries: Charles Darwin, Charles Daubeny, and the Reception of Natural Selection, 1859–1865

Richard Bellon

Abstract Charles Darwin was convinced after the publication of the *Origin of Species* in late 1859 that evolutionists would soon form "a good body of working men." Natural selection, he believed, would appeal to workers because it did not only organize existing knowledge but also inspired novel investigation. His optimism proved correct but premature. In the early 1860s, commentators were far more likely to appropriate natural selection as a weapon for their existing cultural or theological agendas than to engage it as a tool for original research. A few naturalists, however, saw its potential and deliberately avoided charged ideological and theological concerns. Charles Daubeny, Oxford's professor of botany, illustrates this dynamic. In largely overlooked remarks at the 1860 meeting of the British Association for the Advancement of Science, he evaluated natural selection skeptically but with pointed respect. He made a plea for less debate and more scientific inquiry. Although he never gave his full assent to evolution, in 1865 he insisted that even the "most determined opponents" of natural selection ought to admire a theory that had led Darwin to so many invaluable discoveries. Daubeny's engagement with natural selection provides a case study in the ways that Victorians naturalists did not just argue about truth but also searched for tools.

Keywords Agricultural chemistry · British Association for the Advancement of Science · Charles Darwin · Charles Daubeny · Final causes · Joseph Hooker · Manure · Natural selection · Orchids · Oxford · Sexuality in plants · Systematic botany

---

R. Bellon (✉)
Department of History and Lyman Briggs College, Michigan State University, East Lansing, MI, USA
e-mail: bellonr@msu.edu

R. G. Delisle (ed.), *Natural Selection*, Evolutionary Biology – New Perspectives on Its Development 3, https://doi.org/10.1007/978-3-030-65536-5_3

3.1 Introduction

Jane Maienschein (1991: 299) concludes in her study of American biology at the turn of the twentieth century that historians miss pivotal developments by focusing too exclusively on scientific theories. "We must in some case at least look to epistemic claims," she writes, "by which is meant a set of claims about what might count as knowledge about the natural world." Philip J. Pauly (2000: 9) reaches a similar conclusion in his survey of American biology. "Scientists' plans and actions became more important than theories and discoveries," he argued. American discussions on evolution in particular were not simply "arguments about truth, but also—and sometimes most centrally—as a search for tools."

This essay applies Maienschein and Pauly's insights to examine the reception of Charles Darwin's theory of evolution by natural selection in Britain immediately after the publication of *On the Origin of Species*. None of Darwin's contemporaries evaluated natural selection purely on intellectual grounds. They examined the theory based on how they believed it would influence the plans and actions of diverse actors. Obviously, judgments on natural selection's truth value influenced the predictions of its tendencies to influence behavior; but often, people judged it true or false based on how they perceived its tendencies, rather than on the reverse.

The botanist Joseph Hooker, Darwin's closest scientific confidant, insisted privately in 1845 that any theory of species origin must remain subservient to the practical needs of natural history research.[1] A decade later he publicly lamented the "unsatisfactory" and "confused state" of botany. Sloppy practice and conceptual confusion reinforced each other in a vicious circle. Darwin took Hooker's anxieties deeply to heart (Hooker and Thomson 1855: 36; Bellon 2006; Endersby 2008). He aspired for natural selection to do more than organize *existing* knowledge under a general theoretical framework. His theory promoted scientific progress by stabilizing the current research practices of natural history. While he openly acknowledged the "difficulties" of natural selection, these were less severe than those which beset special creation. Only *his* theory clearly suggested hypotheses to resolve the obscurities. Natural selection, he boasted, promised "a grand and almost untrodden field of inquiry" (Darwin 1859: 203–207, 484–488).

Darwin, then, engineered natural selection to aid research. Many allies, and even some erstwhile opponents, approached natural selection on these terms. Others, again both allies and opponents, plugged Darwin's ideas into the rancorous ideological and theological controversies which had rumbled through British life for decades. For the participants in these cultural melees, natural selection's set of interlocking propositions about patterns in the natural world was largely interesting—often exclusively—to the extent that they could be weaponized. Darwin designed natural selection as a scientific scalpel; his contemporaries repeatedly appropriated it as a machete for close-quarter ideological battle.

[1]Burkhardt et al. (1985– : 3: 254–255) [Hooker to Darwin, 14 September 1845]

The notorious 1860 meeting of the British Association for the Advancement of Science in Oxford provides an illuminating case study of the diverse uses found for natural selection, both as a scalpel and a machete. The face-to-face debate between Samuel Wilberforce, Lord Bishop of Oxford, and Thomas Huxley has attracted both mythmaking and sophisticated historical analysis. The most interesting participant in the debates over evolution at the 1860 meeting, however, has attracted little attention in either myth or history. The Rev. Charles Daubeny, professor of botany at Oxford, approached natural selection with a deep religious agenda—one that obligated him to treat the theory with respect. As a man of science, he had a profound duty to uncover the mechanisms of God's creative agency in the physical world. If natural selection advanced the plans and action of serious research, he had a scientific and a religious duty to take it seriously and use it reverently.

3.2 A Good Body of Working Men: Sunday, Christmas Day, 1859. Down House, Downe, Kent

Charles Darwin wrote several letters to friends on Christmas Day in 1859. The *Origin* had been published almost exactly a month earlier, and he felt confident about its prospects. He offered Huxley "hearty thanks & real admiration for your generous zeal" on his behalf. He complained about his persistently poor health to his cousin, William Fox, but also cheered that the *Origin* was selling well and "has already made a few enthusiastic & first-rate converts, viz [Charles] Lyell, Hooker, Huxley, [William] Carpenter &c."[2]

Darwin saved his most personal reflection for Hooker. "Everything which I write about or think of, I long to talk over with you," Darwin had gushed to his friend a few years earlier.[3] In the fortnight before Christmas, he had peppered Hooker with hopeful predictions. On the 14th, he forecast that "we [evolutionists] shall soon be a good body of working men & shall have, I am convinced, all young & rising naturalists on our side." A week later: "The assent & arguments & facts of working naturalists like yourself are far more important than my own Book." He returned to the theme on Christmas.

> There is one thing I am fully convinced of that the future progress (which is the really important point) of the subject will depend on the really good & well-known workers, like yourself Lyell & Huxley, having taken up the subject, than my own work.— I see it plainly it is this that strikes my non-scientific friends.[4]

This focus on "working" naturalists resulted from his recognition that the fate of his theory hinged on his ability to rally a community of expert naturalists around it. It

[2]Burkhardt et al. (1985– : 7: 449–452) [Darwin to Fox, to Hulxey, 25 December 1859]

[3]Burkhardt et al. (1985– : 6: 372) [Darwin to Hooker, 12 April 1857]

[4]Burkhardt et al. (1985– : 7: 431, 441, 450) [Darwin to Hooker, 14, 21, and 25 December 1859]

also demonstrated an awareness that he needed more than passive intellectual acquiescence. The assent had to be reflected in active day-by-day investigation.

Darwin's Christmastime reflections adhered to the principles of scientific method articulated by John Herschel and William Whewell (Hull 2003). The influence of both men ran deep in British science; more than anyone else, they systematized the principles at the heart of the British Association. Whewell insisted that theories earned legitimacy only "through innumerable vicissitudes of confusion, error, and imperfect truth" (1849: 178). He attributed the dramatic conceptual advance of geological theory, for example, to a community of men (Darwin conspicuously included) who tested their ideas "with the hammer in their hands, with the knapsack at their backs, or at their saddlebow" (1831: 643–6). The history of science shows, Whewell said, that the paths opened by scientific theories only become clear after "the dust of the confused march" settles (1837: 1: 12–13).

Whewell drew heavily on Francis Bacon's condemnation of Anticipation—as opposed to Interpretation—of Nature. The anticipator catapulted from a scattering of facts to sweeping general axioms. The interpreter "passes by a continuous and gradual ascent through intermediate generalities" (1831: 53; 1857b: 303). Anticipation was the triumph of vice over virtue. "Laborious observation, narrow and modest inference, caution, slow and gradual advance, limited knowledge, are all unwelcome efforts and restraints to the mind of man, when his speculative spirit is once roused," he moralized; "yet these are the necessary conditions of all advance in the Inductive Sciences" (1857a: 1: 339).

Herschel's *A Preliminary Discourse on the Study of Natural Philosophy* served as one of Darwin's lodestars (Darwin 1958: 67–68; Secord 2014: 80–106). In this landmark book, Herschel identified "the analysis of phenomena" as the immediate object of any physical theory. Well-constructed theories suggest testable hypotheses. Hypotheses in turn lead to "the trial of many curious experiments." The outcomes of these trials allow investigators to "form a theory more in consonance with fact" or to choose between rival theories. Only after extensive trial and difficulty does a theory "truly represent all the facts, and include all the laws, to which observation and induction lead" (1830: 190–220).

Herschel drew upon analogies with building construction to explain the role of theory in science. He described theories as the "corner-stones in the temple of science, a building always altering, always enlarging, and combining in every age, in its several departments, every form of architecture from the rudest to the most refined which that age admits." Switching metaphors in the next paragraph, he imaged theories "like a column built by men, [ascending] amid extraneous apparatus and shapeless masses of materials." Whether cornerstone or column, a theory did not simply shoot up out of the ground like a natural growth. Human labor created theory, and theory, in turn, made further construction possible by giving meaning and unity to what would otherwise be fleeting impressions (1840: 272–274; Janssen 2019: 115–116). Celestial mechanics, the most complete of all scientific fields, acquired conceptual stability by uniting the genius of theorists with the industry of observers (Herschel 1831: 16).

Herschel acknowledged that Isaac Newton did leave a complete system of planetary motion. The inverse-square law explained broad celestial patterns, but did not by itself account for the stability of the solar system amidst orbital wobbles and variations. Herschel emphasized that the gaps in Newton's system of universal gravitation were not embarrassments, because the theory itself provided the tools to fill them. In the nineteenth century, "there is not a single perturbation, great or small, which observation has ever detected, which has not been traced up to its origin in the mutual gravitation of the parts of our system, and been minutely accounted for, in its numerical amount and value, by strict calculation on Newton's principles" (1833: 313–314).

But even the great Newton travelled down blind alleys. Whewell observed that his followers disregarded his theories on optics and on the material cause of gravity (1857b: 302). An honest, well-constructed theory had use even when it failed, however. "When a theory, which has been received on good evidence, appears to fall, the really essential and valuable part of it survives the fall," Whewell observed; "that which has been once discovered continues to be true" (1834: 266).

Darwin, as Michael Ruse observes, "wanted to make his theory as Newtonian as possible" (1979: 176; see also Hodge 2009: 53–54 and Bellon 2014). He constructed the *Origin* carefully to follow Newton's template as consecrated by Whewell and Hershel. This influence is particularly clear in Darwin's understanding of a scientific "revolution." Both Herschel (1851: 4) and Whewell (1834: 267; 1838: 630) accepted conceptual revolutions as necessary for the advancement of scientific knowledge. Darwin confessed in the *Origin* that "when the views entertained in this volume on the origin of species, or when analogous views are generally admitted, we can dimly foresee that there will be a considerable revolution in natural history" (1859: 484).

This invocation of revolution was not celebratory. Darwin acknowledged what he could not deny, that natural selection upended long-cherished assumptions about the nature of life. He recognized the disruptive potential of evolutionary speculations even as a young naturalist contemplating patterns in the geographical and temporal distribution of South American fossils (Brinkman 2009). If the great Newton himself left an incomplete revolution, no wonder Darwin pushed his own to the dimly foreseen future. He never claimed to have constructed a final explanation of species origin. Natural selection was a working principle, a tool that would stand or fall exactly to the extent that it guided future research productively; it was not a speculative "Anticipation" that sought to bypass the arduous collective work of testing, revision, and confirmation.

By pushing his potential revolution to the dim far horizon, he reassured readers that his theory would preserve, or enhance, the core practices of natural history. "Systematists will be able to pursue their labours as at present," Darwin pledged immediately after confessing to revolutionary intent; "but they will not be incessantly haunted by the shadowy doubt whether this or that form be in essence a species."[5] He noted that evolution provided solidity to technical terms like analogy, community of type, morphology, and adaptive characters. New discoveries beckoned (1859: 484–486).

No one better understood Darwin's agenda in the *Origin* than Hooker—he had, as Darwin put it, scrutinized his evidence and reasoning as a "stern & awful judge & sceptic" for nearly 15 years.[6] While Darwin wrote the *Origin*, Hooker toiled simultaneously on a technically dense defense of evolution in his Introductory Essay to *Flora Tasmaniæ*. Drafts and proof sheets flew back and forth between Down House and Kew Gardens. When the published version of Hooker's essay finally arrived, Darwin read it with absorption from Boxing Day until shortly after New Year. "It is by far the grandest & most interesting Essay on subjects of the nature discussed I have ever read," he enthused.[7]

Hooker drew on deep practical experience to argue that "the long serviceable theory of special creations" had become barren for the study of plant systematics, biogeography, and morphology. If biological patterns "may receive a rational explanation under another theory, it is to this latter that the naturalist should look for the means of penetrating the mystery which envelopes the history of species, holding himself ready to lay it down when it shall prove as useless for the further advance of science" (1860: xxvi; Bellon 2006).

Darwin had a good reason for his Christmas Day optimism about the *Origin*. True enough, the first public notice of the book had proved to be a rather nasty affair. The reviewer for *The Athenaeum*, a prominent weekly devoted to science and the arts, savaged Darwin's "poetical" and "bold" views which, he claimed, contradicted both the established scientific knowledge and the wisdom of theologians who had the obligation to condemn "another elaborate theory to exclude Deity from renewed acts of creation." Darwin seethed (Leifchild 1859; Browne 2002: 87–88). But he still felt hopeful that working naturalists would, like Hooker, Huxley, Lyell, and a handful of others, grant him a much fairer hearing. This confidence evaporated over the next half year.

Many workers refused to enter the controversy at all. Hooker briefly discussed evolution and natural selection with George Bentham, a friend and collaborator, until he curtly cut off the conversation. Bentham resolved to keep working on descriptive systematic botany without bothering about the origin of species. "My turn of mind is far too matter of fact to grapple a subject which is full of hypotheses and cannot be dealt with without giving full scope to the imaginative faculty," he informed Hooker (Bellon 2003: 286). At least Bentham read Hooker's essay; many prominent botanists simply ignored it.[8] Other men of science joined the active opposition. By

[5]See Adams (Chap. 8) for a discussion of the persistently complicated relationship between systematic practice and evolutionary theory.

[6]Burkhardt et al. (1985– : 7: 89) [Darwin to Hooker, 6 May 1858]

[7]Burkhardt et al. (1985– : 7: 453, 8: 6–7) [Darwin to Hooker, 26 December 1859 and 3 January 1860]

[8]Burkhardt et al. (1985– : 8: 211) [Darwin to Hooker, 15 May 1860]

summer Darwin noted that most of the reviews in England bitterly opposed him.[9] Perhaps most vexingly, Richard Owen condemned him *personally*—and not just the argument of the *Origin*—without compromise in the April issue of the *Edinburgh Review*. Owen summarily ejected Darwin from the company of naturalists "who trouble the intellectual world little with their beliefs, but enrich it greatly with their proofs" (1860: 529–530; Rupke 1994: 238–242).

The *Saturday Review* reported in May that discussions of the *Origin* had "passed beyond the bounds of the study and lecture-room into the drawing-room and the public street." The writer concluded that most people for and against Darwin "sought for confirmation of their own opinions" (Anon. 1860b). The *Origin* sailed on choppy seas.

3.3 Improving Our Moral and Intellectual Faculties: Wednesday, June 27, 1860. Sheldonian Theatre, University of Oxford

The British Association opened its 30th annual meeting on a sunny day in Oxford on Wednesday, June 27. The morning and early afternoon were devoted to council and general-committee business. The first public meeting began at 4:00 in the afternoon when Prince Albert entered the auditorium of Christopher Wren's magnificent Sheldonian Theatre with a retinue of aristocrats and university dignitaries. The large assembly of men and women rose and cheered the Prince Concert, the one who had served as president of the previous year's meeting in Aberdeen. "My last and pleasing duty is to introduce my successor, Lord Wrottesley," he announced to warm applause (Anon. 1860c).

John, second Baron Wrottesley, was a productive observational astronomer with a long record of service to the Royal and Astronomical societies (Brayley 1867–68). Now in possession of the chair amidst the pomp and ceremony, he undertook one of the most important duties on the social calendar of British science. Each year the association's president made a stately address on the current progress of science, one designed to define and celebrate natural knowledge as integral to the larger sweep of British culture. The president's remarks were not for the moment only. The press covered them widely, and the full address appeared in the meeting's *Report*.

Wrottesley understandably gave astronomy pride of place in his remarks, but he covered advances across the spectrum of scientific activity, applauding, for example, the application of chemistry, physics, and microscopy to biology for fostering an explosion of precise experimental data. He remarked on welcome curricular reforms in Oxford and on the prospects for parliamentary research grants. He described at some length recent geological excavations in France's Somme Valley which had uncovered prehistoric human tools. The address put the weight of authority on the

[9]Burkhardt et al. (1985– : 8: 287) [Darwin to Bronn, 14 July 1860]

side of Lyell's contentious dating of these tools, which pointed toward the deep antiquity of human history (Lyell ghostwrote this portion of the address) (Van Riper 1993: 157). This startlingly reappraisal of the human past was palatable, Wrottesley implied clearly, because it was grounded on new facts and extensive ongoing investigation. The earnest study of nature, he stated directly, "raises us in the scale of human beings and improves every moral and intellectual faculty" (1861: lvii).

Wrottesley ended his address with a paean to "the marvels of the Creator's power" and to "the beauties and prodigies of contrivance" in the living world. Serious scientific study brought one nearer to God, he said in conclusion (1861: lxxv). These sentiments were undoubtedly sincere but also had a pro forma flavor. Statements along the same pious line appeared in nearly every presidential address. Prince Albert had ended his remarks the previous year on an almost identically reverent note (Albert 1860: lxix).

The British Association, from its beginnings in 1831, had tried (not always successfully) to balance freedom of inquiry with reassurances that this independence would draw men and women ever closer to God (Morrell and Thackray 1981: 224–245). The 1839 presidential address by the Rev. William Vernon Harcourt, the founder of the association, shows that this alliance between science and religion was often clumsy and ambiguous. He condemned "the dying embers of . . . obsolete bigotry" which made misguided appeals to scripture to justify "crude and ignorant speculations." Scientific workers must stand resolutely against anyone who aspired "to stir up a worse than civil war between the feelings of piety and the deductions of reason." At the same time, Harcourt insisted that science had conclusively demonstrated "*that all men are the children of one human father, and the handiwork of one Almighty God* [original italics]." He warned his audience against the "misnomered philosopher" who disputed settled evidence of God's handiwork in nature. "Disgrace not the name of Science by throwing stones at the Temple of Truth," he warned (1840: 16–17, 20–21).

As the audience dispersed after Wrottesley's address, a question hung over the meeting. Was opposition to Darwin's theory of natural selection driven by a spasm of obsolete bigotry? Or had he become a misnomered philosopher who dishonored science with wanton speculation?

3.4 The Necessity of Further Inquiries: Thursday, June 28, 1860. Oxford University Museum

Section D—Zoology and Botany, including Physiology—met in Oxford's new university museum. The sections were the working heart of the British Association, venues where a gentlemanly inner circle joined with novice and provincial researchers to share results with a national audience. Sections set standards and policed social hierarchies. The gentlemanly leaders boosted lower-status researchers who followed their lead and disciplined those who did not, all in public view (Morrell and Thackray 1981: 451–460).

Lyell's views on human antiquity which Wrottesley had endorsed on Wednesday were first aired the previous year in geology's Section C (Lyell 1860). A competing paper by the Rev. John Anderson (1860) next attempted to refute geological analysis, pointing to the deep antiquity of human history. Lyell and his allies responded with polite condescension and a fusillade of enough factual detail to "scarify him," as one put it. The goal was less to convince Anderson of his errors, however, and more to warn others of the fate facing anyone foolhardy enough to trod on elite toes (Van Riper 1993: 123–125). Twelve years earlier, Lyell led a similar barrage against a geological paper by Robert Chambers, widely and rightly assumed to be the anonymous author of the evolutionary blockbuster, *Vestiges of the Natural History of Creation*. Lyell openly acknowledged that he and his friends meted out this rough treatment to show "that reasonings in the style of the Author of the Vestiges would not be tolerated among scientific men" (Secord 2000: 434–436). These pummelings could not force critics of mainstream science into silence, but they did send a strong message about elite consensus. Section C's notoriety for drama made it a popular public draw.

In 1860, however, with evolution being a lively controversy, Section D promised the best entertainment. As the local eminence, Daubeny originally agreed to serve as section president but backed out because of his arduous responsibilities as a vice president for the meeting as a whole. John Stevens Henslow—Daubeny's counterpart professor of botany at Cambridge, Darwin's beloved mentor, and Hooker's father-in-law—assumed the chair in his place and gaveled the section open. The first presentation addressed the headway made by a scientific dredging project off the north and east coasts of Scotland (little, it turned out, because Scottish weather refused to cooperate). The second lauded the principles and progress of the Smithsonian Institution in the United States (Anon. 1860a: 25–26). Finally, Daubeny took the floor and addressed the great question of the day in his paper "On the Final Causes of the Sexuality of Plants, with Particular Reference to Mr. Darwin's Work on the Origin of Species."

Daubeny, no less than Darwin, adhered to the same broad methodological principles articulated by his contemporaries Whewell and Herschel. Scientific colleagues valued the Oxford don as a diligent, accurate, and creative experimentalist. John Phillips, who replaced William Buckland in the Oxford geology chair, remembered Daubeny as "always prompt and sagacious in fixing upon the main argument and the right plan for following up successful experiment or retrieving occasional failure." He was also, his friend said, temperamentally disposed to remain "calm amid jarring creeds and contending parties" (Phillips 1868–9: lxxvii–lxxviii).

Daubeny had been a fixture in Oxford for nearly four decades (Goddard 2004; Oldroyd 2004). He assumed the chair of chemistry in 1822 when little interest attached to physical science in the university. From the start, he fought to integrate science into Oxford education with the goal, as he put it later, to accommodate "our medieval institutions to the wants of the present age" (1867b: 2: 41). His inaugural lecture promised a course organized around the general laws of chemical affinity and heat. He reassured students that he would not afflict them with an unphilosophical "assemblage of unconnected facts and obscure formulæ." He tactfully made the case that integrating chemistry into the university curriculum was not an innovation. Rather, chemical study expanded "the higher powers of the mind" and deserved to be considered "as no unimportant part of that noble and extended superstructure, which the founders of our University establishment doubtless meant to erect upon the solid basis of classical education" (1823: 6–9, 18).

Daubeny applied chemistry to a longstanding interest in geology. One result was the reputation-making *Description of Active and Extinct Volcanoes*, which Darwin brought with him on the voyage of HMS *Beagle*.[10] The revised edition, which appeared in 1848, retains enduring value (Daubeny 2011 [1848]). He added the chair of botany to his duties in 1834. From outside the university, the botanist John Lindley (1834) celebrated Daubeny's new position as a sign that "the absurd courses of study pursued in the great English schools" would not long survive. Daubeny, on the inside, crafted a different message. His inaugural lecture in this subject prudently tied the study of plant life to theological education. Botany, he said, provided "no bad exercise of the mental powers." But, far beyond that utility, it afforded "to the Divine some of the most beautiful illustrations of design with which nature can supply him, as well as the most delicate subjects of inquiry connected with the doctrines of final causes" (1834: 27).

These inaugural lectures show how meticulously Daubeny tried to accommodate the university's prevailing conservative attitudes (Rupke 1997: 548–549; Robb-Smith 1997: 571). He attempted to raise the profile of Oxford science by inviting the second meeting of the British Association to the university in 1832 without prior permission from his academic superiors. The gamble paid off, and the university welcomed the event, although the scene of non-Anglican men of science honored in Anglican Oxford generated severe and lasting ill-will among some powerful conservative factions (Morrell and Thackray 1981: 230–233, 240; Bellon 2015: 218–221). During the 1830s, his colleague Baden Powell, the professor of geometry, attracted no students at all to his lectures in some years and only one or two in others (Rupke 1997: 548).

The situation of the physical sciences in Oxford had improved significantly by the early 1850s. John Mozley attributed this progress largely to "the ability, the perseverance, and the disinterested scientific zeal of Dr. Daubeny" (1853: 178). He inspired a circle of younger science colleagues, particularly the professor of medicine Henry Acland, to campaign for a university science museum to serve teaching, collection management, research, and public education. The campaign succeeded. The university's governing Convocation voted final approval of the new building in December 1853.

Daubeny attended the elaborate ceremony to lay the cornerstone on June 20, 1855. The university's chancellor Edward Geoffrey, 14th Earl of Derby, presided. The assembly sung the canticle *Benedicite omnia Opera*—"O All ye Works of

[10] Burkhardt et al. (1985– : 1: 157–158) [Sedgwick to Darwin, 18 September 1831]

the Lord, bless ye the Lord: praise him, and magnify him forever" (Mercer 1855: 106–107)—and the national anthem. Acland composed a prayer for the occasion.

> Avert the evil, and promote the good, O merciful Father. Grant that the building now to be erected on this spot may foster the progress of those Sciences which reveal to us the wonders of Thy creative powers. And do Thou, by Thy heavenly grace, cause the knowledge thus imparted to us to fill us with apprehension of Thy greatness, Thy wisdom, and Thy love. (Atlay 1903: 213)

The ceremony inspired Daubeny to publish reveries about a museum with well-equipped laboratories, lecture theatres, plant and animal specimens arranged in "an assemblage of the whole series of organized beings that had existed upon the earth since the first dawn of creation," and a well-stocked library; "whatever, in short, human industry had from time to time brought to light as illustrative of the past as well as the present history of our globe" (1867b: 2: 143–144). While no museum could live up to this ideal, the Oxford museum was no longer a dream. Science had finally placed an imposing architectural footprint on the ancient university, a magnificent triumph in Daubeny's decades-long battle for science at Oxford (Fox 1997; Yanni 2005: 62–90).

Wrottesley hailed the museum in his presidential address as "substantial proof of [Oxford's] sincere interest in the diffusion of science." He looked forward to it inspiring students to "a closer contemplation and more diligent study of the glorious works of Nature" (1861: lxvii). The British Association meeting marked the inaugural use of the museum, with several sections, including Section D, congregated in the building. The new museum embodied in stone and steel the broad principles that Daubeny brought to the *Origin*. Its design aspired to speak to more than the visitor's reason, Acland said; its Gothic style aspired to warm the heart with "faithful love of the Nature of God" (Acland and Ruskin 1859: 57–58). It was in these architectural surroundings that Daubeny, the clergyman naturalist, brought agendas of faith and reason to natural selection.

Daubeny worked from well-developed principles. He articulated his personal balance of faith and scientific vocation 4 years earlier in his 1856 presidential address to the British Association in Cheltenham. He argued against the "startling consequences" of evolution, drawing on the operation of universal gravitation for an analogy to species fixity.

> The [divine] will, which confines the variations in the vegetable structure within a certain range, lest the order of creation should be disturbed by the introduction of an indefinite number of intermediate forms, is apparently the same in its motive, as that which brings back the celestial Luminaries to their original orbits, after the completion of a cycle of changes induced by their mutual perturbations. (1857b: lxiii)

Daubeny's association of species limits with stable orbits was not simply the product of the natural theological imagination.

Pierre-Simon Laplace, the French mathematical astronomer who, more than anyone else, placed celestial mechanics on an exclusively Newtonian footing, made the same connection 70 years earlier in his legendary paper on the perplexing motions of Saturn and Jupiter.

> So the system of the world merely oscillates around a mean state from which it deviates very little. ... This stability of this system of the world, which assures its duration, is a phenomenon most worthy of our attention because it shows us in the heavens the same intention to maintain order in the universe that nature has so admirably followed on earth to sustain individuals and to perpetuate the species.

In one well-circulated story, Napoleon challenged Laplace on the absence of any reference to God in his masterpiece of Newtonian analysis, *Celestial Mechanics*. "I have no need of that hypothesis," he retorted (Hahn 2005: 78–80, 172). Herschel, Whewell, and other British scientific theists did have this need which they satisfied rather easily by attributing stability to divine forethought. Whewell claimed that Laplace demonstrated God's ability to achieve His ends through natural law (1833: 343–350). God's glory, Herschel wrote in the *Preliminary Discourse*, appeared most majestically in "the calm, energetic regularity of nature," the patient observation of which brings people "nearer to their Creator" (1830: 16–17). The connection of stable orbits to fixed species appears to have been entirely off-hand for Laplace. It was not for his British admirers.

Daubeny's address did more than simply assert species fixity with airy references to divine will. He raised serious scientific questions. If species could vary indefinitely in response to environmental conditions, he asked, what anchored them to a restricted geographical range? If a genus possessed unlimited plasticity, it should have the power "to diffuse itself throughout the world, as easily as it has done over that portion of space within which it is actually circumscribed." He pointed to Hooker's taxonomic and biogeographical work on genera, such as *Coprosma*, a group of shrubs and small trees. The younger botanist showed that *Coprosma* species exhibited such striking regional variation that subgroups are barely recognizable as members of the same species. And yet, the genus remained mostly endemic to New Zealand (1857b: lxii–lxiii).

Daubeny made a thoughtful choice in using *Coprosma* as one of his examples. The genus embodied the acute challenges of bringing taxonomic order to plants. Hooker moaned privately to Darwin that *Coprosma* presented so much variability that "for the life of me I do not know how to draw the line between there being only one species, or 28!"[11] This uncertainty was not an academic issue but a severe headache to someone toiling on *Flora Novæ-Zelandiæ*. Ultimately, he settled on 19 distinct species. *Coprosma* might have been "the most variable and difficult New Zealand genus," but it merely represented the extreme edge of a chronic problem (Hooker 1853–55: 1: 103–111).

Again and again and again, naturalists encountered populations which were both clearly distinct *and* clearly related. Does the taxonomist treat these populations as two varieties of the same species or two closely allied species in the same genus? The doctrine that all individuals in a species traced descent from a single ancestral form with a distinct origin provided no guidance for answering this question in practice. The 19 species of *Coprosma* which Hooker identified in New Zealand were not

[11]Burkhardt et al. (1985– : 5: 67–68) [Hooker to Darwin, November 1851]

obvious facts of nature but the result of his own particular judgment and subject to reconsideration. Few if any fellow experts, even with access to the same materials, would divide *Coprosma* as he had. H.C. Watson observed that where one botanist might see one highly variable species, another might identify up to a dozen. This startling divergence on species delimitation, he noted, resulted from practical rather than conceptual disagreement. Knowledgeable botanists frequently reached dramatically different conclusions about the number of species within a group while still agreeing closely "in their abstract ideas and definitions of the term *species*" (1847–59: 4: 34, 44–52). The problem of delimiting species was not limited to botany. Darwin collided with it innumerable times when classifying barnacles (Stott 2003).

Henslow's *The Principles of Descriptive and Physiological Botany* insisted that botanists could not work together without a global catalog of plants based on uniform standards of nomenclature and classification. Philosophical classification, then, provided an indispensable communal tool in the search for "the laws and principles which regulate the structure and fix the properties of plants" (1835: 136). Ideally, a uniform standard for delimiting species would facilitate research into the laws governing plant form and function; a better conceptual understanding of plant life would in turn improve the practice of naming species. In reality, Hooker lamented to his friend Asa Gray that botanists were wandering further and further apart in their judgments. "I do not say this because my views diverge from others, but because I see others as divergent from one another as I am from them" (Bellon 2006: 19).

Hooker considered the entire contentious situation an embarrassing scandal. Chaotic nomenclature mocked botany's claim to philosophical legitimacy. The confusion made it nearly impossible for naturalists to arrive at general conclusions about global distribution patterns of plants that, as Henslow's textbook pointed out, the goal of the entire laborious systematic enterprise. Hooker aggressively advocated a broad species concept. He lumped allied subgroups under the same species name whenever possible. Most elite naturalists, including Darwin, Lyell, and Henslow, shared Hooker's disdain for the so-called species mongers (typically lower-status naturalists) who tended to multiply species names (McOuat 2001; Endersby 2001, 2008; Allen 2003; Bellon 2006).

Daubeny was not an active systematic botanist, but, as the manager of a large herbarium at the Oxford Botanic Garden, he acutely appreciated the conceptual and practical difficulties of delimiting species on general principles. Daubeny recognized the chaos sowed by different systems of chemical notation (1837: xxix–xxxi); disordered taxonomic practices presented an even more acute challenge to the communal practice of natural history. In his 1856 presidential address, he praised the great service rendered to natural science by Hooker and his colleague Thomas Thomson—"these indefatigable botanists"—in protesting "against that undue multiplication of species" (1857b: lxii).

The analogy Daubeny drew between fixed species and stable orbits raised a glaring problem. Laplace applied the logic of Newtonian mechanics to show how the solar system maintained stability amidst perturbations. The theory of special creation offered no similar support for establishing the limits of species variability. But neither did grandiose invocations "of the universal self-sustaining and self-evolving powers which pervade all nature," which his Oxford colleague Powell placed at the heart of his scientific and religious reasoning (2000 [1860]: 255–256).

Natural selection offered a tool deliberately crafted for biogeographical and taxonomic work. When Daubeny read the *Origin*, he did not abandon his firm conviction that plants manifested God's purpose in nature. But botanists still had the duty to discover the specific contours of God's creative superintendence in plant sexuality or in the natural groupings of plant species. Did natural selection have anything useful to contribute to the study of these topics? Probably, Daubeny acknowledged.

Daubeny did play a bit of bait-and-switch with his remarks. Most of his presentation surveyed current understanding of plant sexuality without any direct reference to natural selection. Anyone who turned up to the session anticipating rhetorical fireworks—what the *Athenaeum* advertised as combatants finding "foemen worthy of their steel, and [making] their charges and countercharges very much to their own satisfaction and the delight of their respective friends" (Anon. 1860a: 19)—were likely disappointed by Daubeny's dry performance. He treated his audience to a dense discussion of plant morphology and physiology. When Daubeny finally *did* mention Darwin, at first he did so to discuss rudimentary organs and pollination mechanisms (1860: 17, 22).

Eventually, about two-thirds finished, Daubeny starting probing natural selection conceptually. Could intricate sexual structures arise in plants *prior to* their physiological role in reproduction? he asked. He doubted it but resolved to leave the question "to Naturalists more equal to the task than myself, to enter the lists against an antagonist furnished with so vast an armoury of facts, and gifted with so singular a power of applying them to the purpose of his theory." He also doubted that the human moral sense could have emerged from anything so crude as the brain of a gorilla. But, in this case, as in all others, questions of "pure science" should proceed in an open spirit, without the impingement of "higher consideration." (A footnote in the published version of the remarks elaborated on this point. Theologians should not interfere in scientific questions, he said, except to act as a type of referee to ensure debates were carried out fairly. And even if it were demonstrated—which he did not concede likely—that the human material form had evolved from apes, this conclusion would not refute the reality of the living human soul.) He turned next, and finally, to the hope that natural selection could help guide the delimitation of species (1860: 22–34).

He finally addressed larger conceptual issues in the final third of his presentation. The audience would have almost certainly struggled to follow him as he jumped from topic to topic. He concluded on a skeptical note.

> There is a great gulf between the act of merging even all the species belonging to a well-defined genus into one, and the assumption that a vertebrated animal could have been developed by natural causes out of an invertebral one, or that the higher faculties of the Human Race could have emanated from the glimmering Intelligence perceived in the Brute. (1860: 34)

So ended a judicious, unapologetically technical, and somewhat disjointed performance. And yet Daubeny's respect for the *Origin* was unmistakable. (Darwin read the pamphlet version and thanked Daubeny for "the *extremely* kind & liberal manner in which you allude to my work.— I can wish & hope for nothing better.")[12] The *Athenaeum* ended its summary of the paper by reporting Daubeny's recommendation for "the necessity of further inquiries" (Anon. 1860a: 26).

Daubeny did not invoke the doctrine of lawful permanency which had figured so prominently in his presidential address at Cheltenham 4 years earlier. The reason is found in the finale to that speech which advocated "a single-minded and reverential pursuit after truth in our present state of trial." He considered tenacious research both as the only path to reliable knowledge of the natural world and as a form of moral discipline "calculated to induce the humility that paves the way to belief" (1857b: lxiii–lxiv, lxx, lxxii–lxxiii). From this perspective, the proper response to Darwin's theory was not to condemn it—or, for that matter, accept it—on doctrinal grounds but to probe its usefulness and its limitations as a tool for contemplating the works of the Creator.

When Daubeny finished his remarks, Henslow as chair asked Huxley for comment. (Hooker or Lyell would have been more obvious spokesmen for Darwin's views, but neither were present.) Huxley demurred, according to the *Athenaeum.*

> Prof. Huxley ... deprecated any discussion on the general question of the truth of Mr. Darwin's theory. He felt that a general audience, in which sentiment would unduly interfere with intellect, was not the public before which such a discussion should be carried on. Dr. Daubeny had brought forth nothing new to demand or require remark.

This lofty attitude was probably driven less by its stated principle and more by a lack of anything intelligent to add in response to a paper anchored by technical issues outside of his interests and expertise. After a few other people commented, Owen stood up, unctuously wrapped himself in "the spirit of the philosopher," and picked up on Daubeny's passing remarks about brain structure. Fundamental anatomical differences exist between human and gorilla brains, Owen announced, and, with the benevolent condescension of unimpeachable Authority, declared that such facts made Darwin's speculations appear rather more courageous than sensible. Huxley's reticence for public controversy immediately evaporated. He stridently disputed Owens's conclusions. This spur of the moment exchange sparked a public controversy over comparative cerebral structure which rumbled for the next several years (Anon. 1860a: 26; Rupke 1994: 259–286; Browne 2002: 119–120).[13]

Under normal circumstances, Daubeny's respectful treatment of natural selection in Section D would have sent an influential message about the proper tone for the larger debate over evolution. Hooker reported to Darwin, however, that it was the fiery confrontation between Huxley and Owen—not Daubeny's circumspect

---

[12]Burkhardt et al. (1985– : 13: 430) [Darwin to Daubeny, 1 August 1860]

[13]Tattersall (Chap. 14) shows that primate brain structure remains a field for debating larger questions about evolutionary processes.

paper—that made "you & your book ... the topic of the day."[14] These were not normal circumstances.

3.5 The Best Weapon for Future Research: Saturday, June 30, 1860. Oxford Botanic Garden

The main entrance to the Oxford Botanic Garden normally remained locked. Regular access was restricted to members of the university. Members of the public gained admittance if they demonstrated that the garden would provide them with interest or improvement (a requirement which explicitly excluded nursery maids and children) (Daubeny 1853 [1864]: i 51). The doors swung open on Saturday evening for Daubeny's conversazione. Special tents were erected in the normally sedate setting for the influx of guests (Anon. 1860a: 18–19).

Social gatherings like this one were as integral to the British Association as addresses, papers, reports, and formal discussions. Members of the extended community of science forged and maintained bonds of affection indispensable for collaborative work. Hooker became engaged to marry Frances Henslow, the eldest daughter of the Cambridge botanist, at the 1847 meeting in Oxford (Bellon 2004). He rarely bothered to attend sessions, preferring to prowl around the hallways and chat with friends.[15] Conversaziones in general were a staple of middle-class social and culture life. Ones at British Association meetings encouraged scientific grandees to mingle and gossip with a cross section of ladies and gentlemen in enlivening settings. On this evening, the garden's plants provided beauty and fragrance, their scientific arrangement intellectual stimulation. The crowd itself became part of the spectacle (Alberti 2003). As host, Daubeny set a genial tone.

The Botanic Garden dated from 1632, making it the oldest in Britain to Daubeny's pride. Despite its antiquity, he fought (and had to fight) to integrate the garden into the intellectual, educational, and religious mission of the university. His guidebook—first published in 1850, revised in 1853, with some supplemental information added in subsequent printings—accentuated the garden's place in the heart of Oxford by highlighting the magnificent views of Merton, Christ Church, and Magdalen architecture from the grounds (Daubeny 1853 [1864]: i 3, 20, ii 3–15). By 1860, the garden, like the new museum, symbolized the progress of physical science in Oxford.

Plants from around the world were laid out in outdoor plots and in greenhouses with an eye to beauty, utility, and systematics. An experimental garden allowed Daubeny to conduct research on plant physiology and agricultural applications. In the 1830s, he demonstrated that, contrary to received wisdom, plants release more

[14] Burkhardt et al. (1985– : 8: 270) [Hooker to Darwin, 2 July 1860]

[15] Burkhardt et al. (1985– : 8: 270–1, 12: 330–1) [Hooker to Darwin, 2 July 1860, 19 September 1864]

oxygen during the day than they absorb at night, which made them ideal for scrubbing air polluted by animal respiration. Experiments conducted over a decade explored the relationship between soil nutrients and plant flourishing, validating plant rotation. In 1847, the last time the British Association met in Oxford, Daubeny joined a committee including Hooker and his father William, director of Kew Gardens, to investigate the influence of carbonic acid on plant growth. Two years later he reported to the association the results of the trials conducted at the Oxford garden (Daubeny 1850, 1853 [1864]: i 2–24, ii 14–15).

Daubeny had inaugurated a new Botanical Museum in 1859, modeled on the Museum of Economic Botany at Kew Gardens. The Oxford garden also housed one of the few herbaria in Britain which approached Kew's in size and usefulness. Thousands of dried specimens (the oldest dated to the seventeenth century) represented species and varieties from around the world. Hooker had recently donated a large cache of specimens collected during his expedition to India from 1848 to 1851 (Daubeny 1853 [1864]: i 56–62, ii 47–48).

Maintaining a large, cosmopolitan herbarium was costly, space-intensive, and time-consuming. Both Daubeny and the Hookers fervently promoted the broad-species concept in part to prevent their herbaria from being overwhelmed with the need to collect supposed species defined by minor regional difference. Each "species" was not just a Latin name in a book but an often fragile and unwieldy physical object—a plant dried on special paper that had to be stored, preserved, and indexed (Bellon 2006; Endersby 2008; Hoquet 2014). Daubeny, like Hooker before him, recognized immediately the relevance of natural selection to the use and maintenance of a comprehensive collection of natural history objects.

For the purposes of the conversazione, the herbarium, experimental garden, and new museum highlighted the productive, tactile work of science. This work did more than produce intellectually and materially useful results, he noted in the guidebook; it revealed God's concern for human well-being written into the structure of soils, animals, and plants. Many of the milling visitors no doubt overlooked one of the most theoretically and practically significant features of the garden—one that helped explain the approach Daubeny took to the *Origin*. He dedicated plots for the homely purpose of testing the efficacy of various manures (Daubeny 1853 [1864]: 22).

Daubeny took a deep interest in manure as Oxford's first professor of rural economy. Better use of manure had improved British agriculture over the past few centuries. He joked in a lecture on Roman husbandry that in the present day, Hercules would not have found much of a labor to perform at the Augean stables. A modern farmer would never allow so much valuable manure to molder used (1857a: 23). But he cautioned his countrymen against feeling too self-satisfied at their progress. In one of his inaugural lectures in the chair, delivered in November 1840, he stressed the need to transform agriculture from "a mere empirical art" into a practice driven by principles which would, ideally, "direct with ease and certainty the operations of the husbandman." Scientific research into manure embodied this sophisticated new approach, in sharp contrast to prevailing trial-and-error wooliness.

> When we inquire in what manner manure acts, we are answered by the most intelligent men, that its action is covered by the veil of Isis; and when we further demand what this means, we discover merely that the excrements of man and animals are supposed to contain an incomprehensible something. (1841: 36–37)

Daubeny's goal was always more ambitious than assigning names to incomprehensible somethings. He wanted to understand as a matter of general law how active chemical agents improve soil fertility. He had no need for the hypothesis that vegetable growth and development must remain a divine mystery.

Daubeny's lodestar was Justus Liebig's *Organic Chemistry in Its Applications to Agriculture and Physiology*. The book had its partial origin in a paper on the decomposition of uric acid which the German chemist presented to the 1837 British Association meeting in Liverpool. He concluded with a call for British men of science to "unite their efforts to those of the chemists of the Continent" (1838: 41). The association routinely charged members to prepare reports on significant scientific topics, and Liebig, in the wake of his instantly celebrated uric-acid paper, agreed to write two of the eight assigned that year, one on isomeric bodies and the other on organic chemistry. He never produced the former. His treatise on agricultural chemistry, written primarily for other purposes, met none of the expectations for such a report, but he tried to paper over this dereliction by dedicating the English edition to the association. The publication coincided exactly with the opening of its 1840 meeting (Brock and Stark 1990).

Liebig placed manure within a natural cycle. A plant draws inorganic components from the soil. An animal ingests those chemicals when it eats the plant and then excretes whatever materials it does not need for its own growth and maintenance. Those materials then leach back into the soil from decomposing manure for the next plant to absorb. This general cycle played out in various complicated ways depending on the nature of the soil, the plant, and the animal. Cattle and horse dung, for example, worked better in some types of soil than human excrement, worse in others. Careful analysis and experimentation would uncover the details of these intricate chemical cycles, long treated as "an unfathomable mystery," and reduce them to rational human command (1840: 174–190).

Daubeny placed *Organic Chemistry in Its Applications to Agriculture and Physiology* at the heart of his new rural economy curriculum. He lamented in the second of his November inaugural lectures that "it is certainly somewhat humiliating" that science had not laid down general principles to explain the action of manure until "Liebig applied himself to its investigation" (1841: 38). Appreciation led to a lasting friendship when Daubeny joined the German chemist on a scientific tour of Britain in 1842, which included a visit to the country house of the Prime Minister, Robert Peel, an ardent public supporter of scientific agriculture with an interest in manures (Brock 1997: 98–99; Peel 1841: 12–16).

Daubeny did not embrace Liebig's principles as self-evidently correct. As Whewell noted privately to a friend in 1843, "the most profound and sagacious physiologists and chemists" could not yet declare which of Liebig's newly published researches "will stand as real discoveries." But the German had provoked that great good, "long thought and study" (Todhunter 1876: 2: 313–314). For Daubeny,

Liebig's general principles laid down the framework for both his small-scale experiments in Oxford and the more ambitious ones carried out across Britain. These principles by definition did not dictate experimental results. More than 15 years later, in his 1856 presidential address to the British Association, Daubeny freely acknowledged that many of Liebig's specific conclusions still remained entangled in legitimate controversy—one of his former students, John Bennet Lawes, led the vanguard against several of Liebig's claims. But, whatever the merits of the competing interpretations, Daubeny insisted that the debates took place using the theoretical tool which Liebig had forged (1857b: lv–lviii; Brock 1997: 173–176).

Visitors to the conversazione would have searched in vain for plots dedicated to the topic of the paper their host had delivered 2 days earlier. His guidebook to the garden, from its first printing in 1853, expressed a "long-cherished" but unrealized intention to carry out experiments "to share in the discovery of the sexuality of plants" (Daubeny 1853 [1864]: 23). Daubeny had lacked principles to guide such investigation, until he read the *Origin*. The book had the potential to serve much the same role for Daubeny as a professor of botany as *Organic Chemistry in Its Applications to Agriculture and Physiology* had played for him as a professor of rural economy. Darwin, like Liebig before him, would be right about some things, wrong about others, but above all, he offered an explanatory toolbox to crack seemingly unfathomable mysteries.

Huxley became the conversazione's center of gravity amidst the garden's intellectually enriching bonanza of sights and smells. Earlier in the day, he had confronted Bishop Wilberforce before a boisterous crowd in the university museum's unfinished library. Now, a few hours later, congratulations washed over him for standing resolutely against such a famously polished orator and slippery controversialist.

The Huxley–Wilberforce debate would later slip into myth. Leonard Huxley in 1900 depicted the event as an "open clash between Science and the Church," with his father rallying "open resistance ... to authority" (Huxley 1900: 1: 194–204). Historians have provided a more sophisticated account that acknowledges the complex agendas at play and the ambiguity of the outcome (Jensen 1988, 1991; Secord 2000: 513–514; Brooke 2001; Browne 2002: 114–125; James 2005a; Hesketh 2009; Desmond and Moore 1991: 485–499; Barton 2018: 173–177), even if the legend has not dislodged easily from popular understanding (James 2005b).

Wilberforce (1860) drew his attack lines from his excoriating review of the *Origin*, due in the upcoming issue of the *Quarterly Review*. Hooker condemned Wilberforce's performance to Darwin as a cavalcade of ignorance, ridicule, and deceit. He knew nothing about science, Hooker grumbled, beyond the poison Owen had whispered in his ear and the calumny he had harvested from hostile reviews.[16] This reaction was understandable, but not entirely fair. Wilberforce had long taken a keen amateur interest in science. He had been an early supporter of plans for the Oxford museum (Atlay 1903: 199). A few years earlier, Edward Forbes assured

[16]Burkhardt et al. (1985– : 8: 270–1) [Hooker to Darwin, 2 July 1860]

readers of the *Westminster Review* that the bishop "knows more of geology than common people give him credit for" (1852: 75). And yet, while Darwin's theory sincerely offended him, Wilberforce had a more dangerous quarry in mind. He had also committed to review another book for the *Quarterly Review* (Browne 2002: 112–114). That work was *Essays and Reviews*, a work of liberal Anglican theology which appeared 4 months after the *Origin* to greater sales and even more bitter recrimination.

The views in *Essays and Reviews* were Wilberforce's ultimate, if unstated, target when he rose to crush Darwin. Of the seven contributions, Powell's "On the Study of the Evidences of Christianity" offended the bishop most grievously. Powell had died a fortnight earlier, or he would likely have been in the room facing Wilberforce. He had held the Savilian chair of geometry at Oxford since 1827. He fought alongside Daubeny to integrate science and mathematics into Oxford's liberal education, although with much less diplomacy (Morrell and Thackray 1981: 392–393). In the decade before his death, he had increasingly concentrated his efforts on theology rather than on the type of technical science he had communicated, including extensively to the British Association, over the past several decades (Corsi 1988).

Powell and his fellow Essayists did not aspire to loosen bonds of faith or weaken the spiritual authority of the established Church of England; quite the opposite, they believed that their theological reforms would fortify the church by aligning the eternal verities of faith with modern thought. Powell argued that the progress of inductive science had demolished miracles as credible evidence for Christianity, at least for "all highly cultivated minds and duly advanced intellects." Science demonstrated God's creation of a law-bound order of nature, and for that reason, when setting the grounds for religious belief, "the argument from *necessity* of miracles is at best a very hazardous one" (2000 [1860]: 239–240, 255).

Powell had aired these inflammatory views months earlier in *The Order of Nature*. Here he scoffed at the "miraculous originations of new species out of nothing." The illegitimate appeal to the supernatural placed much of biology beyond the reach of science, he complained. Species must originate through "ordinary natural causes" (1859: 163, 252–253; Corsi 1988: 209–224). What these causes might be, he did not say. Powell received the proofs to his contribution to *Essays and Reviews* a few weeks after the appearance of the *Origin*. He revised his essay accordingly.

> Mr. Darwin's masterly volume on *The Origin of Species* by the law of 'natural selection,'—which now substantiates on undeniable grounds the very principle so long denounced by the first naturalists,—*the origination of new species by natural causes*: a work which must soon bring about an entire revolution of opinion in favour of the grand principle of the self-evolving powers of nature. (2000 [1860]: 258)

Natural selection, then, became a tool for supporting Powell's wide-ranging arguments about science, nature, miracles, and the foundations of Christian belief. Where Darwin "dimly" foresaw a revolution within the narrow field of natural history, Powell forecast one "soon" in the metaphysical understanding of nature.

Wilberforce nearly choked on his rage at Powell's conception of a self-evolving, miracle-bereft world.[17] Once you dismiss the evidentiary values of scriptural miracles, he believed, you uproot the sacred character of Christian revelation. "Scarcely-veiled Atheism" he hissed in the *Quarterly Review* (1861: 251). Neither Wilberforce nor anyone else needed Powell to draw a line between the *Origin* and explosive theological questions, of course. But the *Origin* did offer Wilberforce a delicious opportunity. He could inflict severe collateral damage on Powell's infidelity by impeaching Darwin. If he could tear down Darwin for scorning "the sober, patient, philosophical courage of [English] philosophy" (1860: 263), he could sink Powell's views as "the misty hieroglyphic of the Atheist" rather than the product of cultured scientific thought (1861: 262).

Huxley appreciated the usefulness of Darwin's theory on practical scientific grounds, but he too had an ideological agenda. "Darwin and Huxley were fighting different battles," Ruth Barton observes. "Huxley saw Darwin as an effective new weapon ... on the side of liberalism. ... The *Origin* mattered because it was a weapon in a larger war" (2018: 173–177). Huxley's aggressive support of evolution was central to his campaign to root science exclusively in natural causes. This scientific naturalism did not require him to embrace natural selection exactly as the *Origin* defined it, and in fact, he dissented from Darwin on many of the fine points of evolutionary theory (Desmond 1994: 262; Chap. 12). In any case, Huxley only felt an incidental pleasure in smashing Wilberforce. His chief target was the bishop's perceived ally, Owen (Rupke 1994: 271).

Hooker joined Huxley on the stage. His defense of the *Origin* unsurprisingly aligned more closely with Darwin's views than Huxley's had. He started by landing a good jibe. Since the bishop asserted all men of science were hostile to the *Origin*, he sarcastically renounced any presumption "to address the audience as a scientific authority." An aggressively detailed analysis of natural selection in light of plant systematics and biogeography followed. He repeated the assurance he made in that Introductory Essay to *Flora Tasmaniæ* that he would "continue to use [Darwin's] hypothesis as the best weapon for future research, holding himself ready to lay it down should a better be forthcoming." Wilberforce misunderstood Darwin's argument so thoroughly, Hooker said acidly, one might think that his Lordship had not even bothered to read it (Anon. 1860a: 65). Lyell was not in the audience, but firsthand accounts convinced him that Hooker had been Darwin's most effective champion (Lyell 1881: 2: 335).

Wilberforce's aggressive strategy did not appeal to everyone, even those who broadly shared his theological commitments. Acland, although devout and friendly with the bishop, recoiled at his treatment of technical matters of science as if they

[17] Wilberforce's attachment to the reality of scriptural miracles does not mean he felt any sympathy with attempts to shackle science to a literal reading of Genesis, which he was known to oppose (Taylor 1847). We should not conflate his religious concerns with evolution, rooted as they were in the theological debates roiling the Church of England in the nineteenth century, with the fundamentalism and biblical literalism which Watts (Chap. 16) identifies as key drivers of modern creationism.

represented an attack on Christian revelation (Atlay 1903: 303). Daubeny's conversazione became a Darwinian rally, even though he himself largely shared Wilberforce's view of miracles. Daubeny conceded that modern thought eviscerated the type of "miracles" which polluted the religious practice of the middle ages. Indeed, the common acceptance of witchcraft and sorcery during these centuries turned his stomach; this revolting superstition caused "a greater amount of misery than any other delusion which ever took hold upon the human mind." But, unlike Powell, he refused to bracket the bright and hopeful miracles described in scripture with gloomy medieval incredulity. God did not depart capriciously from His order of nature but neither did He renounce the freedom to do so. Daubeny agreed with Wilberforce that rejecting the evidentiary values of *all* miracles washed away the foundations of revealed religion. God would not have introduced revelation absent miraculous credentials to verify its truth. The modern world must steer between medieval superstition and a cold, irreligious rationalism, he concluded (Daubeny 1867a). When discussing these matters, Daubeny adopted the tone of someone gently correcting an unfortunate error, in sharp contrast to Wilberforce's booming castigation of sin.

The sense that most people's sympathy at the meeting flowed to Wilberforce created esprit de corps inside the Botanic Garden. Someone gushed to Huxley that they so enjoyed the collision that they wished it could happen again. "Once in a lifetime is enough, if not too much," responded the pugnacious Huxley, disingenuously (Huxley 1900: 1: 202–203).

3.6 Some Most Interesting Work with Orchids: Saturday, June 30, 1860. Sudbrook Park, Richmond

Darwin briefly considered attending the Oxford meeting and even arranged accommodations in Magdalen College. But he hated social travel at the best of times. He had skipped the British Association every year since 1847. The prospect of walking into controversy horrified him. Fresh reports arrived from Cambridge that addresses by Phillips and Adam Sedgwick tied his theory to Powell's views on miracles, denouncing both. He dreaded facing even Sedgwick, an old friend and mentor. Owen would also be hard to sidestep, and Darwin still smoldered over the attack in the *Edinburgh Review*. Two days before the meeting, a bout of severe stomach problems provided him with enough reason to scuttle his tentative plans for Oxford. Instead, he left for Sudbrook Park, the brand-new spa of his hydropathic doctor, Edward Lane, for a regimen of the water cure (Browne 2002: 114–118).

Darwin found time for work amidst the treatments. On Saturday, he wrote a letter full of technical instructions to a young botanist named Alexander Goodman More. He had first asked More for help a week earlier. Correspondence was a central component of natural history practice, and Darwin used it effectively and extensively (Secord 1994; Harvey 2009). In this case, he needed information on orchids in the genus *Ophrys*, but many species did not grow near his home. Darwin had better hope for the Isle of Wight, where More lived. He coaxed the young botanist into making time-consuming observations and collections on his behalf.[18]

Earlier in the month, he had published a general plea for information on these orchids in the *Gardeners' Chronicle*. Most orchids depended absolutely on insects for fertilization, he wrote. The fly orchid (*Ophrys muscifera*) followed this general pattern; the bee orchid (*O. apifera*) did not.

> It is this curious apparent contradiction in the structure of the Bee Orchis—one part, namely the sticky glands, being adapted for fertilisation by insect agency—another part, namely the natural falling out of the pollen-masses, being adapted for self-fertilisation without insect agency—which makes me anxious to hear what happens to the pollen-masses of the Bee Orchis in other districts or parts of England. (Darwin 1860)

As Darwin submitted to Lane's water-cure treatments, he continued to puzzle over these paradoxical plants and to organize collaborative research to answer his questions.

Darwin had been investigating the relationship between insects and flowering plants for two decades. His first article on the subject appeared in the *Gardeners' Chronicle* in 1841. He described bumblebees which bored holes into flowers, a practice that allowed them to collect nectar more quickly than crawling down into the bloom. Unfortunately for the plants, these clever bees neglected to transport pollen from flower to flower. He joked that they had proven themselves "unworthy members of society" by pilfering nectar without fulfilling their role in pollination, "the, so imagined, final cause of their existence!" (1841). In hindsight, a serious subtext ran under sardonic comments about the "wicked bees." The partnership between bees and flowers was not a manifestation of a static, divinely created harmony. Natural selection constantly rewrote the fraught bargains between insects and plants in open-ended coevolutionary relationships (Hoquet 2010; Bellon 2013: 131–132).

In 1860, he returned to plant sexuality with new urgency. The rejection of static final causes, which he had masked nearly 20 years earlier beneath humorous asides, drove a thorough reevaluation of plant reproduction. While he pursued projects on several plant families, orchids provided the model system. The perplexing, delightful diversity in Orchidaceae—botanists had named around 6000 species—made sense when understood as the result of evolutionary diversification driven by coadaptation to an array of local pollinators.

While Darwin and Daubeny arrived independently at an interest in plant sexuality, it is far from a coincidence that the two naturalists converged on the same broad set of problems. Daubeny's Oxford address described the discovery of plant sexuality as unquestionably "the greatest step . . . towards obtaining an insight into the secrets of vegetable organization" (1860: 3); Darwin astutely recognized decades earlier the possibilities of applying natural selection to such a fundamental problem. Daubeny had no path forward for his dream, expressed almost plaintively in his

[18]Burkhardt et al. (1985– : 8: 264–5, 269) [Darwin to More, 24 and 30 June 1860]

guidebook to Oxford's botanic garden, of solving the riddle of the flower. Darwin did—not because he was a better botanist but because he possessed a theory which generated an integrated series of hypotheses.

Darwin engaged in studied understatement when he wrote to Lyell on June 20 that "I have been doing some most interesting work with Orchids."[19] Ten days later, Wilberforce mocked him for profaning the principles of induction, oblivious to the project—soon to be recognized as a model of meticulous inductive research—Darwin was contemplating amidst the water cure, 50 miles away at Sudbrook Park.

3.7 Guiding a Train of Discovery: Wednesday, June 28, 1865. Guildhall, Tiverton, Devon

As Daubeny's health faded, he spent winters in the mild climate of south Devon. There he became active in the Devonshire Association for the Advancement of Science, Literature, and Art. He accepted the presidency in 1865. His presidential address in the Guildhall of Tiverton celebrated the "insight into the Mysteries of Nature" he had witnessed over a long career. His review of scientific progress, much witnessed at first hand, touched on several topics, including Liebig's great if still controversial services to practical agriculture (he stressed that his admiration for the German chemist should not be misunderstood as "partizan" advocacy of his views) (1865: 1, 24–26). When Daubeny's galloping survey touched on the *Origin*, he slowed to examine Darwin's theory at length, which he had never stopped pondering over the past 5 years.

Darwin had written him an appreciative letter full of botanical shoptalk a fortnight after the Oxford meeting. "The whole subject of sexuality seems to me profoundly interesting & mysterious," Darwin enthused; "& the more it is stirred up the better."[20] Daubeny followed Darwin's botany with keen appreciation. This research program proved wildly successful. A torrent of articles and books during the 1860s revolutionized the scientific understanding of flowering-plant form and function (Allan 1977; Ayres 2008; Bellon 2009, 2011, 2013; Hoquet 2010; Tabb 2016; Endersby 2016). Daubeny was particularly impressed by Darwin's ponderously titled monograph, *On the Various Contrivances by Which British and Foreign Orchids Are Fertilised by Insects, and on the Good Effects of Intercrossing*.

Daubeny made little personal progress on his dream of placing plant sexuality on a firmer conceptual foundation. He did not lack knowledge, materials, opportunities, or pertinent questions. He did not know where to start for answers. His fundamental difficulty was captured by a term that figured prominently in his inaugural botany lecture in Oxford and a quarter century later in his British Association remarks on

[19]Burkhardt et al. (1985– : 8: 263) [Darwin to Lyell, 20 June 1860]

[20]Burkhardt et al. (1985– : 13: 428–429) [Darwin to Daubeny, 16 July 1860]

Darwin: "final causes." He adhered to a strong—and ultimately unproductive—knowledge claim about the inherence of stability and purpose in the physical world.

The invocation of goal directedness in the term "final cause" carried a certain amount of ambiguity.[21] It could mean the end achieved by discrete biological contrivances. Darwin himself sometimes used "final cause" in this more limited sense for, say, the egg-laying instincts of the cuckoo (1859: 216–217). But in his view, these types of "final causes" were actually in no sense final; they were merely more or less temporary expedients, subject always to the modifying pressures of natural selection. Daubeny, as he illustrated in his 1834 inaugural botany lecture in Oxford and his 1856 Cheltenham address, adopted a more expansive understanding of "final cause" which encompassed the "unity of design pervading the universe" (1857b: lxiv). Darwin had no use for the hypothesis that a prevailing order and symmetry governed nature. In the *Origin*, he criticized the doctrine of final causes as utterly ineffectual for explaining morphological patterns (1859: 435). By the fourth edition in 1866, he confidently labeled the conviction that God created groups of plants and animals "on a uniformly regulated plan" as not just inconsistent with available evidence but patently unscientific (1866: 513).

Daubeny's understanding of final cause took for granted the fact that the confusing patterns of plant sexuality must reflect harmony and stability as a manifestation of God's deliberate design—somehow. Figuring out the "somehow" was the business of science, and researchers should not presuppose an answer. Daubeny agreed with Whewell: "we are not to assume that we know the objects of the Creator's design, and put this assumed purpose in the place of a physical cause" (Whewell 1833: 352). Astronomers struggled for decades to comprehend the law-bound mechanisms through which God preserved the stability of the solar system. Botanists should not expect to get off any easier in their explication of floral systems. The astronomers, however, worked from Newton's theory of universal gravitation; the botanists had no such tool.

In Oxford, Daubeny publicly challenged Darwin to demonstrate that natural selection could provide a tool to break some of the impasses that bedevilled botany. Darwin met this friendly and fair contest with "boyish delight."[22] He explained the bewildering patterns of floral morphology where Daubeny and other botanists had failed because he approached the problem from an evolutionary perspective that treated "final causes" as "so imagined." *Orchids* bursts with well-demonstrated

[21]Adaptation and teleology continue to provoke contention among biologists to this day, as every essay in this volume shows in some way or another. The disagreement between Darwin and Daubeny on "final causes" was a particularly Victorian manifestation of these persistent disputes—were biological adaptations the manifestations of an inherent tendency towards beauty and stability which God built into natural law or were they volatile and contingent products of blind mechanism? The issue at play here is the overall character of natural law, one that transcends biological adaptation. As Delisle points out (Chap. 4), Darwin was a nineteenth-century scholar addressing issues relevant to his scientific culture.

[22]Burkhardt et al. (1985– : 9: 175) [Darwin to Malden, 15–16 June 1861]

causal explanations for a multitude of biological contrivances. None are "final" because plants, pollinators, and their shared environments remain in endless flux.

Take the peculiar bee orchid which Darwin was pondering on the same day as the so-called Wilberforce–Huxley debate. The doctrine of divine symmetry in nature offered no testable hypotheses on the species' incongruous floral structure. Natural selection did. Observation showed that insects rarely, if ever, visit bee orchid flowers, which itself was odd. The vast majority of orchids relied on insect pollination. Darwin (1876) had no shortage of evidence that showed the fertility benefits of cross-over self-fertilization in orchids, a conclusion that he extended to flowering plants generally. But if the availability of a plant's pollinators declined, for whatever reason, perpetual self-fertilization offered a strategy to stave off extinction. Darwin hypothesized that the bee orchid's relationship with its pollinators must have miscarried. Under selective pressure, it evolved mechanisms for self-fertilization but retained structures adapted to its recently abandoned reproductive strategy (1862: 63–72, 1877: 52–59; Endersby 2016: 201).

Might the bee orchid one day readapt to cross-fertilization? If not, would it decline into extinction? No way to know, Darwin admitted (1877: 58). There are no cosmic final causes, no striving toward symmetry, just life as it struggled to survive in the happenstance of the moment. The kludgy contrivances in bee orchid presented just one of countless puzzles which were unsolvable on the assumption that God individually fabricated thousands of orchid species. "Is it not a more simple and intelligible view that all Orchids owe what they have in common to descent from some monocotyledonous plant?" he asked (1862: 306–307).

Daubeny and his fellow expert naturalists unanimously praised the results of Darwin's evolutionary botany. George Bentham recanted his stubborn refusal to discuss natural selection and became, in his straitlaced way, one of Darwin's most effective champions (Bellon 2003). A few non-specialists tried to criticize the evolutionary perspective in *Orchids* and made fools of themselves in the process. Darwin's friends nominated him in 1862 for the Copley Medal, the highest honor bestowed by the Royal Society. An anti-evolutionary old guard at the society, including the president Edward Sabine and Daubeny's old friend Phillips, blocked the nomination for 2 years until forced to relent because of, as Sabine put it, "Darwin's great services, and recognising that his recent work on Orchids must be classed amongst these" (Bellon 2011: 410–411, 415–416).

Daubeny lectured on Darwin's orchid research in the summer of 1862. "It was the first time that your views had been publickly noticed at Oxford, since the famous discussion in which the Bishop of Oxford & Huxley played so prominent a part," he reported to Darwin. The lecture did not simply describe Darwin's results but defended him from theological criticism in "one the great centres of clerical influence."[23] That October, at the British Association meeting in Cambridge, he pointedly defended Darwin in a crowded meeting of Section D with Huxley in the chair. He recommended to everyone that they should read *Orchids*, "which was important

[23]Burkhardt et al. (1985– : 10: 301–2) [Daubeny to Darwin, 5 July 1862]

to his reputation, as it would dispel many notions which had been wrongly entertained with regard to the [allegedly irreligious] tendency of his writing" (Anon. 1862).

Daubeny violates the most cherished popular myths about the Darwinian Revolution. He was a clergyman who squared up bravely against those who attacked evolution on open or disguised religious ground. He insisted that too many gaps existed in Darwin's reasoning to gain his full assent, but he nonetheless openly respected that reasoning and framed the disagreement as both friendly and scientifically productive. Daubeny was, in short, a skeptic of evolution who nevertheless admired Darwin and argued forcefully for the *Origin*'s scientific legitimacy. He does not fit the grand narrative of metaphysical battle between evolutionists and creationists, so it is not surprising—although, for our understanding of the reception of natural selection, most unfortunate—that his participation in the early debates over the *Origin* has been reduced to footnote status.

Daubeny's address to the Devonshire association in Tiverton would be one of the last before his death in 1867. He had a great deal to say about Darwin. No, he was not ready to subscribe to natural selection as an established *theory*. Even its advocates, he said, had to admit that it was still beset with serious—but not insuperable—difficulties. Nor was he ready to concede that all species originated through the exclusive action of secondary causes. But, nonetheless, he celebrated natural selection as a *tool*.

> Even its most determined opponents ought to entertain an indulgent feeling towards a theory which has guided its author into a train of discoveries, both as to the vegetable and animal kingdom, any one of which would be sufficient to establish the reputation of an ordinary observer. (1865: 10–17)

3.8 Conclusion: Great Principles as Powerful Engines of Discovery: Sunday, February 16, 1838. Somerset House, London

Daubeny was born 7 months after Whewell and died 21 months after him. The careers of these scientific dons shared more than chronological overlap. They worked together for decades as members of the gentlemanly elite which led the British scientific establishment. Both served as president of the British Association. They came into routine contact in London at the meetings of the Geological Society, and it was in the society's Somerset House apartments that, in part, the roots of the debate over natural selection were sunk.

Whewell served as society president in 1838 and dedicated a significant portion of his anniversary address praising a pair of young, up-and-coming naturalists. Owen received commendation for the paleontology which had just won him the society's Wollaston Medal. Darwin received even warmer praises for the results of his *Beagle* work, "one of the most important events for geology which has occurred for many years." Whewell expressed his expectation that "these two naturalists, so fitted by their endowments and character to advance the progress of science," would develop "the great principles which it is given to them to wield, becoming every year more powerful instruments of discovery" (1838: 624–6, 643–6).

Whewell could not have realized that, by jointly anointing Darwin and Owen as the most promising theorists of the rising generation of British naturalists, he was setting up what would become the most consequential rivalry in the nineteenth-century British science. More than 20 years later, Darwin and Owen had both fulfilled their promise and contributed "great principles"—ones which collided with a deafening crash. Whewell's remarks in Somerset House set the ground rules for this unforeseen (if not unforeseeable) conflict. Did the principle of natural selection offer a powerful instrument of discovery?

Owen denied that it did. His notorious attack in the *Edinburgh Review* accused Darwin of repudiating the value of "close and long-continued research." By writing the *Origin*, said Owen, Darwin had lost his rank among the "men who trouble the intellectual world little with their beliefs, but enrich it greatly with their proofs" (1860: 529–30). Owen was motivated by far more than simple jealously. He recognized—in a way that Darwin did not fully appreciate—that natural selection cut painfully against the theoretical commitments at the heart of his research program (Rupke 1994: 238–242; Cosans 2009).

Although the *Athenaeum*'s account of the 1860 British Association meeting explicitly placed Daubeny and Owen shoulder-to-shoulder in opposition to Darwin's theory (Anon. 1860a: 19), the Oxford botanist refused to impugn Darwin's motives or competence. On the contrary, from his first public statement on natural selection in Section D to the end of his life, he encouraged his colleagues to test the usefulness of Darwin's "principle" as a tool of discovery.

In the summer of 1860, and for a few years afterwards, Daubeny's conciliatory approach was drowned out by the din of a culture war exemplified by Owen's denunciation in the *Edinburgh Review*. Owen's censure was more than incendiary and mean-spirited; it was startlingly foolish. Daubeny realized, as Owen should have, that Darwin had ample ability to wield natural selection in the service of new research and deserved the opportunity to try. Daubeny's practical engagement with natural selection as a tool for investigation proved remarkably prescient, down to the suggestion of plant sexuality as a field of investigation. The success of Owen's attacks on the *Origin* proved fleeting once Darwin started accumulating what Daubeny rightly identified as a remarkable "train of discoveries" under the light of natural selection.

Daubeny was never a minor figure in the early debates over natural selection. By the time he offered his warm, if qualified, praise for Darwin's great principle in Tiverton in 1865, the debate over natural selection had shifted in precisely the direction he pointed 5 years earlier at the British Association. He approached natural selection with the same practical consideration he gave to Liebig's agricultural chemistry—as a tool which would prove the extent of its truth through the reach of its usefulness, not as a heresy to stamp out or a creed to embrace.

References

Acland HW, Ruskin J (1859) The Oxford Museum. Smith, Elder, London

Albert, Prince Consort (1860) Presidential address. In: Report of the twenty-ninth meeting of the British Association for the Advancement of Science; held at Aberdeen in September 1859. John Murray, London, pp lix–lxix

Alberti SJMM (2003) Conversaziones and the experience of science in Victorian England. J Victorian Cult 8:208–230

Allan M (1977) Darwin and his flowers: the key to natural selection. Farber and Farber, London

Allen DE (2003) George Bentham's *Handbook for the British flora*: from controversy to cult. Arch Nat Hist 30:224–236

Anderson J (1860) On human remains in superficial drift. In: Report of the twenty-ninth meeting of the British Association for the Advancement of Science; held at Aberdeen in September 1859. John Murray, London, pp 95–97

Anon (1860a) British Association. The Athenaeum no. 1706-1707 (July 7 and 14), pp 18–32, 59–69

Anon (1860b) Professor Owen on the origin of species. Saturday Rev 9:573–574

Anon (1860c) The British Association for the Advancement of Science. Morning Post 26,998 (June 28):4

Anon (1862) The British Association for the Advancement of Science. Daily News 5117(October 3):2

Atlay JB (1903) Sir Henry Wentworth Acland: a memoir. Smith, Elder, London

Ayres P (2008) The aliveness of plants: the Darwins at the dawn of plant science. Pickering & Chatto, London

Barton R (2018) The X-Club: power and authority in Victorian science. University of Chicago Press, Chicago

Bellon R (2003) The great question in agitation: George Bentham and the origin of species. Arch Nat Hist 30:282–297

Bellon R (2004) Hooker, Frances Harriet (née Henslow) (1825-74). In: Lightman B et al (eds) Dictionary of nineteenth-century British scientists, 4 vols, vol 2 pp 993-994 Thoemmes Continuum, Bristol

Bellon R (2006) Joseph Hooker takes a "fixed post": transmutation and the "present unsatisfactory state of systematic botany," 1844-60. J Hist Biol 39:1–39

Bellon R (2009) Charles Darwin solves the "riddle of the flower"; or, why don't historians of biology know about the birds and bees? Hist Sci 47:373–406

Bellon R (2011) Inspiration in the harness of daily labor: Darwin, botany and the triumph of evolution, 1859-1868. Isis 102:392–420

Bellon R (2013) Darwin's evolutionary botany. In: Ruse M (ed) The Cambridge encyclopedia of Darwin and evolutionary thought. Cambridge University Press, Cambridge, pp 131–138

Bellon R (2014) There is grandeur in this view of Newton: Charles Darwin, Isaac Newton and Victorian conceptions of scientific virtue. Endeavour 38:222–234

Bellon R (2015) A sincere and teachable heart: self-denying virtue in British intellectual life, 1736-1859. Brill, Boston

Brayley EW (1867–68) Obituary notice of John, second baron Wrottesley. Mon Notice R Astron Soc 28:64–68

Brinkman PD (2009) Charles Darwin's *Beagle* voyage, fossil vertebrate succession, and "the gradual birth & death of species". J Hist Biol 43:363–399

Brock WH (1997) Justus von Liebig: the chemical gatekeeper. Cambridge University Press, Cambridge

Brock WH, Stark S (1990) Liebig, Gregory and the British Association, 1837-1842. Ambix 37:134–147

Brooke JH (2001) The Wilberforce-Huxley debate: why did it happen. Science & Christian Belief 13:127–141

Browne J (2002) Charles Darwin: the power of place. Alfred A. Knopf, New York

Burkhardt F et al (eds) (1985–) The correspondence of Charles Darwin. Cambridge University Press, Cambridge, 27 vols. Letters also accessible at https://www.darwinproject.ac.uk/
Corsi P (1988) Science and religion: Baden Powell and the Anglican debate, 1800–1860. Cambridge University Press, Cambridge
Cosans CE (2009) Owen's ape & Darwin's bulldog: beyond Darwinism and creationism. Indiana University Press, Bloomington
Darwin C (1841) Humble-bees. Gardeners' Chron 34(21 August):550
Darwin C (1859) On the origin of species by means of natural selection, or the preservation of favoured races in the struggle for life. John Murray, London
Darwin C (1860) Fertilisation of British orchids by insect agency. Gardeners' Chron 23(9 June):528
Darwin C (1862) On the various contrivances by which British and foreign orchids are fertilised by insects, and on the good effects of intercrossing. John Murray, London
Darwin C (1866) On the origin of species by means of natural selection, or the preservation of favoured races in the struggle for life, 4th edn. John Murray, London
Darwin C (1876) The effects of cross and self fertilisation in the vegetable kingdom. John Murray, London
Darwin C (1877) The various contrivances by which orchids are fertilised by insects, 2nd edn. John Murray, London
Darwin C (1958) The autobiography of Charles Darwin, 1809-1882. Barlow N (ed). Collins, London
Daubeny C (1823) Inaugural lecture on the study of chemistry: read at the Ashmolean Museum, November 2, 1822. Oxford University Press, Oxford
Daubeny C (1834) An inaugural lecture on the study of botany: read in the Library of the Botanic Garden, Oxford, May 1, MDCCCXXXIV. H. Parker, Oxford
Daubeny C (1837) Address. In: Report of the sixth meeting of the British Association for the Advancement of Science; held at Bristol in August 1836. John Murray, London, pp xxi–xxxvi
Daubeny C (1841) Three lectures of agriculture. J.H. Parker, Oxford
Daubeny C (1850) On the influence of carbonic acid gas on the health of plants, especially those allied to the fossil remains of the coal formation. In: Report of the nineteenth meeting of the British Association for the Advancement of Science; held at Birmingham in September 1849. John Murray, London, 1884, pp 56–63
Daubeny C (1853 [1864]) Oxford Botanic Garden; or, A popular guide to the Botanic Garden of Oxford, to which is appended an Address to the members of the University, delivered on May 20, 1853, with an appendix, 2nd edn. Oxford Botanic Garden, Oxford [note: the guide (i) has different pagination to the address, appendix, and a supplement added to the 1864 printing (ii) because why make things easy]
Daubeny C (1857a) Lectures on Roman Husbandry. J.H. and Jas. Parker, Oxford
Daubeny C (1857b) Presidential address. In: Report of the twenty-sixth meeting of the British Association for the Advancement of Science; held at Cheltenham in August 1856. John Murray, London, pp xlviii–lxxiii
Daubeny C (1860) Remarks on the final causes of the sexuality of plants, with particular reference to Mr. Darwin's work on the origin of species. J.H. and James Parker, Oxford
Daubeny C (1865) The president's address. Rep Trans Devonshire Assoc Adv Sci Lit Art 4:1–29
Daubeny C (1867a) Christianity and rationalism in their relations to natural science: being a protest against certain principles advocated in Mr. Lecky's History of the rise and influence of the spirit of rationalism in Europe. James Parker, London
Daubeny C (1867b) Miscellanies: being a collection of memoirs and essays on scientific and literary subjects, 2 vols. James Parker, Oxford
Daubeny C (2011 [1848]) A description of active and extinct volcanoes, 2nd ed., Cambridge University Press, Cambridge
Desmond A (1994) Huxley: the devil's disciple. Michael Joseph, London
Desmond A, Moore J (1991) Darwin: the life of a tormented evolutionist. Warner Books, New York

Endersby J (2001) “From having no herbarium”: local knowledge versus metropolitan expertise: Joseph Hooker’s Australasian correspondence with William Colenso and Ronald Gunn. Pac Sci 55:343–358

Endersby J (2008) Imperial nature: Joseph Hooker and the practices of Victorian science. University of Chicago Press, Chicago

Endersby J (2016) Orchid: a cultural history. University of Chicago Press, Chicago

Forbes E (1852) The future of geology. Westminster Rev 58:67–94

Fox R (1997) The University Museum and Oxford science, 1850-1880. In: Brock MG, Curthoys MC (eds) Nineteenth-century Oxford, Part I. In: Aston TH (general ed) Vol 6 of The history of the University of Oxford. Clarendon, Oxford, pp 641–691

Goddard N (2004) Daubeny, Charles Giles Bridle, 1795-1867. In: Matthew HCG, Harrison B (eds) Oxford dictionary of national biography. https://www.oxforddnb.com/

Hahn R (2005) Pierre Simon Laplace, 1749-1827: a determined scientist. Harvard University Press, Cambridge

Harcourt WV (1840) Presidential address. In: Report of the ninth meeting of the British Association for the Advancement of Science; held at Birmingham in August 1839. John Murray, London, pp 3–22

Harvey J (2009) Darwin’s “angels”: the women correspondents of Charles Darwin. Intellect Hist Rev 19:197–210

Henslow JS (1835) The principles of descriptive and physiological botany. Longman, Rees, Orme, Brown, & Green, London

Herschel J (1830) A preliminary discourse on the study of natural philosophy. Longman, Rees, Orme, Brown & Green, London

Herschel J (1831) An address by the president on April 11, 1827. Mon Notices R Astron Soc 1:14–19

Herschel J (1833) A treatise on astronomy. Longman, Rees, Orme, Brown, Green, & Longman, London

Herschel J (1840) Terrestrial magnetism. Q Rev 66:272–274

Herschel J (1851) Outlines of astronomy, 4th edn. Longman, Brown, Green, and Longmans, London

Hesketh I (2009) Of apes and ancestors: evolution, Christianity, and the Oxford Debate. University of Toronto Press, Toronto

Hodge J (2009) The notebook programmes and the projects of Darwin’s London years. In: Hodge J, Radick G (eds) The Cambridge companion to Darwin, 2nd edn. Cambridge University Press, Cambridge, pp 40–68

Hooker J (1853–55) The botany of the Antarctic voyage of H.M. Discovery Ships ‘Erebus’ and ‘Terror’ in the years 1839-1843. Part II: Flora Novæ-Zelandiæ. Lovell Reeve, London, 2 vols

Hooker J (1860) Introductory essay to Flora Tasmaniæ. In: The botany of the Antarctic voyage of H.M. Discovery Ships ‘Erebus’ and ‘Terror’ in the years 1839–1843. Part III: Flora Tasmaniæ, 2 vols. Lovell Reeve, London, vol 2, pp i–cxxviii

Hooker J, Thomson T (1855) Introductory essay. In: Flora Indica: being a systematic Account of the plants of British India, together with observations on the structure and affinities of their natural orders and genera. Pamplin, London

Hoquet T (2010) Darwin teleologist? design in the Orchids. Comptes Rendus Biologies 333:119–128

Hoquet T (2014) Botanical authority: Benjamin Delessert’s collections between travelers and Candolle’s natural method (1803–1847). Isis 105:508–539

Hull DL (2003) Darwin’s science and Victorian philosophy of science. In: Hull DL, Radick G (eds) The Cambridge companion to Darwin. Cambridge University Press, Cambridge, pp 168–191

Huxley L (1900) Life and letters of Thomas Henry Huxley. D. Appleton, New York, 2 vols

James FAJL (2005a) An “open clash between science and the church”? Wilberforce, Huxley and Hooker on Darwin at the British Association, Oxford 1860. In: Knight DM, Eddy MD (eds)

Science and belief: from natural philosophy to natural science, 1700-1900. Ashgate, Aldershot, pp 171–193
James FAJL (2005b) On Wilberforce and Huxley. Astron Geophys 46:1.9
Janssen M (2019) Arches and scaffolds: bridging continuity and discontinuity in theory change. In: Love AC, Wimsatt WC (eds) Beyond the meme: development and structure in cultural evolution. University of Minnesota Press, Minneapolis, pp 95–199
Jensen JV (1988) Return to the Wilberforce–Huxley debate. Br J Hist Sci 21:161–179
Jensen JV (1991) Thomas Henry Huxley: communicating for science. University of Delaware Press, Newark
Leifchild JR (1859) On the origin of species. The Athenaeum 1673 (19 November):659–660.
Liebig C (1838) On the products of the decomposition of uric acid. In: Report of the seventh meeting of the British Association for the Advancement of Science; held at Liverpool in September 1837. John Murray, London, pp 38–41
Liebig L (1840) Organic chemistry in its applications to agriculture and physiology. Playfair L (ed) Taylor and Walton, London
Lindley J (1834) Review of An inaugural lecture on the study of botany by Charles Daubeny. The Athenaeum 344 (May 31):401–402
Lyell C (1860) On the occurrence of works of human art in post-Pliocene deposits. In: Report of the twenty-ninth meeting of the British Association for the Advancement of Science; held at Aberdeen in September 1859. John Murray, London, pp 93–95
Lyell K (1881) Life, letters and journals of Sir Charles Lyell. John Murray, London, 2 vols
Maienschein J (1991) Transforming traditions in American biology, 1880-1915. Johns Hopkins University Press, Baltimore
McOuat G (2001) Cataloguing power: delineating "competent naturalists" and the meaning of species at the British Museum. Br J Hist Sci 34:1–28
Mercer W (1855) The church psalter and hymn book. Jewell and Letchford, London
Morrell J, Thackray A (1981) Gentlemen of science: early years of the British Association for the Advancement of Science. Clarendon, Oxford
Mozley JB (1853) The Oxford Commission. Q Rev 93:152–238
Oldroyd D (2004) Daubeny, Charles Giles Bridle (1795-1867). In: Lightman B et al (eds) Dictionary of nineteenth-century British scientists, vol 2. Thoemmes Continuum, Bristol, 4 vols, pp 543–547
Owen R (1860) Darwin on the origin of species. Edinb Rev 111:487–532
Pauly PJ (2000) Biologists and the promise of American life: from Meriwether Lewis to Alfred Kinsey. Princeton University Press, Princeton
Peel R (1841) An inaugural address delivered by the Right Honourable Sir Robert Peel, Bart. M.P., President of the Tamworth Library and Reading Room, 2nd edn. James Bain, London
Phillips J (1868–9) Charles Giles Bridle Daubeny. Proc R Soc Lond 17:lxxiv–lxxx
Powell B (1859) The order of nature considered in reference to the claims of revelation. Longman, Brown, Green, Longmans, & Roberts, London
Powell B (2000 [1860]) On the study of the evidences of Christianity. In: Shea V Whitla W (eds) Essays and reviews: the 1860 text and its readings. University Press of Virginia, Charlottesville, pp 233–273
Robb-Smith AHT (1997) Medical education. In: Brock MG, Curthoys MC (eds) Nineteenth-century Oxford, Part I. Aston TH (general ed) Vol 6 of The history of the University of Oxford. Clarendon, Oxford, pp 563–582
Rupke NA (1994) Richard Owen: Victorian naturalist. Yale University Press, New Haven
Rupke NA (1997) Oxford's scientific awakening and the role of geology. In: Brock MG, Curthoys MC (eds) Nineteenth-century Oxford, Part I. Aston TH (general ed) Vol 6 of The history of the University of Oxford. Clarendon, Oxford, pp 543–562
Ruse M (1979) The Darwinian revolution: science red in tooth and claw. University of Chicago Press, Chicago

Secord A (1994) Corresponding interests: artisans and gentlemen in nineteenth-century natural history. Br J Hist Sci 27:383–408

Secord JA (2000) Victorian sensation: the extraordinary publication, reception, and secret authorship of Vestiges of the natural history of creation. University of Chicago Press, Chicago

Secord JA (2014) Visions of science: books and readers at the dawn of the Victorian age. University of Chicago Press, Chicago

Stott R (2003) Darwin and the barnacle. W.W. Norton, New York

Tabb K (2016) Darwin at orchis bank: selection after the *Origin*. Stud Hist Philos Sci Part C Stud Hist Philos Biol Biomed Sci 55:11–20

Taylor WC (1847) Review of Pride a hindrance to true knowledge by Samuel Wilberforce. The Athenaeum. 1028 (July 10):727–728

Todhunter I (1876) William Whewell, D.D. Master of Trinity College Cambridge: an account of his writings with selections from his literary and scientific correspondence, 2 vols. Macmillan, London

Van Riper AB (1993) Men among mammoths: Victorian science and the discovery of human prehistory. University of Chicago Press, Chicago

Watson HC (1847–59) Cybele Britannica; or British plants, and their geological relations. Longman, London, 4 vols

Whewell W (1831) Lyell's principles of geology. Br Critic 9:180–206

Whewell W (1833) Astronomy and general physics considered with reference to natural theology. William Pickering, London

Whewell W (1834) Mr. Whewell's reply to the Edinburgh Review. Br Mag 5:263–268

Whewell W (1837) History of the inductive sciences, from the earliest to the present times. John W. Parker, London, 3 vols

Whewell W (1838) Presidential address. Proc Geol Soc Lon 2:624–649

Whewell W (1849) On the fundamental antithesis of philosophy [read 5 February 1844]. Trans Camb Philos Soc 8:170–181

Whewell W (1857a) History of the inductive sciences, from the earliest to the present times, 3rd edn. John W. Parker, London, 3 vols

Whewell W (1857b) Spedding's complete edition of the *Works of Bacon*. Edinb Rev 106:287–322

Wilberforce S (1860) Darwin's Origin of Species. Q Rev 108:225–264

Wilberforce S (1861) Essays and reviews. Q Rev 109:248–305

Wrottesley J (1861) Presidential address. In: Report of the thirtieth meeting of the British Association for the Advancement of Science; held at Oxford in June and July 1860. John Murray, London, pp lv–lxxv

Yanni C (2005) Nature's museums: Victorian science and the architecture of display. Princeton Architectural Press, New York

Chapter 4
Natural Selection as a Mere Auxiliary Hypothesis (Sensu Stricto I. Lakatos) in Charles Darwin's *Origin of Species*

Richard G. Delisle

Abstract It is a widely accepted view that Charles Darwin's theory of biological evolution is founded on the notion of adaptation, a creative evolutionary process driven by natural selection and instituting, in its wake, a vast range of possible evolutionary outcomes. Taking the *Origin of Species* as the basis of our analysis, this chapter will contest the received view which assumes that evolutionary process leads to evolutionary pattern. For Darwin, on the contrary, pattern comes first with process being merely superimposed upon a pre-established pattern. At the core of Darwin's theory, one finds a consistent and robust view of the history of life ("tree of life"), largely neglected among Darwin scholars, which includes (1) a neontological view of life (as against paleontology); (2) the principle of divergence, whose function is to ensure that extinct forms fall in-between extant ones (thus preventing the rise of variations unrecorded among extant forms); (3) the principle of gradation, which levels out discontinuities and novelties by tightly connecting past forms to recent ones; (4) life being organized in a network-like fashion (a pattern not easily undone); (5) the significance of evolutionary stagnation; (6) fixed taxonomic positions since the beginning of life; (7) a segmented tree of life organized around separate and independent classes and phyla; (8) permanent and stable high taxonomic groups deeply entrenched in nature; and (9) a biological modernity acquired in its main features since the Cambrian period. In light of this view of life, Darwin put natural selection in a conceptual straightjacket, depriving it of its evolutionary creativity and reducing it to an auxiliary hypothesis, as defined in Imre Lakatos's view of science.

Keywords Charles Darwin · Darwinism · Natural selection · Evolutionary process · Principle of divergence · Principle of gradation · Tree of life · View of life · Evolutionary pattern · Rhetorical strategies · Imre Lakatos · Theoretical hard core · Auxiliary hypotheses · Methodology of scientific research programmes

---

R. G. Delisle (✉)
School of Liberal Education, University of Lethbridge, Lethbridge, AB, Canada

Department of Philosophy, University of Lethbridge, Lethbridge, AB, Canada
e-mail: richard.delisle@uleth.ca

R. G. Delisle (ed.), *Natural Selection*, Evolutionary Biology – New Perspectives on Its Development 3, https://doi.org/10.1007/978-3-030-65536-5_4

4.1 Introduction

The place of Charles Darwin (1809–1882) in science is a peculiar one: judged from the contemporary understanding of his work, this mid-nineteenth century scholar seems far ahead of his time. This assessment is readily apparent in the following quotations from contemporary scholars: "Among contemporary evolutionary theorists Darwin functions as an icon, an image against which theories may receive approbation or reprobation. To select from the historical Darwin those features that best comport with one's own predilections in the contemporary scientific debate is to have those predilections sanctioned by the master" (Richards 1988: 146). And again: "No other field of science is as burdened by its past as is evolutionary biology... [It] can be defined to a large degree as the ongoing attempt of Darwin's intellectual descendants to come to terms with his overwhelming influence" (Horgan 1996: 114). Finally, one also reads: "One way of coming to terms with Darwin's influence is to appropriate him in support of one's own views" (Shanahan 2004: 293). Of course, the cardinal notion here is "natural selection" and the key book the *Origin of Species* (1859). To postulate that the process of biological evolution is organized around the mechanism of natural selection is, under a prevailing view, to walk in the footsteps of Darwin.

The received view is today so firmly entrenched in evolutionary science that it is seen as well founded by proponents of standard Darwinism, reformers of Darwinism, and opponents of Darwinism. For instance:

1. One view places natural selection as a key explanatory component at the core of evolutionary biology, thus implicitly or explicitly building upon Darwin's own theory (Alcock 2001, 2017; Dawkins 1976, 1996, 2009; Dennett 1995; Mayr 1980, 1983, 1991; Depew 2017; Depew and Weber 1989, 1995; Wilson 1975; Wray et al. 2014).
2. Another view suggests that while we should retain some essential elements of Darwin's theory, such as natural selection, a new theory should be elaborated by leaving a significant place for other explanatory elements designed to address phenomena Darwin himself had neglected (Brooks and Wiley 1988; Eldredge 1985; Gould 1980, 2002; Gould and Eldredge 1977; Gould and Lewontin 1979; Jablonka and Lamb 2006; Kimura 1983; Laland et al. 2014; Pigliucci and Müller 2010).
3. Yet another view considerably reduces the significance of Darwin's theory by proposing a more radical alternative, one in which natural selection is confined to a relatively minor explanatory role (Goodwin 1994, 2009; Kauffman 1993, 1995; Reid 1985, 2007; Ulanowicz 1997, 2009).

Evolutionary biology as it appears today consists in a range of overlapping theories falling between two extremes, with one pole emphasizing the strong, creative role of natural selection, the other confining it to a weak and merely eliminative role. Adherents at both ends of the spectrum agree that Darwin's contribution to science is definitively established: a theory explaining the process and pattern of evolution under the action of natural selection. As William Provine explains (1985: 827): "Natural selection of individual differences… was the primary mechanism of evolution at every level of the evolutionary process. Certainly this is the dominant view neo-Darwinians today attribute to Darwin." This understanding would appear to be confirmed by additional authority from the analyses of historians and philosophers of biology, who concur that Darwin himself positioned natural selection at the explanatory core of his theory (Hodge 1977, 2013; Ruse 1975, 1979, 2013).

This chapter will argue that the received view lacks a solid foundation. At the outset, it must be acknowledged that there is something annoying about questioning Darwin's contribution, an achievement that enjoys nearly universal consensus. I will stop short, however, of sheepishness in drawing attention to significant aspects of Darwin's thinking impossible to ignore. At a fairly superficial level, the content of the *Origin of Species* looks fairly straightforward to a modern evolutionist. On closer analysis, however, the book shows itself to be tremendously complex, composed of distinct explanatory levels and organized around incommensurable tensions and theories (Delisle 2019). When these lacunae come into view, Darwin appears different from the picture created in the post-1959 era, following the celebrations of the 100th anniversary of the *Origin of Species* (Smocovitis 1999) and subsequently appropriated by modern thinkers who refashioned his work to suit their immediate needs. This chapter will take the reader through some of these explanatory levels in search of Darwin's ultimate theoretical core, a quest that will also require going beyond his rhetorical strategies.

At the heart of Darwin's *magnum opus*, one does not find a pattern of evolution explained by the open or uninhibited action of natural selection. Instead, one encounters a pre-established view of life—a pan-divergent pattern—thwarting the rise of genuine evolutionary novelties and the expression of true historicity. Darwin is not the founding figure of "evolutionary contingency" in biology, as it sometimes claimed. Under this other understanding, "natural selection" can be seen as constituting an auxiliary hypothesis whose function consists in protecting the core of the theory from refutation, to borrow Imre Lakatos's (1970) view of science. It will be seen that Darwin modifies the role of natural selection at will, exploiting it when it seems to justify a pan-divergent view and entirely disposing of it when it seems to threaten it. In short, it is held that natural selection fills more a justificatory function than an explanatory one.

The issue raised here extends well beyond the debate that raged in 1970s and 1980s over whether "process" or "pattern" should have explanatory precedence in evolutionary explanations (Bock 1974, 1977, 1981; Bretsky 1979; Cracraft 1981a, b; Delisle 2001; Delson et al. 1977; Eldredge and Tattersall 1975, 1982; Halstead 1981; Hennig 1981 [1969]: 38–42; Hull 1979, 1988: 232–276; Mayr 1974 ; Mayr and Ashlock 1991; Platnick 1979; Rosen et al. 1979; Schafersman 1985; Simpson 1975; Szalay 1977, 1981, 1991). For Darwin, not only does "pattern" always comes first, but it is also envisioned in such a rigid way as to leave no room for evolutionary change other than "divergence." Conceptually speaking, the principle of divergence at the heart of Darwin's theory cannot be equated with its modern and more recent counterpart. There is something profoundly unmodern about Darwin's so-called tree of life. This chapter is an attempt to put the *Origin of Species* back into the intellectual context of its time, when scholars were struggling to catch up with the implications of a fast-moving revolution called Transformism.

4.2 Rhetorical Strategies in the *Origin of Species*

The last decades have seen a number of scholars converging on the view that the argumentative structure of the *Origin of Species* (1859) is organized around few key features (Hodge 1977, 1989, 2013; Recker 1987; Ruse 1975, 1979, 2000, 2013; Waters 2009). The resulting picture goes as follows: (1) the *first* chapters focus on the action of a selective process on the rise of variation in the context of life forms facing the conditions of the economy of nature; (2) the *last* chapters review the evolutionary effects of such evolutionary principles in fields such as paleontology, biogeography, embryology, morphology, and systematics; and (3) the *middle* chapters are designed to rebut possible objections. In terms of epistemology, it is argued that the *Origin* includes a section concerned with the *causes* of evolution and one interested in its *effects*. It is not uncommon to read in the recent literature that Darwin's theory is, therefore, about a process (natural selection) whose evolutionary outcome is reflected in a specific pattern (divergence) (Browne 1983: 215; Kohn 1985, 2009; Limoges 1968; Ruse 1999: 191). This view is summarized by Peter J. Bowler (1988: 7–9):

> The most radical aspect of Darwin's approach was his reliance on adaptation as the sole driving agent of evolution. Species change because they must adapt to new environments or because they become more specialized for their existing life styles... Once adaptation is accepted as the sole directing agent, evolution has to be seen as an irregularly branching tree... If samples from a single original species migrate to several new locations, each will adapt to its new environment in its own way, and eventually a group of distinct but related species will be formed. Each of the descendant species will then undergo its own evolutionary process, depending on the opportunities for further migration and adaptation that open up to it. Evolution must thus be regarded as a pattern of haphazard branching, with the branches constantly diverging further apart and redividing where possible... Darwin devised the theory of natural selection as an explanation of how populations adapt to changes in their environment... The variations from which the environment selects are essentially random and undirected... The fact that variations are not directed along fixed lines emphasizes the haphazard or open-ended nature of a process governed only by adaptation.

Conceived in this way, the notion of natural selection can legitimately claim to constitute an essential cause and core component of a theory devoted to explaining biological evolution.

While the epistemological distinction employed in the analyses above is no doubt useful to a certain extent, that distinction lacks the capacity to penetrate the deep explanatory structure of the *Origin of Species*. Darwin may in fact have tried to pursue *some* of the epistemological ideals presented in these analyses, but he fails in large measure simply because he *himself* offers the reader something quite different.

Fortunately, some of Darwin's contemporaries are there to remind us that we have introduced more order into the *Origin* than is really there and that we may be reading our own modern view back into it. Reviewing the publication of the first edition of the *Origin*, Darwin's friend and supporter, Thomas Henry Huxley (1860: 542), came up with this rather harsh comment:

> But this very superabundance of matter must have been embarrassing to a writer who, for the present, can only put forward an abstract of his views; and thence it arises, perhaps, that notwithstanding the clearness of the style, those who attempt fairly to digest the book find much of it a sort of intellectual pemmican—a mass of facts crushed and pounded into shape, rather than held together by the ordinary medium of an obvious logical bond: due attention will, without doubt, discover this bond, but it is often hard to find.

Other contemporaries of Darwin also noted other problems with its argumentation. As expressed by Alvar Ellegard (1990 [1958]: 189) in his overview of the literature of the period: "A host of press voices accused Darwin of mingling together 'facts and assumptions, probabilities and speculations. . . in most illogical confusion'... and of using his imagination to supply the lack of factual observation." Apparently, studying the work of Charles Darwin comes with the double peril of working our way through his rhetorical apparatus and simultaneously resisting the temptation of reading modern evolutionary biology back into the *Origin of Species*.[1]

At this junction, let us enter into the specifics of our analysis by turning to Darwin's own words and ideas. For the sake of simplicity, I will largely rely on the 1859 edition of the *Origin*, although what is said there is present throughout all six editions.[2] The main thread of our analysis will be guided by the nature of the relationship between "process" and "pattern." Among other things, this analysis will show that, far from being "one long argument" (1859: 459), the *Origin* is a profoundly dislocated book offering a series of competing views.

4.3 The Network of Life Resisting Natural Selection

According to the received view, the first chapters of the *Origin* contain the explanatory core of Darwin's theory. Here, Darwin presumably explains how the limited number of places in nature forces life forms to compete against each other, the

---

[1]On a personal note, when I first read the *Origin of Species*, I accepted the received view without much question. This is unsurprising given my training under the auspices of the so-called Modern Synthesis. To this day, I teach a yearly course entirely devoted to the *Origin* in the spirit of the received view. Only after many years of agonizing work have I been able to break free from the double peril mentioned above. While doing so has opened up for me a new paradigm in research, in practical terms, it has left me somewhat schizophrenic owing to my teaching. See also a similar comment made by Mark Adams (Chap. 8 in this book) when he realized that the received view concerning the history of genetics was problematic.

[2]The reader will find many relevant pages to the first (1859) and last (1872) editions of the *Origin of Species* that are analyzed in this chapter in Delisle (2019).

outcome being a selection process for natural variations that provide an advantage to some carriers (survival) and disadvantage to others (extinction). The action of natural selection (nature) is conceived as analogous with artificial selection (breeders). Over time, the argument continues, evolutionary change and adaptation are generated, a process reflected in the evolutionary pattern of forms moving away from each other (divergence).

What the received view largely overlooks, however, is the persistence of an underlying yet dominant narrative running through those chapters that focuses on the nature of life itself. Moving away from issues concerned with *hypothetical* evolutionary processes, this underlying narrative highlights the *factuality* of life being organized in a network-like fashion. At this explanatory level, the anticipated relationship between process (natural selection) and pattern (divergence)—the action of natural selection instigating divergence—is shattered and replaced by the inherent connectiveness of life, whose natural pattern imposes itself upon a natural selection that is powerless at undoing it. Darwin offers the reader a long list of statements to the effect that life is, indeed, network-like. In Chapters 1–5 and 8 of the *Origin of Species* (1859), one finds a consistent argument:

1. Life forms resisting change over several generations (p. 7).
2. Life forms having the tendency to reverse to the parental condition (pp. 14, 15, 25, 26, 152–154, 159).
3. Life forms being conceived as non-monolithic taxonomic entities called "sub-breeds" or "strains"; entities characterized by their inherent ability to bind by reproducing together (pp. 31, 32, 96, 267).
4. Life forms naturally favoring interbreeding over inbreeding by reproductively reaching out to forms that are not too closely related (pp. 70, 71, 96, 104–106, 248–250, 253).
5. Life forms characterized by different levels of fertility below the family level, thus involving the following taxonomic groups: genera, species, varieties, and races (pp. 22, 23, 248, 253, 255–257, 261, 267–272).
6. Breeders able to create perfectly fertile forms even though they are separated at the generic level (pp. 22, 23, 26).
7. Darwin's heredity theory based on blending, by which intermediate states in offspring are generated by mixing up together the features of both parents (p. 108).
8. The recognition that polyphyletism—the derivation from more than one ancestral line—is a common feature in domesticated forms (pp. 17, 28, 29, 40).

Yet, when Charles Darwin first introduces his Diagram in the middle of Chapter 4 of the *Origin of Species* (see Fig. 4.1), the reader is asked to read it through the background assumption of evolutionary divergence. Is Darwin's demand justified by his own writing? No, the reader is confronted with a *rhetorical strategy* presented in visual form. Speaking of Chapter 8, David Morrison (2015: 547, 548) justly observes:

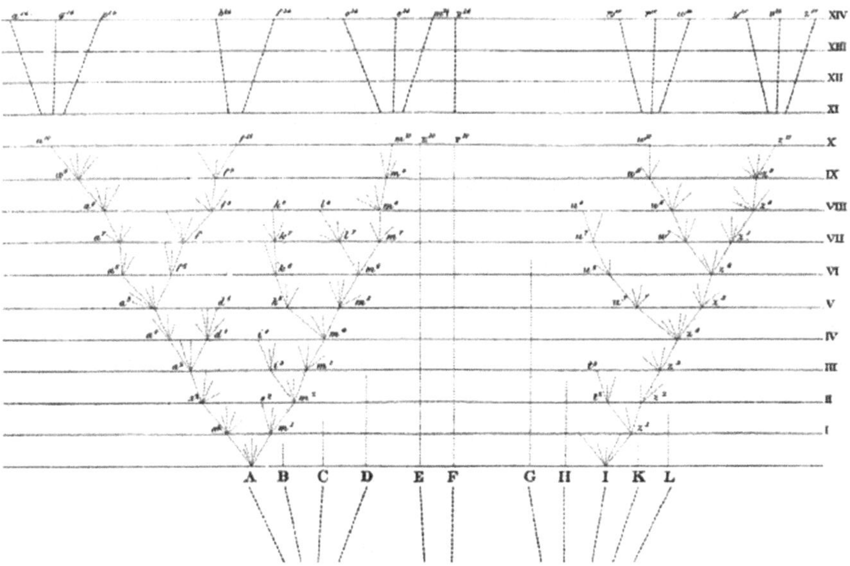

Fig. 4.1 *Darwin's diagram.* This sole illustration in the *Origin of Species* is a genealogy of extant forms and not a phylogenetic tree

> Darwin's original book has a chapter on hybridization (Chapter 8), which he uses as part of his argument that evolutionary divergence is continuous, so that all degrees of breeding should exist, from complete interbreeding within a species, through occasional cross-species hybridization, and on to complete isolation (e.g., between phyla). However, he concurrently fails to note that his Tree of Life must, therefore, have reticulations (representing the hybridization events).

What is said here of Chapter 8, I hold, applies equally to Chapters 1 through 5. And Morrison (2016: 459) concludes: "[s]o, even Darwin's Tree is actually a network." In those chapters, natural selection is too weak a causal force to organize life under any significant evolutionary pattern. In the very best case, Darwin should obtain a huge amount of reticulate evolution, not divergence. This explains why Darwin (1859: 102–108) could not come up with an even remotely convincing explanation of full speciation (Wagner 1868; Romanes 1886; Vorzimmer 1970; Sulloway 1979; Mayr 1982: 411–417, 1991: 31–34; Bulmer 2004). Darwin struggles to impose a purely theoretical agenda—natural selection and divergence—upon a robust network-like reality he himself so consistently presented. Apparently, "process" and "pattern" share in the *Origin of Species* an epistemological relationship more complicated than anticipated.

4.4 A Process (Natural Selection) Producing an Imaginary Pattern (Divergence)

A cursory reading of Chapters 6 and 7 of the *Origin of Species* (1859) gives the impression that Darwin is hard at work strengthening his theory by providing proof of the existence of intermediate forms in evolution. His method of investigation is made explicit in these words in the 6th edition:

> In searching for the gradations through which an organ in any species has been perfected, we ought to look exclusively to its lineal progenitors; but this is scarcely ever possible, and we are forced to look to other species and genera of the same group, that is to the collateral descendants from the same parent-form, in order to see what gradations are possible, and for the chance of some gradations having being transmitted in an unaltered or little altered condition (Darwin 1872: 144)

Irrespective of Darwin's difficulties with life being organized in a network-like fashion, his proposed method is at least coherent with his attempt to organize evolution under the principle of divergence. Some phylogenetic hypotheses involving cases such as the squirrels, the petrels, the ants, and the bees might even be *consistent* with his proposed theory (Darwin 1859: 180, 184–185, 219–235), although not to the exclusion of other competing theories such as reticulate evolution and polyphyletism.

Soon enough, however, the reader discovers that an entirely different world is being described by Darwin once she or he looks behind his rhetorical agenda. Suddenly, in a kind of paradigm or gestalt shift, the overall picture of evolution is no longer cast in terms of "divergence" but rather "parallel evolution," as if separate and independent evolutionary lineages were rushing to acquire similar features across a wide range of taxonomic groups (1859: 181–182, 186–189, 193–194, 216–224):

1. The various manifestations of parasitic behaviors in ants, bees, and birds.
2. The rise of flying capacities in three different orders of land mammals (flying lemurs, bats, and lemurs).
3. The co-occurrence of luminous organs in a few insects only remotely related.
4. The development of electricity-producing organs in dozens of fish that are also distantly related.
5. The appearance of variously complex eyes in separate phyla (reptiles, insects, and starfish).

In Chapters 6 and 7 of the *Origin* (1859), the underlying discourse is strikingly "un-Darwinian." To understand why, it will first be necessary to pay attention to one fundamental feature of the theory Darwin hopes to put before us: the particular nature of the relationship between natural selection (process) and divergence (pattern). Elliott Sober (2011: 35) captures this relationship nicely:

> [T]he relationship between common ancestry and natural selection in Darwin's theory has a special feature. Natural selection and common ancestry fit together in the theory, but only if selection has not been all-powerful. If *all* the traits that organisms have are present because

> there was selection for them, Darwin's Principle will conclude that none of them provides strong evidence for common ancestry. What is needed is that selection causes branching and extinction but that some traits persist in lineages for *non*adaptive reasons... The conjunction of common ancestry *and* natural selection would be unknowable, according to Darwin's Principle, if the second conjunct described the *only* cause of trait evolution. Hyperadaptationism has no place in this two-part theory. [italics original]

To put it another way, to be well founded, Darwin's theory had to maintain a delicate balance between the action of natural selection and the traces of affiliation binding related forms (Delisle 2019: 115–130). If the power of natural selection is too strong, traces of affiliation will entirely be erased under adaptation; if the power of natural selection is too weak, affiliated forms will remain bounded in the network of life. We have already seen Darwin confronted with this last issue in Chapters 1–5 of the *Origin* (1859). This time, in Chapters 6 and 7, Darwin faced another problem by moving to the other extreme, suggesting that the force of natural selection is so overwhelming as to generate massive amounts of parallel evolution or analogies, thus shattering the principle of divergence. Apparently, confronted with the inadequacies of Chapters 1–5, Darwin felt the need to overcompensate for the weakness of his case, thus undermining the method of investigation he explicitly outlined in the remarks quoted above (Darwin 1859: 187; 1872: 144). Indeed, Darwin became ever more ambitious in his quest for intermediate connections between forms—first looking for traits in presumably very closely related species—then extending his search to taxonomic groups as distinct as those belonging to different orders, classes, and even phyla. The more Darwin widened the taxonomic scope of his analysis, however, the more he risked reaching the breaking point of his theory by entering the territory of evolutionary parallelism and analogies. This breaking point is reached several times in the *Origin*, although Darwin tries to conceal it by resorting to several argumentative expedients in Chapters 6 and 7:

1. Different taxonomic levels are mixed up from case to case within the same page or few pages.
2. Many examples contain no indication as to the exact taxonomic levels discussed.
3. The nature of the relationship between "natural selection" and "divergence (descent)" is modified at will, with cases overstressing natural selection over descent and other cases delicately balancing the two.
4. Homologies and analogies are indiscriminately and confusingly exploited.

In addition to the confusion created by these expedients, the reader has to face Darwin's significant use of imaginary scenarios. A number of so-called case studies simply go beyond the confines of empirical reality. The paradigmatic example of this tendency concerns the complexification of the eye. Looking for intermediate stages in widely separated taxonomic groups, Darwin writes:

> The simplest organ which can be called an eye consists of an optic nerve, surrounded by pigment-cells and covered by translucent skin, but without any lens or other refractive body... In certain starfishes, small depressions in the layer of pigment which surrounds the nerve are filled... with transparent gelatinous matter, projecting with a convex surface, like the cornea in the higher animals... In the great class of the Articulata, we may start from an

> optic nerve simply coated with pigment, the latter sometimes forming a sort of pupil, but destitute of a lens or other optical contrivance. With insects it is now known that the numerous facets on the cornea of their compound eyes form true lenses, and that the cones include curiously modified nervous filaments. But these organs in the Articulata are so much diversified that Müller formerly made three main classes with seven subdivisions, besides a fourth main class of aggregated simple eyes. . . Within the highest division of the animal kingdom, namely, the Vertebrata, we can start from an eye so simple, that it consists, as in the lancelet, of a little sack of transparent skin, furnished with a nerve and lined with pigment, but destitute of any other apparatus. In fishes and reptiles, as Owen remarked, 'the range of gradations of dioptric structures is very great.' It is a significant fact that even in man... the beautiful crystalline lens is formed in the embryo by an accumulation of epidermic cells, lying in a sack-like fold of the skin; and the vitreous body is formed from embryonic sub-cutaneous tissue (Darwin 1872: 144–146).

What is Darwin trying to achieve here? Is he collecting here and there across the taxonomic spectrum traces suggesting an evolutionary road going from a simple eye to a complex one? Is he aware that doing so implies fusing sources of information supporting evolutionary parallelism and analogies? The importance of these phenomena for the case study of eyes caught the attention of St. George Mivart in his *On the Genesis of Species* (1871: 76–78). The reader cannot but be confused by Darwin's approach in Chapters 6 and 7 which blurs the explanatory lines between actual phylogenetic propositions, possible phylogenetic hypotheses, and information gathered merely to throw indirect light on evolutionary steps taken during evolution.

Darwin then pursues his case study by resorting to an analogy between the eye and the telescope, fusing the art of nature with that of human engineering. In a thought experiment, Darwin's abstract reasoning moves away from the separate evolutionary lines that gave rise to more or less complex eyes in order to put forward the case for a unique and ideal evolutionary pathway in eye evolution:

> It is scarcely possible to avoid comparing the eye to a telescope. We know that this instrument has been perfected by the long-continued efforts of the highest human intellects; and we naturally infer that the eye has been formed by a somewhat analogous process. . . If we must compare the eye to an optical instrument, we ought in imagination to take a thick layer of transparent tissue, with a nerve sensitive to light beneath, and then suppose every part of this layer to be continually changing slowly in density, so as to separate into layers of different densities and thicknesses, and with the surfaces of each layer slowly changing in form. Further we must suppose that there is a power always intently watching each slight accidental alteration in the transparent layers; and carefully selecting each alteration which. . . tend to produce a distincter image. . . [N]atural selection will pick out with unerring skill each improvement (Darwin 1859: 188–189; see also 1872: 146)

No surprise that William Whewell (1864: 15; see also Hull 2009: 188–189) commented that the eye-telescope analogy merely allows us to *imagine* evolutionary transitions, and not document actual transitions.

After having been confronted in Chapters 1–5 of the *Origin of Species* (1859) to the network-like organization of life resisting the action of natural selection, Charles Darwin inverts the polarity of his explanation in Chapters 6 and 7 by tearing apart his divergent tree of life in a flight of fancy while simultaneously boosting the power of natural selection by presenting it as an all-powerful agent generating a massive quantity of evolutionary analogies. By so doing, of course, Darwin shatters the delicate balance his theory would have required between "natural selection" and "divergence." Evolutionary process and evolutionary pattern are, for Darwin, two distinct and independent explanatory variables whose relationship he modifies at will from case to case.

4.5 Explanatory Precedence of Pattern Over Process: Putting the Broad Framework in Place

We now know that it would be futile to try to identify a single coherent narrative in the *Origin of Species* (see also Delisle 2017a, 2019). Nevertheless, it is at least possible to identify a predominating narrative in terms of its consistency: this is seen in Darwin's Diagram (see Fig. 4.1) and its accompanying conceptual cluster. Here lies the explanatory core of Darwin's theory, I hold, if one is willing to ignore the other incompatible components. At least as far as Darwin's Diagram is concerned, I subscribe to the view of Mary P. Winsor (2009: 47) when she says that "divergence" and "natural selection" are uncoupled in the *Origin of Species*:

> The leaders of the Modern Synthesis and their epigones have convinced us that this means that belief in evolution could not be scientific unless it included a cause, which they have achieved by overlooking Darwin's claim that a reasonable person ought to accept branching evolution 'even if it were unsupported by other facts or arguments' (Darwin 1859: 458)... Yet after the Modern Synthesis, many writers treat natural selection and evolution as interdependent and inextricable parts of one unified theory. This flies in the face of the historical facts.

The principle of divergence lies at the heart of a rich conceptual bundle that severely constrains the role of natural selection. For Darwin, there is much more to "divergence" than just a divergent pattern. Before unpacking this bundle, a preliminary obstacle to our understanding of the principle of divergence should be removed. A cursory look at Darwin's Diagram leaves the reader with the impression that change is characterized by evolutionary opportunism, with life forms evolving in a haphazard fashion: (a) many lineages are diverging from each other, (b) some lineages belonging to A and I converge toward each other to a degree, (c) a number of lineages remain stable for various amounts of geological times, and (d) extinction characterizes the fate of many lineages.

These differential evolutionary outcomes would seem to be the product of the encounter between, on the one hand, the rise of unpredictable variations in a host of life forms and, on the other, equally unpredictable biotic and abiotic changes. This explains, the argument continues, the indeterministic nature of the field of evolutionary biology and justifies its classification among historical disciplines: Darwinism thus recognizes the historical nature of biological evolution, "evolutionary contingency" being its key characteristic (Simpson 1964: 176–189; Mayr 1982: 21–82; Gould 1986, 1989: 277–291; 2002: 1333–1336). Of course, Darwin's oft-cited remarks supporting this seemingly incontestable view are the following:

> These several facts accord well with my theory. I believe in *no fixed law of development*, causing all the inhabitants of a country to change abruptly, or simultaneously, or to an equal degree. The process of modification must be extremely slow. The variability of each species is quite independent of that of all others. Whether such variability be taken advantage of by natural selection, and whether the variations be accumulated to a greater or lesser amount, thus causing a greater or lesser amount of modification in the varying species, depends on many *complex contingencies*, – on the variability being of a beneficial nature, on the power of intercrossing, on the rate of breeding, on the slowly changing physical conditions of the country, and more especially on the nature of the other inhabitants with which the varying species comes into competition. Hence it is by no means surprising that one species should retain the same identical form much longer than others; or, if changing, that it should change less (Darwin 1859: 314). [my emphasis]

Yet, this quote is deceiving. Darwin did not have the same understanding of such notions as "no fixed law of development" and "complex contingencies" as do modern evolutionists. Let us see how he coupled his view of the tree of life with a long list of explanatory devices to effectively put evolution in a straightjacket, seriously depriving the evolutionary process of room to maneuver.

1. *When Neontology Overshadows Paleontology*. Darwin belongs to a group of nineteenth-century scholars who confronted the rise of evolutionism by working their way through an *assimilation process*: forms alive today (extant forms) were used as a gauge with which to classify past and extinct forms. As the annals of life were being gradually revealed to the world, Darwin rushed to assimilate extinct forms to extant ones. "To travel in space is to travel in time" is certainly the tacit motto of the *Origin of Species*. By travelling in geographical space, extant forms presenting similarities between themselves were reunited under the assumption that they are connected through common ancestors (1859: 349–350). Providing a long list of reasons as to why the field of paleontology is of limited value in our inquiry of the past (1859: 280, 287, 289, 293–301, 306), Darwin felt justified at building an evolutionary biology very largely based on the field of neontology.
2. *When Extinct Forms Fall In-Between Extant Forms*. For Darwin, the principle of divergence plays the precise function of bracketing all the past under the present (see, for instance, Darwin 1859: 303–306). The idea of common ancestry extends beyond the mere notion of kinship also to include the assumption that the present always contains more biological variability than at any previous time in the history of life, which explains Darwin's commitment to what Stephen Jay Gould (1989: 39–52) calls the "cone of increasing diversity." It has not been sufficiently stressed that Darwin's principle of divergence constitutes a conceptual tool that prevents the past from overflowing the present with variations unknown among extant forms, as is the case when evolution is envisioned using a "decimation model" (compare Figs. 4.2 and 4.3). For Darwin, the past remains under the epistemological control of the present at all times; nothing truly new is expected to emerge from an unknown past. Can a history of life deprived of evolutionary novelties really be, then, characterized by "evolutionary contingency"? If so, what explanatory room is left for a creative force called "natural

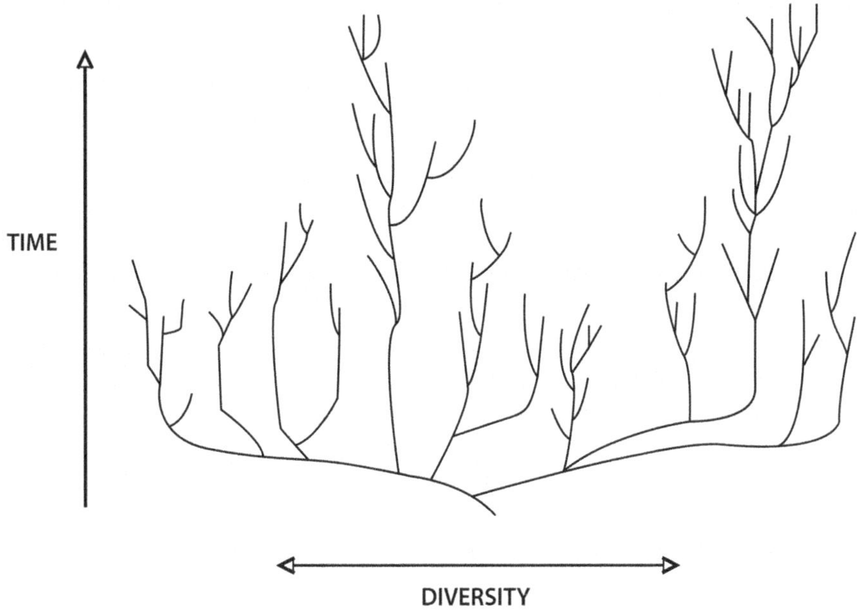

Fig. 4.2 *Decimation model*. This evolutionary model is based on a rapid multiplication of types followed by a significant amount of decimation (extinction). Many body plans that existed in the past have left no traces today

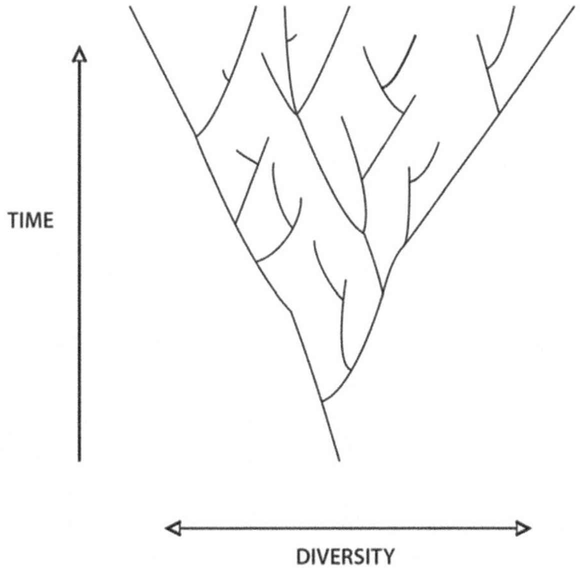

Fig. 4.3 *Cone of increasing diversity*. This evolutionary model assumes that variability observed among extant forms entirely captures that of the past, since all fossils are either aligned with or fall in-between extant forms. This is Charles Darwin's view, which he based on the notion of a slow deployment of increasing diversity within a class

selection"? The following quote is extraordinary for its epistemological implications:

> Let us now look to the mutual affinities of extinct and living species. They all fall into one grand natural system; and this fact is at once explained on the principle of descent. The more ancient any form is, the more, as a general rule, it differs from living forms. *But... all fossils can be classed either in still existing groups, or between them.* That the extinct forms of life help to fill up the wide intervals between existing genera, families, and orders, cannot be disputed. For if we confine our attention either to the living or to the extinct alone, the series is far less perfect than if we combine both into one general system. With respect to the Vertebrata, whole pages could be filled with striking illustrations... showing how extinct animals fall in between existing groups... (Darwin 1859: 329). [my emphasis]

3. *Fixed Taxonomic Positions Ever Since the Beginning of Life.* As if Darwin's tree of life was not already rigid enough, he postulated another notion that would further undermine the openness of the evolutionary process. In presenting his Diagram (see Fig. 4.1), Darwin argues that the relative taxonomic positions of the main extant groups were preserved with respect to one another throughout all of evolutionary history. It is worth quoting this passage in full:

> The reader will best understand what is meant, if he will take the trouble of referring to the diagram... We will suppose the letters A to L to represent allied genera, which lived during the Silurian epoch, and these have descended from a species which existed at an unknown anterior period. Species of three of these genera (A, F, and I) have transmitted modified descendants to the present day, represented by the fifteen genera (a^{14} to z^{14}) on the uppermost horizontal line. Now all these modified descendants from a single species, are represented as related in blood or descent to the same degree... yet they differ widely and in different degrees from each other. The forms descended from A, now broken up into two or three families, constitute a distinct order from those descended from I, also broken up into two families. Nor can the existing species, descended from A, be ranked in the same genus with the parent A; or those from I, with the parent I. But the existing genus F^{14} may be supposed to have been but slightly modified; and it will then rank with the parent-genus F; just as some few still living organic beings belong to Silurian genera. So that the amount or value of the differences between organic beings all related to each other in the same degree of blood, has come to be widely different. *Nevertheless their genealogical arrangement remains strictly true, not only at the present time, but at each successive period of descent.* All the modified descendants from A will have inherited something in common from their common parent, as will all the descendants from I... All the descendants of the genus F... are supposed to have been but little modified, and they yet form a single genus. But this genus, though much isolated, will still occupy its proper intermediate position; for F originally was intermediate in character between A and I, and the several genera descended from these two genera [A and I] will have inherited to a certain extent their characters (Darwin 1859: 420–422; see also 1872: 369–370). [my emphasis]

By so arguing, Darwin ensured that no major evolutionary events would be lost to the taxonomic study of extant forms: it is argued that the position of lineages A, F, and I (including their descendants) has persisted ever since the beginning of life, with F remaining in the middle, I on the right side, and A on the left side. Why, then, bother studying an uncertain evolutionary past, if this only confirms what the present already reveals in full?

4. *A Modern-Looking World as Far Back as We Can Geologically Probe*. In an evolutionary world that seems to be characterized by its ever-eternal modernity, it is perfectly consistent to see Darwin imagining common ancestors that are merely downsized forms when compared with their extant counterparts:

> In the vertebrata, we see a series of internal vertebrae bearing certain processes and appendages; in the articulata, we see the body divided into a series of segments, bearing external appendages; and in the flowering plants, we see a series of successive spiral whorls of leaves. An *indefinite repetition of the same part or organ* is the common characteristic... of all low or little-modified forms; therefore we may readily believe that the unknown progenitor of the vertebrata possessed many vertebrae; the unknown progenitor of the articulata, many segments; and the unknown progenitor of flowering plants, many spiral whorls of leaves (Darwin 1859: 437). [my emphasis]

One cannot but be struck by the preformationist tone of this passage which appeals to the "indefinite repetition of the same part or organ." Here, Darwin is hard at work recycling variations observed among extant forms by projecting them backward in time, thus creating imaginary ancestors who are merely pale echoes of the present.

5. *The Principle of Gradation: Tightly Riveting the Past to the Present*. Epistemologically speaking, the principle of divergence is so intimately connected with the "principle of gradation" that the two fuse together at the core of Darwin's theory under a dual principle whose nature is simultaneously about "divergence" and "gradation." As Darwin repeatedly remarks in the *Origin of Species* (1859: 194, 206, 210, 243): *Natura non facit saltum* ("nature makes no jumps"). Of course, Darwin's gradualistic view of matters requires him to postulate that evolution is filled with intermediate stages. More importantly for present purposes, however, this view fills a precise function in his explanatory apparatus: whereas "divergence" prevents the past from overwhelming the biological diversity of the present by bracketing it, "gradation" has the role of ensuring that life constitutes a perfectly smooth and uninterrupted network organized in a tightly knit fabric, so that no such disruption is possible. As Darwin (1859: 489) writes: "As all the living forms of life are the lineal descendants of those which lived long before the Silurian epoch, we may feel certain that the ordinary succession by generation has never once been broken, and that no cataclysm has desolated the whole world." The world is thus perfectly continuous, deprived of breaks or discontinuities. The world is one and united, built as a single block with the past interpenetrating the present or, even better, being literally incorporated into the present. Thus, the explanatory function of the principle of divergence-gradation is at the core of Darwin's theory.

Because Darwin deprived the evolutionary past of any significant existence for itself, he becomes an easy target of the same epistemological criticism William Whewell leveled against Charles Lyell in 1832:

> It seems to us somewhat rash to suppose, as the uniformitarian does, that the information which we at present possess concerning the course of physical occurrences, affecting the

> earth and its inhabitants, is sufficient to enable us to construct classifications, which shall include all that is past under the categories of the present. Limited as our knowledge is in time, in space, in kind, it would be very wonderful if it should have suggested to us all the laws and causes by which the natural history of the globe. . . is influenced—it would be strange, if it should not even have left us ignorant of some of the most important of the agents which, since the beginning of time, have been in action. . . (Whewell 1832: 126)

As Marjorie Grene and David Depew (2004: 171) have observed: "In light of mass extinctions spread across quite different geological eras, Whewell thought it decidedly unscientific for Lyell to shoehorn the history of life into categories cobbled up to fit contemporary life forms." Whereas Whewell anticipated that the present would be blind to yet hidden novelties in the past, Darwin would not seriously contemplate that possibility.

4.6 Channeling the Action of Natural Selection

Evolutionary contingency as we envision it today is not what Darwin had in mind. A closer look at how Darwin explains the *process* of evolutionary divergence will make this point clearer still. In a nutshell, evolutionary change does exist for Darwin, but only insofar as it occurs according to a divergent pattern. His view on the matter is so unyielding that he refuses to contemplate the possibility of other kinds of evolutionary change, such as convergent evolution or reticulate evolution in their full expression.[3] Darwin's arguments imply an evolutionary process that is *channeled*. At this point, a defender of the received view might reply that divergence constitutes the overall evolutionary pattern for Darwin only because it is a by-product of the action of natural selection: it is the process that generates the pattern, not the reverse. This argument could then be developed by gathering the following explanatory components:

1. Places in the economy of nature are limited in a world filled to near capacity.
2. Demographic increases or overpopulation presses organisms and species against each other, creating the conditions for competition.
3. Evolutionary change or adaptation is the way taken to outcompete other forms, the best variations carrying the day at the expense of others which are doomed to extinction.
4. Adaptive pressure is not dissipated in a random fashion by pushing lineages in all sort of directions; instead, this pressure is released along a divergent evolutionary path.

[3]Darwin's Diagram shows that he allows for a *very limited amount* of evolutionary convergence when, for instance, some lines of lineages A and I are compared with each other. What Darwin wants to avoid at all costs is losing sight of the relative position of the main taxonomic groups during the entire history of life.

5. It is by evolving away from competitors through divergence that some forms manage to successfully carve out a place for themselves in the economy of nature. This principle is so prevalent in Darwin that it acquires a near law-like status under the appellation of a "general rule":

> But I must remark that I do not suppose that the process ever goes on so regularly as is represented in the diagram, though in itself made somewhat irregular. I am far from thinking that the *most divergent varieties* will invariably prevail and multiply: a medium form may often long endure, and may or may not produce more than one modified descendant; for natural selection will always act according to the nature of the places which are either unoccupied or not perfectly occupied by other beings; and this will depend on infinitely complex relations. But as a *general rule*, the more diversified in structure the descendants from any one species can be rendered, the more places they will be enabled to seize on, and the more their modified progeny will be increased (1859: 118–119). [my emphasis]

The received view just described is based on a selection of components that apparently lend support to the view that "natural selection" and "divergence" are closely interrelated at the core of Darwin's theory. But this interpretation can only be maintained at the expense of ignoring the significant explanatory apparatus put in place by Darwin around his Diagram. For him, divergence has explanatory precedence over the action of natural selection. This is also seen in the case of the "competitive exclusion principle," a concept used by Darwin (although this terminology is not his). Darwin explains that competition between two forms is most fierce between closely related forms, if only because they are so similar in their constitution that they are bound to exploit the same resources in nature:

> But the struggle almost invariably will be most severe between the individuals of the same species, for they frequent the same districts, require the same food, and are exposed to the same dangers. In the case of varieties of the same species, the struggle will generally be almost equally severe... As species of the same genus have usually, though by no means invariably, some similarity in habits and constitution, and always in structure, the struggle will generally be more severe between species of the same genus, when they come into competition with each other, than between species of distinct genera... We can dimly see why the competition should be most severe between allied forms, which fill nearly the same place in the economy of nature (1859: 75–76).

On this understanding, a strong incentive exists for closely related forms to diverge away from each other. So far, so good, for the thesis that *pattern* is a by-product of *process*. Things fall apart, however, when we follow Darwin's reasoning a little further. In the quote above, Darwin states that such an incentive to diverge is reduced when forms are more distantly related: "the struggle will generally be more severe between species of the same genus... than between species of distinct genera" (p. 76). This claim repeated elsewhere in the *Origin* (1859: 110, 321) begs the question: "What is driving the process of divergence all the way up the taxonomic scale?," a question that makes explicit what is only implicit in Darwin. Should not the process be thwarted when two related forms become separated at the level of a family, an order, or a class, by which time they have become so modified as to exploit different resources? Darwin avoided asking this embarrassing question. Yet, by his own logic, the competitive exclusion principle should gradually be deactivated in reverse proportion to the taxonomic ascent of two related forms. Why assume, then, that divergence will continue under taxonomic ascent? Darwin is not relying on the *process* of divergence to support his overall view; he is merely extrapolating the divergent *pattern* he takes for granted across all taxonomic levels. Pattern has explanatory precedence over process.

Should we really be surprised by this? An inspection of Darwin's Diagram reveals that, at whatever taxonomic level he reads it (1859: 116–126, 331–332, 412–413), an overall pan-divergent pattern emerges, a recurrence that recalls "fractal geometry." How could it have been otherwise for someone who claims that all extinct forms fall in-between extant ones, with the "cone of increasing diversity" implying that divergence today is at a maximum relative to the entire evolutionary past? Rather than exhibiting any explanatory force of its own, natural selection is a process merely superimposed upon the pattern of divergence.

Expanding upon the fractal-like configuration of Darwin's view of life, it should also be noted how his pan-divergent view further deprives natural selection of its creative role by confining it to divergence. Indeed, as the competitive exclusion principle exhausts itself during taxonomic ascent, it would have been consistent on Darwin's part to hold that such a relaxed selective context permitted the manifestation of other evolutionary patterns, like convergent evolution. This, however, Darwin would not allow. To do so would have blurred or confused the taxonomic positions of the various groups under consideration. In his Diagram, it should be recalled, Darwin wanted to believe he had a clear view on the relative position of lineages A, F, and I ever since the beginning of life. The homogeneity or fractal-like configuration of Darwin's pan-divergent Diagram—at a low taxonomic level, at a high taxonomic level, in the present time, and in the past time—is guaranteed by his commitment to integral uniformitarianism under the principle of gradation (*Natura non facit saltum*). Darwin's unique and repetitive geometry (divergence) annihilates the creativity of evolutionary time by strongly channeling the action of natural selection, making evolution somewhat predictable and deterministic.[4]

[4]Darwin's pan-divergent view of evolution makes it somewhat predictable on a number of other manifestations: (a) the most divergent forms are more likely to survive; (b) the less divergent ones are more susceptible to go extinct; (c) more often than not, adaptive trends will be organized around a divergent pattern, etc. I thank David Ceccarelli for having required of me more precision to my thoughts on such an issue.

4.7 Blocking the Action of Natural Selection: Stability, Equilibrium, and Cyclicity

We all too often assume that Darwin's Diagram applies to the entire tree of life. This, however, is not what Darwin had in mind. He was quite explicit that his reflection about evolution was confined to *distinct and separate classes* (1859: 128–129; 207, 450, 483–484).[5] It seems innocuous enough for us as modern scholars to simply extend the analysis to the whole tree of life, but to do so on Darwin's behalf is a significant distortion. Knowing that Darwin worked his way by projecting extant forms backward in time, the overall picture that emerges is a series of separate and independent classes evolving in parallel since the beginning of life. Darwin's (1859: 484, 488, 490) occasional suggestion that life arose from a limited number of progenitors is more rhetoric than argument. As long as one conceives so-called "common" ancestors as pale reflections of extant forms, as Darwin does, one always ends up with forms looking strangely similar to currently existing ones. The following quote captures the essence of Darwin's approach: "I must premise, that I have nothing to do with the origin of the primary mental powers, any more than I have with that of life itself. We are concerned only with the diversities of instinct and of the other mental qualities of animals within the same class" (1859: 207). Darwin's Diagram is not, properly speaking, a "tree of life"; instead, it is a genealogy of extant forms confined to a class. And Darwin was not alone among his contemporaries and countrymen in shoehorning extinct forms into extant ones. For instance, Richard Owen (1860: 515) wrote:

> Every known fossil belongs to someone or other of the existing classes, and that the organic remains of the most ancient fossiliferous strata do not indicate or suggest that any earlier and different class of being remains to be discovered, or has been irretrievably lost in the universal metamorphism of the oldest rocks.

Similarly, Thomas Henry Huxley (1862: xlviii) claims:

> No fossil animal is so distinct from those now living as to require to be arranged even in a separate class from those which contain existing forms. It is only when we come to the orders, which may be roughly estimated at about a hundred and thirty, that we meet with fossil animals so distinct from those now living as to require orders for themselves; and these do not amount, on the most liberal estimate, to more than about ten per cent. of the whole.

---

[5]Robert J. Richards (1999: 129) rightly stresses the neglected fact that Darwin's biology is actively constructed around what the former calls "archetypes," such as the vertebrate archetype. I argue that Darwin's focus on biological types is connected to his commitment to the uniformitarian-neontological approach which consists in projecting separate extant classes backward in geological times. In so doing, Darwin embraced an approach that is also shared with his contemporaries and countrymen Charles Lyell, Richard Owen, and Thomas Henry Huxley (see also main text below).

Like Owen and Huxley, Charles Darwin took the class and phylum levels to be perfectly stable entities in the history of life, restricting extinction events to the order level and below. This is implicitly stated in these words:

> From the first growth of the tree, many a limb and branch has decayed and dropped off; and these lost branches of various sizes may represent those whole *orders*, *families*, and *genera* which have now no living representatives, and which are known to us only from having been found in a fossil state (Darwin 1859: 129–130). [my emphasis]

Darwin, Owen, and Huxley worked their way backward in time, using extant classes as absolute reference points for classifying extinct forms.[6] Darwin's vision of life's history in the *Origin of Species* is much more consistent and robust than usually acknowledged by the received view, weighing heavily on how the process-pattern issue is envisioned. Let us continue unpacking the conceptual cluster accompanying Darwin's view of life.

For Darwin, the Silurian period is an impermeable temporal barrier, before which nothing is known of life, after which life is largely organized around well-defined groups as seen among extant forms, and at which point life seems to have suddenly sprung full-fledged out of nothing (1859: 302–310, 338). Reading him leaves the reader with the strange impression that life is largely at a standstill, with its main features having being in place for a very long time. Again, T.H. Huxley (1862: I) managed to express this sentiment:

> [E]nough has been said to justify the statement that, in view of the immense diversity of known animal and vegetable forms, and the enormous lapse of time indicated by the accumulation of fossiliferous strata, the only circumstance to be wondered at is, not that changes of life, as exhibited by positive evidence, have been so great, but that they have been so small.

Furthermore, the Silurian (or Cambrian in 1872: 285–286) constitutes, for Darwin, a genuine epistemological divide: (a) during the Silurian and after, one is confronted with life in all its factuality, fully constituted and organized around extant classes; (b) during the pre-Silurian or pre-Cambrian period, one contemplates entirely hypothetical ideas about the past. This situation puts substantial explanatory stress on this earlier period. Indeed, Darwin (1859: 307; 1872: 286) argues that half the time allotted to the entire history of life is found before the Silurian or Cambrian, implying that the constitution of classes and phyla must have had to occur in that first half. Darwin is therefore being completely consistent in arguing that throughout the history of life, Earth has been populated by just as many different taxonomic groups as we encounter today and as far back as we can geologically probe, including distinct classes: "at the most remote geological period, the earth may have been as well-populated with many species of many genera, families, orders, and classes, as at the present day" (1859: 126; see also 1872: 97).

[6]By focusing on the differences between the views of Charles Darwin, Thomas Henry Huxley, and Richard Owen, the historiography overlooked the significant ideas that unite them. See Delisle (2019: 209–210).

Charles Darwin's Silurian or Cambrian world is nearly as modern as ours today; a world structured around separate extant classes has been in place since the Silurian or Cambrian period. In addition, these classes have been spared the fate of extinction.[7] To put it another way, higher-level taxonomic groups like classes and phyla are, according to Darwin, deeply entrenched in nature. This is explained by ever more dominant groups driving less dominant ones to extinction, ultimately favoring the entrenchment of a limited number of high taxonomic entities: "Thus we can account for the fact that all organisms, recent and extinct, are included under a few great orders, under still fewer classes" (1859: 428, 429); or again: "we can understand how it is that there exist but very few classes in each main division of the animal and vegetable kingdoms" (1859: 126). Darwin (1859: 51–52) even hints that among the lower-level taxonomic entities, some varieties and species are more permanent than others.

One encounters, therefore, ever more entrenched entities as one gradually climbs the taxonomic scale from bottom to top, with classes and phyla approaching near-total domination, permanence, and stability. This notion is even consistent with Darwin's "competitive exclusion principle" when applied at higher taxonomic levels. As seen above, Darwin (1859: 76, 110, 321) holds that the incentive for two related forms to diverge away from each other is reduced as they climb the taxonomic scale. Extending Darwin's explicit formulation for low taxonomic levels to its implicit and logical consequence at higher taxonomic levels, one sees that classes and phyla are, by virtue of their inherent dominant position, without direct competitors because of having carved out for themselves exclusive places in the economy of nature. Within this adaptive void, the competitive exclusion principle is eventually entirely neutralized during taxonomic ascent.

The idea of permanent and nearly static classes and phyla is sharply at odds with the received view, which takes for granted that "evolutionary contingency" is a fundamental feature of Darwin's theory. Naturally, this assumption is also accompanied by the notion of an open-ended evolution, evolutionary opportunism being explained by the action of natural selection in the context of unpredictable biotic and abiotic changes. It is now easy to see why this interpretation lacks a solid exegetical foundation. We have seen how Darwin deployed a long list of explanatory devices in order to flesh out the key features of life's history. One such feature concerns the permanence and near-stability of classes and phyla. Once this is recognized, all that was left to explain for Darwin were the evolutionary events at lower taxonomic levels. This Darwin did by outlining a theory for which the notion of "evolutionary

[7]For Darwin, classes avoided extinction. This can be seen from deducing Darwin's position by juxtaposing two sources of information: (1) separate evolutionary groups belonging to the class level existed as far back as we can geologically probe (1859: 126; 1872: 97); (2) all known fossils can be classified either among the still-extant groups or between them; none fall outside the variational range of extant forms, as guaranteed by the principle of divergence which has been effective since the beginning of life (1859: 329; 1872: 301). Darwin did recognize the extinction of ammonites and trilobites, but these groups were, for him, to be placed at either the family- or order-level only (1859: 321; 1872: 297).

equilibrium" is essential, restricting evolution to the rise and fall of varieties, species, genera, families, and orders, confining them within permanent and entrenched classes and phyla.

The Darwinian world *blocked* the action of natural selection by also closing it on itself on the number of species: Darwin argues that the number of species in the history of life remained roughly the same:

> Thus the appearance of new forms and the disappearance of old forms. . . are bound together. In certain flourishing groups, the number of new specific forms which have been produced within a given time is probably greater than that of the old forms which have been exterminated; but we know that the number of species has not gone on indefinitely increasing, at least during the later geological periods, so that looking to later times we may believe that the production of new forms has caused the extinction of about the same number of old forms (1859: 320; see also 1872: 296).

Darwin's world is based on a surprisingly rigid mechanical equilibrium, wherein the evolutionary fate of population A^1 depends on the evolutionary fate of population A^2 (1859: 63–64, 109, 172). Expansion in space and time of the former is directly proportionate to the decline of the latter. In a world filled to capacity, as Darwin's world manifestly is, expansion of a form can only be achieved at the expense of another. Conversely, if A^2 stands its evolutionary ground, A^1 will be prevented from expanding. Evolutionary gates are opening and closing on such geometrical relationships, a mechanical equilibrium not obviously congenial to the notion of "evolutionary contingency."

Of course, the received view would have it that Darwin required forms to be pressed against each other in order to create the conditions for competition and, ultimately, evolutionary change through the search of adaptive responses. Yet, a rather neglected feature of Darwin's Diagram is the evolutionary stability and stagnation of no fewer than 9 lineages out of 11 (see Fig. 4.1). Stagnation is a notion discussed in several places in the *Origin of Species* (1859: 281, 314–315, 351, 402, 429). Darwin's expedient was to claim that many forms are involved in a sort of "adaptive equilibrium," a relationship involving the action of natural selection, but one that is blocked from generating evolutionary change under mutually entrenched forms, wherein adaptive forces cancel each other out: "Battle within battle must ever be recurring with varying success; and yet in the long-run the forces are so nicely balanced, that the face of nature remains uniform for long periods of time..." (1859: 73). Here again, we see the explanatory precedence of evolutionary pattern over evolutionary process: evolutionary stability or stagnation is recognized as a fact or pattern, requiring an evolutionary process in order to justify it; in this case, "adaptive equilibrium" is that process.[8]

---

[8]It may be tempting to argue that Charles Darwin believed in biological progress, contrary to the view advanced here in this chapter. Darwin did believe in progress; however, a close analysis reveals that for it he offers rhetoric rather than argument in support of this belief. At a more fundamental level, Darwin developed a robust conceptual cluster concerning the annals of life that thwarted the expression of progress. This question has been treated at some length in Delisle (2019: 197–228).

4.8 Dismissing the Action of Natural Selection

One last case will help us highlight the nature of the relationship between evolutionary pattern and evolutionary process in Darwin's mind. This time, we see him protecting the principle of divergence by entirely *dismissing* the role of natural selection in order to avoid undermining it. Forced to choose between "divergence" or "natural selection," he selected divergence. Two examples will be provided. The first concerns the classic example of the Galapagos Islands (1859: 397–403), in which Darwin tries to explain the biogeographical facts concerning the migration of faunas and floras to a group of 10 islands in the Pacific Ocean 500 or 600 miles off the shores of Ecuador (South America). Darwin's main argument is the following: various life forms on these islands migrated from the near mainland where their common ancestors originally resided, only to diverge from one another through their differential adaptation on separate islands.

> Thus the several islands of the Galapagos Archipelago are tenanted... in a quite marvellous manner, by very closely related species; so that the inhabitants of each separate island, though mostly distinct, are related in an incomparably closer degree to each other than to the inhabitants of any other part of the world. And this is just what might have been expected on my view, for the islands are situated so near each other that they would almost certainly receive immigrants from the same original source, or from each other (Darwin 1859: 400).

This description apparently suggests that divergence and natural selection are intertwined. But this fails to take into account a more challenging situation in which Darwin is forced to explain why specific forms residing on one island did not, in general, invade other islands within sight of each other:

> But we often take, I think, an erroneous view of the probability of closely allied species invading each other's territory, when put into free intercommunication. Undoubtedly if one species has any advantage whatever over another, it will in a very brief time wholly or in part supplant it; but if both are equally well fitted for their own places in nature, both probably will hold their own places and keep separate for almost any length of time... In the Galapagos Archipelago, many even the birds, though so well adapted for flying from island to island, are distinct on each; thus there are three closely-allied species of mocking-thrush, each confined to its own island. . . I think we need not greatly marvel at the endemic and representative species, which inhabit the several islands of the Galapagos Archipelago, not having universally spread from island to island. In many other instances, as in the several districts of the same continent, pre-occupation has probably played an important part in checking the commingling of species under the same conditions of life (Darwin 1859: 402–403).

What Darwin does here is simply accept the overall pattern of divergence—in fact, he accepts a series of divergences—accompanying the invasion of the Galapagos Islands while ultimately dismissing the process of natural selection within its geographical confines by resorting to the unusual explanation of "pre-occupation." Indeed, whereas the force of natural selection is actively appealed to during the first wave of divergences when life forms migrated from the mainland, it is no longer the case when the separate forms are established on their respective islands. Explanatorily speaking, "divergence" remains steady throughout this scenario, while "natural selection" is either invoked or dismissed during its different phases. Darwin thus preferred to take liberties with the notion of natural selection, leaving the core of his pan-divergent theory intact. Darwin subscribed to this explanation even though the *Origin of Species* is filled with allusions to the effect that (a) sooner or later the rise of new variations will tip the balance by favoring one species over another and that (b) there is always room for improvement or change, no matter how many places in the economy of nature are apparently full.

Our second example will see Darwin developing a migration scenario in the Northern Hemisphere (1859: 365–369). Under the impetus of an expanding glaciation, alpine plants adapted for a cold climate migrated southward, both in Europe and North America. This scenario postulates that these plants originally shared common ancestors in the Arctic Region before their southward migration. The profound similarities binding this flora in both regions of the world—organisms belonging to the same variety, to two related varieties, or to two closely related species—are thus explained by an overall pattern of divergence. With the end of the glaciation, continues Darwin, this flora retreated back northward. The similarities uniting this flora are explained as follows:

> Thus we can understand the identity of many plants at points so immensely remote as on the mountains of the United States and of Europe. We can thus also understand the fact that the Alpine plants of each mountain-range are more especially related to the arctic forms living due north or nearly due north of them: for the migration as the cold came on, and the re-migration on the returning warmth, will generally have been due south to north. The Alpine plants, for example, of Scotland. . . and those of the Pyrenees. . . are more especially allied to the plants of northern Scandinavia; those of the United States to Labrador; those of the mountains of Siberia to the arctic regions of that country. . . In illustrating what, as I believe, actually took place during the Glacial period, I assumed that at its commencement the arctic productions were as uniform round the polar regions as they are at the present day. (Darwin 1859: 367–369)

Darwin's main point in this scenario consists in arguing that the alpine plants in the Northern Hemisphere are similar because they have diverged from common Arctic ancestors. The challenge, however, arises out of the need to explain how that flora managed to retain its degree of similarity, irrespective both of the extent of the geographical space covered and the considerable motion involved during the southward migration and its retreat northward. Should we not have expected that these conditions were ideal for the activation of more important adaptive responses and greater degrees of evolutionary change than are illustrated here? Precisely because Darwin wanted to protect the principle of divergence, this limited diversification event involving close varieties or species required the dismissal of the influence of natural selection. This time, he does so by appealing to the notion of a group of forms migrating together as a "single body":

> The arctic forms, during their long southern migration and re-migration northward, will have been exposed to nearly the same climate, and, as is especially to be noticed, they will have kept in a body together; consequently their mutual relations will not have been much disturbed, and, in accordance with the principles inculcated in this volume, they will not have been liable to much modification (Darwin 1859: 368).

Whenever Darwin is confronted with a specific evolutionary pattern, he prefers to meddle with the evolutionary process in order to find a way to justify the former.

4.9 Natural Selection as an Auxiliary Hypothesis

To assume that natural selection is the explanatory core of the *Origin of Species* is to overlook its dominant narrative (among other competing narratives). Charles Darwin deployed a fleshed-out and consistent view of the history of life ("tree of life"), one that created an epistemological imbalance in favor of evolutionary pattern, and very much to the detriment of evolutionary process. The fact that the principle of divergence came first in his intellectual development suggests that it may have played a role in Darwin's conception of the separateness of pattern and process (Browne 1983; Kohn 1985, 2009; Limoges 1968; Mayr 1985, 1991; Sober 2011; Winsor 2009). Furthermore, as I have argued elsewhere (Delisle 2014, 2019), the ontological underpinnings of Darwin's view of the history of life rest on several notions rooted in the seventeenth and eighteenth centuries, the principle of divergence merely being the tip of a conceptual iceberg (or cluster) that can be summarized in the following way:

1. *Behind the Principle of Divergence.* For Darwin, the principle of divergence is, above all, an extension of his epistemology of assimilation: past and extinct forms are shoehorned under extant ones, thus preventing the overflow of unknown biological variation by ensuring that all the past is bracketed or fall in-between currently existing forms. This is expressed in Darwin's "cone of increasing diversity," which reveals that divergence is now at a high point relative to the entire evolutionary past. Darwin's biology is essentially a neontological science that frames the findings of paleontology. He holds that the main extant groups retained their position relative to one another ever since the beginning of life. On this view, no major evolutionary events are lost to the taxonomic investigation of groups. The longstanding modernity of the world constitutes a key feature for Darwin, with the Silurian/Cambrian temporal barrier delimiting the time when the main extant taxonomic groups sprung forth fully fledged. On the earlier side of that barrier, hypothetical common ancestors are imagined by projecting backward in geological times variations borne by extant forms. These presumed ancestors are merely downsized forms when compared with their extant counterparts.
2. *Behind the Principle of Gradation.* Another core component of Darwin's theory is the principle of gradation, which presents life as a tightly knit network lacking significant evolutionary disruptions: *Natura non facit saltum.* This unified and continuous world ensures that the past is enclosed within the present. The homogeneity of this network is such that it shares the repetitive features of fractal geometry, with the same pan-divergent evolutionary pattern recurring throughout: at low taxonomic levels, at high taxonomic levels, in the past and in the present. Yet, by insisting on the principle of gradation, Darwin faced the problem

of disentangling the network of life. It is no surprise that in Chapters 1–5 and 8 of the *Origin of Species* (1859), Darwin was confronted to the inherent connectiveness of reproductive networks, a pattern he could not undo with the power of natural selection. That is why in Chapters 6 and 7, he had to resort to the subterfuge of showing off an all-powerful natural selection in the guise of evolutionary analogies, upsetting the delicate balance between "natural selection" and "divergence" his theory would have required.

3. *A Modern and Fully Constituted World.* The core of the *Origin* is built around notions like divergence, gradation, the network of life, and a long-established modernity. The potential tensions between these components are, at least in part, resolved by constraints having been imposed upon the evolutionary dynamics or process. Among the explanatory devices used by Darwin to reach this end, one finds the following: (a) evolutionary stagnation or a lack of evolutionary change is recognized as a common phenomenon; (b) the action of the competitive exclusion principle among related forms exhausts itself in reverse proportion to their taxonomic ascent; (c) the more taxonomic groups climb the taxonomic scale, the more stable and entrenched in nature they become; (d) evolutionary motion is largely restricted to the rise and fall of varieties, species, genera, families, and orders, entities confined within permanent extant classes and phyla.

These are the numerous features Darwin saw in the history of life, many of which have been neglected by the proponents of the received view. In fleshing out this vision, Darwin alternatively exploited different strategies when the time came to offer a justification (i.e., a causal explanation) for the postulated evolutionary pattern: the action of natural selection was either channeled, blocked, or entirely dismissed. In no way was natural selection conceived as a driving force capable of generating a full range of possible evolutionary outcomes. In order to protect the core of his theory, Darwin disposed at will of the evolutionary process.

Following the view of the nature of science proposed by Imre Lakatos (1970: 132–138; see also Chalmers 1999: 130–148; Larvor 1998: 50–57; Laudan et al. 1986: 199–207), I argue that it is useful to distinguish for Darwin the core components (his view of life) from the peripheral components (natural selection) (Delisle 2017a).[9] The "hard core" captures what is most essential and stable, for made of a set of fundamental assumptions. Questioning that core is to move beyond what Darwin had in mind. Under Lakatos's "methodology of scientific research programmes," this core must be treated as irrefutable by methodological decision, becoming the "negative heuristic" that forbids any attack on it. Separate from the hard core, "auxiliary hypotheses" serve the function of protecting the hard core from refutation through a series of adjustments. These auxiliary hypotheses act as a protective belt: their modification offers flexibility at creating a match between the empirical world and the view propounded under a research entity. If necessary, auxiliary hypotheses can be disposed of and replaced by others: "It is this protective belt of auxiliary

[9]The opposition in Darwin's theory between a hard core and auxiliary hypotheses is presented from a somewhat different perspective in Delisle (2019: 131–160).

hypotheses which has to bear the brunt of tests and get adjusted and re-adjusted, or even completely replaced, to defend the thus-hardened core" (Lakatos 1970: 133). In our proposal, natural selection plays precisely this role in Darwin's account of evolution. He continually adjusted its role from case to case, helping him to keep his view of the history of life unchanged. In fact, natural selection constitutes the perfect auxiliary hypothesis since it provides endless explanatory flexibility whenever a postulated evolutionary pattern requires justification.[10]

4.10 Conclusion

In light of our analysis, it becomes extremely difficult to provide a definition of "Darwinism" that is operational (Delisle 2017a, b, 2019; Chap. 1 in this book). According to the received view, Charles Darwin was committed to a strong selective theory: natural selection drives evolution, an open-ended process that is reflected in a whole range of evolutionary patterns. Can the label "Darwinism" be maintained for a theory that either channels, blocks, or dismisses natural selection, thus confining its explanatory role to that of an auxiliary hypothesis? The most consistent or robust contribution Darwin made in the *Origin of Species* is to a *pan-divergent theory of evolution that generates a limited amount of evolutionary change*. This is far from what modern scholars usually have in mind when they invoke Darwin in their work.

The characterization of Darwin presented here may be surprising, but this is because of the distorting lens generally used to look at him. Indeed, the reductionistic and atomistic research tradition which sees genes, traits, organisms, or populations as the basic entities being carried on in the flow of evolutionary time under a strong selective impetus (as one sees in the writings of Fisher, Mayr, and Dawkins) has been retroactively applied to Darwin. By so doing, it was overlooked that Darwin belongs to a different research tradition, one which attempts to synthesize several key explanatory components: the evolutionary mechanisms per se, an interpretation of the tree of life or the chain of being, embryology-development, and an ontology of morphology, among other issues, all of which imply relevant methodological choices. In this context, it is to be expected that mechanisms will rarely be envisioned as central to evolutionary explanations, being confined to a secondary or subsidiary role. The theory of punctuated equilibrium of the 1970s and 1980s succeeded in convincing many scientists that the mechanism-centered view of evolutionary biology is incomplete, forcing evolutionists to confront other issues

[10]As recognized by Gould and Lewontin (1979), the "adaptationist programme" can be exploited to explain all sorts of evolutionary phenomena generated by an adaptive process, whether this is justified or not. But the same explanatory flexibility can be used by resorting to its negative application, as is sometimes done by Darwin when he postulated a lack of selective forces or the establishment of an "adaptive equilibrium" thwarting the expression of evolutionary change. Of course, it is tempting to argue, as I do, that using the adaptationist programme in such a negative way is to move outside that program altogether.

as well. This "revival" reconnected the field with the long-established tradition which consists in putting together the various explanatory pieces of a complex puzzle called "evolution," an exercise which is rarely deprived of internal explanatory tensions and contradictions.

Georges Cuvier, for instance, could not have conceived of biological evolution in light of his understanding of both a discontinuous fossil record and a holistic morphology based on biological types. For evolutionists, however, there was no other way but to seek compromises and calibrate the various explanatory components, giving priority to some over others. Darwin did this by emphasizing his overall pan-divergent view of life (based on a cluster of ideas) over the action of natural selection. Ernst Haeckel instrumentalized the evolutionary mechanisms by arranging for variations to be oriented (see Chap. 5 in this book) in the service of more important explanatory components like the tree of life, embryological development, deterministic factors at the physicochemical level, and an overall strong progressionism (the latter being merely a superficial feature of Darwin's thinking). Of course, paleontologists like Henry Fairfield Osborn (see Chap. 7 in this book) and Otto Schindewolf, among many others, gave precedence to the message they deciphered from the fossil record, thus looking for evolutionary mechanisms compatible with it (and not the other way around). And, while largely neglected under a mechanism-centered view, the positions of the architects of the "Modern Synthesis" like Julian S. Huxley, Theodosius Dobzhansky, and Bernhard Rensch remain largely indecipherable unless their ultimate underpinnings are taken into consideration (Delisle 2008, 2009a, b, 2011). For Huxley and Dobzhansky, for instance, their understanding of the evolutionary mechanisms was directly connected to a holistic and directional view of the history of life. For Rensch, it also depended on directionalism, but one founded on a reductionistic approach which appeals to a physicochemical-based monism (see Chap. 9 in this book). Even George Gaylord Simpson—irrespective of his compatibility rhetoric between population genetics and paleontology—could not entirely escape directionalism in the history of life (see Chap. 10 in this book). Glossing that complexity of thinking systems under a narrow view of science (positivistic or otherwise) can only provide an incomplete picture. What is at stake here, apparently, is not so much how science should be envisioned, but rather how we shall deal with the very nature of evolutionary biology.

When fields such as paleobiology and evo-devo developed in the 1970s and 1990s, respectively, their proponents knowingly or unknowingly worked within what is likely the most robust and dominant research tradition to have existed so far in the short history of evolutionary biology.

Acknowledgments I am grateful to David Ceccarelli and Barbara Continenza for their comments and suggestions to improve this chapter. I thank James Tierney (Yale University) for his assistance in improving the quality of my English.

References

Alcock J (2001) The triumph of sociobiology. Oxford University Press, Oxford

Alcock J (2017) The behavioral sciences and sociobiology: a Darwinian approach. In: Delisle RG (ed) The Darwinian tradition in context: research programs in evolutionary biology. Springer, Cham, pp 37–59

Bock W (1974) Philosophical foundations of classical evolutionary classification. Syst Zool 22:375–392

Bock W (1977) Foundations and methods of evolutionary classification. In: Hecht M, Goody P, Hecht B (eds) Major patterns in vertebrate evolution. Plenum, New York, pp 851–895

Bock W (1981) Functional-adaptive analysis in evolutionary classification. Am Zool 21:5–20

Bowler PJ (1988) The non-Darwinian revolution. Johns Hopkins University Press, Baltimore

Bretsky S (1979) Recognition of ancestor-descendant relationships in invertebrate paleontology. In: Cracraft J, Eldredge N (eds) Phylogenetic analysis and paleontology. Columbia University Press, New York, pp 113–163

Brooks D, Wiley E (1988) Evolution as entropy: toward a unified theory of biology, 2nd edn. University of Chicago Press, Chicago

Browne J (1983) The secular ark: studies in the history of biogeography. Yale University Press, New Haven

Bulmer M (2004) Did Jenkin's swamping argument invalidate Darwin's theory of natural selection? Br J Hist Sci 37:281–297

Chalmers AF (1999) What is this thing called science? 3rd edn. Hackett, Indianapolis

Cracraft J (1981a) The use of functional and adaptive criteria in phylogenetic systematics. Am Zool 21:21–36

Cracraft J (1981b) A pattern and process in paleobiology: the role of cladistic analysis in systematic paleontology. Paleobiology 7:456–468

Darwin C (1859) On the origin of species. John Murray, London

Darwin C (1872) The origin of species, 6th edn. John Murray, London

Dawkins R (1976) The selfish gene. Oxford University Press, Oxford

Dawkins R (1996) Climbing Mount Improbable. WW Norton, New York

Dawkins R (2009) The greatest show on earth: the evidence for evolution. Free, New York

Delisle RG (2001) Adaptationism versus cladism in human evolution studies. In: Corbey R, Roebroeks W (eds) Studying human origins: disciplinary history and epistemology. Amsterdam University Press, Amsterdam, pp 107–121

Delisle RG (2008) Expanding the framework of the holism/reductionism debate in neo-Darwinism: the case of Theodosius Dobzkansky and Bernhard Rensch. Hist Philos Life Sci 30:207–226

Delisle RG (2009a) Les philosophies du néo-darwinisme. Presses Universitaires de France, Paris

Delisle RG (2009b) The uncertain foundation of neo-Darwinism: metaphysical and epistemological pluralism in the evolutionary synthesis. Stud Hist Philos Biol Biomed Sci 40:119–132

Delisle RG (2011) What was really synthesized during the evolutionary synthesis? A historiographic proposal. Stud Hist Philos Biol Biomed Sci 42:50–59

Delisle RG (2014) Evolution in a fully constituted world: Charles Darwin's debts towards a static world in the Origin of Species (1859). Endeavour 38:197–210

Delisle RG (2017a) From Charles Darwin to the Evolutionary Synthesis: weak and diffused connections only. In: Delisle RG (ed) The Darwinian tradition in context: research programs in evolutionary biology. Springer, Cham, pp 133–168

Delisle RG (2017b) Introduction: Darwinism or a kaleidoscope of research programs and ideas? In: Delisle RG (ed) The Darwinian tradition in context: research programs in evolutionary biology. Springer, Cham, pp 1–8

Delisle RG (2019) Charles Darwin's incomplete revolution: the Origin of Species and the static worldview. Springer, Cham

Delson E, Eldredge N, Tattersall I (1977) Reconstruction of hominid phylogeny: a testable framework based on cladistic analysis. J Hum Evol 6:263–278

Dennett D (1995) Darwin's dangerous idea. Simon & Schuster, New York
Depew D (2017) Darwinism in the twentieth century: productive encounters with saltation, acquired characteristics, and development. In: Delisle RG (ed) The Darwinian tradition in context: research programs in evolutionary biology. Springer, Cham, pp 61–88
Depew D, Weber B (1989) The evolution of the Darwinian research tradition. Syst Res 6:255–263
Depew D, Weber B (1995) Darwinism evolving. MIT Press, Cambridge
Eldredge N (1985) Unfinished synthesis. Oxford University Press, Oxford
Eldredge N, Tattersall I (1975) Evolutionary models, phylogenetic reconstruction, and another look at hominid phylogeny. In: Szalay F (ed) Approaches to primate paleobiology, vol 5. Karger, Basel, pp 218–242
Eldredge N, Tattersall I (1982) The myths of human evolution. Columbia University Press, New York
Ellegard A (1990 [1958]) Darwin and the general reader: the reception of Darwin's theory of evolution in the British periodical press, 1859–1872. University of Chicago Press, Chicago
Goodwin B (1994) How the leopard changed its spots: the evolution of complexity. Princeton University Press, Princeton
Goodwin B (2009) Beyond the Darwinian paradigm: understanding biological forms. In: Ruse M, Travis J (eds) Evolution: the first four billion years. Belknap Press, Cambridge, pp 299–312
Gould SJ (1980) Is a new and general theory of evolution emerging? Paleobiology 6:119–130
Gould SJ (1986) Evolution and the triumph of homology, or why history matters. Am Sci 74:60–69
Gould SJ (1989) Wonderful life: the Burgess Shale and the nature of history. W.W. Norton, New York
Gould SJ (2002) The structure of evolutionary theory. Belknap, Cambridge
Gould SJ, Eldredge N (1977) Punctuated equilibria: the tempo and mode of evolution reconsidered. Paleobiology 3:115–151
Gould SJ, Lewontin R (1979) The spandrels of San Marco and the Panglossian paradigm: a critique of the adaptationist programme. Proc R Soc Lond 205(B):581–598
Grene M, Depew D (2004) The philosophy of biology: an episodic history. Cambridge University Press, Cambridge
Halstead LB (1981) Halstead replies. Nature 289:106–107
Hennig W (1981 [1969]) Insect phylogeny. Wiley, New York
Hodge MJS (1977) The structure and strategy of Darwin's 'long argument'. Br J Hist Sci 10:237–246
Hodge MJS (1989) Darwin's theory and Darwin's arguments. In: Ruse M (ed) What the philosophy of biology is. Kluwer, Dordrecht, pp 163–182
Hodge MJS (2013) Darwin's book: On the Origin of Species. Sci Educ 22(9):2267–2294
Horgan J (1996) The end of science: facing the limits of knowledge in the twilight of the scientific age. Addison-Wesley, Reading
Hull DL (1979) The limits of cladism. Syst Zool 28:416–440
Hull DL (1988) Science as a process. University of Chicago Press, Chicago
Hull DL (2009) Darwin's science and Victorian philosophy of science. In: Hodge MJS, Radick G (eds) The Cambridge companion to Darwin, 2nd edn. Cambridge University Press, Cambridge, pp 173–196
Huxley TH (1860) Darwin on the Origin of Species. Westminster Rev 17:541–570
Huxley TH (1862) The anniversary address. Q J Geol Soc Lond 18:xl–liv
Jablonka E, Lamb M (2006) Evolution in four dimensions. MIT Press, Cambridge
Kauffman S (1993) The origins of order: self-organization and selection in evolution. Oxford University Press, Oxford
Kauffman S (1995) At home in the universe: the search for the laws of self-organization and complexity. Oxford University Press, Oxford
Kimura M (1983) The neutral theory of molecular evolution. Cambridge University Press, Cambridge

Kohn D (1985) Darwin's principle of divergence as internal dialogue. In: Kohn D (ed) The Darwinian heritage. Princeton University Press, Princeton, pp 245–257
Kohn D (2009) Darwin's keystone: the principle of divergence. In: Ruse M, Richards RJ (eds) The Cambridge companion to the 'Origin of Species'. Cambridge University Press, Cambridge, pp 87–108
Lakatos I (1970) Falsification and the methodology of scientific research programmes. In: Lakatos I, Musgrave A (eds) Criticism and the growth of knowledge. Cambridge University Press, Cambridge, pp 91–196
Laland K, Uller T, Feldman M, Sterelny K, Müller G, Moczek A, Jablonska E, Odling-Smee J (2014) Does evolutionary theory need a rethink? Yes, urgently. Nature 514:161–164
Larvor B (1998) Lakatos: an introduction. Routledge, New York
Laudan L, Donovan A, Laudan R, Barker P, Brown H, Leplin J, Thagard P, Wykstra S (1986) Scientific change: philosophical models and historical research. Synthese 69:141–223
Limoges C (1968) Darwin, Milne-Edwards et le principe de divergence. Actes du XIIe Congrès International d'Histoire des Sciences 8:11–115
Mayr E (1974) Cladistic analysis or cladistic classification? Zeitschrift für Zoologische Systematik und Evolutionforschung 12:94–128
Mayr E (1980) Prologue: some thoughts on the history of the Evolutionary Synthesis. In: Mayr E, Provine WB (eds) The Evolutionary Synthesis: perspectives on the unification of biology. Harvard University Press, Cambridge, pp 1–48
Mayr E (1982) The growth of biological thought. Belknap, Cambridge
Mayr E (1983) How to carry out the adaptationist program? Am Nat 121:324–334
Mayr E (1985) Darwin's five theories of evolution. In: Kohn D (ed) The Darwinian heritage. Princeton University Press, Princeton, pp 755–772
Mayr E (1991) One long argument: Charles Darwin and the genesis of modern evolutionary thought. Harvard University Press, Cambridge
Mayr E, Ashlock PD (1991) Principles of systematic zoology, 2nd edn. McGraw-Hill, New York
Mivart SG (1871) On the genesis of species. Macmillan, London
Morrison DA (2015) [Book review] The tree of life: evolution and classification of living organisms. Syst Biol 64:546–548
Morrison DA (2016) Genealogies: pedigrees and phylogenies are reticulating networks not just divergent trees. Evol Biol 43:456–473
Owen R (1860) [Darwin on the Origin of Species]. Edinb Rev 111:487–532
Pigliucci M, Müller G (2010) Elements of an extended evolutionary synthesis. In: Pigliucci M, Müller G (eds) Evolution – The extended synthesis. MIT Press, Cambridge, pp 3–17
Platnick N (1979) Philosophy and the transformation of cladistics. Syst Zool 28:537–546
Provine W (1985) Adaptation and mechanisms of evolution after Darwin: a study in persistent controversies. In: Kohn D (ed) The Darwinian heritage. Princeton University Press, Princeton, pp 825–866
Recker D (1987) Causal efficacy: the structure of Darwin's argument strategy in the Origin of Species. Philos Biol 54:147–175
Reid RGB (1985) Evolutionary theory: the unfinished synthesis. Croom Helm, Beckenham
Reid RGB (2007) Biological emergences: evolution by natural experiment. MIT Press, Cambridge
Richards RJ (1988) The moral foundations of the idea of evolutionary progress: Darwin, Spencer, and the neo-Darwinians. In: Nitecki MH (ed) Evolutionary progress. University of Chicago Press, Chicago, pp 129–148
Richards RJ (1999) Darwin's romantic biology: the foundation of his evolutionary ethics. In: Maienchein J, Ruse M (eds) Biology and the foundation of ethics. Cambridge University Press, Cambridge, pp 113–153
Romanes GJ (1886) Physiological selection: an additional suggestion on the Origin of Species. J Linn Soc 19:337–411
Rosen D, Nelson G, Patterson C (1979) Foreword to a reprint of W. Hennig's phylogenetic systematics [1966]. University of Illinois Press, Urbana

Ruse M (1975) Darwin's debt to philosophy: an examination of the influence of the philosophical ideas of John FW Herschel and William Whewell on the development of Charles Darwin's theory of evolution. Stud Hist Philos Sci 6:159–181
Ruse M (1979) The Darwinian revolution. University of Chicago Press, Chicago
Ruse M (1999) The Darwinian revolution, 2nd edn. University of Chicago Press, Chicago
Ruse M (2000) Darwin and the philosophers: epistemological factors in the development and reception of the theory of the Origin of Species. In: Creath R, Maienschein J (eds) Biology and epistemology. Cambridge University Press, Cambridge, pp 3–26
Ruse M (2013) The origin of species. In: Ruse M (ed) The Cambridge Encyclopedia of Darwin and evolutionary thought. Cambridge University Press, Cambridge, pp 95–102
Schafersman S (1985) Anatomy of a controversy: Halstead vs. the British Museum (Natural History). In: Godfrey L (ed) What Darwin Began: modern Darwinism and Non-Darwinian perspectives on evolution. Allyn and Bacon, Boston, pp 186–219
Shanahan T (2004) The evolution of Darwinism: selection, adaptation, and progress in evolutionary biology. Cambridge University Press, Cambridge
Simpson GG (1964) This view of life. Harcourt, Brace & WorldBrace & World, New York
Simpson GG (1975) Recent advances in methods of phylogenetic inference. In: Luckett W, Szalay F (eds) Phylogeny of the primates. Plenum, New York, pp 3–19
Smocovitis VB (1999) The 1959 Darwin centennial celebration in America. Osiris 14:274–323
Sober E (2011) Did Darwin write the *Origin* backward? Philosophical essays on Darwin's theory. Prometheus, Amherst
Sulloway FJ (1979) Geographic isolation in Darwin's thinking: the vicissitudes of a crucial Idea. Stud Hist Biol 3:23–65
Szalay F (1977) Ancestors, descendants, sister groups and testing of phylogenetic hypothesis. Syst Zool 26:12–18
Szalay F (1981) Functional analysis and the practice of the phylogenetic method as reflected by some mammalian studies. Am Zool 21:37–45
Szalay F (1991) The unresolved world between taxonomy and population biology: what is, and what is not, macroevolution. J Hum Evol 20:271–280
Ulanowicz R (1997) Ecology: the ascendant perspective. Columbia University Press, New York
Ulanowicz R (2009) A third window: natural life beyond Newton and Darwin. Templeton Foundation Press, West Conshohocken, PA
Vorzimmer PJ (1970) Charles Darwin: the years of controversy. Temple University Press, Philadelphia
Wagner M (1868) Die Darwin'sche Theorie und das Migrationgesetz der Organismen. Duncker & Humblot, Leipzig
Waters CK (2009) The arguments in the Origin of Species. In: Hodge MJS, Radick G (eds) The Cambridge companion to Darwin, 2nd edn. Cambridge University Press, Cambridge, pp 120–143
Whewell W (1832) [Review of volume 2 of Lyell's] principles of geology. Q Rev 47:103–132
Whewell W (1864) Astronomy and general physics considered with reference to natural theology, 7th edn. Pickering, London
Wilson EO (1975) Sociobiology: the new synthesis. Harvard University Press, Cambridge
Winsor MP (2009) Taxonomy was the foundation of Darwin's evolution. Taxon 58:43–49
Wray G, Hoekstra H, Futuyma D, Lenski R, Mackay T, Schluter D, Strassmann J (2014) Does evolutionary theory need a rethink? No, all is well. Nature 514:161–164

Chapter 5
Natural Selection in Ernst Haeckel's Legacy

Georgy S. Levit and Uwe Hossfeld

Abstract The "German Darwin" Ernst Haeckel was very influential not only in Germany but in non-German-speaking countries as well. He was a key figure during the "first Darwinian revolution," a period when the foundation for the modern evolutionary theory was laid. Haeckel defended and developed the Darwinian theory with unparalleled passion and energy. He created a conceptual framework within which the majority of Darwinians worldwide worked over decades. Contemporary evolutionary theory is unthinkable without notions coined by Haeckel such as "phylogeny," "ontogeny," "phylogenetic tree," or "ecology." Moreover, his theories were encouraged and admired by Darwin himself. It was Haeckel who crucially contributed to the visualization of the Darwinian theory and who tried to convert Darwinism into a universal worldview. Yet it remains controversial to what extent Haeckel's view of evolutionary mechanisms corresponded to those in Darwin's own theory. In this chapter, we will examine this issue and demonstrate that although Haeckel championed natural selection throughout his whole career, his neo-Lamarckian concept of variation made his grasp of natural selection different from that of Darwin. As paradoxical as it may sound from the modern viewpoint, Haeckel made these neo-Lamarckian adjustments in order to render the Darwinian theoretical system more straightforward.

Keywords Ernst Haeckel · Darwinian theory · Natural selection · Lamarckism · Evolutionary mechanisms

5.1 Introduction

Ernst Haeckel is, without doubt, a crucial figure in the growth of Darwinian thought in the nineteenth century (Hossfeld et al. 2019; Levit and Hossfeld 2017, 2019). As Robert Richards has emphasized, "More people at the turn of the century learned of

G. S. Levit (✉) · U. Hossfeld
Biology Education Research Group, Friedrich-Schiller-University Jena, Jena, Germany

R. G. Delisle (ed.), *Natural Selection*, Evolutionary Biology – New Perspectives on Its Development 3, https://doi.org/10.1007/978-3-030-65536-5_5

Fig. 5.1 Ernst Haeckel (portrait) from "Challenger Report" (1884, Courtesy: Günther-Berlin)

evolutionary theory from his pen than from any other source, including Darwin's own writings" (Richards 2018). Considering Haeckel's significance and level of influence, the question whether his version of Darwinism differed from that of the British genius himself is of key importance (Fig. 5.1). There is a widespread tendency among contemporary historians of biology to regard Haeckel as Darwin's supporter in terms of general evolutionism but to deprive him the priority of being one of the earliest champions of natural selection. Richards has made a clear statement on the role of natural selection in Haeckel's theoretical system: "His *Generelle Morphologie der Organismen* (1866) sought to explain those relationships through the devices that Darwin had advanced: namely, natural selection and the inheritance of acquired characteristics. Depending on the traits and the situation of the organism, one of these devices might be emphasized more than the other. Through the course of Haeckel's career, he tilted to the Lamarckian notion, but kept natural selection at the ready" (Richards 2018). Ernst Mayr has argued along similar lines, stating that Haeckel accepted natural selection only "in part" (Mayr 1991, p. 37). At the same time, Mayr was convinced that the place of natural selection in Haeckel's thought was under-researched. In a letter to one of us (UH), Mayr emphasized the necessity of estimating the exact role of natural selection and

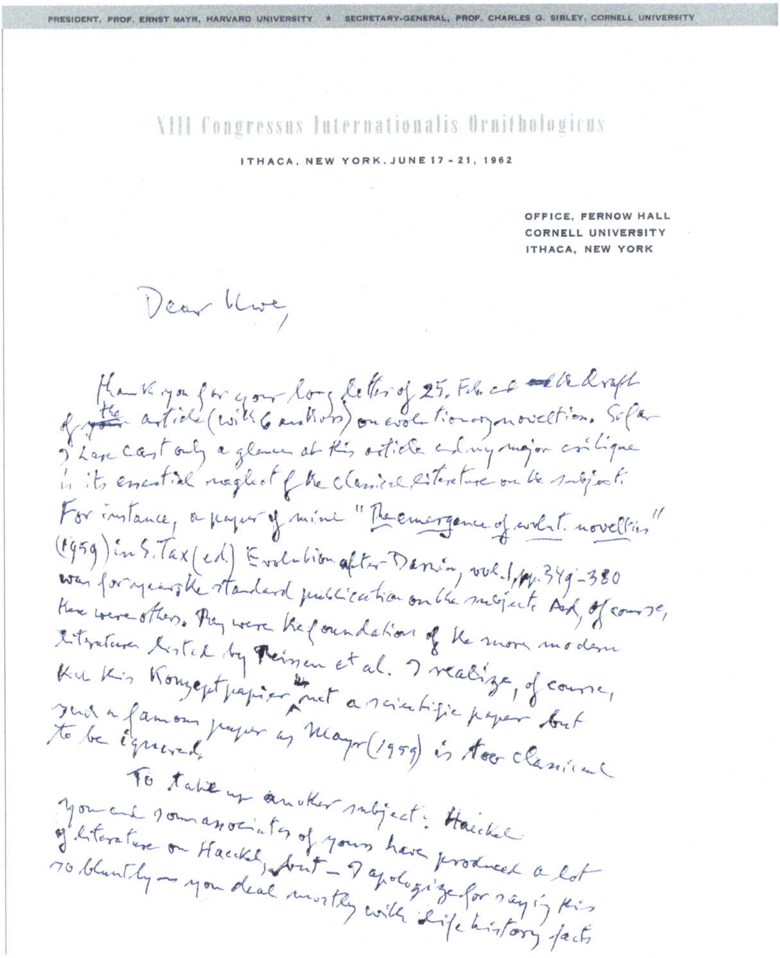
PRESIDENT, PROF. ERNST MAYR, HARVARD UNIVERSITY ★ SECRETARY-GENERAL, PROF. CHARLES G. SIBLEY, CORNELL UNIVERSITY

XIII Congressus Internationalis Ornithologicus

ITHACA, NEW YORK, JUNE 17 - 21, 1962

OFFICE, FERNOW HALL
CORNELL UNIVERSITY
ITHACA, NEW YORK

Dear Uwe,

Thank you for your long letter of 25. Feb. and the draft of the article (with 6 authors) on evolutionary novelties. So far I have cast only a glance at this article and my major critique is its essential neglect of the classical literature on the subject. For instance, a paper of mine "The emergence of evolut. novelties" (1959) in S. Tax (ed.) Evolution after Darwin, vol. 1, pp. 349–380 was for years the standard publication on the subject. And, of course, there were others. They were the foundation of the more modern literature listed by Reinen et al. I realize, of course, that this Konzeptpapier is not a scientific paper but such a famous paper as Mayr (1959) is too classical to be ignored.

To take up another subject: Haeckel. You and some associates of yours have produced a lot of literature on Haeckel, but — I apologize for saying this so bluntly — you deal mostly with life history facts

Fig. 5.2 Ernst Mayr's letter, Bedford, Massachusetts, 5 March 2004

alternative evolutionary mechanisms in Haeckel's works, especially in his early publications: "What we now need is a careful analysis of all of Haeckel's statements on natural selection. Did he accept it without reservations? Did he suggest that natural selection was not sufficient to explain all evolutionary phenomena? [. . .] Was selection for Haeckel a selection of the best or an elimination of the worst? Did Haeckel appreciate the importance of sexual selection?"[1] (Fig. 5.2).

Mario di Gregorio has investigated these issues extensively and formulated original views on Haeckel's understanding and use of natural selection. Haeckel, he claimed, hardly applied natural selection to individuals, but rather to groups,

[1]E. Mayr to U. Hossfeld, Bedford, MA, 5 March 2004.

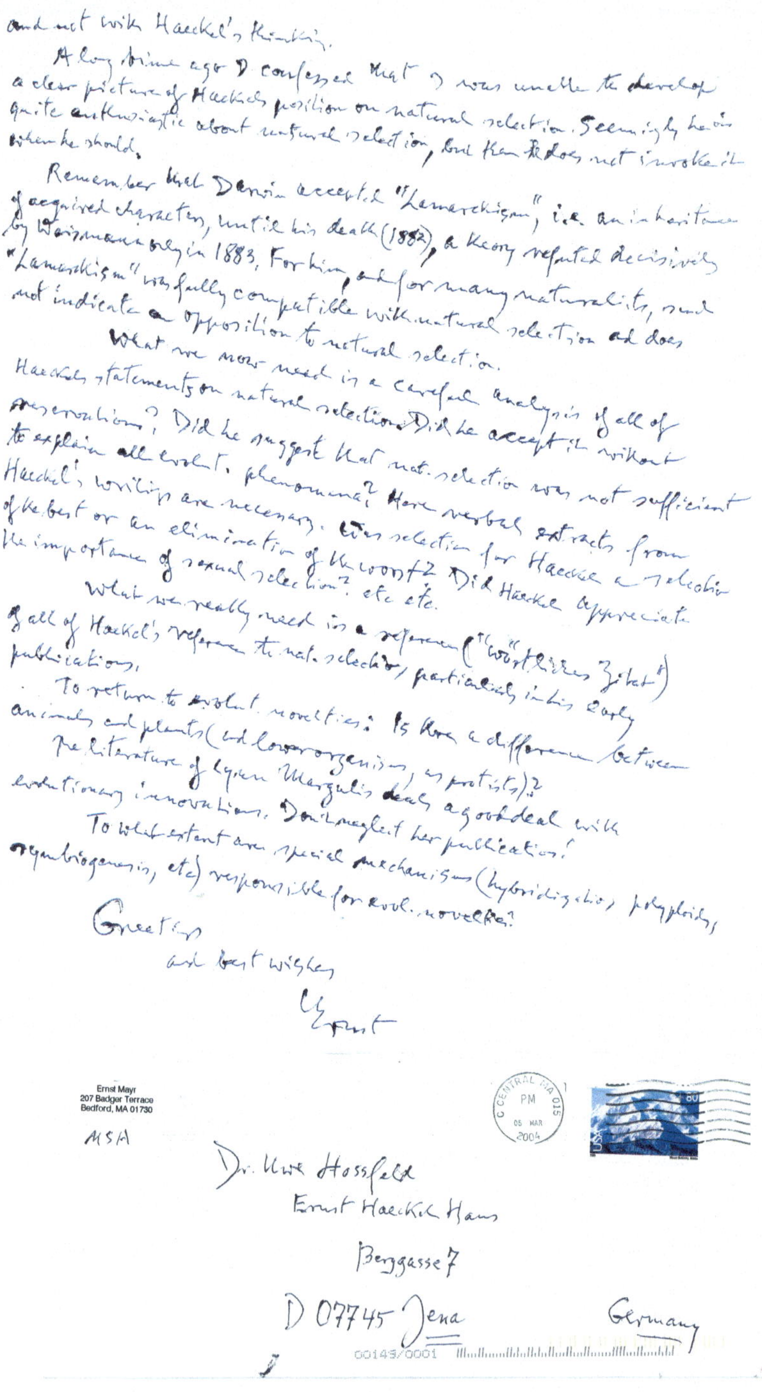

and not with Haeckel's thinking.

A long time ago I confessed that I was unable to develop a clear picture of Haeckel's position on natural selection. Seemingly he is quite enthusiastic about natural selection, but then he does not invoke it when he should.

Remember that Darwin accepted "Lamarckism", i.e. an inheritance of acquired characters, until his death (1882), a theory refuted decisively by Weismann only in 1883. For him, and for many naturalists, "Lamarckism" was fully compatible with natural selection and does not indicate an opposition to natural selection.

What we now need is a careful analysis of all of Haeckel's statements on natural selection. Did he accept it without reservations? Did he suggest that nat. selection was not sufficient to explain all evolut. phenomena? Mere verbal extracts from Haeckel's writings are necessary. Was selection for Haeckel a selection of the best or an elimination of the worst? Did Haeckel appreciate the importance of sexual selection? etc. etc.

What we really need is a reference ("wörtliches Zitat") of all of Haeckel's references to nat. selection, particularly in his early publications.

To return to evolut. novelties: Is there a difference between animals and plants (and lower organisms, as protists)?

The literature of Lynn Margulis deals a good deal with evolutionary innovations. Don't neglect her publications!

To what extent are special mechanisms (hybridization, polyploidy, symbiogenesis, etc) responsible for evol. novelties?

Greetings
and best wishes
Ernst

Ernst Mayr
207 Badger Terrace
Bedford, MA 01730

MSA

CENTRAL MA 015
PM
05 MAR
2004

Dr. Uwe Hossfeld
Ernst Haeckel Haus
Berggasse 7
D 07745 Jena Germany

00149/0001

Fig. 5.2 (continued)

"because groups were better suited to his concept of morphological levels of individuality" (Di Gregorio 2008). All these views are at odds with Darwin's own estimation of Haeckel. In as early as 1864, after receiving Haeckel's copy of *Neue Stettiner Zeitung* (20 September 1863) with a report on Haeckel's "Speech of Stettin" and Haeckel's paper on marine zooplankton (Haeckel 1864), Darwin wrote: "I am delighted that so distinguished a naturalist should confirm & expound my views; and I can clearly see that you are one of the few who clearly understands Natural Selection."[2] Darwin was delighted not only by Haeckel's grasp of natural selection; right in the introduction to the first edition of *The Descent of Man*, Darwin claimed that Haeckel was the only major evolutionist who immediately realized the significance of sexual selection: "Prof. Häckel is the sole author who, since the publication of the 'Origin,' has discussed, in his various works, in a very able manner, the subject of sexual selection, and has seen its full importance" (Darwin 1871, p. 5). In the 1882 edition of the book, Darwin repeated this passage even in a little bit extended form (Darwin 1882, p. 3). Certainly, as noted by Richards elsewhere, some historians of science would argue that if Darwin's German were better, "he would have detected deviant tendencies in the work of his new disciple" (Richards 2004). But Darwin was not as quick and superficial in his judgments as one might suppose. His letters suggest that when having difficulties with German (as he did, for example, in the work of Fritz Müller) he employed assistants to help him acquire an accurate grasp of German-language concepts.[3]

The major objective of this chapter is to evaluate Richards's thesis (Richards 2018) which can be broken down to two statements: (1) In the *Generelle Morphologie*, Haeckel advocated both natural selection and the inheritance of acquired characters; (2) over the course of his career, Haeckel became more and more neo-Lamarckian, tending to downplay the role of natural selection as an actual explanatory pattern, although he did not abandon it completely. We will proceed as follows. First, we will explicate Darwin's own views on evolutionary mechanisms as Darwin was Haeckel's major inspiration and Haeckel's initial objective was to substantiate Darwin's insights by doing extensive empirical research. Second, we ask whether Haeckel gives fair due to all evolutionary mechanisms suggested by Darwin, as Darwin never understood natural selection to be the sole mechanism of evolutionary advancement. Third, we will demonstrate the role of natural selection in Haeckel's works along with auxiliary evolutionary mechanisms beginning with *Generelle Morphologie (GM)* and ending with his final publications. We will try to concentrate on Haeckel's purely scientific, technical (as opposed to popular) works as they remained untranslated and are little known to English-speaking audiences.

Our evidence suggests that Haeckel tried to exactly follow Darwin and to make his doctrine more straightforward. Haeckel introduced into his works all the major

[2]Darwin to Ernst Haeckel, 9 March 1864. https://www.darwinproject.ac.uk/letter/DCP-LETT-4422.xml.

[3]Darwin to Haeckel, 21 November 1864. https://www.darwinproject.ac.uk/letter/?docId=letters/DCP-LETT-4676.xml.

evolutionary mechanisms Darwin mentioned, i.e., use/disuse, direct impact of the environment, macromutations, and natural selection. As our qualitative and quantitative studies demonstrate, natural selection never disappeared from Haeckel's vocabulary, yet his understanding of it differed crucially from that of Darwin, with the core issue being Haeckel's interpretation of variation. If Darwin's variation was predominantly random (isotropic), Haeckel's variation was predominantly definitely directed, i.e., immediately adaptive. Therefore, Haeckel embedded neo-Lamarckian inheritance, by which we mean the inheritance of acquired characters, right into his theory of natural selection.[4] Haeckel's successor in Jena (and prominent "old-school-Darwinian"), Ludwig Plate, saw no problem in this approach: "Definitely directed variation and selection are not mutually exclusive, but can work together. It does not matter to selection, if a certain change is in the same direction as the one before or not, if the change continues in the same direction or not" (Plate 1913, p. 510; Levit and Hossfeld 2006).

Haeckel's theory of heredity (perigenesis), which he coined in the mid-1870s, only strengthened his neo-Lamarckian view of variation. Thus, Haeckelian natural selection was not opposed to neo-Lamarckism but peacefully coexisted with it. Furthermore, we argue that Darwin's own theory of pangenesis is better suited to neo-Lamarckian interpretations than to strict Darwinian selectionism (as paradoxical as this may sound).

5.2 Darwin on Evolutionary Mechanisms

Darwinism is a dynamic and complex theoretical system consisting of several tightly interconnected postulates and numerous auxiliary hypotheses. The difficulty with defining Darwinism is connected with the fact that the theory of natural selection only achieved logical consistency and conceptual maturity decades after Darwin's death (Levit et al. 2008b, 2011). The first step toward Darwinism as we know it came with the synthetic theory of evolution (STE), which proposed a logically coherent and empirically well-substantiated theoretical system which become "Darwinian" in the modern sense (Reif et al. 2000; Granovitch 2021). That is why an appeal to Darwin's own writings is not the best argument in favor of the "Darwinian" character of a concept.[5] Contemporary "post-synthetic" Darwinism possesses a prerogative of retrospective classification of concepts as "Darwinian" or "anti-Darwinian" irrespectively of Darwin's own views. This prerogative has been legitimized by the almost perfect logical consistency of this complex theoretical system,

[4]Although the genuine nature of Lamarck's original theory is a topic of debate, the term neo-Lamarckism (Lamarckism) "has come to mean the inheritance of acquired characteristics" (Bowler 2003, p. 90).

[5]Comparatively, an appeal to Einstein's writings on general relativity constitutes a strong argument in favor of the "Einsteinianism" of a particular concept.

elaborated by generations of experimental biologists and theoreticians during the second half of the twentieth century. That is why Haeckel's Darwinism must be compared to Darwin's very own Darwinism without references to "synthetic" or "post-synthetic" doctrines. This is the only way to determine the "Darwinian" nature of Haeckel's theory as viewed through the "Down House" window (Levit and Hossfeld 2019).

Darwin's own theory, which advanced the notions of organic evolution and common descent, introduced the principle of natural selection within a broader theoretical context. First of all, Darwin championed sexual selection which drastically increased the explanatory power of his theoretical system. Furthermore, already in the first edition of the *Origin* (1859), this context encompassed a multiplicity of evolutionary mechanisms including neo-Lamarckian variation: "Lamarckian variation is there, under the terms 'direct action of the conditions of life' and the inheritance of 'habit,' and hence 'use and disuse'" (Olby 2009). Although these neo-Lamarckian alternatives to natural selection played a minor role in his explanatory pattern, Darwin always accepted these alternative mechanisms (Olby 2017). The same is true for Darwin's mutationism (famous Darwin's "sports," i.e., macromutations). The founder of mutationism, Hugo de Vries, who claimed that sudden and non-reverting saltations bring about new species, even "portrayed Darwin as the father of macromutationism" (Gould 1983). Besides, this initial version of Darwinism incorporated the germs of a concept of evolutionary constraints expressed, for example, as the concept of correlation. These ideas later gave rise to a theory of orthogenesis, i.e., a theory of directed evolution (Levit and Olsson 2006; Levit et al. 2008b).

The role of these "alternative" evolutionary mechanisms in Darwin's theory probably even increased over time, a fact which becomes apparent when comparing the first and sixth editions of *On the Origin of Species* (Darwin 1859, 1872) as "in the last two editions non-selective forces come into play" (Liepmann 1981). Some scholars have opposed this "Darwin's Lamarckianization" thesis: "From Darwin's own perspective, nothing has changed: he is just trying to make clearer a point that he has always made but that has been constantly overlooked" (Hoquet 2017). In Hoquet's view, instead of becoming more and more Lamarckian, "Darwin stresses the power of variations, something acting simultaneously with the power of natural selection" (Hoquet 2017). If it is so, it fits our hypothesis that Haeckel decided to coin his own theory of variation and heredity in order to make the whole theoretical system more straightforward even better than the "Lamarckianization" thesis.

Historians of science disagree about the role of various evolutionary mechanisms in Darwin's explanatory paradigm. For example, Winther claims that "Darwin was caught in the logical bind [...]" as "he attempted to champion the importance of natural selection in producing adaptations while also accentuating systematic and necessarily adaptive somatically-mediated variations" (Winther 2000). Any emphasis he placed on non-selectionist mechanisms of adaptation necessarily decreased the standing of his major discovery, namely, the importance of natural selection. By contrast, Ernst Mayr saw no conflict between various evolutionary mechanisms within Darwin's own theoretical system: "For Darwin inheritance of acquired characters and a direct effect of the environment were compatible with natural selection" (Mayr 1997). Winther (2000) seems to be right if we view Darwin's theory through the lens of the modern synthesis, whereas Mayr is absolutely right if we see the theory with Darwin's own eyes. Twenty-three years after the first publication of *Origin* (1859), in the last edition of *The Descent of Man* prepared during his lifetime, Darwin listed all the major evolutionary mechanisms practically on one page (Darwin 1882, pp. 607–608). Expectedly, he considered the primary mechanism to be the struggle for existence and natural selection. Sexual selection is another powerful driving force of evolution. Darwin's neo-Lamarckian "long-continued use or disuse of parts will have done much in the same direction with natural selection" (the view passionately supported by Haeckel, as we will see below). At the same time, Darwin also asserted that various parts of the organism are modified in accord with the "principle of correlation," i.e., he admitted some constraints on the independent variation of features. Finally, Darwin claimed that "something may be attributed to the direct and definite action of the surrounding conditions of life" (the second neo-Lamarckian mechanism). Darwin also included in his list of evolutionary mechanisms the "occasional modifications" bringing about "structures" which "cannot be accounted for by any form of selection, or by the inherited effects of use and disuse of parts" (Darwin 1882, p. 608). Darwin described macromutations (sports) as one more auxiliary evolutionary mechanism, but did so only tentatively, therefore excluding it from the list of "regular" evolutionary mechanisms. "Sports" are more commonly found under domestication, Darwin argued, than in nature, where they are "extremely rare" (Largent 2009). As to the general dynamics of all these factors in Darwin's own theoretical system, his tendency toward either admitting the increasingly important role of non-selectionist mechanisms or to better articulating their role (depending on our interpretation of Darwinian texts) is clearly detectable through the six editions of the *Origin* and in some of his other publications. Independently of how exceptional non-selectionist mechanisms were from Darwin's own viewpoint, external observers were entitled to take them seriously and to consider their role in the general concert of evolution.

Darwin was Haeckel's scientific icon, and our analysis suggests that Haeckel followed Darwin quite closely in his description of evolutionary mechanisms (Levit and Hossfeld 2019). Indeed, it would be surprising if he did not. Haeckel elaborated on both natural selection and non-selectionist mechanisms and tried to build them into a noncontradictory theoretical system. But he demonstrated a tendency to convert Darwin's tentative concepts and cautious assumptions into strong beliefs. As Darwin remarked in his typical gentle manner after reading *Generelle Morphologie*, "Your reviewer will say you have spoken much too strongly."[6] This strength of conviction is the principal difference between Haeckel and Darwin.

[6]Letter 5293 (Darwin to Haeckel, 18 August 1866).

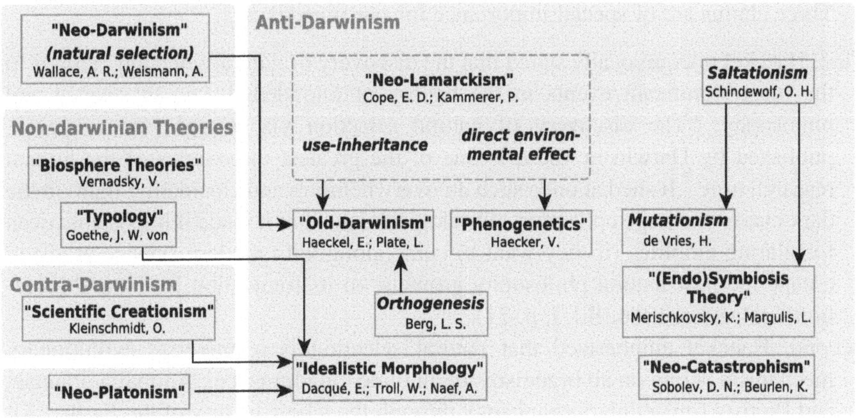

Fig. 5.3 Natural selection and the alternative evolutionary mechanisms at the turn of the centuries (nineteenth to twentieth)

One's interpretation of Darwin's texts depended much on the school one belonged to. Haeckel (old-school-Darwinism) and Weismann (neo-Darwinism) picked up different ideas from Darwin's heritage. This split into two major rival selectionist movements—namely, old- and neo-Darwinism—was determined by each movement's position toward the supposed pluralism of evolutionary mechanisms in Darwin's texts (Levit and Hossfeld 2006). Haeckel was certainly on the site of the pluralists and therefore followed Darwin's intention to create an "open-ended" theory escaping monolithic explanations (Fig. 5.3).

5.3 Natural Selection and Sexual Selection in the *Generelle Morphologie*

The two-volume *Generelle Morphologie* (*General Morphology of Organisms*) (hereafter *GM*) is Haeckel's first major Darwinian work, which appeared in 1866 and embraced almost all aspects of Darwin's theoretical system beginning with detailed phylogenetic trees and ending with the philosophy of monism (Haeckel 1866). Significant space is also devoted to natural selection. Below we outline Haeckel's understanding of natural selection in the *GM*, a task aided by the authors' new translations of several full-length quotations (as *GM* is unavailable in English).

Haeckel devoted specifically to natural selection sections III and VII of the 19th chapter of the so-called fifth book, which is a part of the second volume. But strong claims regarding natural selection can be found also in other parts of the text.

Three claims are of special importance for our objectives.

First, Haeckel unequivocally stated that the discovery of natural selection belongs to the most significant events in the history of knowledge, i.e., in science and philosophy: "The discovery of natural selection via struggle for existence, published by Darwin in 1859, is one of the greatest discoveries of the human research urge.[7] It shed at once such an overwhelming and elucidating light on the dark chaos of a huge collection of biological data, that it made it impossible even for glaring empirics (if they want to come along with modern science at all) to escape the new natural philosophy growing on its foundation [of natural selection]" (Haeckel 1866, Bd. I, p. 71).[8]

Second, Haeckel emphasized that natural selection is a universal evolutionary mechanism acting on all organisms of all three kingdoms, i.e., Animalia, Plantae, and Protista (unicellular organisms), through the whole history of the Earth: "All the large numbers of species of all three kingdoms, which ever existed on our Earth, came about this way, originating from a few autogenous species, under the influence of natural selection discovered by Darwin" (Haeckel 1866, Bd. II, p. 30).[9]

Third, Haeckel insisted that natural selection constitutes the crucial argument "for the exclusive validity of mechanically acting causes in the whole field of biology," i.e., the ultimate evidence for the necessity of naturalistic-causal explanations, thereby whisking away all kinds of teleology (Haeckel 1866, Bd. I, p. 100). In other words, natural selection was for Haeckel not only an ultimate explanatory pattern of biological phenomena but the cornerstone of the new naturalistic worldview opposed to religious prejudices.

Haeckel gave examples of adaptations which could appear only by natural selection. In the already mentioned section VII of the 19th chapter, he discussed the "pelagic fauna of crystal animals" (Haeckel 1866, Bd. II, p. 242). By "crystal animals" or "glass-like animals" (Glasthiere), Haeckel was referring to transparent marine fauna like *Ctenophora* or *Pyrosomatida*. He proceeded from the assumption that "crystal animals" existed in the past in varieties with different levels of transparency and colorlessness (Fig. 5.5). The most transparent and colorless had an advantage in the struggle for existence in clear waters. These individually

[7]"Forschungstrieb," a term akin to Blumenbach's *Nisus Formativus*.

[8]*All citations from GM are given in our translation.* German original: "Die 1859 von Charles Darwin veröffentlichte Entdeckung der natürlichen Zuchtwahl im Kampfe ums Dasein, eine der grössten Entdeckungen des menschlichen Forschungstriebes, hat mit einem Male ein so gewaltiges und klärendes Licht in das dunkle Chaos der haufenweis gesammelten biologischen Thatsachen geworfen, dass es auch den crassesten Empirikern fernerhin, wenn sie überhaupt mit der Wissenschaft fortschreiten wollen, nicht mehr möglich sein wird, sich der daraus emporwachsenden neuen Naturphilosophie zu entziehen."

[9]"Alle die zahlreichen Arten der drei Reiche, welche jemals auf unserer Erde gelebt haben, sind in dieser Weise, unter dem Einflüsse der von Darwin entdeckten natürlichen Zuchtwahl, im Laufe der Zeit aus einer geringen Anzahl autogener Species hervorgegangen."

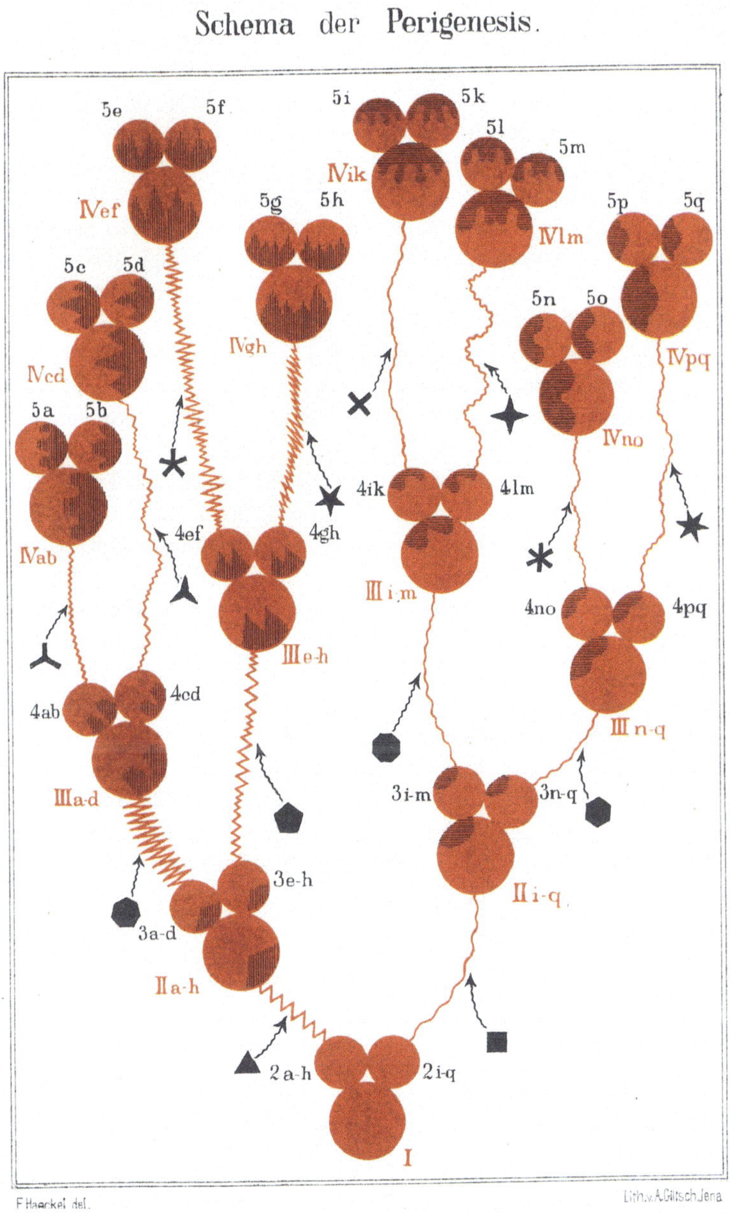

Fig. 5.4 Haeckel's illustration of his perigenesis hypothesis (from Haeckel 1876, pp. 80–81). Haeckel's scheme of perigenesis reflects his idea of interaction between inheritance and adaptation. It presents a transmission of hereditary information in five generations. Four generations indicated as (I–IV) and the last fifth generation as (5a–5q). The plastidule movements transmitting hereditary information are depicted as the red wave lines (pure hereditary information without external influences). Environmental influences are presented as the black wave lines. Environmental

advantageous features were strengthened and secured over many generations so that eventually, completely transparent (glass-like) organic structures came into being. There can be no doubts, Haeckel argued, that these glass-like structures appeared as a result of natural selection as their close non-pelagic relatives dwelling on the seabed or on the coast demonstrate no glass-like body composition, but are opaque and colored (Ibid., p. 243). The transparency of the "crystal animals" gave advantages to both predators and prey, which in the hypothetical past existed in different "varieties." Haeckel brings in this connection also a "very special argument" that many sea animals are colorless and transparent only as pelagic living larvae, but later, when dwelling on the sea bottom or on the coast acquire colors and become opaque as with most Echinodermata (Ibid., pp. 243–244).

Thus, for the three above reasons, natural selection appears to be the most important evolutionary mechanism for Haeckel. This view is also underpinned by our quantitative analysis as, in the *GM*, Haeckel employed the term "natural selection" in various word combinations (see Table 5.1) more than 160 times.[10] If we consider that Haeckel mentioned sexual selection 20 times (see Table 5.2), along with an intensive discussion of artificial selection (which he mentions 32 times; see Table 5.3), we come to more than 200 references to selectionist mechanisms in both volumes of *GM*.

The above three claims suggest the following. First, natural selection was for Haeckel not only a purely biological concept but the foundation of a new "natural philosophy." To fully appreciate this claim, one should consider that for Haeckel, "all true natural science is philosophy, and all true philosophy is natural science. All true science (*Wissenschaft*), therefore, is natural philosophy" (Haeckel 1866, Bd. II, p. 447). Natural selection therefore was crucial for Haeckel's "universalism," i.e., for his attempts to offer all-embracing "scientific" explanations of the universe driven by an aspiration to convert Darwinism into a worldview. Natural selection was *the* universal evolutionary mechanism within a universalist theoretical system, which Haeckel presented already in the *GM*.

Fig. 5.4 (continued) influences modify plastidules' movements and in that way function as adaptations: "In this way, variations are introduced into the germ line in accordance with Haeckel's neo-Lamarckian theory" (Allen 2014). Various black geometric figures symbolize a diversity of environmental conditions an organism is exposed to. These varying environmental conditions cause various modifications in the internal motions within the plastidules, which is graphically reflected as a black hatching on the red balls. The "Micky-Maus-like" triple-balls consist each of one big circle and two small circles. The two small circles are the result of the division of the big one, i.e., they symbolize the division of the plastidules. The scheme is valid for both unicellular and multicellular organisms

[10]German allows many synonymous expressions for natural selection—*natürliche Auslese, natürliche Zuchtwahl, natürliche Selektion*, etc.—and therefore it is difficult to conduct very exact calculations, but the number we give provides a rough estimate (see Table 5.1).

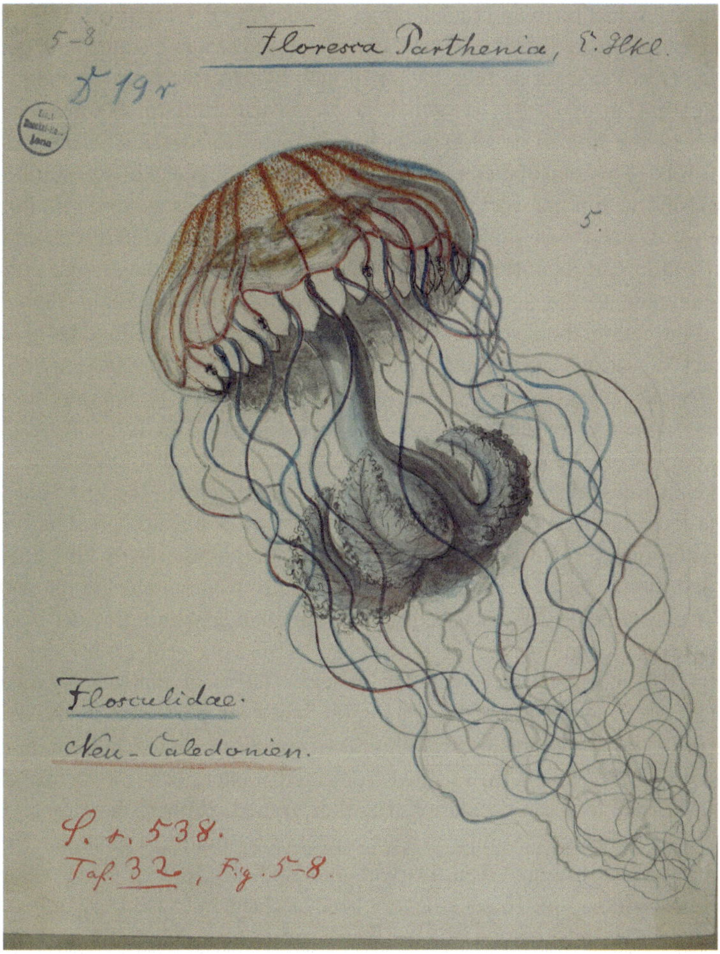

Fig. 5.5 Haeckel's own drawing of Medusae (*Floresca parthenia*) (Ernst-Haeckel-Haus, Jena)

Second, Haeckel considered natural selection to be the most important element within Darwin's own theory. It is not an accident that Haeckel named section II of the *GM*, "The Theory of Selection," and equated the latter to Darwinism writ large (Ibid., p. 166). And, again, one should bear in mind that Haeckel considered himself a true Darwinian.

Third, natural selection was, for Haeckel, a pattern-making *method* of approaching natural phenomena. His threefold parallelism of ontogeny, phylogeny, and classification not only generally "provided the strongest proof for Darwin's theory of descent with modification" (Rieppel 2019) but coupled with natural selection established the core of the causal explanation of evolution.

So natural selection was crucial for Haeckel. As Di Gregorio (2008) claims, Haeckel clearly distinguished various theories within Darwin's conceptual construction. Thus, Haeckel separated the theory of natural selection from a general evolutionism (evolution as such), for which he suggested the term "Lamarckism," as Lamarck was the first to develop a scientific "theory of descent" (Haeckel 1866, Bd. II, p. 166). This must be considered when analyzing Haeckel's position toward "Lamarckism" within his works, as it has little to do with its current meaning.

Some words need to be said about the context in which Haeckel discussed natural selection within *GM*. Natural selection is mostly analyzed in the second volume of the *GM* devoted to the "general theory of evolution [Entwickelungslehre]" and organized into eight "books" (major chapters). He begins the volume by presenting systematics of various organismic groups and then goes on to ontogeny and phylogeny. The last two "books" of the volume are devoted to the significance of evolutionary theory for anthropology and cosmology. The fifth "book," titled, "First part of the general introduction to evolutionary theory [Entwickelungsgeschichte]" is divided into five subchapters (16–20): "The notion and the tasks of ontogeny" (16), "evolutionary history of physiological individuals" (17), "evolutionary history of morphological individuals" (18), "the theory of descent and the theory of selection" (19), and finally "the ontogenetic theses" (20). Natural selection is mostly discussed in the 19th chapter within a wider discussion about the overall significance of the theory of descent (Haeckel 1866, Bd. II, pp. 166–170). Haeckel starts with the general statement that all organisms on Earth originate from a few (or even a single) "most simple" species, the so-called Monera (unicellular organisms, roughly corresponding to the current notion of bacteria), which, in their turn, evolved from inert matter. This theory was not the sudden discovery of a single author, Haeckel argued. Rather, there were several scholars who advanced similar views, most importantly Lamarck, Geoffroy St. Hilaire, Goethe, Lorenz Oken, Darwin, and Wallace (Ibid., p. 153). Within this pantheon, Darwin had for Haeckel a very special place, because Darwin had begun an unprecedented "reformation" of the theory of descent and 1859 marked a new period in the history of life sciences. Haeckel considers Darwin a true "hero," combining broad empirical knowledge and a deep philosophical grasp of nature (Ibid., p. 163). Darwin's merits, Haeckel argues, include two major elements. First, Darwin made the "doctrine of descent" into a strict and deep theory embracing all biological disciplines. Second, Darwin invented the theory of natural selection which provided a causal explanation of evolution (the theory of transformation—*Umwandlungslehre*). Haeckel also mentions Wallace who independently came to "the same basic ideas" (*zu den selben Grundideen*), foremost among these being the concept of natural selection. For Haeckel, natural selection "follows with necessity the natural tendency of organisms to multiply in geometrical progression, whereas their required existence conditions (and especially nourishments) grow only in arithmetic progression" (Ibid., p. 164). Under these circumstances, the struggle for existence operates as "a breeder" (*züchtend*) and brings about new species. The significance of the discovery of natural selection, Haeckel continued, cannot be overestimated as evolutionary theory (*Transmutationstheorie*) becomes a true and complete theory on the top of all biological sciences only due to the theory of natural selection. Natural selection is *causae efficientes* of evolution (Haeckel opposed *causae efficientes* as natural acting causes (*Werk-Ursachen*) to *causae finales* corresponding to supranatural teleology (Zweck-Ursachen) (Ibid., p. 26) based on the interaction of two functions immanent to all living organisms: heredity and adaptation (Ibid., pp. 167–168). All organismic features are either the result of adaptation or heredity. As organisms adapt to various environments and inherit new features, one can observe a process of differentiation having no constraints because variability has no limits. In this way "from one and the same species, due to adaptation to very different life conditions, occur very different species" (Ibid., p. 168). Describing the process of adaptation, Haeckel uses the term "the inheritance of characters acquired through adaptation [*Anpassung*]."[11] Now, since the individuals of the same species do not vary in exactly the same manner, the struggle for existence comes into being. Thus, natural selection selects individuals best adapted to their environments. These better adapted organisms are also more perfect (*vollkommen*) than their predecessors, and therefore the whole process of evolution runs toward perfection or progress (*Vervollkommnung*, *Fortschritt*) in the Earth's organisms organization. Perfecting (getting more perfect, advanced, or complete) is for Haeckel an everlasting process and one of the major characteristics of evolution. The struggle for existence ceases by an increasing divergence of traits.

All the above sounds quite Darwinian, maybe with the exception of Haeckel's straightforward progressionism as, for Darwin, natural selection was not a universal law of advancement (Hoquet 2017).[12]

The difference between Darwin and Haeckel becomes apparent when the latter explicates his understanding of heredity. Haeckel provided a detailed account of his concept of heredity and adaptation already in the *GM*. He distinguished "progressive" and "conservative" heredity as organisms may inherit ancestral (conservative) or newly acquired (progressive) features. In this context, he introduced "the law of progressive heredity" or "the law of inheritance of acquired characters," proving that "all descendants of their parents inherit not only old characters inherited by these parents [from their ancestors], but (as a minimum partly) also new characters acquired by them [parents] during their lifetimes" (Haeckel 1866, Bd. II, p. 178). Haeckel clearly stated that these new characters are acquired via adaptation. These adaptive features may be due to use/disuse, habit, or developmental interaction with an environment, but ultimately can be all reduced to nutrition (Ibid., p. 192), since the persistence of all organisms is only possible on the foundation of nutrition-based metabolism. This metabolism is *the* cause and the fundamental condition of changes leading to adaptation (Ibid., p. 193). New substances (molecules) can also be assimilated from the external world. In this way, environmental substances (*Materien der Aussenwelt*) have a chance to influence the protein structures (*Eiweiss-Verbindungen*) of changing organisms. At the same time, Haeckel distinguished direct and indirect adaption (Ibid., p. 196). A direct (or actual) adaptation is

[11]"Vererbung der durch Anpassung erworbenen Abänderungen."

[12]Cf. "Haeckel was indeed a progressionist, but then so was Darwin" (Richards 2009, p. 147).

an immediate adaptation of an organism to its environment during its lifetime. Indirect (or potential) adaptation means that only the next generation will enjoy the fruits of adaptive changes. In this theory, individual changes are, however, never completely congruent and, depending on the character of environmental influences they experience, may even look random. So there is always space for natural selection to act on subtle differences.

To sum up, in the *GM*, Haeckel gave natural selection an absolutely central role in his evolutionary theory. He considered natural selection the most essential Darwinian concept and even equated the term "Darwinism" with the term "natural selection." Furthermore, he maintained that the discovery of natural selection was one of the most significant events in the history of knowledge since it offered a pattern-making method of establishing scientific worldviews.

Yet Haeckelian natural selection is not equivalent to Darwinian natural selection. The crucial difference between the two authors' theories lies in their respective interpretations of variation.

Darwin held that the environment was necessary for both adaptation and variation. He provided an external mechanism for adaptation: natural selection. He also proposed an external mechanism for variation: changes in the environment (Winther 2000). Darwin as well as Haeckel believed that the environment produces variation by acting both on the reproductive system and soma: "[...] the conditions of life appear to act in two ways, - directly on the whole organisation or on certain parts alone, and indirectly by affecting the reproductive system" (Darwin 1872, p. 5). Variations induced by environmental "disturbance" of soma could sometimes be definitely directed and adaptive in Darwin's view. But as a rule, Darwin's variation is random (isotropic):

> With respect to the direct action, we must bear in mind that in every case, as Professor Weismann has lately insisted, and as I have incidentally shown in my work on 'Variation under Domestication,' there are two factors: namely, the nature of the organism, and the nature of the conditions. The former seems to be much the more important; for nearly similar variations sometimes arise under, as far as we can judge, dissimilar conditions; and, on the other hand, dissimilar variations under conditions which appear to be nearly uniform. The effects on the offspring are either definite or indefinite. (Darwin 1872, p. 6)

Haeckel's variation, by contrast, is predominantly definitely directed.[13] Thus, the principal difference between Darwin and Haeckel here is that for Darwin, "the nature of the organism" was stronger than "the nature of the conditions," whereas for Haeckel, the environment was an immediate force directing "the nature of the organism." Consequently, Haeckelian natural selection is even more severe than Darwin's as organisms vary in a similar direction and thereby compete for similar resources. Haeckel's understanding of variation creates a certain theoretical difficulty in separating his "Darwinism" from his "neo-Lamarckism," because his

[13]Richards maintains that Haeckel advocated "accidental variations" to a much greater extent than we think as *Anpassung* in Haeckel's texts does not necessarily mean (pre)adaptation to environmental circumstances (Richards 2009, p. 145).

neo-Lamarckism appears to be built-in into his concept of natural selection as a necessary component.

5.4 Haeckel on Natural Selection in the Late Publications

In the period post-*GM*, Haeckel's interest in biological anthropology and embryology visibly strengthened, though sometimes at the cost of detailed discussions of evolutionary mechanisms. In the 1870s and thereafter, Haeckel continued to publish manuscripts with both low and high frequency of references to natural selection, sometimes ignoring this concept in the works where it would be highly expected and mentioning it in seemingly surprising contexts. For example, in 1872, he published a seminal three-volume monograph on calcareous sponges, *Kalkschwämme*, which included two volumes of text and one atlas with illustrations (Haeckel 1872). In this monograph, Haeckel coined his famous "biogenetic law" (ontogeny recapitulates phylogeny) as sponges expressed in their whole being "the profound meaning of this biogenetic fundamental law [*biogenetischen Grundgesetzes*]" (Haeckel 1872, Bd. I, p. 215; Olsson et al. 2017; Porges et al. 2019). The entire organization of these animals, Haeckel emphasized, becomes clear to us only through their ontogeny, "through which we are led directly to their phylogeny" (Haeckel 1872, Bd. I, p. 215). The biogenetic law, being in the *GM* only a hypothesis, gained a strong empirical foundation in the *Kalkschwämme* (Reynolds 2019). Since the "biogenetic law" initially evolved within the pages of *GM* (Porges et al. 2019) and proved central to Haeckel's evolutionary theory, one would expect to see natural selection in the *Kalkschwämme* as well. Yet there is not a single reference to this evolutionary mechanism in the entire monograph.

By contrast, in a 100-page, rather ideologically motivated paper called *Freedom in Science and Teaching*, originally published in German in 1878 and a year later appearing in English translation, Haeckel referenced natural selection 20 times (Haeckel 1879). In this paper, Haeckel argued against his teacher, Rudolf Virchow, who maintained that science was essentially esoteric in character and who reasoned against the attempts of Darwinians to appeal directly to a broad audience, i.e., to popularize Darwinism and "materialism." Virchow placed special attention on Darwinian biological anthropology, which was, according to him, only a vague hypothesis. Contrarily, Haeckel advocated a limitless freedom of science and, as one would expect, argued in favor of the descent of man from apes. In this paper, he repeated his thesis that Darwinism equals natural selection: "This theory of elimination was first clearly recognized and appreciated in its full significance by Charles Darwin in 1859, and the selection-hypothesis which he founded on it is Darwinism properly so called" (Haeckel 1879). He emphasized that the theory of natural selection was "the immeasurable step" in establishing the idea of evolution and in combating the immutability of species (a fundamental teleological idea, in Haeckel's terms) and creationism. It was the theory of selection that annihilated the doctrine of isolated creation "in one blow" and "suddenly opened" Haeckel's eyes "to a comprehension of that greatest of all biological riddles," the origin of specific biological forms. He also emphasized that the theory of selection is applicable to human evolution: "The theory of selection teaches that in human life, as in animal and plant life everywhere, and at all times, only a small and chosen minority can exist and flourish, while the enormous majority starve and perish miserably and more or less prematurely" (Haeckel 1879). Phrases like "animal and plant life everywhere" and "at all times" in the above quotation suggest that Haeckel still believed in the universal character of natural selection as an evolutionary mechanism. In fact, he stressed in the paper that "most organic species have originated by a process of selection." In sum, there is nothing to suggest that Haeckel had begun to have doubts about natural selection at the time of writing this paper, which was actually a public address. Surely, Haeckel's emphasis on the significance of natural selection can be explained by the polemic nature of the paper as its major objective was protecting Darwinism from accusations of it being just a hypothesis. But, from another side, in these polemics, Haeckel was forced to outline the immutable hard core of Darwinism, to use Lakatosian terms, and this hard core was for him constituted by the theory of natural selection.

In more technical publications of the mid-1870s, Haeckel continued to praise natural selection as the core of evolutionary theory. For instance, in a pamphlet devoted to Haeckel's very own theory of heredity (see next chapter for details), Haeckel emphasized that "Darwin's theory of selection and the theory of descent justified by it, in accord with my deepest conviction, remains unshakable" (Haeckel 1876, p. 18).

Haeckel continued to employ the concept of natural selection also in very late publications. In his major late three-volume technical work, *Systematische Phylogenie* (*Systematic Phylogeny*) (Haeckel 1894–1896), he reiterated his commitment to natural selection as a major evolutionary mechanism and referenced the term natural selection 39 times, accompanied by several references to sexual selection. He also made several strong claims about natural selection; thus, in the first volume, devoted to the phylogeny of protists and plants, Haeckel claimed that both major evolutionary laws, "the law of progressive differentiation" and "the law of perfectioning" (Vervollkommnung), logically follow from the theory of natural selection.

The following extensive quotation effectively illustrates both Haeckel's grasp of phylogeny and his understanding of the role of natural selection in it:

> The natural process of phylogeny is on the whole a process of progressive development.[14] In the history of the organic world, the number, the diversity and the perfection of organic forms increases in the course of time; this historical progress is the more evident, the more we approach the present day. The principal fact of this progressive development finds its explanation in the theory of selection. As natural selection, due to the struggle for existence, takes an effect incessantly and anytime by way of adaptation and inheritance, it has as its

[14]Haeckel used the term "Entwickelung" (development) in the sense of contemporary term "evolution."

necessary result a continual multiplication, differentiation and perfecting of organisms. (Haeckel 1894, p. 11)[15]

It is easy to see that in this last voluminous technical work Haeckel is still a champion of natural selection and regards it as a major evolutionary mechanism. He still believed that progressive development "generally occurred through the continuous operations of natural selection," a belief highlighted by Richards in his writings on the *Generelle Morphologie* (Richards 2009, p. 147). As in the early works, Haeckel insisted that natural selection is a universal evolutionary mechanism acting at all times and in all places. Elsewhere, when discussing the age of the Earth, he mentioned natural selection as the major directing evolutionary force (Haeckel 1894, p. 17).

Haeckel's popular writings of the end of the nineteenth century reflect his adherence to natural selection as well. Thus, in what is probably his best-known book, *The Riddle of the Universe*, originally published in German as *Die Welträthsel* in 1899 and thus belonging to his very late publications, one can find 32 mentions of selectionist mechanisms (27 directly to natural selection). Most importantly, he repeats the idea initially formulated in the *GM* that the core of Darwin's discovery is the theory of natural selection: "Darwin, however, had not only the signal merit of bringing all the results of the various biological sciences to a common focus in the principle of descent, and thus giving them a harmonious interpretation, but he also discovered, in the principle of selection, that direct cause of transformism which Lamarck had missed" (Haeckel 1934, p. 64). Haeckel would certainly oppose the view presented by Richard Delisle in this volume that natural selection was only an "auxiliary hypothesis" for Darwin (Delisle 2021).

In other words, in his late and very late publications, Haeckel emphasized the role of natural selection as strongly as he did in the early works. A relatively low frequency of references to natural selection in *Systematische Phylogenie* can be explained by the overall lower level of attention he gives to the mechanisms of evolution in the text, instead devoting himself to lengthy discussions of the exact pathways of evolution (phylogenetic reconstructions). The generally lower frequency of natural selection in later works compared to *Generelle Morphologie* is due to the plain fact that Haeckel never again composed a work of comparable comprehensiveness by keeping a very high level of technicality (as opposed to popular publications with less technical details). There was no second edition of the *GM* either, whereas his popular works usually had many editions.

[15]German original: "Der phylogenetische Naturprocess ist im Grossen und Ganzen ein Process der fortschreitenden Entwickelung. In der Geschichte der organischen Welt nimmt von Periode zu Periode die Zahl, Mannichfaltigkeit und Vollkommenheit der organischen Formen zu; dieser historische Fortschritt wird in der Palaeontologie um so auffallender, je mehr wir uns der Gegenwart nähern. Die grosse Thatsache dieser progressiven Entwickelung findet ihre Erklärung durch die Selections-Theorie; denn die natürliche Zuchtwahl durch den Kampf um's Dasein, welche jederzeit und unaufhörlich mittelst der Anpassung und Vererbung wirksam ist, hat zur notwendigen Folge eine beständige Vermehrung, Differenzirung und Vervollkommnung der Organismen."

5.5 Non-selectionist Mechanisms in Haeckel's Works

As already mentioned, Haeckel advocated the whole range of non-selectionist mechanisms as well. The major difficulty with detecting and statistically analyzing these mechanisms in his texts is the great number of synonymous expressions he used for the same phenomenon.

Haeckel undoubtedly championed the best-known mechanism of neo-Lamarckian evolution, i.e., the "use and disuse of organs" (*Gebrauch und Nichtgebrauch der Organe*) accompanied by the inheritance of acquired characters. He also believed that the environment had a direct impact on organisms' heredity (*direkte oder universelle oder actuelle Anpassung*). Both of these neo-Lamarckian evolutionary mechanisms, advocated by Darwin as well, appear already in the *Generelle Morphologie* (e.g., Haeckel 1866, Bd. II, pp. 168, 196, 205, 364) and then never disappear from Haeckel's publications.

The idea that large mutations or "sports" bring about evolutionary "monsters" (*monströse Abänderung, sprungweise Abänderung, plötzliche Ausartung, monströse Entwickelung*) (terminologically close to the famous "hopeful monsters" of the twentieth-century geneticist Richard Goldschmidt) is also to be found already in the *Generelle Morphologie* (Haeckel 1866, pp. 204, 205). These large-scale, one-step mutations obey "the law of monstrous variation" (*lex variationis monstrosae*) and apply to all known kinds of organisms: "All organisms, under certain very unusual and deviant nutritional conditions, are able to produce offspring deviating from the characters of their parent organisms to such an extraordinary and unusual extent (and not in a usual low degree) that man can label them monsters or malformations" (Haeckel 1866, Bd. II, p. 204). Haeckel was of the opinion that "monsters" are relatively common among the offspring of humans, domesticated animals, and plants. These monster mutations can beget not only new species but also new families and orders. Although Darwin also admitted the existence of "sports" and monsters, Haeckel went much further than his scientific idol and advocated for the validity of this concept "too strongly."

Haeckel also championed Darwin's idea of correlation in both his early and late publications (e.g., Haeckel 1866, p. 218; 1874, p. 133; 1896, p. 231). Darwin defined correlated variation as an interconnectedness of various parts of an organism from the evolutionary perspective: "I mean by this expression that the whole organisation is so tied together, during its growth and development, that when slight variations in any one part occur and are accumulated through natural selection, other parts become modified" (Darwin 1872, p. 114). Haeckel, as in other cases, elevated Darwin's idea of correlation to the rank of a law (Haeckel 1866, Bd. II, p. 216). Although the idea of correlation is not necessarily non-selectionist in nature, it can be associated with both orthogenesis (the idea that macroevolution is linear and determined by definite variations; Ceccarelli 2021) and neo-Lamarckism.

Although Haeckel traced the law of correlation back to Goethe, he defined it in terms reminiscent of Darwin's original concept and emphasized that correlation can lead to nonadaptive organismic features. All organismic changes occur as adaptive reactions, Haeckel argued, but considering that an alteration of an individual character may affect the whole organism, the latter may demonstrate nonadaptive features as well (*Abänderungen welche nicht unmittelbar durch jene Anpassung bedingt sind*) (Haeckel 1866, Bd. II, p. 216). In that sense, Haeckel revolted against the adaptationist fallacy (the idea that nearly all evolutionary changes are adaptive) early on. Remarkably, he saw the ultimate cause of correlation in the "nutritive interaction" between all parts of an organism.

Haeckel advocated for non-selectionist mechanisms also in the later works, but he had no clearly detectable bias toward them. For Haeckel, it was important to demonstrate that evolution is a consistently causal process, that no "wonder" is in play, and that all evolutionary events occur in the continuous theoretical space, where various phenomena and their explanations are not opposed to each other, but complement each other. For example, in *The Evolution of Man*, he devoted several pages to "sudden variation" bringing about new fixed species (e.g., a common two-horned he-goat begets a four-horned goat) but insisted that this phenomenon can be ultimately reduced to the Darwinian idea of gradual evolutionary changes: "All these functions of evolution which 'suddenly and by a leap' produce this four-horned form of goat are in reality perfectly 'gradual and continuous' changes in the evolution of those masses of cells of which we have spoken: they depend on a change in the nutrition of the tissue at these two points in the frontal bone and skin" (Haeckel 1879, p. 169). Haeckel also narrowed the scope of possible evolutionary changes caused by sudden mutations over the course of his scholarly career. If in early works sudden mutations could bring about new families in one jump, in later works he tended to limit their range to the level of species.

Haeckel subordinated other non-selectionist evolutionary mechanisms to Darwinian ones as well. In *The Evolution of Man* (1879), one can find all the major mechanisms Haeckel suggested in the *Generelle Morphologie* (1866): use and disuse, direct adaptation, saltations, and correlative adaptation (e.g., Haeckel 1879, pp. 86, 158). At the same time, the frequency of their use is comparable with his early publications. Thus, he employed the notion of natural selection 25 times in *The Evolution*, accompanied by 11 references to sexual selection. By contrast, the expression "direct or actual adaptation" is used only once, and "use or disuse of organs" is employed two times. Most importantly, Haeckel labels "use and disuse" (which he also calls "adaptation through practice and habit") as "Lamarck's theory" and asserts that Lamarck failed to grasp the most important factor of evolution, namely, natural selection:

> [. . .] but Lamarck did not reach the principle which Darwin subsequently introduced as the most important factor in the Theory of Transmutation, namely, the principle of Natural Selection in the Struggle for Existence. Lamarck failed to discover this most important causal relation, and this, together with the low condition of all biological sciences at that time, prevented him from more firmly establishing his theory of the common descent of animals and man. (Haeckel 1879, p. 86)

In other words, in *The Evolution of Man* (1879), Haeckel advocated neo-Lamarckian mechanisms but clearly subordinated them to the "major causal principle," i.e., to the principle of natural selection.

The same tendency can be found in the two volumes of *The History of Creation* (1880, 1887), although Haeckel mentions neo-Lamarckian mechanisms far more frequently therein (in comparison to the *GM* and *The Evolution*). Direct (or actual) adaptation is mentioned in the text 12 times, and use/disuse 5 times (e.g., Haeckel 1880, pp. 225, 225, 227, 231, 231, 245). Comparatively, however, "natural selection" appears quite frequently—a total of 69 times in the text. Altogether there are 146 references to selectionist processes in both volumes of *The History of Creation* (Haeckel 1880, 1887). The preference for selectionism is evident in this quite late Haeckelian text.

In the three volumes of his *Systematische Phylogenie* (1894, 1895, and 1896), which were not primarily devoted to the discussion of evolutionary mechanisms, selectionist processes, as already mentioned, were referenced 39 times altogether (Table 5.4). The same three volumes contain only three references to direct (or actual) adaptation and four references to use/disuse, i.e., only seven total references to Lamarckian mechanisms. One can certainly find less direct discussion of neo-Lamarckian mechanisms in this text, but his tendency of highlighting selectionism and bringing it into conceptual communication with neo-Lamarckism is apparent.

Haeckel does not neglect natural selection in his post-*Generelle Morphologie* texts, nor does he increasingly prefer non-selectionist evolutionary mechanisms, even considering his very late fundamental treatise, *Systematische Phylogenie*. The core issue with Haeckel's original interpretation of evolutionary mechanisms was not his alleged departure from natural selection over the course of time, but his view of the nature of variation, which rendered his understanding of natural selection partly incompatible with Darwin's and completely incompatible with that proposed by the modern synthesis.

5.6 Haeckel's View of Variation and His Theory of Heredity

Haeckel remained a champion of natural selection throughout his whole career as a Darwinian biologist, yet his version of the theory of natural selection was incongruent with Darwin's. The major difference between the two theories laid in their interpretations of the nature of variation. Darwinian variation was mostly isotropic and only partly definitely directed (Winther 2000). Haeckel's variation was mostly directed and adaptive and only isotropic to a minor extent. The directedness of variation was due to the ability of an environment to influence developmental processes on a molecular-biological level.

As to the question why Haeckel tilted toward definitely directed variation, our analysis proves that Haeckel's approach made the whole explanatory pattern more logical and consistent than Darwin's conceptual system. Darwin, as is well known, tentatively proposed the hypothesis of pangenesis, according to which animal and plant cells threw off so-called gemmules (particles) circulating within the organism (Darwin 1868). These gemmules were supposed to be the bearers of hereditary information, including information on an organism's environment, and to transmit this information from parents to offspring. The blending of gemmules occurring in the progeny (blending inheritance) guaranteed the transmission of information from both parents in the case of sexual reproduction. Darwin's pangenesis hypothesis, first coined in *The Variation of Animals and Plants Under Domestication* in 1868 (2 years later than Haeckel's *GM*), was quite a Lamarckian one (Darwin 1868). As Olby puts it, "As for the inheritance of acquired characters, Pangenesis accommodates it, for altered organs will send their kind of gemmules to the reproductive system" (Olby 2017). Independently of the question of how Darwin himself combined the ideas of isotropic variation and pangenesis, there are numerous places in *The Variation* where he appears to allow a Haeckelian interpretation of variation:

> In variations caused by the direct action of changed conditions, whether of a definite or indefinite nature, as with the fleeces of sheep in hot countries, with maize grown in cold countries, with inherited gout, &c. conditions, whether acting on the embryo, the young or adult animal, can cause inherited modifications. It is equally or even more unintelligible on any ordinary view, how the effects of the long-continued use or disuse of any part, or of changed habits of body or mind, can be inherited. A more perplexing problem can hardly be proposed; but on our view we have only to suppose that certain cells become at last not only functionally but structurally modified; and that these throw off similarly modified gemmules. (Darwin 1868, pp. 394–395)

Darwin was certainly far from the idea of making variation completely neo-Lamarckian. Rather, he just "wanted to include in his scheme the possibility of the inheritance of some limited acquired characters" and "pangenesis gave him the chance to be Lamarckian without any of Lamarck's inner strivings" (Browne 2002, pp. 283–284). Haeckel's hypothesis of definitely directed variation reflected on this part of Darwin's intuition regarding the nature of heredity and rendered the whole theory more straightforward and less self-contradictory for an external viewer. Haeckel, as always, spoke "too strongly." He argued that the very possibility of natural selection was guaranteed by the unequal abilities of various organisms (of the same population) to immediately adapt to certain environmental conditions. Commenting on the "causes" of heritability (*Erblichkeit*) as a "virtual power" (as opposed to heredity as an actual characteristic), Haeckel asserted that although little is known about its mechanisms, with all probability some material particles are transferred from the parental organism to the descendants and these particles transmit both parental and environmental information. In that sense, Haeckel followed Darwin's "pangenesis" theory in his early publications (Levit et al. 2008a).

In an 1875 lecture presented to the *Medical-Scientific Society of Jena*, Haeckel propounded his own theory of heredity, the so-called perigenesis theory, which was consequently published as a separate pamphlet (Haeckel 1876; Di Gregorio 2005, p. 224; Reynolds 2008) (Fig. 5.4). Haeckel emphasized that the rudiments of this hypothesis were to be found already in the *GM* (Haeckel 1876, p. 17). The objective of the theory was to explicate the predominantly neo-Lamarckian nature of variation: "Like Darwin, Haeckel maintained an important role for the inheritance of acquired characteristics, and his theory of heredity provided a mechanism for how it might work" (Allen 2014). At the core of the theory was the idea of "the plastidules," molecules constituting protoplasm and consisting only of atoms. In accordance with his monism, Haeckel argued that all atoms are ensouled (*beseelt*) (Haeckel 1876, p. 39) and therefore that the plastidules possess an "unconscious memory" determining their wavelike motions. These motions Haeckel described as "ramified undulation" (*verzweigte Wellenbewegung*) (Ibid., p. 65). This "ramified undulation" is the essence of the "perigenesis theory" and the *causa efficiens* of any biogenetic process. The perigenesis hypothesis was, for Haeckel, a "genetic molecular theory" (*genetische Molekular-Theorie*) (Ibid., p. 17) bringing together inheritance as an internal organismic characteristic and adaption as modification in accordance with immediate external conditions. As the plastidules do have a "memory," they transmit hereditary information from one generation to another, at the same time "being liable to have their undulations affected by external forces" (Di Gregorio 2005, p. 226). In other words, the wave dynamics of the plastidules guaranteed the intergenerational transmission of hereditary information including newly acquired environmental information. As Fig. 5.4 explicates, various external influences induce various wavelike motions of the plastidules, and these changes are inherited by following generations (e.g., IVef to 5e, 5f). Since, for example, IVpq saved a memory of different external conditions than IVef, it gave different hereditary information to the next generation: 5p, 5q.

Despite the neo-Lamarckian nature of the perigenesis theory, there was a room for natural selection within it: "Like Darwin's gemmules, Haeckel's theory of perigenesis provided a mechanism—in his case a quasi-molecular one—for the origin of variations, and thus for creating the raw material on which selection could act" (Allen 2014).

It was evident already to Haeckel's contemporaries that his theory "does not furnish a clearer explanation than does Mr. Darwin's pangenesis" (Lankester 1876). Some of Haeckel's and Darwin's colleagues such as George John Romanes and Edwin Ray Lankester (Lankester 1876) saw great similarities between Darwin's and Haeckel's views. Romanes maintained that Darwin's theory was superior to Haeckel's because Haeckel's "terms are so much more general" (Burkhardt 2017, p. 26). At the same time, Lankester argued that both Darwin and Haeckel were "students of Mr. Herbert Spencer's works" and that both theories had their roots in Spencer's neo-Lamarckian concept of "life units." Lankester concluded that Darwin and Haeckel's theories were similar but equally obscure. Reacting to news of the perigenesis theory, Darwin himself wrote to Haeckel in 1876: "With respect to Pangenesis, I am sorry that you dissent so strongly from it, as it has lately risen in my estimation; but you are thoroughly right to explain in the clearest & strongest terms, your dissent."[16] Although Darwin himself was unhappy with Haeckel's departure from pangenesis, and although Haeckel stressed that perigenesis is his only fundamental disagreement with Darwin, both theories are fundamentally similar insofar as they both posit the existence of hypothetical "particles" having a kind of a "memory" which are able to transmit information in a neo-Lamarckian way.

[16]Letter no. DCP-LETT-10506 (Darwin to Haeckel, 14 May 1876).

Indeed, both theories are neo-Lamarckian enough to be more compatible with the idea of directed variation than with isotropic variation. Making his theory of natural selection fully compatible with neo-Lamarckism, Haeckel rendered the whole theoretical system more consistent with the proposed mechanisms of heredity. Haeckel's model enabled improved congruity between selectionist and neo-Lamarckian elements of the theory.

5.7 Conclusion

Considering Haeckel's extraordinary importance for the development of continental Darwinism, and that the level of his influence sometimes surpassed that of Darwin himself up until the end of the nineteenth century, Haeckel's understanding of natural selection sheds light not only on his own theory but also on the very essence of Darwinism around the time when Darwin was working. One should not forget that Haeckel and Darwin were in contact and that Darwin vividly reacted to Haeckel's achievements, either encouraging them or remaining reserved, as in the case of the perigenesis theory.

Our quantitative analysis of Haeckel's Darwinian texts and the review of his selected theoretical claims demonstrates that Haeckel considered natural selection through his whole career. The crucial issue was, however, his grasp of the very nature of natural selection. Insofar as Haeckel championed the idea of predominantly definitely directed variation, his version of natural selection differed essentially from that of Darwin. Although Darwin also developed a fairly neo-Lamarckian theory of heredity (pangenesis), he simultaneously championed predominantly isotropic variation and only accepted adaptive (directed) variation in a limited way. Haeckelian variation is, on the contrary, predominantly adaptive, and this provides the Haeckelian concept of natural selection with a different emphasis. Paradoxically, for a contemporary reader, the concept of adaptive variation makes Haeckelian natural selection seem more harsh as, if it were true, competing individuals would vary in a similar direction and experience selective pressure on their very subtle differences.

Our analysis also demonstrates that Haeckel advocated (purely) non-selectionist evolutionary mechanisms as well. Along with neo-Lamarckian direct environmental impact and use/disuse, Haeckel considered large mutations (jumps, sports, monstrosities) to be legitimate mechanisms of evolution, and he also supported the Darwinian idea of correlation.

In sum, Haeckel used Darwin as an inspiration and his understanding of evolutionary mechanisms can be traced back to Darwin's texts. However, Haeckel provided Darwin's ideas with new content and reformulated them within a different theoretical context. Although this reformulation did not necessarily bring him into intractable conflict with Darwin's own Darwinism (as Darwin was not in principle against neo-Lamarckian explanations), it renders his version of Darwinian selectionism even further from contemporary understandings thereof.

Haeckel's consistent support of natural selection as an evolutionary mechanism can be explained by his fidelity to Darwin's ideas, but more importantly by natural selection's ability to explain the whole range of biological phenomena from *monera* to man. Natural selection was, for Haeckel, *the* causal mechanism of evolution which allow us to explain evolutionary progress in naturalistic terms. Paradoxically, it was a neo-Lamarckian natural selection.

Acknowledgments We are very thankful to Wiebke Eichhorn (Wolgast, Germany) for composing the reference tables.

Dr. Thomas Bach (Ernst-Haeckel-Haus, Jena, Germany) was very kind to supply us with archival materials and advices. Cameron Yetman (Halifax, Canada) kindly assisted in improving the English version of this chapter.

Archival Sources

Letters of Ernst Mayr to Uwe Hossfeld (private archive of Uwe Hossfeld)
Letters of Darwin to Haeckel (www.darwinproject.ac.uk)

Supplementary Materials

Table 5.1 Selectionist terms in Haeckel's *Generelle Morphologie* (1866)

Notions	Frequency	English translation
Selections-Theorie	37	Selection theory
Selection	9	Selection
Selection-Gesetz	1	Law of selection
Natural selection	3	Natural selection
Natürliche Selection	2	Natural selection
Auslese	11	Selection
Natürliche Auslese	4	Natural selection
Auswahl	9	Selection
Natürlicher Auswahl-Process	1	Process of natural selection
Natürlichre Züchtungs-Process	1	Process of natural selection
Zweckmäßige Auswahl	1	Purposeful selection
Zuchtwahl	8	Selection
Zuchtwahl-Lehre	2	Selection doctrine
Zuchtwahllehre	1 (3)	Selection teaching
Natürliche Zuchtwahl	14	Natural selection
Gleichfarbige Zuchtwahl (selectio concolor)	1	Selection of a certain color to adapt to the environment
Natürliche Züchtung	60	Natural selection
In sum	= 165	

Table 5.2 "Sexual selection" in the *Generelle Morphologie*

Notions	Frequency	English translation
Sexuelle Selection	2	Sexual selection
Geschlechtliche Auslese	1	Sexual selection
Sexuelle Auslese	3	Sexual selection
Weibliche Zuchtwahl	3	Female choice
Männliche Zuchtwahl	3	Male choice
Sexuelle Zuchtwahl	7	Sexual selection
Sexuelle Züchtung	1	Sexual selection
In sum	= 20	

Table 5.3 "Artificial selection" and "the process of (artificial) selection" in the *Generelle Morphologie*

Notions	Frequency
Künstliche Züchtung	26
Künstliche Auslese	1
Künstliche Zuchtwahl	1
Züchtungs-Vorgang	2
In sum	= 30

Table 5.4 References to the selectionist mechanisms in three volumes of Haeckel's *Systematische Phylogenie* (1894, 1895, and 1896)

Notions	Frequency	English translation
Selection	4	Selection
Selections-Theorie	13	The theory of selection
Selections-Princip	1	The principle of selection
Selection-Process	1	Selection process
Natural-Selection	6	Natural selection
Cellular-Selection	3	Cellular selection
Personal-Selection	1	Individual selection
Kosmetische Selection	1	Cosmetic selection
Natürliche Zuchtwahl	5	Natural selection
Züchtung	2	Selection
Natürliche Züchtung	2	Natural selection
In sum	= 39	

References

Allen G (2014) Origins of the classical gene concept, 1900–1950: genetics, mechanistic, philosophy, and the capitalization of agriculture. Perspect Biol Med 57(1):8–39

Bowler P (2003) Evolution: the history of idea. University of California Press, Berkeley

Browne J (2002) Charles Darwin. The power of place. Princeton University Press, Princeton

Burkhardt F (ed) (2017) The correspondence of Charles Darwin, vol 25. Cambridge University Press, Cambridge

Ceccarelli D (2021) Recasting natural selection: Osborn and the pluralistic view of life. In: Delisle RG (ed) Natural selection: revisiting its explanatory role in evolutionary biology. Springer, Cham, pp 171–194

Darwin C (1859) On the origin of species by means of natural selection, or the preservation of favoured races in the struggle for life. John Murray, London

Darwin C (1868) The variation of animals and plants under domestication. John Murray, London

Darwin C (1871) The descent of man, and selection in relation to sex, 1st edn. John Murray, London

Darwin C (1872) On the origin of species, 6th edn. John Murray, London

Darwin C (1882) The descent of man, and selection in relation to sex, 2nd edn. John Murray, London

Delisle RG (2021) Natural selection as a mere auxiliary hypothesis (Sensu Stricto I. Lakatos) in Charles Darwin's origin of species. In: Delisle RG (ed) Natural selection: revisiting its explanatory role in evolutionary biology. Springer, Cham, pp 73–104

Di Gregorio M (2005) From here to eternity: Ernst Haeckel and scientific faith, religion, theology, and natural science. Vandenhoeck & Ruprecht, Göttingen

Di Gregorio M (2008) Under Darwin's Banner: Ernst Haeckel, Carl Gegenbaur, and evolutionary morphology. In: Engels EM, Glick T (eds) The reception of Darwin in Europe. Continuum, London, pp 79–97

Gould SJ (1983) Irrelevance, submission and partnership: the changing role of paleontology in Darwin's three centennials and a modest proposal for macroevolution. In: Bendall DS (ed) Evolution from molecules to men. Cambridge University Press, Cambridge, pp 347–366

Granovitch AI (2021) Natural selection, morphoprocess and a logical field of evolutionary process. In: Delisle RG (ed) Natural selection: revisiting its explanatory role in evolutionary biology. Springer, Cham, pp 391–418

Haeckel E (1864) Beiträge zur Kenntniss der Corycaeiden. Jenaische Zeitschrift für Medicin und Naturwissenschaft 1:61–112

Haeckel E (1866) Generelle Morphologie der Organismen. Bd. I. Allgemeine Anatomie. Bd. II. Allgemeine Entwickelungsgeschichte. Georg Reimer, Berlin

Haeckel E (1872) Monographie der Kalkschwämme. Erster Band (Genereller Theil). In: Biologie der Kalkschwämme. Georg Reimer, Berlin

Haeckel E (1874) Anthropogenie oder Entwickelungsgeschichte des Menschen. Verlag von W, Engelmann, Leipzig

Haeckel E (1876) Die Perigenesis der Plastidule oder die Wellenerzeugung der Lebenstheilchen. Georg Reimer, Berlin

Haeckel E (1879) Freedom in science and teaching. D. Appleton, New York

Haeckel E (1880, 1887) The history of creation, vols I & II. D. Appleton, New York

Haeckel E (1894–1896) Systematische Phylogenie. Georg Reimer, Berlin

Haeckel E (1934) The riddle of the universe. Watts, London

Hoquet T (2017) The evolution of the origin (1859-1872). In: Ruse M (ed) The Cambridge encyclopedia of Darwin and evolutionary thought. Cambridge University Press, Cambridge, pp 158–164

Hossfeld U, Levit GS, Kutschera U (eds) (2019) Ernst Haeckel (1843–1919): the German Darwin and his impact on modern biology. Special Issue Theory Biosci 138(1):1–202

Lankester ER (1876) Perigenesis vs. pangenesis: Ernst Haeckel new theory of heredity. Nature July 13:235–237

Largent MA (2009) Darwin's analogy between artificial and natural selection in the *Origin of Species*. In: Ruse M, Richards R (eds) The Cambridge companion to the "Origin of Species". Cambridge University Press, Cambridge

Levit GS, Hossfeld U (2006) The forgotten "Old Darwinian" synthesis: the evolutionary theory of Ludwig H. Plate (1862-1937). NTM Int J Hist Ethics Nat Sci Technol Med 14:9–25

Levit GS, Hossfeld U (2017) Major research traditions in twentieth-century evolutionary biology: the relations of Germany's Darwinism with them. In: Delisle R (ed) The Darwinian tradition in context. Springer, Cham, pp 169–193

Levit GS, Hossfeld U (2019) Ernst Haeckel in the history of biology. Curr Biol 29:R1269–R1300
Levit GS, Olsson L (2006) "Evolution on Rails": mechanisms and levels of orthogenesis. Ann Hist Philos Biol 11:97–136
Levit GS, Simunek M, Hossfeld U (2008a) Psychoontogeny and psychophylogeny: Bernhard Rensch's (1900–1990) selectionist turn through the prism of panpsychistic identism. Theor Biosc 127:297–322
Levit GS, Meister K, Hossfeld U (2008b) Alternative evolutionary theories: a historical survey. J Bioecon 10(1):71–96
Levit GS, Hossfeld U, Witt U (2011) Can Darwinism be "generalized" and of what use would this be? J Evol Econ 21:545–562
Liepmann H (1981) The six editions of the "Origin of Species": a comparative study. Acta Biotheoretica 30:199–214
Mayr E (1991) One long argument. Harvard University Press, Cambridge
Mayr E (1997) The objects of selection. PNAS 94:2091–2094
Olby R (2009) Variation and Inheritance. In: Ruse M, Richards RJ (eds) The Cambridge companion to the "Origin of Species". Cambridge University Press, Cambridge, pp 30–46
Olby R (2017) Darwin and heredity. In: Ruse M (ed) The Cambridge encyclopedia of Darwin and evolutionary thought. Cambridge University Press, Cambridge, pp 116–123
Olsson L, Levit GS, Hossfeld U (2017) The "Biogenetic Law" in zoology: from Ernst Haeckel's formulation to current approaches. Theory Biosci 136:19–29
Plate L (1913) Selektionsprinzip und Probleme der Artbildung. Verlag Wilhelm Engelmann, Leipzig
Porges K, Stewart I, Hossfeld U, Levit GS (2019) From idea to law: theory, concept and terminological formation in Ernst Haeckel's works. Russ J Dev Biol 50(6):290–302
Reif W-E, Junker T, Hossfeld U (2000) The synthetic theory of evolution: general problems and the German contribution to the synthesis. Theory Biosci 119(1):41–91
Reynolds A (2008) Ernst Haeckel and the theory of the cell state: remarks on the history of a bio-political metaphor. Hist Sci xlvi:123–152
Reynolds A (2019) Ernst Haeckel and the philosophy of sponges. Theory Biosci 138:133–146
Richards RJ (2004) If this be heresy: Haeckel's conversion to Darwinism. In: Lustig A et al (eds) Darwinian heresies. Cambridge University Press, Cambridge, pp 101–130
Richards RJ (2009) The tragic sense of life. The University of Chicago Press, Chicago
Richards RJ (2018) Ernst Haeckel: a dream transformed. In: Harman O, Dietrich M (eds) Dreamers, visionaries, and revolutionaries in the life sciences. The University of Chicago Press, Chicago, pp 35–50
Rieppel O (2019) The concept of the "organic individual" in Haeckel's writings. Theory Biosci 138 (1):147–157
Winther RG (2000) Darwin on variation and heredity. J Hist Biol 33:425–455

Part III
Rethinking a So-Called Intermediary Period

Chapter 6
The Origins of Theoretical Developmental Genetics: Reinterpreting William Bateson's Role in the History of Evolutionary Thought

Carlos Ochoa

> *Selection is a true phenomenon; but its function is to select, not to create.*
> *—William Bateson (1904)*

Abstract By considering the traditional history of evolutionary thought, William Bateson has been portrayed as an irrational anti-Darwinist who found in Mendel's laws the basis of heredity that supported his belief in saltational evolution; in so doing, he delayed the modern synthesis. However, this chapter suggests a very different interpretation. Bateson was one of the first geneticists whose ideas are closely associated with current developmental biology. I demonstrate that Bateson's studies were focused on how morphological structures arise and are maintained and not just the research of hereditary transmission of characters. From this framework, he attempted to understand how complex structures can emerge from transmission elements (which we know today as genes) during their ontogeny. In other words, he was one of the first scientists who conceived what we know today as the action of genes in controlling development. This view allowed him to criticize the model of gradual evolution, challenging natural selection as a creative force.

Keywords William Bateson · Developmental genetics · Variation · Heredity · Evo-devo · Homology · Mendelism · Modern synthesis · Natural selection

6.1 Introduction

The modern synthesis was a social movement of great significance for the biological sciences that contributed to the general agreement with the theory of natural selection, and it became part of evolutionary orthodoxy. The central commitments of this movement have persisted to this day, and they are part of contemporary education

C. Ochoa (✉)
Department of Evolutionary Biology, Faculty of Science, Universidad Nacional Autónoma de México, Coyoacán, Mexico City, México
e-mail: carlos008a@ciencias.unam.mx

R. G. Delisle (ed.), *Natural Selection*, Evolutionary Biology – New Perspectives on Its Development 3, https://doi.org/10.1007/978-3-030-65536-5_6

and research programs. Some authors even consider that a modern theory of evolution must obey the constructed foundation by this movement (e.g., the extended synthesis). However, it is important to note that the embryology was not considered in this movement, even though it has been vital to scale up our understanding of evolution. One of the consequences of this negligence is that the traditional narrative of the synthesis lacks those chapters that display the importance of embryology in evolution; rather, they focused on highlighting the achievements of population genetics. This leads us to revisit the origins of the modern synthesis and develop a new historiographic narrative (e.g., Amundson 2005; Delisle 2009, 2011, 2017; Stoltzfus and Cable 2014; Ochoa 2017, 2021; Stoltzfus 2017; Adams 2021; Schwartz 2021; van der Meer 2021).

According to most historians, the modern synthesis emerged in two stages—the first was the union of Mendelian inheritance with the theory of natural selection, while the second was the agreement and integration of the traditional disciplines of biology to endorse the assertion of the first stage. As we can see, this first stage was crucial for the progress of the modern synthesis which is related to the foundation of population genetics. This stage will be our main source of research.

Following the traditional narrative, before the twentieth century, Darwin and his followers argued that natural selection acted upon small continuous variation, then they believed that evolution proceeds gradually, but their ignorance with respect to the causes of heredity and variation delayed the general acceptance of Darwinism. This knowledge of inheritance came in 1900 when Mendel's laws were rediscovered. However, William Bateson, a champion of saltational evolution, grabbed Mendel's laws as a support for his belief that natural selection worked rather in discontinuous variation. Subsequently, Bateson and his followers, known as Mendelians, debated with mathematical biologists, called biometricians, who endorsed Darwin's original claim. However, the Mendelians did not realize that Mendelian inheritance was the "missing piece" to complement the Darwinian theory. Thus, the early Mendelians were mistaken in not realizing that their views were complementary to Darwinian ideas of gradual evolution. This delayed the reception of Darwinism by geneticists. According to this interpretation, the conflict was solved when it was recognized that Mendel's laws and Darwin's theory were complementary rather than contradictory, this assumption allowing the development of population genetics (e.g., see Provine 1971; Mayr 1980, 1982).

The "missing piece" theory has been embraced by different historians and philosophers who in turn have thrown up a large series of speculations about why the synthesis between Mendelism and Darwinism was delayed. For example, Mayr (1973, 1980) points out that the early Mendelians failed because they supported typological thinking[1] which prevented them to think "populationally"; besides, they

[1]For Mayr, typological thinking traced back to Platonic *eidos* whose fallout swayed Linnaeus' notion of the fixity of species, and it was established in its evolutionary version with the saltational evolution. In contrast, population thinking, which is related to Darwinism, emphasizes that the unit of evolution is the population and not the individual. However, this dichotomy between "typological thinking and population thinking" is nothing more than a rhetorical resource that Mayr used to

had conceptual muddles about variation and heredity. For other authors, the conflict was due to philosophical, methodological, or sociopolitical differences that shunned an agreement between their opinions (Farrall 1975; Norton 1975; MacKenzie and Barnes 1975; Hull 1985, pp. 806–807; Olby 1988; Kim 1994). Others argue that the debate was simply due to personal conflicts that if they had not existed, the synthesis would have occurred earlier (Sturtevant 1965, p. 58; Froggatt and Nevin 1971; Provine 1971). For Hull (1985, p. 805), for example, the fact that this debate has lasted is "an inexplicable embarrassment."

Nevertheless, the missing piece theory has been challenged by some historians who have pointed out several irregularities in standard historiography (e.g., Nordmann 1992; Gayon 1998; Stoltzfus and Cable 2014; Ochoa 2017, pp. 35–85; Adams 2021). From a skeptical point of view, we might say that the synthesis between Mendelism and Darwinism was incomplete since relying on a reductionist approach to Mendelism that had set aside embryology as a field of study (Amundson 2005; Ochoa 2017, pp. 35–85). For this reason, I challenge the proposal of Provine (1971) when he said that if biometricians and Mendelians had worked together, instead of attacking each other, population genetics would have emerged 15 years earlier than expected. Unlike Provine who was looking for the origins of population genetics, I consider that the present work corresponds to a pursuit for the origins of developmental genetics whereby some distinctive topics of evo-devo such as developmental repatterning (e.g., heterochrony, heterometry, etc.), modularity, phenotypic plasticity, developmental constraints, etc., might have occurred earlier if population genetics had not been successful. Particularly, I shall analyze the ideas of William Bateson who seems to be a crucial character in this endeavor; however, his figure has been tarnished by the traditional narrative.

William Bateson, an English naturalist, is best known as the main popularizer of Mendel's law during the early twentieth century. However, in the traditional view, Bateson is portrayed as an irrational anti-Darwinist who found in Mendel's assumptions the basis of heredity that supported his perspective on saltational evolution. In so doing, Bateson supposedly failed to associate Mendelian inheritance with his earlier work about discontinuous variation, not realizing that Mendel's assumptions were complementary to the Darwinian argument about continuous variation (Froggatt and Nevin 1971; Provine 1971; Mayr 1982).

The origins of the tarnished reputation of William Bateson as an enemy of the synthesis and progress of science can be found in the same book that inaugurated the first stage of the modern synthesis, *The Genetical Theory of Natural Selection*, by Ronald A. Fisher, written in 1930, just 4 years after Bateson's death. In the book's preface, after referring to Bateson as one of the main defenders of Mendel, Fisher stated:

> Unfortunately he was unprepared to recognize the mathematical or statistical aspects of biology, and from this and other causes he was not only incapable of framing an evolutionary

---

justify his participation in the development of the modern synthesis (see Winsor 2003; Chung 2003; Amundson 2005; Witteveen 2015, 2016; Ochoa 2017).

> theory himself, but entirely failed to see how Mendelism supplied the *missing parts* of the structure first erected by Darwin. His interpretation of Mendelian facts was from the first too exclusively coloured by his earlier belief in the discontinuous origin of specific forms. Though his influence upon evolutionary theory was thus chiefly retrogressive, the mighty body of Mendelian researches throughout the world has evidently outgrown the fallacies with which it was at first fostered. (Fisher 1930, pp. ix–x; my emphasis)

Nonetheless, curiously these were not Fisher's words, but his mentor's, Major Leonard Darwin, one of Darwin's sons. In a letter dated the 16th of January 1929, just when Fisher was writing his famous book, Darwin wrote to Fisher about his opinion of the figure of William Bateson:

> As to Bateson, If I had to write, I should write something like the following. But I am not well up in what he did do, and may well blunder....
>
> [...] The merit for this discovery must mainly rest with Mendel, whilst amongst our countryman, Bateson played the leading part in its rediscovery. Unfortunately he was unable to grasp the mathematical or statistical aspects of biology, and from this and other causes, he was not only incapable of framing an evolutionary theory himself, but entirely failed to see how Mendelism supplied the missing parts of the structure first erected by Darwin. Nothing but harm can come from following Bateson in regard to evolutionary theory, though his name will come to be honoured for his pioneer work in Mendelism when what he failed to do as regards theory has been accomplished.
>
> Having written it, I daresay I should tear it up, and advise you to do ditto....

Fisher replied gratefully, sharing with him a similar discourse: "it puts the point admirably, and though I have already altered the wording somewhat, it seems to me just what was wanted." He also added: "The only thing to do is to commend Bateson's enthusiasm for genetics, without saying, which would rather comfort my conscience, 'while greatly retarding its progress in his own country'" (Bennett 1983, pp. 95–96). In any case, Fisher ignored Leonard Darwin's suggestion and reproduced it in his book, thereby initiating the well-known myth about the synthesis between Darwinism and Mendelism.

According to Peterson (2008), this standpoint initiated by Fisher was perpetuated by Julian Huxley and Ernst Mayr at the *Origin of the Species* centennial conference held at the University of Chicago in 1959. Consecutively, it was firmly restated in Mayr and Provine's 1980 publication, *The Evolutionary Synthesis*, that addressed 1974 workshops organized by the *American Academy of Arts and Sciences Committee on the Recent History of Science and Technology*, particularly by Mayr's contributed works. Obviously, among these episodes, one of the accounts to promote the synthesis discourse was to spread out the figure of Bateson as an opponent of Darwinism who had suitable ideas about how inheritance works, but with flawed notions of how evolution operates.

However, as Stoltzfus and Cable (2014, p. 537) have noted, this myth of the missing piece was drawn on a framework based on complete ignorance of Bateson's ideas by Leonard Darwin and Ronald Fisher. This inconsistency had been pointed out by Cock (1973, p. 19) in his analysis of Fisher's criticism of Bateson: "Fisher's criticisms are, it must be concluded, unfair and, in large part, based on a misunderstanding of Bateson's views. Examination of them has... perhaps enabled us to see the Mendelism-Biometry controversy in a new light." Bearing this in mind, it is necessary to reexamine Bateson's thought as suggested by some authors (Nordmann 1992; Bateson 2002; Peterson 2008; Joshi 2010; Rushton 2014), because although his arguments are incompatible with the Darwinian orthodox perspective, this does not mean that they are false. A fresh look at history can give us further clarification concerning the conflict (Cock 1973; Nordmann 1992). Furthermore, his view on variation and heredity might be interpreted in light of the current progress of evo-devo (Bateson 2002; Joshi 2010).

So my primary aim of this chapter is to vindicate William Bateson's figure in the history of evolutionary thought. He should not be represented as a character associated with the popularization of Mendel's laws, and that he had a mistaken view of evolution that delayed the progress of science. Rather, he should be one of the first geneticists whose notions are closely associated with those later developed by evolutionary developmental biology. Even his criticism regarding natural selection is also valid from a contemporary point of view.

I shall analyze in Sects. 6.2 and 6.3 his conceptions on variation and heredity before and after embracing Mendelism. Afterward, I will explore his general criticism of Darwinism in Sect. 6.4. We shall observe that the main point of Bateson's research was how morphological structures arise and are maintained. From this framework, he attempted to understand how well-defined structures can emerge from transmission elements (genes) during their ontogeny. In other words, he was one of the first scientists who conceived what we know today as the action of genes in controlling development. This view led him to criticize the model of gradual evolution, challenging natural selection as a creative force.

6.2 Materials for the Study of Variation and Heredity

In this section, we will see that Bateson's first strategy was to demonstrate that discontinuity is a common phenomenon, while the transitional forms were rare. Further on he concluded that the best way to understand the gaps among species was through the study of the origin and maintaining of the morphological structures which are discontinuous by themselves. Consecutively, we will notice that Bateson became interested in the study of homology as a causal product that requires an explanation grounded in embryology. On this basis, I conclude that his research was focused on the origin of morphological units and not just on hereditary transmission of characters.

6.2.1 Bateson on Variation

Although we can represent William Bateson as either the main popularizer of Mendelism in the early twentieth century or one of the founders of the study of heredity, it is important to say that his vocation was developed in the field of morphology (Cock and Forsdyke 2008). His treatise, *Materials for the Study of Variation* (1894), is generally used as a reference of his belief in discontinuity before converting to Mendelism, but it is the culmination of his previous work linking morphology with evolution. One of the main achievements of this treatise was to demonstrate that discontinuity in organic forms is a common phenomenon in nature and that the best method to understand the origin of species was through the study of the origin and maintaining of these discontinuous forms and not the study of the adaptation of individuals in their environment as generally assumed.

Bateson noticed that most characters, which make up the bodies of organisms, are repetitions of the same fundamental structure, and hence he thought adaptation to external conditions has little to do with the origin of those characters. That is, the study of evolutionary processes should be focused on the emergence of nonadaptive structures that constitute the framework of animals (and plants), and this, for him, must be the study of variation. In so doing, he broadly suggested that the study of variation offers a means whereby we might know some important processes of evolution: "variation, whatever may be its cause, and however it may be limited, is the essential phenomenon of evolution. Variation, in fact, is evolution. The readiest way, then, of solving the problem of evolution is to study the facts of variation" (Bateson 1894, p. 6). To achieve this goal, he distinguished two types of variation: discontinuous variation, which pertains to the differences between the number of structures or organs in distinct species as well as anomalies in particular individuals; and continuous variation or small differences between individuals which form a series of intermediates. Considering this, he claimed to show evidence of the importance of the first type of variation.

For Bateson, continuous variation is displayed as the smooth differences among individuals that hypothetically are the outcome of the adaptation by natural selection. However, the knowledge of discontinuous variation looked more promising than continuous variation because it could provide more data about the origin of structures and hence the origin of the discontinuity among species. In other words, Bateson stressed that the origin of species does not rely upon the knowledge of organisms in their environment; on the contrary, it is their intrinsic properties that should be investigated. And this explains why he preferred discontinuous variation as the main research issue.[2] But what exactly did Bateson mean by discontinuous variation?

Recall the title of his book: *Materials for the Study of Variation.* This treatise mainly describes cases of specific individuals showing certain teratological anomalies, for instance, horses with extra fingers, organisms with supernumerary segments, flies with legs instead of antennas, etc. To account for these phenomena, he makes a

[2]Peterson and Müller (2016) argue that many morphological novelties cannot solely be explained by small continuous variations but also by discontinuous variations: "microevolutionary events are insufficient for explaining discontinuous forms of change and phenotypic novelties. The idea that small, continuous, incremental variational change is the sole cause of phenotypic evolution continues to be challenged by qualitatively discontinuous changes that also need to be accounted for by evolutionary theory" (pp. 328–329).

distinction among three main kinds of discontinuous variation: meristic variation, substantive variation, and homeotic variation. *Meristic* variation is the phenomenon of repeated structures that tend to display in individuals as a series of similar parts. *Substantive* variation is the phenomenon that presents the qualitative condition of a given character, for example, size, proportion, and color. Finally, *homeotic* variation is the phenomenon of abrupt and profound change of one structure into another, such as the modification of an antenna into a leg.[3] Let us take a closer look at each of these kinds of variations.

Meristic Variation According to Bateson, all the structures that make up organisms can be simplified into a fundamental unit; in other words, the parts are nothing more than a repetition of a basic structure. As an illustration, the phenomenon of segmentation is a well-known condition seen in vertebrates, annelids, and arthropods. These are repeated elements along the entire axis of the body, repetitions in any part of the body such as the axis of an insect leg. Moreover, such repetitions might also be exhibited as symmetry, this is, a structure that can be formed as a twin image of the other, but inverted. This is observed in both animals and plants, and these numerous examples of repetition of parts are not isolated cases but universal biological phenomena: "This phenomenon of repetition of parts, generally occurring in such a way as to form a symmetry or pattern, comes near to being a universal character of the bodies of living things. It will in cases which follow be often convenient to employ a single term to denote this phenomenon wherever and however occurring. For this purpose the term **merism** will be used" (Bateson 1894, p. 20; bold original).

Taking repetitions into account, he showed throughout his book many cases of individuals displaying some type of abnormalities. Considering the insect leg of *Periplaneta americana*, the tarsus of this cockroach typically comprises five parts, but in approximately 25% of the population, the tarsus is composed of only four parts (Fig. 6.1a). In the case of vertebrates, we observe examples such as an individual of the snake species, *Python molurus*, that exhibits an anomalous repetition of the vertebrae and ribs during its development (Fig. 6.1b). Besides, an individual of the frog *Rana temporaria* presents ten vertebrae instead of nine in its normal state (Fig. 6.1c, III). However, Bateson showed that in all these cases we can see that the phenomenon of repetition of parts is common and it has been important for the rise of new types. Yet if evolution proceeds by such discontinuous steps, how could their occurrence be accounted for?

Substantive Variation Repetition and symmetry are very common despite they always come along with other differences regarding qualitative value. These differences are most important in order to distinguish organisms from each other. Furthermore, they are essential to separate large divisions such as genera and families.

[3] We may associate meristic variation with heterometry; some cases of substantive variation would correspond to heterochrony; and homeotic variation with either heterotopy or heterotypy. Currently, Diogo et al. (2017) claim that some developmental anomalies such as loss and addition (meristic variation), proportion (substantive variation), and identity transformation (homeotic variation) of elements play a fairly important role in evolution.

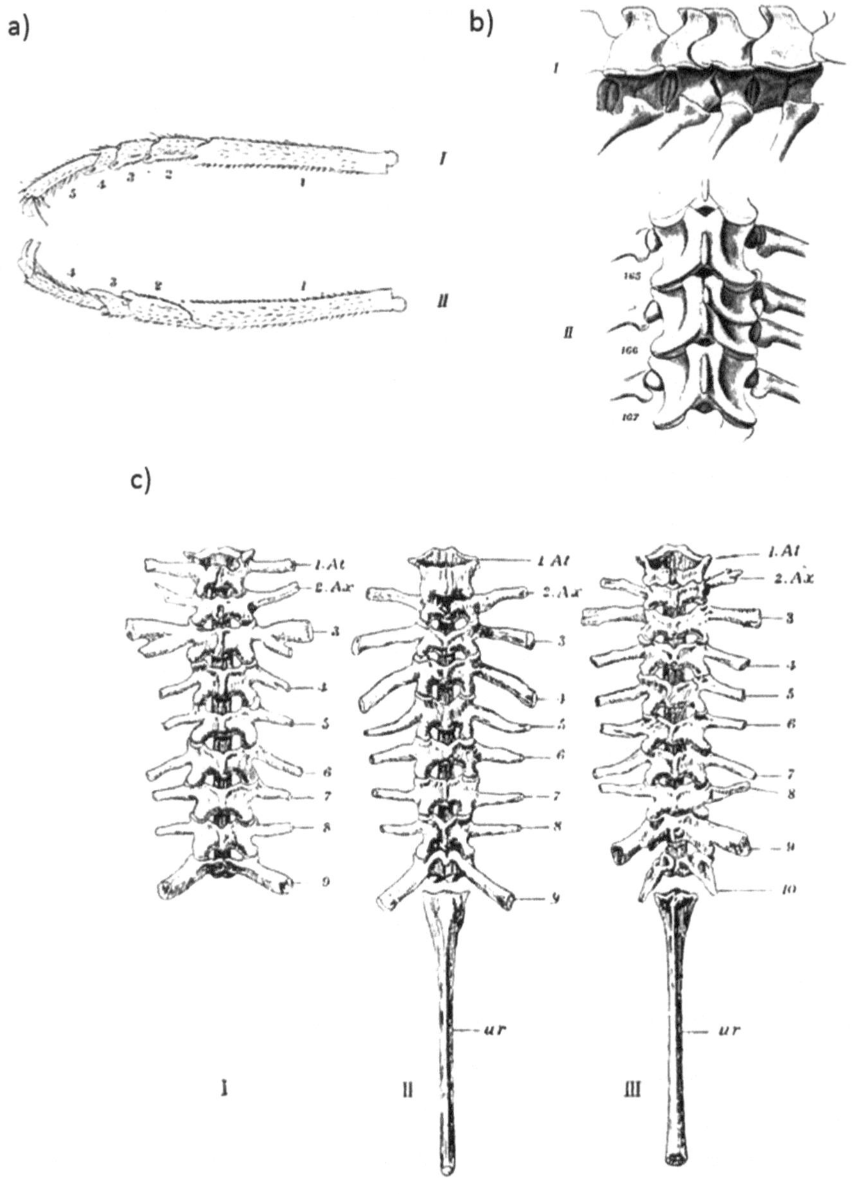

Fig. 6.1 Cases of meristic variation. (**a**) (**I**) The normal left tarsus of a cockroach consisting of five parts. (**II**) The abnormal right tarsus that has four parts instead of five. (**b**) The spinal column of *Python molurus*. (**I**) View of the right side of vertebrae 147–150—see the development abnormality of vertebrae 148th and 149th. Dorsal surface of vertebrae 166–167—a duplication of vertebra 166th is seen on the right side. (**c**) Spinal columns of the species *Rana temporaria*. (**I**) An abnormal individual that presents the atlas with tubercle unlike the other vertebrae. (**II**) A normal individual with nine vertebrae and one urostil. (**III**) A further abnormal individual showing an additional vertebra (from Bateson 1894)

Such forms present variations of the parts that are represented by the differences in shape, proportion, size, color, etc.: "From these phenomena of division may be distinguished variations in the actual constitution or substance of the parts themselves. To these variations the name **substantive** will be given" (Bateson 1894, p. 23; bold original).

Although we might use this type of variation to represent intermediate forms (e.g., continuous differences in size, proportion, color, etc.), the fact remains that these forms appear in a discontinuous fashion. As an illustration, having numerical data of the different sizes within a population, we can note that the intermediates are either rare or do not exist, and we only observe the size distribution between the two extremes (e.g., tall or short). For instance, the beetle *Xylotrupes gideon* possesses big and small males without intermediate forms, as is the case for size variations between the males of the common earwig *Forficula auricularia* whose differences are even more evident. Similarly, the growth of the skull in bulldogs is less proportioned to the normal growth of the lower jaw. This variation is parallel in other breeds such as the pug and the Japanese pug, and it has also been observed in cattle, pigs, and fish. Interestingly, a fisherman who found a trout specimen with this abnormality thought that this was a new species (Fig. 6.2a). This parallelism evidences that "the variations were from the first definite and striking." Similarly, variations in color such as color patterns of animals and plants are due to chemical stability. In such cases, we can observe a few intermediates.

It is important to note that Bateson stressed what matters here is not if the intermediates exist, but whether any of these stable forms passed through an intermediate state or not. To account for the absence of continuous forms, Bateson resorted to Galton's polyhedron metaphor, which was used to explain that some forms are more stable than others. In modern terms, we would argue that some forms are more possible than others because of certain rules of embryonic stability within the morphospace: "To employ the metaphor which Galton has used so well—and which may prove hereafter to be more than a metaphor—we are concerned with the question of the positions of Organic Stability; and in so far as the intermediate forms are not or have not been positions of Organic Stability, in so far is the variation discontinuous" (Bateson 1894, p. 42).[4]

Homeotic Variation For Bateson, this kind of variation is so peculiar, because it shows an abrupt and profound transformation of one organ into another. For example, an individual of *Palaeocimbex quadrimaculatus* developed a leg instead of antennas in the same position (Fig. 6.2b). Another example is a crab specimen that has a well-formed cheliped on the third maxilliped (jaw piece) (Fig. 6.2c). Additionally, Bateson pointed out the atlas of a mutant frog of *Rana temporaria* that exhibits unusual tubers (Fig. 6.1c, I). Besides, he asserted that plant experts had

[4]Regarding the model of developmental constraints, Pere Alberch (1982) argues that embryologic interactions drastically constrain evolutionary change by producing a discontinuous and directional distribution of forms. Under this model, some forms are impossible (intermediate) and others are likely to appear (discontinuous).

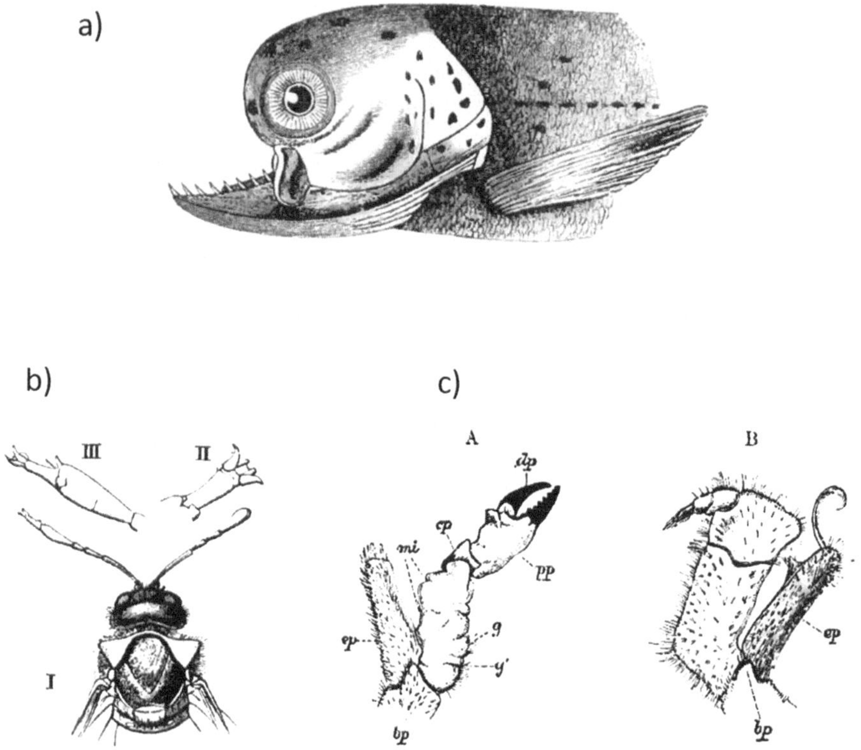

Fig. 6.2 Cases of substantive and homeotic variation. (**a**) Bulldog-headed trout. (**b**) (**I**) *Palaeocimbex quadrimaculatus* specimen presenting the left antenna as a leg. (**II**) The antenna seen from the front. (**III**) The antenna seen from above. (**c**) (**A**) The left and (**B**) right side of the third maxilliped of *Cancer pagurus*—see (**A**) the abnormal development of what looks like a cheliped (from Bateson 1894)

previously noticed this phenomenon, denominated as "metamorphy," but since the use of this word might have different meanings, he stated: "I therefore propose to substitute the term **homeosis**, which is also more correct" (Bateson 1894, p. 85; bold original).

Bateson was the first to describe this phenomenon that, as we know today, lies basically at the foundation of modern genetics and contemporary developmental biology. Although he was unaware of the underlying mechanisms that produce such kinds of abnormalities, he highlighted the fact that this phenomenon is the key to understand the secrets of evolution: "In this study, besides, this particular kind of variation will be found to be especially important and I believe that in the future its significance and the mode of its occurrence will become an object of high interest" (Bateson 1894, p. 84).

But why these phenomena of repetition, discontinuity, and replacement were so attractive to him? As we shall see below, Bateson was interested in the underlying mechanisms of embryonic development. The repetition of parts of organs shows that there is an ontogenetic process by which we can understand how animals and plants develop. Why is this process important for understanding evolution? If some individuals present an anomaly in the number, proportion, or identity of their elements during development, then gaps between species must occur promptly, abruptly, and discontinuously: "Is it not then possible that the discontinuity of species may be a consequence and expression of the discontinuity of variation?" (Bateson 1894, p. 69). Therefore, "It suggests in brief *that the discontinuity of species results from the discontinuity of variation*" (Bateson 1894, p. 568; italics original).

So Bateson was interested in individuals that show teratologic variations since they were important to understand underlying mechanisms that operate during embryogenesis. Therefore, such analysis would give us data about the origin of species. In short, Bateson argued in favor of discontinuous variation because he had in mind the central role of embryonic developmental processes for explaining evolution. The gaps between species could be explained not by the gradual and adaptive change but by the study of discontinuous variation: "the discontinuity of which species is an expression has its origin not in the environment, nor in any phenomenon of adaptation, but in the intrinsic nature of organisms themselves, manifested in the original discontinuity of variation" (Bateson 1894, p. 567).

Something important must be insisted upon here: Bateson did not write *Materials for the Study of Variation* to describe the underlying mechanisms by which we would account for the origin of structures or species; he rather described a compendium of cases that suggested that discontinuity is a common phenomenon in nature. Bateson's works were developed in two steps. The first step was to gather cases of discontinuity as much as possible; the next step was to explain the process involved in the origin of both structures and species.

6.2.2 Bateson and Heredity Before Mendel's Law

Was the problem of inheritance what primarily fretted Bateson? What did this concept mean for him? Did he get from Mendel the key to grasp the facts of heredity and evolution? According to the traditional narrative, while Bateson was traveling by train to give a lecture at the *Royal Horticultural Society* on May 8, 1900, he read the foundation of Mendel's heredity, throughout Hugo de Vries's paper, and noted that this theory of heredity fitted perfectly with his studies on discontinuity (Mayr 1982, p. 733). Since then, Bateson became a champion of Mendelism in the early twentieth century as well as an expert about heredity. Subsequently, he debated with the biometricians to argue in favor of Mendel's laws as evidence for natural selection acting upon discontinuous variation (Provine 1971). But why did Bateson adopt

Mendelism? Before answering all these questions, let us take a closer look at the development of his notion of inheritance before he adopted Mendelism.

Interestingly, Bateson disagreed with the term heredity, because he thought it was not an adequate metaphor. Such notion gave the idea that the body of offspring depended entirely on the traits transmitted by parents.[5] For Bateson, superficial similarities between offspring and parents were irrelevant; therefore, this concept of heredity he had challenged highlighted the individual genealogical relationship in the transmission of characters. Parents could transmit physical features typical of a family, and offspring might innovate traits with their superficial differences. However, he disagreed with that because, although we cannot deny that offspring resemble their parents, each individual is morphologically distinct: "For in nature the body of one individual has never been the body of its parent, and is not formed by a plastic operation from it [i.e., as a copy]; but the new body is made again new from the beginning, just as if the wax model had gone back into the melting-pot before the new model was begun" (Bateson 1894, p. 33).

It is necessary to emphasize that the study of the origin of variation as the units comprising the bodies of organisms was the main goal of his research rather than the hereditary transmission of characters: "The phenomena of variation and the origin of a variety must necessarily be studied first, while the question of the perpetuation of the variety properly forms a distinct subject" (Bateson 1894, p. 76). So what did he mean by the origin of variation if we do not consider hereditary transmission? If the use of the concept of heredity (transmission) was inappropriate, then what alternatives did he present as an explanation of the perpetuation of structures?

Bateson wrote *Materials for the Study of Variation* to provide the necessary material for evolutionists to work on the problem of variation. This might have helped them to solve the issue of the origin of the species. Thus, although if we had understood how hereditary elements are transmitted, this would not have provided a solution to the actual problem of variation (see below). If we consider that hereditary transmission represents a different issue other than the origin and development of the structures that make up the body of organisms, then we can understand why Bateson's main aim was the study of form and not the hereditary transmission. Even if we had associated ancestry with hereditary transmission, ancestry would nonetheless be a notion related to embryology: "the facts of embryology do support the doctrine of community of descent" (Bateson 1894, p. 8).

Amundson (2005) claims that before the modern synthesis was established, inheritance had been related to embryology. However, the notion of heredity,

[5]Once he became a Mendelian, he adopted the term and included the notion of the transmission of characters by parents to offspring in his explanatory domain; however, he advises careful use of the terms: "the terms 'Heredity' and 'Inheritance' are founded on a misapplication of metaphor, and in the light of our present knowledge it is becoming clearer that the ideas of 'transmission' of a character by parent to offspring, or of there being any 'contribution' made by an ancestor to its posterity, must only be admitted under the strictest reserve, and merely as descriptive terms" (Bateson 1902, p. 26).

which was embraced further on by neo-Darwinists, entailed the transmission of likeness from parents to offspring, regardless of the embryonic processes. But what kinds of problems might occur if we consider only the hereditary transmission and not the embryonic development of structure? Since *Materials for the Study of Variation*, Bateson had already grasped this difficulty, so he pointed out that this conception would ignore one of the principal topics of the study of morphology, that is, homology:

> Many times in the course of this work we have had occasion to consider the modifications in the conception of homology demanded by the facts of variation[...] It is not so much that to enlarge the conception of homology so as to include the phenomena of meristic variation is a direct help, as that to maintain the old view is a hindrance and keeps up an obstacle in the way of any attempt to apprehend the real nature of the phenomena of division, and hence of meristic variation. (Bateson 1894, p. 570)

The problem of homology lies in the origin of characters, which depends on the rise of repeated structures. The transmission of continuous characters was of no concern to him. The analysis of the origin of organs and their development was the most reliable test of homology revealing, therefore, the causes of the repetition of parts as the most important thing for understanding the real meaning of homology.[6] From this point of view, he stated that homology should be evaluated under two central conditions:

1. *Taking into account homology between members of the same series.* The repetition of parts shows a similar organization, which supports a common underlying developmental mechanism.
2. *Considering the individuality of the members of a series.* If two series are compared that have the same number of repeated units, we will have no problems for evaluating homology between those species. We can make equivalences of each unit since they occupy the same position in the series. But if one series differs from another in number, we will have difficulties to make those equivalences since we would not know what the matching parts are in one series when compared to the other. With this in mind, Bateson asserted that the problem of phylogeny was not so simple as far as the issue of individuality on homology is concerned: "no one has proposed to homologize single leaves of one tree with single leaves on another [...] Which vertebra of a pigeon, which has 15 cervical vertebrae, is homologous with the first dorsal of a swan which has 26 cervicals?" (Bateson 1894, pp. 32–33).

Bateson embraced a different conception of homology. The emergence of those morphological units is not through a continuous process of transmission that goes from parents to offspring or from ancestors to descendants; rather, it constitutes a

[6]Recent studies have highlighted the importance of defining homology regarding the biological causes that originate and preserve those morphological units that we call homologues (e.g., Müller 2003; Brigandt 2007; Wagner 2007, 2014; Kuratani 2009; Ochoa and Rasskin-Gutman 2015).

causal outcome that demands an explanation based on embryology.[7] As a result, he concluded that our perception has settled misconceptions about the interpretation of nature, because homology seen as continuous transmission of individual traits may confuse us about how nature works: "In these subjective conceptions of homology and of variation, we have allowed ourselves to judge too much by human criterions of difficulty, and we have let ourselves fancy that nature has produced the forms of life from each other in the ways" (Bateson 1894, p. 33).

Before Bateson had encountered Mendel's assumptions, he already displayed a great interest in the origin of morphological structures. The causes of repeated parts are difficult to deduce by means of the transmission of continuous characters. The preservation of the underlying embryonic mechanisms is evidence of common ancestry. These processes produce homologous structures whose consecutive modifications are the source of new variation. That is why Bateson's subsequent notion of "heredity" relates to a dynamical process of embryology, one grounded on the knowledge of the origin of form.

6.3 Bateson as a Developmental Geneticist

In this section, we will see that Bateson's adoption of Mendel's principle of segregation rests on several reasons. One of them was to understand how transmission of characters is related to the origin and development of structures. I emphasize that the principle of segregation alone does not encompass completely the domain of heredity since this principle accounts for how hereditary units (which we call today "genes") are transmitted, but it does not explain how the characters are developed from those elements. In other words, the principle of segregation by itself did not solve the problems of the mechanisms of generation.

We will see that Bateson attempted to integrate the knowledge of the transmission of hereditary units with his previous study on the origin and development of morphological structures. I argue that he conceived a process not much different from what we know today as the action of genes controlling ontogeny. In some examples of substantive variation, this model applied very well; however, in some others, such as color patterns and meristic variation of repetition of parts, this model was more difficult to sustain. Given the lack of information in his time, Bateson assumed that these biological mechanisms could be analogous to those observed in the physics of inorganic objects. In this context, he developed his famous vibratory theory of repeated parts.

[7]Müller (2003) stresses that the study of homology in morphology (biological homology concept) focuses on the underlying biological mechanisms rather than on the genealogical terms used by systematists (historical homology concept).

6.3.1 Bateson on Mendelism

Bateson has been remembered as one of the great popularizers of Mendelism in the early twentieth century. By defending the principle of segregation,[8] he suggested alternative methods for breeding experiments that settled his research programs on the study of inheritance. But if the origin of form, and not the hereditary transmission, was the primary aim of Bateson's studies, what did he find in Mendel's work that led him to the problem of hereditary transmission?

In a previous work (Ochoa 2017, pp. 55–63), I argued that Bateson's adoption of Mendelian principles was related to the attacks of the biometricians, which had intensified since 1895, particularly by Raphael Weldon. Likewise, biometricians such as Karl Pearson had embraced a mathematically reliable model of inheritance, the so-called Galton's law of ancestral heredity, by which they traced continuous characters across generations. Although the law of ancestral heredity was not in itself a hindrance for Bateson, this proposal captured the interest of the scientific community. In the principle of segregation, Bateson had found an empirically powerful hereditary transmission system, as well as a robust alternative to the law of ancestral heredity.[9] The principle of segregation explains the constancy of characters resulting from hereditary factors. According to the observations made during breeding experiments, hereditary factors segregate and predict the ratios of characters in the second generation: three dominant phenotypes and one recessive. Thus, if characters are transmitted unchanged from one generation to the next, maintaining the observed ratios as expected, given their dominance and recessiveness, there will neither be contributions from the individual characters of all ancestors, nor will there be equal probabilities of expression of the ancestral characters, as the law of ancestral heredity states.

But there were yet more specific reasons why Bateson adopted Mendel's principle of segregation. Darden (1977) emphasized six. First, Bateson observed that the data from old breeding experiments could be correctly interpreted under the new scheme. Breeding experiments in the past reaffirmed the predictions of the Mendelian system. Second, this principle was related to the problem of the origin of species. Given that discontinuous variation prevails through Mendelian factors, it is possible to account for the discontinuity of species through the transmission of these elements. Third, discontinuous variation also avoided the problem posed by the swamping effect, which states that many continuous characters can be swamped by other characters due to interbreeding across generations, while discontinuous characters are transferred entirely and prevail for generations. Fourth, Mendelism inaugurated a new experimental method of artificial breeding using statistical

[8]Mendel's laws were not assigned by early Mendelians; rather, they only referred to a single law—namely, the principle of segregation (see Marks 2008).

[9]As we will discuss below, Bateson originally thought that the principle of segregation as a modification to the law of ancestral heredity applied to some cases of prepotency, but in later publications, he argued that the principle of segregation comprises all situations.

analyses to generate precise data that strengthened the law. Fifth, Bateson suggested that the principle of segregation should be of general application and does not, therefore, only apply to discontinuous variation but also to continuous variation.[10]

However, Darden's sixth reason (1977), which is the most important for this analysis, consists of two related arguments. First, Bateson linked the principle of segregation to embryonic development, and second, he indicated that this notion was not necessarily based on the assumption that hereditary units had a material basis. This is a notable point in his analysis: Darden noted that while Bateson believed segregation to be a central piece in understanding inheritance, this was not the only element comprising the domain of heredity: "He insisted that a theory of heredity should account for development as well as transmission of characters" (Darden 1977, p. 98). Various authors have pointed this out when discussing Bateson's objection to the chromosome theory of heredity (e.g., Cock 1983; Bateson 2002; Rushton 2014).

Based on the above, we can distinguish between two types of issue regarding the domain of heredity. One corresponds to the transmission of characters by hereditary factors and the other to how characters develop from these factors. The principle of segregation was a theory that accounted for the transmission of hereditary units, but did not solve the general problem of heredity, which concerned the mechanisms involved in embryonic development (Darden 1991; Amundson 2005; Ochoa 2017). In Bateson's eyes, the term "heredity" was not restricted to only the transmission of characters from parents to offspring, but also included the identity of characters, or the repetition of similar traits during the development of an individual (see Bateson 1909b, pp. 275–277). Furthermore, we must include another important element with these ideas: the study of variation. Since heredity was not separate from the mechanisms that produce variation, Bateson concentrated on understanding these processes concurrently. This was one of the prime aims of his research prior to his embracing Mendelism (see above). In this lies the original meaning of the term "genetics," which is the study of variation and heredity.[11]

Bateson wrote *Mendel's Principles of Heredity* (1902, 1909) as a powerful defense and affirmation of Mendel's principle of segregation. However, when writing some of the chapters in *Problems of Genetics* (1913)—where he incorporated his previous work on variation into the issue of hereditary transmission of characters—Bateson's frustration with trying to solve the problems of variation and heredity is clearly noticeable. Disappointed with the facts gathered on variation, he admitted, for example:

[10]This assumption refutes the traditional narrative account that Bateson had not realized that Mendel's principles could also be applied to continuity. In fact, Bateson thought that many continuous characters fell within the domain of the principle of segregation and the "blending" characters were due to processes that may also be explained by this scheme; see below.

[11]Bateson first used the word "genetics" in 1905 in a letter to the zoologist Adam Sedgwick (see Bateson 1928, p. 34).

> Now since variation consists as much in meristic change as in alteration in substance or material, there is one great range of problems of causation from which we are as yet entirely cut off. We know nothing of the causation of division, and we have scarcely an observation, experiment or surmise touching the causes by which the meristic processes may be altered. (Bateson 1913, p. 86)

Undaunted by these drawbacks, Bateson's goal remained to deduce the mechanisms underlying the phenomena of heredity and variation. In order to harmonize the Mendelian system with his earlier view on variation, Bateson conceived that hereditary factors were not static material entities that were simply transmitted across generations. And as Darden (1977, p. 98) comments, "He could not adopt such factors because he thought independent, static, material units could not account for the organized, dynamic changes in development."

Nonetheless, Darden (1991) argues that embracing Mendelism did not mean that early geneticists believed that characters were literally passed on from parents to offspring by germ cells. This was because the only way they had to carry out research was by inferring the existence of hereditary units from the study of visible characters during breeding experiments. Hence, they recognized the need for an obvious distinction between visible characters and the hereditary factors that caused the appearance of those characters. The question then became, how could they account for the development of these characters from these transmission factors? In a sense, Bateson's explanation is a modern one since he had perceived that such transmission factors, whatever their physical nature, would trigger the development of a character. In other words, Bateson had a glimpse of what, in modern terms, we understand as the role that genes have in controlling development.

6.3.2 *The Action of Unit-Characters in Controlling Development*

Inspired by Wilhelm Roux's "causal embryology" or *Entwicklungsmechanik*, Bateson borrowed the notion of the "fermentative effect" from Hans Driesch, who argued that the cell's nucleus induced the development of morphological structures through the production of chemical changes via an enzymatic effect. According to Bateson, the presence of a hereditary factor could activate the production of specific substances leading to the generation of a certain outcome of organization, while the absence of this factor would yield an alternative pathway of development. In *Mendel's Principles of Heredity* (1909), he wrote:

> The round seed of peas, or of maize, is one which contains something possessing the power of turning most of the reserve-materials into starch. If the dominant factor endowed with this power is absent, much of the sugar remains sugar, and the seed wrinkles on ripening. The actual physiological processes involved are doubtless more complex than this, but there is no mistaking the essential nature of the distinction between the round and the wrinkled seed. So also it is easy to understand that an albino is an organism from which a ferment responsible for the production of colour has been omitted... Again it is easy to imagine that the presence

> or absence of a ferment can confer a greater power of resistance to the attack of a fungus. I suppose also that the dominant whites met with in some animals and plants may reasonably be represented as organisms possessing a substance which has the power of suppressing the development of pigment, whether by preventing its excretion or by destroying it when formed... the heredity of characters consists in the transmission of the power to produce something with properties resembling those of ferments. (Bateson 1909b, pp. 267–268)

In his *Problems of Genetics* (1913), Bateson laid out a more comprehensive outline. As mentioned above, much of the book is devoted to unifying his early work on variation and heredity with the principle of segregation. Concerning both substantive variation and meristic variation, he suggested that the study of genetics involved understanding the underlying mechanisms that produced such types of variations. These processes could either be mechanical "which relates to the manner in which material is divided and distributed" or chemical "which relates to the constitution of the materials themselves" (p. 32). Regarding substantive variation (e.g., color), he explained that germ cells transmitted the hereditary factor that produced a chemical substance (chromogen), allowing the development of color in a character: "The facts of colour-inheritance for instance prove that germ-cells, otherwise identical, may be formed *possessing* the chromogen-factor which is necessary to the formation of colour in the flowers, or *destitute* of that factor. Similarly the germ-cells may possess the ferment which, by its action on the chromogenic substance, produces the colour, or they may be without that ferment" (p. 33; italics original).

In the case of animals, the typical purple color of the Silky fowl is due to the pigment-producing factor, which in Mendelian terms is denoted by the letter "P." This factor influences the distribution of black pigment in the mesoblastic membranes across the body: "This variation must undoubtedly have been one of *addition*." In contrast, the Brown Leghorn variety "possesses a factor which has the power of partially or completely restricting the operation of the pigment-producing factor," this factor being denoted by the letter "D." "Thus in order that the black-skinned breed could be evolved from such a type as a Brown Leghorn it must be necessary *both* that P should be added *and* that D should drop out" (Bateson 1913, p. 85; his italics).[12] For other cases, Bateson conceded that the hereditary factors also function as a kind of switch that activates or suppresses development by cell division: "We want especially to know whether these divisions are due to the *addition* of some factor or power which enables the part to divide, or whether the division results from the *absence* of something which in the normal body prevents the part from dividing" (Bateson 1913, p. 46; his italics).[13]

---

[12]Nowadays, many studies show that the regulation of some genes causes variation in the color. For example, the variation in the melanocortin 1 receptor (MC1R) gene, which participates in the production of melanin in plumage, provides some color differences between birds (see, e.g., Doucet et al. 2004; Hoque et al. 2013).

[13]Many authors have currently stressed the importance of gene regulation to control the change in developmental timing (see Richardson 1995; Smith 2003; Arthur 2011).

The misunderstanding of the "action of unit-characters" may possibly lie in a historiographical confusion that has long been repeated. Although many authors have recognized that Bateson's conception of heredity was associated with embryology, most have commented that he chose a theory of heredity based on vibratory forces rather than proposing a particulate theory of inheritance (e.g., Coleman 1970; Gould 2002, p. 403; Rushton 2014).[14] Albeit his objection to hereditary units as particles is not called into question, I must emphasize that this "vibratory theory" was not a theory of hereditary transmission, but rather an assumption that attempted to explain the dynamics of the interactions that exist between the hereditary transmission factors and their subsequent expression during embryonic development, particularly for cases such as color patterns and the repetition of parts (meristic variation). Moreover, neither is it a "theory," since Bateson admitted that there was no way to visualize these processes. As a result, all that could be done was to use analogies of the mechanical forces observed in the physics of inorganic objects: "Many attempts have been made to interpret the processes of division and repetition, in terms of mechanics, or at least to refer them to their nearest mechanical analogies, so far with little success... In the preliminary stage of ignorance, having no more trustworthy clue, I do not think it wholly unprofitable to consider the applicability of this analogy somewhat more fully" (Bateson 1913, pp. 61–62).

Patterns are elements that regularly repeat themselves, for example, color patterns seen in animals such as spots, stripes, etc. Bateson speculated that the underlying mechanisms in such biological patterns could be analogous to some physical mechanisms, for instance, the pattern produced by the physical force of the wind when it makes ripples in a mackerel sky or on a flat sandy beach. So the striped pattern observed in zebras does not only emerge through the chemical reaction produced by the mere presence of the factor but also because of the mechanical forces operating throughout the body. More specifically, once the factor triggers the pigment production, it is subsequently distributed throughout the body tissue during cell division (or independently of it) by a rhythmic force analogous to the action of the wind, generating this typical striped pattern similar to ripples in the sand: "We

[14]Various authors quoted a letter from Bateson sent to his sister Anna on September 14, 1891: "Did I tell you anything about my new VIBRATORY THEORY of REPETITION of PARTS in Animals and Plants? I have been turning it over again lately and feel sure there is something in it. It is the best idea I ever had or am likely to have— Do you see what I mean?— Divisions between segments, petals, etc. are *internodal* lines like those in sand figures made by sound, i.e. lines of maximum vibratory strain, while the mid-segmental lines and the petals, etc. are the *nodal* lines, or places of minimum movement. Hence all the *patterns* and *recurrence of patterns* in animals and plants—hence the perfection of symmetry—hence bilaterally symmetrical variation, and the *completeness* of repetition whether of a part repeated in a radial or linear series etc. etc. I am, as you see, in a great fluster... P.S. Of course, Heredity becomes quite a simple phenomenon in light of this" (Bateson 1928, pp. 42–43; his italics). Here it is interesting to note that Bateson spoke of these processes as a "vibratory theory of repetition of parts" and not as a theory of heredity. In other letters, he refers to this as the "undulatory hypothesis," but finally in *Problems of Genetics* (1913), he argued that it was a hypothetical model of mechanisms to be discovered that may be analogous to the physical mechanisms acting upon the production of non-biological patterns; see below.

cannot tell what in the zebra corresponds to the wind or the flow of the current, but we can perceive that in the distribution of the pigments, that is to say, of the chromogen-substances or of the ferments which act upon them, a rhythmical disturbance has been set up which has produced the pattern we see" (Bateson 1913, p. 36).[15]

Regarding meristic variation or the repetition of parts (e.g., vertebrae, leaves, appendages, etc.), the mechanisms appear to be more complicated than indicated above. In spite of this, Bateson continued with the analogy and added other models for a better understanding of the physics behind these processes. As an example, he mentioned Ernst Chladni's vibrating plates forming shapes, the patterns seen inside of the bowing of a tube when lycopodium powder is added, and the phenomenon of osmotic growths developed by Stéphane Leduc. According to Bateson, a rhythmical mechanical force, analogous to that observed in Chladni's plates or Leduc's osmotic growths, is responsible for separating the tissues that constitute the segmentation. During this process, differentiation occurs in potentially identical structures as the result of many chemical reactions: "But in the segmentations of living things the nodes and internodes, once determined by the dividing forces, would each become the seat of appropriate and distinct chemical processes leading to the differentiation of the parts, and the deposition of the bones, petals, spines, hairs, and other organs in relation to the meristic ground-plan" (Bateson 1913, p. 60).

We should keep in mind that this "theory" does account for the origin of structure from an embryological point of view. It explains how morphological units emerge and whose causes are independent of hereditary transmission. By adopting Mendelism, Bateson did not abandon the primary aims of his initial research, which were to understand the origin of variation. The problem Bateson addressed was how characters are developed from hereditary factors and not the issue of hereditary transmission per se. Nevertheless, the question of hereditary transmission was not resolved and, from Bateson's point of view, the same difficulties remained as regards the integration of this model of the generation of repetition of parts through rhythmic forces with the application of the principle of segregation in the other models: "We may conceive such differences as due to change of form in the successive 'waves' of division, but we cannot yet imagine segregation otherwise than as acting by the removal or retention of a material element. Future observation by some novel method may suggest some other possibility, but such cases bring before us very clearly the difficulties by which the problem is beset" (Bateson 1913, p. 70).

In summary, while we can portray Bateson as the main popularizer of Mendelism in the early twentieth century and one of the founders of the study of heredity, we must also recognize that he was one of the first researchers to study the action of unit-

[15]Interestingly, Caballero et al. (2012) suggest a model that explains the color patterns seen in animals through physical and mechanical processes as well as cellular interactions and not only by the participation of genes alone. Likewise, Newman (2017) argues that some structural arrangements such as tissue layers, segments, appendages "can be readily generated by physical organizing forces acting on tissue masses, with minimal programming by the genome."

characters (now called genes) during embryonic development. In short, Bateson could be considered as the founder of modern theoretical developmental genetics.

6.4 Bateson's Critique of Darwinism

In this section, we will look at Bateson's critique of Darwinism in two crucial stages. The first stage corresponds to the period before his debate with the biometricians. Bateson criticized strict adaptationism as a method of evolutionary research and gradualism as a model to account for the origin of species. For Bateson, studying the origin of structures through their discontinuity was the best strategy for understanding the gaps between species rather than gradual change, as established by the theory of natural selection.

The second stage begins with the debate with the biometricians. Bateson's theory of the action of unit-characters in controlling development allowed him to explain the existence of intermediate forms between two populations. However, his conclusions were not promising for gradualism. First of all, intermediate traits stem from the action and interaction of unit-characters during ontogeny, for example, a dominant factor might incompletely suppress a character during development. Second, many individual variations arise as a result of the influence of the environment. In both these cases, the characters were not wholly inherited. In consequence, natural selection could not operate efficiently at this level. For Bateson, creativity laid in the development of morphological structures because they emerge with a specific organization from the outset. The role of natural selection, therefore, was to preserve a character, but not create it.

6.4.1 The First Clash with Darwinism

According to Gayon (1998), three essential elements comprise Darwinism: heredity, variation, and natural selection. For Darwin, heredity represented common ancestry, and its subsequent modifications were due to variation and natural selection. According to Darwin, this model of descent with modification was a gradual and adaptive process: natural selection operates upon small variations in the members of a population. The selected new variations are transmitted to the next generation, and as a result, these individuals are better adapted to local environments. This gradual process through geological time accounts for the origin of species. Gayon (1998) stresses that during Darwin's time, studies of variation and heredity were scarce. Darwin had advanced in conjecture, but the theory needed empirical support. As a result, the criticism and ensuing discussion focused on these issues.

We may note that Bateson's first clash with Darwinism was not initially with the biometricians, but with the preliminary assumptions made by Darwin and some of his followers. Bateson was influenced by *Natural Inheritance* (1889) by Francis

Galton, who argued in favor of saltational evolution (Gillham 2015). Bateson wrote *Materials for the Study of Variation* (1894) as an anti-Darwinian thesis of the study of variation. There are two themes to his criticism: (1) adaptationism as a method of research into natural history and (2) gradualism as a process that involves selection acting upon small variants across time.

Regarding adaptationism, Bateson criticized a method used by English naturalists that consisted in designating immediate utility to a given structure. He claimed that this method, despite it being inefficient, had attracted the attention of many naturalists interested in the topic of evolution. According to Bateson, although we cannot deny the existence of adaptations, such a method was no longer useful once we consider the origin of variation as well as the differences between species. Likewise, he saw that there were many characters that should not have an adaptive value. Even if we were supporters of Darwinism, it must not be assumed that every part of an animal is the product of adaptation, and when we do consider adaptations, we might ask ourselves: how can we test the functional purpose of a character in an environment? The method of inferring adaptations is inefficient, and despite the attractiveness of such research, it should not be taken as a primary basis for the study of the origin of structures and species: "But in the absence of such correct and final estimates of utility, we must never use the utility of a structure as a point of departure in considering the manner of its origin... Hence, though the study of adaptation will always remain a fascinating branch of Natural History, it is not and cannot be a means of directly solving the problem of the origin of species" (Bateson 1894, pp. 12–13).[16]

Concerning gradualism, Darwinism states that natural selection takes advantage of each small variation of individuals in a population. This process takes time and it involves constant gradual change. Given the above, Bateson wondered why we do not see such a chain of continuity from species to species if, hypothetically, according to this theory, there is a continuous series of adapted intermediates. In other words, why do we always find gaps between species? Bateson thought there was little evidence for gradualism so far and mentioned that "In the early days of the theory of natural selection it was hoped that with searching the direct utility of such small differences would be found, but time has been running now and the hope is unfulfilled" (Bateson 1894, p. 11).

If the intermediate forms were scarce, then there was no evidence that evolutionary change had been continuous. This is the reason why Bateson became interested in the study of variation, and he thought the study of what he called discontinuous variation was the best way to explain the gaps between species. The evidence for discontinuous forms was overwhelming, while intermediate forms were rare.

---

[16]Similarly, during the late twentieth century, conceptions such as "exaptation" and "spandrel" attempted to exclude the teleonomic aspect of biology (Gould and Lewontin 1979; Gould and Vrba 1982; Gould 2002). Diogo (2017) recently incorporated into his evolutionary framework the assumption that the structures making up an organism do not have an optimal match with their environment.

As regards the foregoing, a number of different points of view on his critique of Darwinism in *Materials for the Study of Variation* should be kept in mind. (1) Continuous variation is related to continuous change between species; the assumptions made in this process lead to the invention of adaptive stories. Bateson claimed that this type of variation is neither needed for the origin of structures nor the origin of the species; rather discontinuous variations (such as meristic and homeotic) were the best way to account for the gaps among species. (2) Substantive variation did not represent a model of gradual change because there was little evidence of intermediate forms. Even if they were continuous, those small differences between individuals did not guarantee subsequent inheritance across generations. (3) Considering the above, each individual differs from its parents, that is, they are not a copy of them. Moreover, if there is a similarity between parents and offspring, this transmission process is not equivalent to the origin and maintenance of homologous structures. Rather, they result from underlying mechanisms operating during ontogeny (such as the repetition of parts and their subsequent differentiation).

As far as we can see, Bateson argued for discontinuous variation since the evidence for discontinuity in nature was so great. If forms are discontinuous, with no intermediate steps, then it was difficult to sustain the gradual change of species. Bateson emphasized that the origin of structures did not lie in the study of adaptation, but in the study of the internal causes that produce variation. In the following section, we turn to his criticism of Darwinism when he had become a Mendelian and when he debated with the biometricians.

6.4.2 The Study of Developmental Genetics Is Opposed to Gradual Evolution

According to Gayon (1998), the problems faced by the post-Darwinian theory of natural selection lie in the phenomena of variation and heredity: How can the study of gradual change be quantifiable? What hypotheses of hereditary variation might be compatible with such a theory? The first scientists who attempted to answer these questions were the biometricians, and they did it through statistical methods applied to populations. But what did biometricians suggest regarding the facts of variation and heredity?

Biometricians were interested in individual variations (i.e., the small differences perceived in each individual in a population). For them, inheritance represented the contribution of the traits from all ancestors to their offspring. The importance of individual variations relies on their ability to be passed down from one generation to the next. If variations had that ability, natural selection would act upon those slight modifications. Under this scenario, evolution might be gradual and adaptive at each step of transformation. Galton's law of ancestral heredity made it possible to calculate this process. According to this principle, the two parents contribute one half of the total heritage of their offspring; the four grandparents contribute one quarter; the eight great grandparents one eighth, and so on. Therefore, the sum of the ancestral contribution results in the total heritage of the individual (Ochoa 2017, pp. 35–36).

Interestingly, the law of ancestral heredity had only been applied to one single case, this being the change in frequency of color patterns in basset hounds. Whatever the original interpretation of Galton's principle may have been, the statistician Pearson took the equation, changed it, and created his own version. For Pearson, and even for Weldon, the law of ancestral heredity was not exclusively associated with any physiological mechanism of heredity, it simply represented a mathematical expression of the inherited continuous variants. In principle, the fundamental idea was to establish a statistical method that would quantify the change. It would subsequently be complemented by a hypothesis of biological inheritance that might correspond with such results (Froggatt and Nevin 1971, pp. 7–8).

At this stage, one might expect a synthesis between law of ancestral heredity and Mendel's principle of segregation. The former would provide the mathematical basis for population change, whereas the latter would complement this by supplying the biological hypothesis of inheritance that the mathematicians hoped to find. But curiously, it was neither George Udny Yule nor Ronald Fisher who first suggested such a reconciliation, but Bateson himself (Cock 1973, p. 10; Olby 1987). When Bateson learned of the principle of segregation, through the publication of Hugo de Vries's work, he originally thought it could be interpreted as a modification of the law of ancestral heredity as regards cases where the prepotency of characters is observed. However, Olby (1987, p. 417) argues that after reading Mendel directly, Bateson changed his mind and saw the two principles as distinct systems. One applied to blending (continuous) characters, and the other to non-blending (discontinuous) characters. He concluded that the discontinuity at the gamete level was not related to the discontinuity at the observed level of characters.

One clarification is required here before proceeding with this discussion. Although we can say that the law of ancestral heredity leads to the assumption of blending inheritance, considering that the sum of total heritage is the contribution of the traits of all ancestors, it was merely a statistical tool that did not contemplate any biological mechanism for blending inheritance. In fact, Porter (2014) points out that blending inheritance was never considered to be a basic mechanism of hereditary transmission. Neither Darwin nor biometricians included a true mechanism of blending inheritance in their considerations; rather, once again, it was Bateson who made this attempt. According to Porter, the term "blending inheritance" is misleading. It is not an alternative biological mechanism to "particulate inheritance," but it stems from Galton's writings who described the continuous characters observed in the crosses between some individuals. Darwinism merely required it to justify the transmission of hereditary variations via continuity.

Returning to the point at issue, Bateson objected to the law of ancestral heredity because the principle of segregation could predict all the outcomes in breeding experiments, and it also had a general application regarding the magnitude of size in the characters studied (Darden 1977). Bateson linked the Mendelian principle to discontinuous characters, and in this sense, these characters avoided the "swamping effect." However, despite the traditional narrative, Bateson had in fact been able to explain the phenomenon of continuous variation in Mendelian terms. In doing so, he developed extremely strong arguments for rejecting the model of gradual evolution.

In Chapter XIII, "Intermediates Between Varieties and the 'Pure Lines' of Johannsen," of *Mendel's Principles of Heredity* (1909), Bateson clarified that there was a great deal of confusion among biologists regarding the "physiological significance of intermediates." He stressed that the biometricians used intermediate forms to demonstrate that the principle of segregation could not apply to individual variations "towards which the gradational forms seem to lead, and the misuse of statistical method so frequent in biometrical attempts to investigate heredity has come about chiefly through misinterpretation of the nature of such gradational forms" (Bateson 1909b, p. 235). Ignoring the facts of discontinuous variations, Bateson added that "Biologists committed to the proposition that varieties arise through the transformation of masses of individuals by the selective accumulation of minute differences saw that with each new case in which discontinuous variation could be proved to occur, the scope for their views was reduced, and the existence of intermediates constituted the most promising line of defence" (Bateson 1909b, p. 235).

Bateson subsequently argued that the study of continuous characters was a complicated task. However, he suggested that these were only the superficial results of the effects produced by the interaction of hereditary transmission factors. Later, he listed the categories of explanations whereby the phenomenon is expected:

1. *Intermediates as heterozygous forms*. In some cases, it is possible to distinguish groups that have homozygous alleles from those that possess heterozygous alleles through the observation of pure characters and intermediate characters. For example, in *Lychnis*, when the white variety *vespertina* is crossed with the red variety *dioica*, the offspring shows a pink color. In *Pisum*, crosses between the "half-dwarf" variety, which is about 3–4 feet tall, with the 1-foot-tall variety of "dwarfs" result in hybrids with an intermediate height: "In all these the one 'dose' of a dominant factor is sufficient to produce the full effect" (Bateson 1909b, p. 237).
2. *Intermediates due to subtraction stages of dominant factors*. We should not consider individuals who possess half of a certain number of characters in their bodies as evidence of a transition stage. Rather, the total pattern represents a pure type by itself. For instance, we should not regard the Dutch rabbit which has a body that is half white and half black as being an intermediate state of a cross between a dark rabbit with a white one. This pattern is due to the restriction of a pigment factor that is expressed in certain areas of their bodies. Moreover, there is a variety of "half-hoary" in the pea plant which shows a hoary state in the ventral surface of the leaves, while the dorsal surface is glabrous. Nevertheless, "no such 'intermediates' have occurred among the thousands of plants raised... except when the definite 'half-hoary' type was originally introduced as a parent" (Bateson 1909b, p. 238).

3. *Intermediates produced by the interference of other factors*. Following the model of action of the unit-characters in controlling development, the dominant factor suppresses the development of certain characters; however, in some cases, this suppression is not always total. As a result, it is possible to perceive an intermediate state in the development of these characters; i.e., "we have seen that a dominant factor may exist which has the power, for example, of suppressing the development of colour, leaving those parts white which in the absence of that factor would otherwise be coloured. This suppressing or inhibiting effect may be total, but when it is not total some puzzling intermediate types may be constituted" (Bateson 1909b, p. 238).
4. *Intermediates as fluctuational forms*. Many continuous variations are produced by the influence exerted by the environment during development and not by heredity itself: "there are intermediates due to the disturbing effects of many small causes not of genetic but presumably of environmental origin" (Bateson 1909b, p. 239).[17]

When considering these explanations, Bateson thought the Darwinian mode of evolutionary change—gradual evolution through the selection of small differences over time—suffered from significant problems. Varieties with intermediate characters are not proof of a population that has gradually changed into another through a series of intermediate forms. Rather, such forms exist because of the effect produced by the action of unit-characters in controlling the development of a character. For example, individuals with intermediate characters, such as the Dutch rabbit, are not a blending of characters caused by crosses between two pure lines. These patterns constitute a pure line that stems from the action of a specific hereditary factor that activates (or suppresses) the production of the pigment in certain areas of the body. In other cases, dominant hereditary units suppress the development of a character incompletely:

> For it would be found that the so-called intermediates consisted of numbers of genetically distinct types with distinct genetic properties depending on the factors which constituted them. Some would carry the colour-factors for the Dutch pattern, others those for more complete or less complete pigmentation, while others would owe their partial whiteness to the presence of the dominant factor which can suppress pigmentation in several stages of completeness. (Bateson 1909b, p. 241)

Now regarding this last example, while selection might potentially operate on intermediate forms—considering that the intermediate form might grant some advantage in its environment—it is difficult to show how such intermediate forms could be transmitted completely to the next generation:

> In such cases again there is every likelihood that careful selection might succeed in isolating subordinate types in which the suppression attains particular degrees of completeness ranging within well defined limits, but the evidence distinctly negatives any attempt to treat these several forms as a continuous series in which any member is capable of reproducing any other among its offspring. (Bateson 1909b, p. 239)

[17]This phenomenon would fall into the domain of what we know today as phenotypic plasticity.

Given that not all these individual variations are hereditary but also result from the environmental stimulus, Bateson could not perceive that there was a selective effect on such characters. He referred to them as "fluctuations" and distinguished them from genetic variations. He made this distinction based on the work of Hugo de Vries[18] and found support for doing so in the empirical work of the Danish geneticist Wilhelm Johannsen. In a famous experiment with beans, Johannsen had proved that artificial selection of certain characters does not change the initial average distribution of variation in a population. He selected the heaviest and lightest seeds from seeds with a variety of weights. By self-fertilizing them through cultivation techniques, he attempted to achieve a pure line of each one in order to fix both characters. However, after measuring the weight of the seeds from the progeny of each of these lines, he observed that the average distribution of the different weights was re-established rather than obtaining seeds of the selected weight. This allowed him to conclude that many individual variations resulted from collateral effects ensuing from the cell's response during development as well as the influence of the environment. As a result, he determined that such variations could not be inherited, and consequently natural selection would be inefficient when acting upon such variations[19]: "*Selection inside the family raised from a single seed did not alter the modal weight, which went back or regressed to that of the individual common parent*... Within the genetically pure line there are fluctuations in weight, but these fluctuations are due to interference which is external, or environmental in the wide sense, and selection of those extremes which are due to such interference produces no effect on the result" (Bateson 1909b, p. 240; his italics).

For Bateson, the lack of agreement about the real mode of evolutionary change derives from our ignorance of the causes behind the origin of variation: "Misconception of the nature and significance of intermediates has deprived the work of the biometrical school of scientific value as a contribution to the study of heredity" (Bateson 1909b, p. 241). Once we had attained such knowledge, however, it would only be a matter of time before we realized that this new branch of study, genetics, challenged the Darwinian mode of evolutionary change:

> The conception of evolution as proceeding through the gradual transformation of masses of individuals by the accumulation of impalpable changes is one that the study of genetics shows immediately to be false. Once for all, that burden so gratuitously undertaken in ignorance of genetic physiology by the evolutionists of the last century may be cast into oblivion. (Bateson 1909b, p. 289)

In brief, his position was that, once the causes of variation are known, we must abandon the claim that natural selection operated upon individual variations over a long period of time: "The scope of natural selection is closely limited by the laws of variation" (Bateson 1909b, p. 289). Individual (continuous) variations also had a

---

[18]However, Stoltzfus and Cable (2014) point out that de Vries's fluctuations also included partially hereditary variations.

[19]On this interpretation of Johannsen's experiments, see Roll-Hansen (2009, 2014) and Ochoa (2017, pp. 64–74).

causal explanation in embryology: they emerge either through the action of unit-characters when controlling development or due to the environment's influence during development. In both cases, direct inheritance is impossible, and thus at this level, natural selection can do little or nothing to affect outcomes.

Many problems arise in the study of genetics when accounting for intermediate forms as proof of gradualism. We may use this knowledge to grasp how natural selection really works. Let us take a closer look at some essential points of his evolutionary approach. Bateson's primary strategy was to find enough data to show that discontinuous variation is a common phenomenon. He then discovered the source of the transmission of these characters by the principle of segregation but then went even further. He attempted to explain the function of hereditary factors: he thought that they caused an enzymatic reaction that resulted in defined embryological pathways. He observed that distinct types of emergent variation did not show intermediate steps but definite and discontinuous products. For example, remember his attempt to explain the pattern of the zebra's stripes:

> When the essential analogy between these various classes of phenomena is perceived, no one will be astonished at, or reluctant to admit, the reality of discontinuity in variation, and if we are as far as ever from knowing the actual causation of pattern we ought not to feel surprised that it may arise suddenly or be suddenly modified in descent. Biologists have felt it easier to conceive the evolution of a striped animal like a zebra from a self-coloured type like a horse (or of the self-coloured from the striped) as a process involving many intergradational steps; but so far as the pattern is concerned, the change may have been decided by a single event, just as the multitudinous and ordered rippling of a beach may be created or obliterated at one tide. (Bateson 1913, pp. 36–38)

Taking this into consideration, Bateson's critique of gradualism is, at the same time, a critique of natural selection as a creative factor, when in fact creativity is in every respect internal to the organism. The origin and establishment of morphological structures are defined in a specific order from the outset. Consider color patterns, the serial repetition of parts, the appearance of legs rather than antennas in flies, the bulldog head in cattle, pigs, and fish, etc. None of these forms owe their presence to the action of natural selection working upon smooth, small, and random differences. For Darwin's centennial celebration, he wrote:

> This order cannot by the nature of the case be dependent on natural selection for its existence, but must be a consequence of the fundamental chemical and physical nature of living things. The study of variation had from the first shown that an orderliness of this kind was present. The bodies and the properties of living things are cosmic, not chaotic. No matter how low in the scale we go, never do we find the slightest hint of a diminution in that all-pervading orderliness, nor can we conceive an organism existing for a moment in any other state. (Bateson 1909a, p. 92)

Given this agenda, this does not mean that natural selection is unimportant in evolution, but "its function is to select, not to create" (Bateson 1904, p. 527). The characters are determined, discontinuous, and possess a certain logic during their development. The deduction is obvious: the value of a character lies in how it is generated and not in how it is preserved.[20] In other words, natural selection determines which characters remain, but it does not decide which characters are generated.[21]

> To begin with, we must relegate Selection to its proper place. Selection permits the viable to continue and decides that the nonviable shall perish... So again, as the course of descent branches in the successive generations, Selection determines along which branch Evolution shall proceed, but it does not decide what novelties that branch shall bring forth. (Bateson 1909a, p. 96)

In summary, before embracing Mendelism and his later debate with the biometricians, Bateson criticized the preliminary model of Darwinism. On the one hand, he demurred on the adaptive method that involves making inferences about the functional value of characters when they arise; and, on the other hand, he rejected the gradualistic mode of evolutionary change. Given that intermediate forms are rare and discontinuous states are common, he proposed that the study of discontinuous variation was the best way to understand gaps between species.

When the biometricians made their appearance, Bateson developed his theory of the action of the unit-characters in controlling development. He argued that the development of intermediate forms by either the action of hereditary factors or the influence of the environment during ontogeny posed many problems to sustain the gradual change of species. He also concluded that natural selection could not operate efficiently at this level and that, in consequence, the creativity of evolutionary change was to be found in the development of discontinuous characters. Under this framework, he relegated natural selection to a simple mechanism that preserved well-formed structures and challenged its role as a creative force.

6.5 Conclusion: The Delay of the Unmodern Synthesis

As discussed above, William Bateson's work studied not only the hereditary transmission of characters but also the origin and preservation of morphological structures. This was one of the primary goals of his research. From this starting point, Bateson attempted to explain how well-defined structures could be generated during ontogeny through the existence of transmission unit-characters (which we know today as genes). As a result, we can say that Bateson was one of the first naturalists who envisioned what we know today as the action of genes on controlling

[20] In similar lines, Linde-Medina (2011) argues that the role of natural selection as a creative force is challenged when we consider how characters develop. She concludes that selection is the cause of the prevalence of the characters, but when development is considered, the explanation behind the character's existence lies in how they are generated. Granovitch (2021, p. 409) also claims that the role of natural selection "consists in decreasing the morphogenetic potential of a species. Viewed from this angle, selectogenesis should be considered as a conservative force."

[21] As an analogy with a soccer game, Pere Alberch said that natural selection can determine which team wins, but developmental biology determines which teams play.

development. For Bateson, the model of activation or suppression of embryological pathways through hereditary factors was an adequate explanation for the development of characters such as color, proportion, and size. However, an additional explanation was required for other characters such as color pattern (e.g., spots, stripes, etc.) and the repetition of parts (e.g., vertebrae, leaves, appendages, etc.). To that end, he proposed his well-known vibratory theory, which has been erroneously presented as a theory of hereditary transmission of characters by historians.

Bateson's models could also account for the causes underlying intermediate forms; however, his conclusions conflicted with Darwinism. First, the interaction of unit-characters and their subsequent expression during development presents highly variable intermediate forms. That is, they are not genetically pure. Second, such individual variations also resulted from the influence of the environment and not from heredity. In both cases, the selection of these variants is inefficient. Given this, Bateson concluded that evolutionary creativity lies in the development of the structures themselves, not in natural selection acting on continuous variants.

With this conclusion in mind, I would like to stress that William Bateson should not only be remembered for his contributions to demonstrating Mendel's principle of segregation but also for the similarity of his ideas to subsequent theoretical advances on evo-devo. Topics such as developmental repatterning, modularity, phenotypic plasticity, developmental constraints, and particularly developmental genetics would have developed many years earlier if population genetics had not been so successful. In this sense, the issue is not that the early Mendelians like Bateson impeded the synthesis between Darwinism and genetics but that population geneticists such as Fisher are far more to blame for delaying the synthesis between developmental biology and evolution.

Here, it is necessary to emphasize that population genetics was developed on inadequate grounds regarding the acquired knowledge of variation and heredity. Ronald Fisher's "synthetic" works should not be considered as a solution to the conflict between Mendelians and biometricians, as many historians have suggested for years. Admittedly, his model did provide a mathematical-statistical solution of evolutionary change through continuous variants in accordance with the principle of segregation. However, his view lacked a biological approach to the causes of variation and heredity from an embryological perspective. Fisher had a reductionist view of evolutionary change, and as a result, he completely ignored embryological development and comparative anatomy (Ochoa 2017).

Amundson (2005) highlights one consequence of this lack that the term "heredity," spread by the modern synthesis, meant the transmission of genes from parents to offspring with no consideration for how characters emerge during ontogeny. Since population genetics only requires the transmission of genes, it does not need to account for the assumptions of developmental genetics. In short, the modern synthesis from its inception was incomplete because it ignored the processes of embryonic development. It propagated a concept of "narrow heredity" which stated that the problems had been solved when Mendel's laws were discovered. Amundson argues that Mendel's proposal as a mechanism of heredity was not a discovery but a convention for what inheritance should be.

From this point of view, it seems to me "an inexplicable embarrassment" that today many educators and students of evolution continue to set out the modern synthesis in its early stages as a scientific revolution that stems from an agreement between two contradictory perspectives. In fact, this "synthesis" was possible only because population geneticists ignored the processes of embryonic development and used Mendel's principle of segregation in their favor. Even worse, the discourse of the "missing piece" had gained general acceptance from the tarnished reputation of Bateson and the ignorance of his ideas. I would therefore urge all evolutionary biologists who identify as Darwinists to question where their beliefs stem from in evolution (see also in this volume Adams 2021; Delisle 2021; Esposito 2021; Granovitch 2021; Ochoa 2021; Schwartz 2021; van der Meer 2021).

During the 1990s, the advances made in evolutionary developmental biology were remarkable. Gilbert, Opitz, and Raff (1996) announced the birth of a new synthesis in the three areas of biology that had been eclipsed by the modern synthesis of genetics: embryology, macroevolution, and homology. In their conclusions, they wrote:

> The role of natural selection in evolution, however, is seen to play less an important role. It is merely a filter for unsuccessful morphologies generated by development. Population genetics is destined to change if it is not to become as irrelevant to evolution as Newtonian mechanics is to contemporary physics... Developmental biology is reclaiming its appropriate place in evolutionary theory. (Gilbert et al. 1996, p. 368)

Acknowledgments Thanks especially to Aaron Thompson and Omar Ochoa-Olmos for their invaluable support. I thank Sarah Debbie Wilson Barrera and Philip Daniels for the English editing. Many thanks to Richard Delisle for the invitation to contribute to this volume and for his suggestions and corrections that improve this work. I also want to thank the reviewers for their suggestions.

References

Adams MB (2021) Little evolution, BIG evolution: rethinking the history of Darwinism, population genetics, and the "synthesis". In: Delisle RG (ed) Natural selection: revisiting its explanatory role in evolutionary biology. Springer, Cham, pp 195–230

Alberch P (1982) Developmental constraints in evolutionary processes. In: Bonner JT (ed) Evolution and development. Springer, New York, pp 313–332

Amundson R (2005) The changing role of the embryo in evolutionary thought: roots of evo-devo. Cambridge University Press, Cambridge

Arthur W (2011) Evolution: a developmental approach. Wiley-Blackwell, Oxford

Bateson W (1894) Material for the study of variation: treated with especial regard to discontinuity in the origin of species. Macmillan, London

Bateson W (1902) Mendel's principle of heredity: a defense. Cambridge University Press, Cambridge

Bateson W (1904) Heredity and evolution. Popular Sci Mon 65:522–531

Bateson W (1909a) Heredity and variation in modern lights. In: Seward AC (ed) Darwin and modern science: essays in commemoration of the centenary of the birth of Charles Darwin and of the fiftieth anniversary of the publication of the origin of species. Cambridge University Press, Cambridge, pp 85–101

Bateson W (1909b) Mendel's principle of heredity. Cambridge University Press, Cambridge
Bateson W (1913) Problems of genetics. Yale University Press, New Haven
Bateson B (1928) William Bateson, F.R.S., naturalist: his essays and addresses, together with a short account of his life. Cambridge University Press, Cambridge
Bateson P (2002) William Bateson: a biologist ahead of his time. J Genet 81:49–58
Bennett JH (ed) (1983) Natural selection, heredity, and eugenics: including selected correspondence of R.A. Fisher with Leonard Darwin and others. Oxford University, Oxford
Brigandt I (2007) Typology now: homology and developmental constraints explain evolvability. Biol Philos 22:709–725
Caballero L, Benítez M, Alvarez-Buylla ER, Hernández S, Arzola AV, Cocho G (2012) An epigenetic model for pigment patterning based on mechanical and cellular interactions. J Exp Zool (Mol Dev Evol) 318:209–223
Chung C (2003) On the origin of the typological/population distinction in Ernst Mayr's changing views of species, 1942-1959. Stud Hist Philos Biol Biomed Sci 34:277–296
Cock AG (1973) William Bateson, Mendelism and biometry. J Hist Biol 6:1–36
Cock AG (1983) William Bateson's rejection and eventual acceptance of chromosome theory. Ann Sci 40:19–59
Cock AG, Forsdyke DR (2008) Treasure your exceptions: the science and life of William Bateson. Springer, New York
Coleman W (1970) Bateson and chromosomes: conservative thought in science. Centaurus 15:228–314
Darden L (1977) William Bateson and the promise of Mendelism. J Hist Biol 10:87–106
Darden L (1991) Theory change in science: strategies from Mendelian genetics. Oxford University Press, New York
Delisle RG (2009) Les philosophies du néo-darwinisme: conceptions divergentes sur l'homme et le sens de l'évolution. PUF, Paris
Delisle RG (2011) What was really synthesized during the evolutionary synthesis? A historiographic proposal. Stud Hist Philos Biol Biomed Sci 42:50–59
Delisle RG (2017) From Charles Darwin to the evolutionary synthesis: weak and diffused connections only. In: Delisle RG (ed) The Darwinian tradition in context: research programs in evolutionary biology. Springer, Cham, pp 133–167
Delisle RG (2021) Natural selection as a mere auxiliary hypothesis in Charles Darwin's *Origin of Species*. In: Delisle RG (ed) Natural selection: revisiting its explanatory role in evolutionary biology. Springer, Cham, pp 73–104
Diogo R (2017) Evolution driven by organismal behavior: a unifying view of life, function, form, mismatches, and trends. Springer, Cham
Diogo R, Guinard G, Diaz RE Jr (2017) Dinosaurs, chameleons, humans, and evo-devo path: linking Étienne Geoffroy's teratology, Waddington's homeorhesis, Alberch's logic of "monsters," and Goldschmidt hopeful "monsters". J Exp Zool (Mol Dev Evol) 328B:207–229
Doucet SM, Shawkey MD, Rathburn MK, Mays HL Jr, Montgomerie R (2004) Concordant evolution of plumage colour, feather microstructure and a melanocortin receptor gene between mainland and island populations of a fairy-wren. Proc R Soc Lond B 271:1663–1670
Esposito M (2021) Cathedrals, corals and mycelia: three analogies for the history of evolutionary biology. In: Delisle RG (ed) Natural selection: revisiting its explanatory role in evolutionary biology. Springer, Cham, pp 11–38
Farrall LA (1975) Controversy and conflict in science: a case study– the English biometric school and Mendel's laws. Soc Stud Sci 5:269–301
Fisher RA (1930) The genetical theory of natural selection. Oxford University Press, Oxford
Froggatt P, Nevin NC (1971) The "Law of Ancestral Heredity" and the Mendelian-ancestrian controversy in England, 1889–1906. J Med Genet 8:1–36
Gayon J (1998) Darwinism's struggle for survival: heredity and the hypothesis of natural selection. Cambridge University Press, Cambridge

Gilbert SF, Opitz JM, Raff RA (1996) Resynthesizing evolutionary and developmental biology. Dev Biol 173:357–372

Gillham NW (2015) The battle between the biometricians and the Mendelians: how Sir Francis Galton's work caused his disciples to reach conflicting conclusions about the hereditary mechanism. Sci Educ (Dordr) 24:61–75

Gould SJ (2002) The structure of evolutionary theory. Harvard University Press, Cambridge

Gould SJ, Lewontin RC (1979) The spandrels of San Marco and the Panglossian paradigm: a critique of the adaptationist programme. Proc R Soc Lond B Biol Sci 205:581–598

Gould SJ, Vrba ES (1982) Exaptation—a missing term in the science of form. Paleobiology 8:4–15

Granovitch AI (2021) Natural selection, morphoprocess and a logical field of evolutionary concepts. In: Delisle RG (ed) Natural selection: revisiting its explanatory role in evolutionary biology. Springer, Cham, pp 391–418

Hoque MR, Jin S, Heo KN, Kang BS, Jo C, Lee JH (2013) Investigation of MC1R SNPs and their relationships with plumage colors in Korean native chicken. Asian-Australas J Anim Sci 26:625–629

Hull DL (1985) Darwinism as a historical entity: a historiographic proposal. In: Kohn D (ed) The Darwinian heritage. Princeton University Press, Princeton, pp 773–812

Joshi A (2010) The deep roots of evo-devo and the "origins" question. J Genet 89:555–557

Kim K-M (1994) Explaining scientific consensus: the case of Mendelian genetics. Guilford, New York

Kuratani S (2009) Modularity, comparative embryology and evo-devo: developmental dissection of evolving body plans. Dev Biol 332:61–69

Linde-Medina M (2011) Adaptation or exaptation? The case of the human hand. J Biosci 36:575–585

MacKenzie D, Barnes B (1975) Biometriker versus Mendelianer: Eine Kontroverse und ihre Erklärung. In: Stehr N, König R (eds) Wissenschaftssoziologie: Studien und Materialien (Kölner Zeitschriff für Soziologie und Sozialpsychologie, Sonderheff 18). Westdeutscher, Wiesbaden, pp 165–196

Marks J (2008) The construction of Mendel's laws. Evol Anthropol 17:250–253

Mayr E (1973) The recent historiography of genetics. J Hist Biol 6:125–154

Mayr E (1980) Prologue: some thoughts on the history of the evolutionary synthesis. In: Mayr E, Provine WB (eds) The evolutionary synthesis: perspectives on the unification of biology. Harvard University Press, Cambridge, pp 1–48

Mayr E (1982) The growth of biological thought: diversity, evolution, and inheritance. Belknap Press of Harvard University Press, Cambridge

Müller GB (2003) Homology: the evolution of morphological organization. In: Müller GB, Newman SA (eds) Origination of organismal form: beyond the gene in developmental and evolutionary biology. MIT Press, Cambridge, MA, pp 51–69

Newman SA (2017) Inherency. In: Nuño de la Rosa L, Müller GB (eds) Evolutionary developmental biology. Springer. https://doi.org/10.1007/978-3-319-33038-9_78-1

Nordmann A (1992) Darwinians at war: Bateson's place in histories of Darwinism. Synthese 91:53–72

Norton BJ (1975) Biology and philosophy: the methodological foundations of Biometry. J Hist Biol 8:85–93

Ochoa C (2017) El eclipse del antidarwinismo: la historia detrás de la síntesis moderna. Centro de Estudios Filosóficos, Políticos y Sociales Vicente Lombardo Toledano, México

Ochoa C (2021) Inertia, trend, and momentum reconsidered: G. G. Simpson—An orthogeneticist? In: Delisle RG (ed) Natural selection: revisiting its explanatory role in evolutionary biology. Springer, Cham, pp 261–290

Ochoa C, Rasskin-Gutman D (2015) Evo-devo mechanisms underlying the continuum between homology and homoplasy. J Exp Zool B (Mol Dev Evol) 324:91–103

Olby R (1987) William Bateson's introduction of Mendelism to England: a reassessment. Brit J Hist Sci 20:399–420

Olby R (1988) The dimensions of scientific controversy: the biometric—Mendelian debate. Brit J Hist Sci 22:299–320
Peterson EL (2008) William Bateson from *Balanoglossus* to *Materials for the Study of Variation*: the transatlantic roots of discontinuity and the (un)naturalness of selection. J Hist Biol 41:267–305
Peterson T, Müller GB (2016) Phenotypic novelty in evodevo: the distinction between continuous and discontinuous variation and its importance in evolutionary theory. Evol Biol 43:314–335
Porter TM (2014) The curious case of blending inheritance. Stud Hist Philos Biol Biomed Sci 46:125–132
Provine W (1971) The origins of theoretical population genetics. University of Chicago Press, Chicago
Richardson MK (1995) Heterochrony and the phylotypic period. Dev Biol 172:412–421
Roll-Hansen N (2009) Sources of Wilhelm Johannsen's genotype theory. J Hist Biol 42:457–493
Roll-Hansen N (2014) Commentary: Wilhelm Johannsen and the problem of heredity at the turn of the 19th century. Int J Epidemiol 43:1007–1013
Rushton AR (2014) William Bateson and the chromosome theory of heredity: a reappraisal. Brit J Hist Sci 47:147–171
Schwartz JH (2021) What's natural about natural selection? In: Delisle RG (ed) Natural selection: revisiting its explanatory role in evolutionary biology. Springer, Cham, pp 329–390
Smith KK (2003) Time's arrow: heterochrony and the evolution of development. Int J Dev Biol 47:613–621
Stoltzfus A (2017) Why we don't want another "synthesis" Biol Direct 12. https://doi.org/10.1186/s13062-017-0194-1
Stoltzfus A, Cable K (2014) Mendelian-mutationism: the forgotten evolutionary synthesis. J Hist Biol 47:501–546
Sturtevant AH (1965) A history of genetics. Harper and Row, New York
van der Meer JM (2021) The concept of natural selection in Theodosius Dobzhansky. Its development and interpretation. In: Delisle RG (ed) Natural selection: revisiting its explanatory role in evolutionary biology. Springer, Cham, pp 291–328
Wagner GP (2007) The developmental genetics of homology. Nat Rev Genet 8:473–479
Wagner GP (2014) Homology, genes, and evolutionary innovation. Princeton University Press, Princeton
Winsor MP (2003) Non-essentialist methods in pre-Darwinian taxonomy. Biol Philos 18:387–400
Witteveen J (2015) "A temporary oversimplification": Mayr, Simpson, Dobzhansky, and the origins of the typology/population dichotomy (part 1 of 2). Stud Hist Philos Biol Biomed Sci 54:20–33
Witteveen J (2016) "A temporary oversimplification": Mayr, Simpson, Dobzhansky, and the origins of the typology/population dichotomy (part 2 of 2). Stud Hist Philos Biol Biomed Sci 57:96–105

Chapter 7
Recasting Natural Selection: Osborn and the Pluralistic View of Life

David Ceccarelli

Abstract Historians have almost overwhelmingly considered Henry Fairfield Osborn (1857–1935) an authority of twentieth-century American science. Behind his political and institutional clout is the parabola of a scientist whose work embodied the complexity of the debate in evolutionary biology at the turn of the century. Considered to be the leading proposer of orthogenesis in American paleontology, throughout his career Osborn denied natural selection the power of producing novel evolutionary characteristics and rather explained evolution as the result of the use-inheritance theory, organic selection, and internally directed variations. Osborn's theoretical shifts reflected all the disputes that characterized the so-called eclipse of Darwinism. At the same time, his late-career effort toward a synthesis among different evolutionary factors and approaches seems a noteworthy aspect of his work too often minimized. This chapter aims at exploring the research program in evolutionary biology that Osborn outlined between the 1910s and the 1920s. I will consider in particular how Osborn aimed at overcoming the specialization of biological studies through a renewed holistic approach in life sciences, outlining a synthesis (the theory of "tetraplasy") among the evolutionary factors proposed in the years of the eclipse of Darwinism, and recasting the role of natural selection within a pluralistic view of life. Though Osborn's agenda gained little acceptance among his contemporaries and even provoked harsh criticism by his Columbia colleague Thomas Hunt Morgan, his efforts seemed to respond to the epistemological necessity that would further motivate George Gaylord Simpson's work.

Keywords Orthogenesis · Natural selection · Paleontology · Henry Fairfield Osborn · Thomas Hunt Morgan · George Gaylord Simpson · Eclipse of Darwinism · Modern synthesis

---

D. Ceccarelli (✉)
Department of History, Humanities and Society, University of Rome Tor Vergata, Rome, Italy
e-mail: david.ceccarelli@uniroma2.it

R. G. Delisle (ed.), *Natural Selection*, Evolutionary Biology – New Perspectives on Its Development 3, https://doi.org/10.1007/978-3-030-65536-5_7

7.1 Introduction

Historians have almost overwhelmingly considered Henry Fairfield Osborn (1857–1935) an authority of American science (Bowler 1983; Rainger 1991; Ruse 1996; Gould 2002). As the foremost paleontologist at Columbia University and 25 years president of the American Museum of Natural History of New York (1908–1933), Osborn was, according to historian Brian Regal, "second only to Albert Einstein as the most popular and well-known scientist in America" (Regal 2002: 12). Opinions on his scientific legacy varied greatly, showing appreciation and scorn at the same time. Osborn has often been depicted as an autocratic, powerful, and capricious figure. Stephen Jay Gould deemed him the patrician of American paleontology who owned all and did as he pleased for about three decades (Gould 2002). According to the anthropologist Margaret Mead, Osborn was "the magnificent old devil" of American science, "an arbitrary and opinionated scientist" whose incredible efforts had however made the museum one of the greatest scientific institutions in the world (Alexander 1997: 28).

Osborn owes his controversial reputation to various motives. As the president of the AMNH for a quarter of century, he obtained large support by a number of American financiers, philanthropists, and politicians,[1] exploiting the museum almost at will. There is no doubt that under his leadership the American Museum's exhibitions were vastly improved (Gregory 1938: 77). As a matter of fact, in the early twentieth century, the museum became one of the main institutions in America to make the deep past imaginable (Clark 2008). Just as significantly, it reflected Osborn's views on racialism, eugenics, and immigration policies (Rainger 1991; Regal 2002; Spiro 2009). In his own words, the museum had to be exploited as a "positive engine for the propagation of socially desirable views" (Fowler 2003: 18).

Above all, Osborn is famous for having been an orthogenesist and anti-Darwinian evolutionist. He was probably the most authoritative figure of what historian Peter J. Bowler called the "American School" of anti-Darwinian evolutionism (Bowler 1983). Together with a number of American naturalists who were born between the 1840s and the 1850s,[2] Osborn opposed the neo-Darwinian agenda inspired by the German biologist August Weismann (1834–1914), according to which the union between gametes (*amphimixis*) was the only source of variability for natural selection to act upon (Depew 2017: 62n). Between the nineteenth and the twentieth

[1]Osborn was born in a wealthy family from Connecticut. In 1861, Henry's aunt married the well-known banker J.P. Morgan, who then became part of the Osborn family. His financial support, together with that of his son and Henry's cousin J.P. Morgan Jr., was essential to the expansion of the American Museum. Besides, the museum counted on many other trustees and supporters such as the eugenicist Madison Grant, the American politician Seth Low, the president of the First National Bank Childs Frick, John Rockefeller, and Theodore Roosevelt (Rainger 1991: 65).

[2]Among the major figures were also Alpheus Hyatt (1838–1902), paleontologist and zoologist at MIT and Boston University; Edward Drinker Cope (1840–1897), paleontologist at the Academy of Natural Sciences of Philadelphia; Alpheus Packard (1839–1905), zoologist at Brown University; and William Berryman Scott (1858–1947), paleontologist at Princeton University.

centuries, neo-Darwinism was more or less properly interpreted as a strictly reductionist view of evolution that claimed the all-sufficiency of natural selection operating on random variations. For his part, Osborn criticized neo-Darwinism throughout his career, maintaining that macroevolution was mainly determined by definite variations and that chance variations and natural selection had no proper creative power.

Osborn's critique of selection has been thoroughly explored by scholars (Bowler 1983; Rainger 1985, 1991; Regal 2002; Gould 2002; Adams 2021). Nevertheless, the research agenda he carried out in the early twentieth century reveals an attitude of pre-synthetic evolutionary studies that the classical historiography has sometimes underrated. As a matter of fact, between the 1910s and the 1920s, Osborn endeavored to outline a theoretical synthesis that aimed at recasting natural selection within a broader view of evolution. Much of his ambitious project turned out to be a failure in the eyes of his contemporaries. Furthermore, his scientific contribution underwent severe criticisms by the architects of the evolutionary synthesis. The aim of this chapter is to critically reassess Osborn's parabola. I argue that, despite his speculative attitudes and biased reconstructions (Ruse 1996: 272), Osborn's efforts reflected the need to provide solutions to the outstanding issues raised by the neo-Darwinian paradigm between the two centuries. Most importantly, his attempt to integrate approaches in evolutionary biology emerged as a response to the increasing partitioning of biological studies that would further motivate the synthesis between paleontology and population genetics in the 1940s. This theoretical and epistemological scenario reasonably blurs the lines of historiographical categories such as "eclipse of Darwinism" and "evolutionary synthesis."

The chapter will consist of four sections. In the first section, I will show how Osborn's evolutionism went through three distinct phases, along which he explained definite variations through the use-inheritance theory (1889–1895), organic selection (1895–1897), and, lastly, internally directed variations (1897–1935). Such periodization will allow us to see how Osborn's critical attitude toward natural selection underwent several adjustments over time, which led him to recast its role within a pluralistic view of evolutionary causation. The second section will examine Osborn's poorly known theory of "tetraplasy," which aimed at outlining a synthesis among the evolutionary factors proposed in the years of the so-called eclipse of Darwinism. This analysis will help highlight, in the third section, how Osborn tried to overcome the overspecialization of biological studies advocating an organicist and holistic approach in life sciences. The fourth section will address Osborn's ultimate step in his evolutionary synthesis. Indeed, his volume *The Origin and Evolution of Life* (1917) proposed a revised version of the theory of tetraplasy that further reconfigured the role of natural selection. Special attention will be devoted to the controversial reception of Osborn's volume, which stirred up a heated discussion between the 1910s and the 1920s.

7.2 Forms and Phases of Anti-Darwinism

Osborn's early works were overtly "neo-Lamarckian." At least until 1890, when he published the paper *The Paleontological Evidence for the Transmission of Acquired Characters*, Osborn was a staunch proponent of the theory of inheritance of acquired characters, following in the footsteps of his master Edward Drinker Cope (1840–1897), the leader of American neo-Lamarckism.

In the 1880s, Osborn had embraced Cope's "tritubercular" theory of mammal dental evolution, according to which three-cone Eocene mammals were the ancestors of all the modern placental mammals (Osborn 1887, 1888; Scott 1892). Cope and Osborn explained the origin and arrangement of the cones, from reptiles' single-cone teeth to the multituberculate structure, as the outcome of animals' chewing habits and diet (Gregory 1934). At first, cones had to be arisen from the contact zone between upper and lower molars as the consequence of the vertical movement of the mandible. Then the horizontal movement of the jaw molded the new cusps. Once it emerged, the three-cone pattern represented a tipping point in mammal evolution. Its supposed mechanical superiority and adaptability, even in comparison with the overspecialized descendants, made it a tremendously widespread structure in the history of mammals (Osborn 1888: 1078–1079).

Similarly to Cope, Osborn frequently used the case of parallel evolution to counter selectionism. The idea that similar bone structures had independently evolved through the contingent accumulation of identical random variations in distinct groups was epistemologically untenable. Considering the parallel evolution of horse species in North America and Europe, Osborn remarked:

> What must be assumed in the strict application of the selection-theory? (*a*) that variations in the lower molars correlated with coincident variations of reversed patterns in the upper molars, these with metamorphosis in the premolars and pocketing of the incisor enamel all new elements and forms at first so minute as to be barely visible, immediately selected and accumulated; (*b*) in the same individuals favorable variations in the proportions of the digits involving readjustments in the entire limbs and skeleton, all coincident with those in the teeth; (*c*) finally, all the above new variations, correlations, and readjustments not found in the hereditary germ-plasm of one period, but arising fortuitously by the union of different strains, observed to occur simultaneously and to be selected at the same rate in the species of the Rocky Mountains, the Thames Valley, and Switzerland! (Osborn 1890: 111)

Natural selection could preserve slight adaptive changes, yet, as already espoused by George Jackson Mivart, the rise of novel structures "from their apparently useless condition" represented "the most vulnerable point in the pure selection theory" (ivi: 110; see also Osborn 1889: 562). The Lamarckian principle offered an adequate explanation for such cases and further enriched the status of the evolutionary theory: "this [Lamarckian] school stands for Lamarckism *plus*—not *versus*—Darwinism, as Lankester has recently put it. There is naturally a diversity of opinion as to how far each of these principles is operative, not that they conflict" (Osborn 1890: 111).

Actually, opinions on the role of natural selection varied greatly among neo-Lamarckians. In a private communication, for instance, Cope openly dissented with Osborn's concession to natural selection: "I agree with your position in the main, but I do not admit the Natural Selection ever originated anything" (Osborn 1931: 393). As a matter of fact, Osborn's critical target was Weismann's "pure selectionism." Not only did it detach lines of reasoning but also had detrimental social and cultural effects, since denying the transmissibility of ontogenetic variations implied denying the cumulative effect of education from generation to generation (Osborn 1891: 363–364).

In the early 1890s, Osborn started to question the experimental evidence of the inheritance of acquired characters, calling into question Cope's theory of mechanical adaptation. The British biologist Edward Poulton[3] persuaded him that nonplastic structures could undergo corrosion by mechanical impact and friction (Osborn 1907a: 235). Whereas Cope dismissed Poulton's criticism pointing out that the deposit of dentine as a process of repair was generally acknowledged by paleontologists, and that "repair" had to be intended "in direct relation to use" (Cope 1896: 381), Osborn slowly moved away from his master's interpretation and started to search for an "unknown" factor of evolution that escaped the dead ends of neo-Lamarckism. This materialized in 1895, when Osborn became one of the three proponents of the theory of organic selection, together with James Mark Baldwin (1861–1934) and Lloyd Morgan (1852–1936). As Osborn stated at the meeting of the New York Academy of Sciences on March 9, 1896, organic selection allowed ontogeny to affect phylogeny without implying the transmission of acquired characters. Indeed, habits could induce ontogenetic variations; thereafter, it was possible for natural selection "to work very slowly and gradually upon predispositions to useful correlated variations, and thus what are primarily ontogenetic variations become slowly apparent as phylogenic variations or congenital characters of the race" (Osborn 1897a: 584). Yet Osborn's infatuation with organic selection lasted no longer than 2 years. Perhaps this is because his paleontological work was rooted in an orthogenetic view of evolutionary progress and thus "organic selection" still sounded too Darwinian (Ceccarelli 2018). In particular, in the article "The Limits of Organic Selection" (1897), Osborn pinpointed that such hypothesis retained some weak points of neo-Lamarckian and neo-Darwinian theories. First, when applied to nonplastic structures such as horns and teeth, accommodations and habits could bring about deterioration rather than development. Furthermore, Baldwin and Morgan considered individual plasticity and adaptability an outcome of natural selection, which was still far from being demonstrated (Osborn 1897b: 950). In light of this, at the turn of the century, Osborn adjusted his view of directional evolution for the second time in the course of a few short years, this time advancing the hypothesis that some variations were inherently predetermined and oriented, an idea he will later call "aristogenesis" (Osborn 1934).

Osborn developed his theory of directed evolution for over 30 years. Much of his work was closely tied to that of his Princeton colleague and friend William

[3]Osborn had first met Poulton during his stay in London between 1879 and 1880. In the 1880s, the two scientists exchanged many letters discussing how Weismannism had irreversibly challenged the theory of the inheritance of acquired characters.

Berryman Scott (1859–1947). During the 1890s, Scott had indeed focused his paleontological researches on parallelism in macroevolution (Ochoa 2021), exploring the works of the German paleontologist Wilhelm H. Waagen (1841–1900) and his hypothesis of orthogenetic evolution through gradual and oriented mutations (Gould 2002).

In *Die Formenreihe des Ammonites Subradiatus* (1869), Waagen pointed out that fossil trends were driven by an inner tendency which was allegedly independent from environmental conditions. By studying Ammonites' parallel evolution in India and Europe, Waagen argued that parallelism did not result from the adaptation to the environment, since these organisms had gradually mutated in the same direction although the ground and the external conditions they lived in were different. Over time, a number of German naturalists further developed the hypothesis of directed variation. In 1893, the zoologist Wilhelm Haacke traced back directed variations to the very mechanical arrangements of the units of the germplasm (*Gemmaria*). Haacke defined this process "orthogenesis," a term which came to be known under Theodor Eimer's definition: evolution by "definitely directed variation" (Bowler 1979).[4]

Osborn showed interest in Waagen's work around 1895 (Osborn 1895: 433). Since then, he embarked on a path of constant theoretical processing and expansion that lasted until his last days. The main core of Osborn's orthogenetic theory remained however unchanged and entailed two fundamental modes of evolution that elicited two types of variations: numerical characters, or "rectigradations," i.e., additions to the organism "which did not exist before," and allometrons, namely, changes of proportion and of modeling (Osborn 1907b: 748). At first, ecological conditions caused what Osborn first called "adaptive radiation" (Osborn 1900). New environmental circumstances triggered allometric variations, such as changes of body size or cranial shape, which led to divergence. Following such phase of differentiation, where branches spring off in all directions "to take advantage of every possible opportunity of securing food" (Osborn 1900: 48), there was a second phase. Indeed, when subject to similar environmental conditions, the descendants of the original group evolved in the same direction. Such form of parallel evolution was no longer explainable through the neo-Lamarckian principle of mechanical adaptation and appeared rather due to the fact that each branch shared "latent potential variations" which, in the right circumstances, became manifest, leading to the independent evolution of similar types in widely separated regions (Osborn 1900, 1902a, b).

In 1905, Osborn compared these variations to Waagen's oriented mutations and further called them "rectigradations" (Osborn 1905a: 457). To some extent, both these modes of change reflected two key aspects of orthogenesis: gradualism and, most importantly, inherent directionality (Grehan, Ainsworth 1985: 186). However, Waagen's mutations and Osborn's rectigradations denoted different objects.

[4]Further historical analysis of orthogenesis focused on non-English-speaking contexts can be found in Levit and Olsson (2006) and Popov (2018).

Whereas Waagen's mutations described "stages of transitions in direct lines of phyletic ascent" with a poor differentiation of characters, rectigradations indicated the stages of "single new characters occurring at definite points" (Osborn 1911: 328). Such process was gradual and "in the direction of future adaptation" (Osborn 1934: 210). Somehow, rectigradations resulted from "some entirely unknown and at the present time inconceivable relation between the forces of heredity and those of ontogeny and environment" (Osborn 1912a: 305). The examination of this relation became one of Osborn's main scientific scopes.

7.3 Tetraplasy as an Alternative to Epistemological Reductionism in Life Sciences

Osborn's writings on the status of the scientific discussion of evolution display his careful attention to the theoretical side of the subject, as well as his concern for the way scientists were pursuing their arguments. As early as 1895, he had pointed out that most of the disputes on inheritance made frequent use of "negative reasoning," which led to the proliferation of "theories which are less grounded upon fact than upon the logical improbabilities of rival theories" (Osborn 1895: 418). Furthermore, both neo-Darwinians and neo-Lamarckians used to conceive evolutionary factors as separable:

> Yet a survey of recent discussion among biologists as to the theory of evolution has shown broad lines of division into several schools of opinion strictly according to the factor from which the subject has been approached. It is true that, conceiving any one of these principal factors as separable, we become involved in endless difficulties; conceiving them as inseparable and continuously interacting under natural conditions, we reach the only true conception of the evolution process. (Osborn 1907b: 744)

Many of the difficulties that had long characterized the post-Darwinian debate were due to a misunderstanding of the mutual causal relationships among the factors of evolution. The evolutionary process had to be studied from a multifactorial perspective that tackled its causes as intimately connected and interdependent. In saying that, Osborn often recalled John Stuart Mill's *System of Logic* remark "the cause is the sum total of the conditions, positive and negative, taken together" (Mill 1843: 366), and indeed stated: "Mill's doctrine is the one on which the Law of the Four Inseparable causes or Factors of Evolution is founded" (Osborn 1912a: 278).

According to Osborn, the factors, or causes of evolution, were essentially four: heredity, ontogeny, environment, and, finally, natural selection. Each of these terms embedded decades of debates as to their definition and theoretical boundaries. The notion of environment, for instance, embraced the old conceptions of *milieu* and *monde ambiente* that emerged within the French natural philosophy during the eighteenth century, which scientists had now to regard as the complex of biotic and abiotic conditions that surrounded the organism (Osborn 1912a: 284). Ontogeny was instead the expression of heredity and its reaction to the environment. During

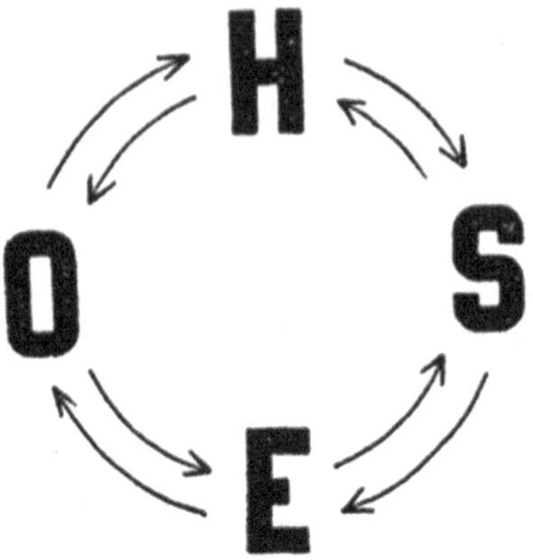

Fig. 7.1 Causal circularity as represented in "Tetraplasy: the Law of the Four Inseparable Factors of Evolution" (1912)

the individual development, numerous changes occurred, including the socially acquired accommodations that played a major role in the process of organic selection. Osborn did not fail to emphasize his doubts about the inheritance of ontogenetic modifications and nurture acquisitions (Osborn 1912a: 291) and actually pointed out that the only inheritable variations were those at the germinal level (heredity). As opposed to the other factors, Osborn remarked that natural selection did not play any "creative" role in evolution, acting upon the results of the interactions of environment, ontogeny, and heredity (Osborn 1912a: p. 294).

In the article "Tetraplasy: the Law of the Four Inseparable Factors of Evolution" (1912a), Osborn tried to sum up the connections among the four factors using the formulation: E (evolution) = H (heredity) × O (ontogeny) × E (environment) × S (selection). The difficulties that evolutionists had faced in the attempt to pinpoint the origin of new characters were to a large extent the consequence of the causal circularity among the four factors (Fig. 7.1). As a matter of fact, it was necessary to discriminate the causes that precede the genesis of character and its actual origin (Osborn 1912a: 300).

By means of sequential numbering, Osborn used his formulation to express various evolutionary scenarios in some sort of formal language.[5] For instance, the sequence H × O × E × S represented the mere condition of stability between the organism and the environment, whereas $H \times O \times E^1 \times S$ expressed the simple occurrence of environmental changes. By that rationale, the sequence $H^3 \times O^2 \times E^1 \times S^4$ could be used to express the far more complex scenario of organic selection: environmental changes create new selective pressure, which is followed by organisms' change of habits. The congenital variation then matches the ontogenetic modification already achieved, and finally natural selection operates on the population.

[5]Osborn wrote: "This formula roughly expresses the intimate nexus which exists between all these processes, a nexus which is quite consistent with the fact that each has also its separate part in life and in evolution. The multiplication sign, X, is to be interpreted in the active and passive sense of *influencing and influenced by*" (Osborn 1912a: 298).

Despite the sequential and rather linear character of his thematization, Osborn's theory of tetraplasy, from τετρα (four) and πλάσις (to form), was meant to highlight how the origin of new structures (rectigradations) and the transformation of existing characters (allometrons) were produced by the continuous interaction among heredity, ontogeny, environment, and selection. However, under certain conditions, each factor could become a leading one (Osborn 1908: 149, 1912a: 306). Most importantly, in complex organisms, each factor may be initiative of different groups of characters, whose combination was always subject to endless readjustments. In this regard, even the slightest alteration in the sphere of action of one factor could result in an overall disturbance.

This multifactorial conception was of great value for paleontological inquiry. By studying the fossils of titanoters, the giant mammals that had populated North America in the Eocene period, Osborn, for instance, observed that (a) the new cusps in the teeth of these animals had emerged through latent homology; (b) the dimension and form of their cranium was mostly controlled by ontogeny; (c) the environment, besides its indirect action through heredity and ontogeny, had determined the increase in body size; and (d) selection operated especially on the fluctuations in skull breadth or skull length respectively (Osborn 1908: 150). In other terms, rectigradations emerged from the hereditary constitution of the individual within the tetraplastic dynamic, whereas allometrons pertained ontogeny, environment, and, finally, natural selection.

Osborn's account of the evolution of titanoters bears witness to his efforts to address macroevolution as a multifactorial and integrated process. Just as importantly, it shows that, however interdependent, evolutionary factors acted according to a certain hierarchy of processes. While heredity was conditioned by ontogeny and environment, and ontogeny initiated changes that were taken up by heredity, natural selection represented an outer arbiter, a regulative mechanism that operated on the product of heredity, ontogeny, and environment. Such thematization of natural selection shows that, during the so-called eclipse of Darwinism, anti-Darwinian evolutionists, instead of denying natural selection, rather pushed it to the edge of the evolutionary explanation. As the paleontologist George Gaylord Simpson admitted, "no theorist, however radically non-Darwinian, has denied the fact that natural selection has some effect on evolution" (Simpson 1944a: 74).

7.4 Experimentalism and Paleontological Observation

Osborn's search for a pluralist explanation of evolution was on equal terms with his reflection on the future of evolutionary biology. He was aware of the technical and methodological changes that were affecting the debate on inheritance and evolution between the nineteenth and the twentieth centuries, especially because of the rise of experimental biology and genetics.

From the early 1890s, a new generation of European and American scientists laid the foundation for the redefinition of the epistemological and methodological boundaries of life sciences. According to Garland Allen's classical account, most of this transformation resulted from the impact of Wilhelm Roux's developmental mechanics (*Entwicklungsmechanik*). Whereas the early accounts of ontogenetic differentiation rested on phylogenetic history and the correspondence between phyletic and developmental stages, Roux focused his explanation on the physical and chemical interactions (Allen 1978: 34). Such approach gained remarkable attention within the American scientific community. On the model of the Zoological Laboratory at Naples (Italy), the Marine Biological Laboratory of Woods Hole (Massachusetts) was founded in 1888 and soon became a key center for the institutionalization of experimental methodology in life sciences. It was here that embryologists such as Thomas Morgan, E.B. Wilson, and C.O. Whitman moved their first steps in the field of experimental embryology, embracing Roux's *Entwicklungsmechanik*.

Along the lines of this transformation, experimentalists often charged classic morphologists with speculation. According to W. Johannsen, the collections of the museums fostered the work of morphologists who merely "operated with phenotypes in phylogenetic speculation," which made it "a science of phenotypes" with no value in genetic inquires (Johannsen 1911: 134). In his book *The Mechanistic Conception of Life* (1912), the German physiologist Jacques Loeb famously despised the tendency to mysticism and metaphysics by "romantic" biologists and evolutionists (Pauly 1987; Esposito 2017). "Biology," Loeb affirmed together with T.H. Morgan and W.J.V. Osterhout in the introduction of the series of books *Monographs of Experimental Biology*, "which not long ago was purely descriptive and speculative, has begun to adopt the methods of exact sciences, recognizing that for permanent progress not only experiments are required but quantitative experiments" (Loeb et al. 1919: 5).

Osborn could not remain indifferent to such claims, showing all his concern about the future of evolutionary biology and, most importantly, paleontology. As early as 1891, he stated the post-Darwinian debate had turned into a fratricidal war among biologists (Osborn 1891: 354). Evolutionists were increasingly working separately, and the debate on evolution could not but reflect such an epistemological attitude. Albeit experimental studies were essential for the understanding of hereditary mechanisms, Osborn felt that the study of nature in laboratories could not give any further information about the organisms' ecology. When pushed to the extreme, experimentalism diverged from the canonical, and somehow authentic, way of doing science (Regal 2002: 72).

Between 1909 and 1912, Osborn publicly charged experimentalists with gathering data from artificial investigation. On the occasion of the American centennial of Charles Darwin's birth in Baltimore (1909), and during the conference held at the Harvey Society in January 1912, Osborn highlighted the perils of experimentalism. A typical example of how experimentalists could fail to interpret data was William Bateson's *Materials for the Study of Variation* (1894). Here, the morphologist William Bateson had analyzed 323 cases of discontinuous variation in mammals' vertebrae, teeth, and skull. Of these, Osborn's assistant at the American Museum

W. D. Matthew examined (Osborn 1912b: 193), 286 were just deformities and reversional characters with no significance in evolution:

> Bateson cites 323 cases of discontinuity in vertebrae, teeth and skull. Of these 286 are teratological, or reversional, and have absolutely no significance in evolution; ten cases supernumerary (or fourth molar) teeth are possibly significant because among the mammals there are a few genera with fourth molars which may possibly have arisen by saltation. There remain only thirty-seven cases which may be ranked as "probably significant," and these are the meristic additions or reductions of vertebrae in the spinal column, significant because of the well-known variations in the vertebral formulae of different mammals, and secondly because vertebrae can be added or subtracted only discontinuously. (Osborn 1912b: 194)

According to Osborn, Bateson's misconception showed that it was necessary to preserve the naturalists' gaze on nature to fully comprehend organisms' variability. However important, the experimental investigation could not provide such information since, Osborn remarked during his speech in Baltimore, their observations took place in a synchronic dimension and thus were "too short-lived to grasp those changes in the hard parts (if they exist) which are so excessively slow as to be invisible and immeasurable by mortal eye" (Osborn 1909: 228). As he would state some years later, "paleontology is the acid test; paleontology is evolution" (Osborn 1932: 56), therefore evolutionists had to strive for a methodological eclecticism, merging data, approaches, and tools of analysis. In many ways, the research agenda he carried out at the American Museum reflected his aspiration. Sided by the researchers Gregory and Matthew, vertebrate paleontology "flourished as a field of biology" at the AMNH (Rainger 1988: 221).

7.5 The Elysian Fields of Mystery

Despite Osborn's popularity and institutional power, his claims for a pluralistic approach to evolutionary biology were mostly vain. The reasons for this are surely multiple. Perhaps, the publication of the volume *The Origin and Evolution of Life. On the Theory of Action, Reaction and Interaction of Energy* (1917) played a major role in the process.

The book presented a synthesis of Osborn's view of evolution and was written for a general audience. Chapter after chapter, Osborn focused on the different scales of biological interaction, ranging from the relationship between inorganic and organic matter to the role of endocrine system in bones' development and trophic relations. The difference between the organic and the inorganic world, he affirmed in the book, laid in the continuous reactions and interactions among energies in organisms. From the point of view of the paleontologist, each skeleton was indeed the result of multilevel energetic interactions. The endocrine response to the physiochemical conditions of the environment was, for instance, among the chief causes of allometric variation. In a similar manner, structural atrophies were always compensated by other parts, as in the case of his best-known fossil reconstruction of *Tyrannosaurus rex*,[6] where "the feeble grasping power and consequent degeneration of the fore limb and hand are more than compensated for by the development of the tail and the hind claws" (Osborn 1917b: 215).

Tetraplasy still represented the benchmark of Osborn's theoretical discussion. In this regard, in *The Origin and Evolution of Life*, he made a seemingly minor adjustment renaming the "four factors" as "physiochemical energies." Unlike the article published in 1912, Osborn now distinguished between life and inorganic environment, namely, between the "physiochemical energies of organisms" and the "physiochemical energies of the space, sun, earth, air and water" (Osborn 1917b: 22). This renewed schematization compelled him to exclude natural selection from the sets of energies acting on the individual. As he declared at the National Academy of Sciences in 1916:

> During the past two years I have been engaged in working out the aspects of this law [Tetraplasy] from the standpoint of physics and chemistry, that is, interchange of energy, in preparation for the Hale Lectures before the National Academy of Sciences on "The Origin and Evolution of Life upon the Earth." I perceive that it was an error to regard Selection as one of the four inseparable factors because it is not a form of energy. Consequently the law should be restated in the following terms: *In each organism the phenomena of life represent the action, reaction, and interaction of four complexes of physico-chemical energy, namely, those of (1) the inorganic environment, (2) the developing individual (cytoplasm and somatic chromatin), (3) the germinal or heredity chromatin, 4) the organic environment. Upon the resultant actions, reactions, and interactions of each organism Selection is constantly operating whenever there is competition with corresponding actions, reactions, and interactions in other organisms*. (Osborn 1917a: 8)

As far as this may be seen as a fatal blow to natural selection, from a certain point of view, such reformulation of tetraplasy has meant that selection ended up having an even more global control on the process. Though it could not originate new characters, as Osborn had maintained since 1889, natural selection was the only law of evolution the other factors and energies were always subject to. An example from the second chapter of *The Origin and Evolution of Life* could be revealing in this regard. By speculating on the primary stages of life, Osborn put forward five hypothesis: that the early steps in the organization of living matter consisted in the assemblage of ten essential elements (hydrogen, oxygen, nitrogen, carbon, phosphorus, sulfur, potassium, calcium, magnesium, and iron); that some kind of mutual attraction may have created a novel binding among these elements; that such arrangement took place by means of colloidal structures; that even the evolution of the simplest organism had proceeded in parallel with the evolution of chemical messengers; that natural selection constantly acted on the whole process. "Was there any stage in this grouping, assemblage, and organization of life forms, however and rudimentary, when the law of natural selection did not operate between different

[6]Barnum Brown found the first fossils of *T. rex* between 1902 and 1907. Osborn described the specimen as "tyrant lizard king" (Osborn 1905b). In *The Origin and Evolution of Life*, he further wrote: "This king of the tyrant saurians is in respect to speed, size, power, and ferocity the most destructive life engine which has ever evolved" (Osborn 1917b: 214).

Organism A		*Organisms B—X*
Physico-chemical Actions, Reactions, and Interactions of the		Physico-Chemical Actions, Reactions and Interactions of the
LIFELESS ENVIRONMENT	NATURAL SELECTION (Competition, Survival, Elimination,	LIFELESS ENVIRONMENT
DEVELOPING ORGANISM Cytoplasm and Chromatin HEREDITY CHROMATIN	Cessation of Selection, Reversed Selection, Organic Selection, etc.)	DEVELOPING ORGANISM Cytoplasm and Chromatin HEREDITY CHROMATIN
LIFE ENVIRONMENT		LIFE ENVIRONMENT

Fig. 7.2 The relation between the complexes of energies and natural selection as represented in "Application of the Laws of Action, Reaction and Interaction in Life Evolution" (1917)

aggregations of matter?" Osborn asked himself questioning the scope of natural selection. "Probably not, because each of the chemical life elements possesses its peculiar properties which in living compounds best serve certain functions" (Osborn 1917b: 68–69).

Darwin's principle, as well as organic selection, could be now understood as "one of the causes of evolution in making the lifeless and living worlds what they now appear to be," determining which combination of energies should survive (ivi: 20). Tetraplasy, now renamed "tetrakinetic theory," could thus be represented as given in Fig. 7.2.

In Osborn's intent, the tetrakinetic theory was not meant to outline an alternative to selectionism but rather to escape the naive unidirectional conceptions of evolutionary causation that at least three generations of naturalists had formerly put forward. Evolutionists such as Buffon, Lamarck, Cope, and Spencer had generally advocated the idea that the causes of evolution were from without inward, namely, beginning in the somatic cells and extending into germ. On the other hand, Wallace, Weismann, Bateson, and De Vries, notwithstanding their difference of opinion about natural selection, had maintained that causes begin in the germ and then extend outward (Osborn 1917b: XIII). To dispel the errors and biases of both the "centripetal" and "centrifugal" conceptions of evolution, it was necessary to embrace an interactionist view.

Osborn's *The Origin and Evolution of Life* gained a controversial success. The cutting-edge graphic reproductions of fossils, paintings, and photographs that feature the entire volume made it quickly popular among general readers (Davidson 2008: 137). However, its murky terminology, alongside the allusion to outdated cosmic philosophies, raised doubts. Though Osborn had distanced himself from any metaphysical or anti-scientific interpretation of evolution, he was blamed for his barely contained mysticism (Clark 2008).

Thomas Hunt Morgan became by far one the harshest critics of *The Origin and Evolution of Life*. Paradoxically, in 1903 it was Osborn, at the time head of the Biology Department at Columbia University, who had helped Morgan obtain his assignment at Columbia University, claiming to be ready to reduce his salary in order to hire Morgan (Regal 2002: 73). One year later, Morgan was appointed professor of experimental zoology.

As Morgan set out his research agenda, the friction between the two scientists became far more manifest. Between 1916 and 1918, the confrontation reached its peak, fostered by Morgan intellectual hostility against any form of speculative biology. One year before *The Origin and Evolution of Life* was published, Morgan released the volume *A Critique of the Theory of Evolution* (1916), where he charged the paleontologists who leave the description and attempt explanations with indulging in the "bad habit of wandering" (Morgan 1916: 26). Though Morgan made no specific mention to Osborn in the text, it is no difficult to figure out whom he was referring to. As a matter of fact, in 1916 Morgan and Osborn had started to debate on the epistemic value of experimental observations. Morgan's famous experiments on *Drosophilae*, Osborn stated on August 3, had little to do with natural history since they were conducted under "unnatural conditions" (Regal 2002: 73).

On December 26, 1917, Morgan sent to Osborn a long detailed letter in order to highlight the weak spots of *The Origin and Evolution of Life*. Morgan's point was clear: Osborn was not worthy of addressing the issues he had raised in the book. Paleontology could not investigate the causes of evolutionary processes. Furthermore, Osborn's language was everything but clear and consistent from an epistemological point of view. His claims that nature was intrinsically "creative" hid too many conceptual pitfalls, leaving space for irrationalism. Osborn's very use of the notion of "energy" was nowhere near that employed by modern scientists. In Morgan's eyes, Osborn was one of those old-fashioned evolutionists such as Cope, Haeckel, Driesch, and Bergson who made up new terminologies instead of explaining things. All in all, this was nothing but the indication of the same intellectual condition: the attitude to cover scientific ignorance with neologisms and terms that give "the appearance of profundity," making mysteries out of mechanisms (Morgan 1916: 59). As he wrote to Osborn on December 26, 1917:

> [...] here and elsewhere you intimate very strongly that there is some innate property (chemical, if you like) of living matter that makes it transcend all properties of other matter. Here, then we come to the "heart" of the subject. Even by intimating the possibility of such a conclusion you leave the safe fields of the energy conception and roam abroad in the Elysian fields of mystery. (Reingold 1962: 125)

In *Tempo and Mode in Evolution* (1944), the American paleontologist George Gaylord Simpson brilliantly summed up Morgan's feeling about Osborn when he wrote that geneticists used to consider the paleontologist as the "man who undertakes to study the principles of the internal combustion engine by standing on a street corner and watching the motor cars whiz by" (Simpson 1944a: XV).

7.6 Conclusion: Simpson, Osborn, and the "Premature Synthesis"

Osborn retired from his president position at the American Museum in 1933. He was more than aware that his views of evolution gained little acceptance by modern evolutionists. As he admitted in his autobiography, "The new principles which I have enunciated from fifty-two years of paleontological research [. . .] have, so far as I know, gained no acceptance in the current realm of either biologic or paleontologic thought" (Osborn 1930: 73). Historians R. Rainger (1991) and B. Regal (2002) have thoroughly documented that even Osborn's closest friends and collaborators distanced themselves from many of his theoretical conclusions on evolution.[7] Emblematically, when Edward Poulton asked him to join the meeting at the British Association for the Advancement of Science called "One Hundred Years of Evolution" in 1931, he recommended Osborn not to touch on evolutionary principles or laws and rather to focus on the evidence of evolution (Regal 2002: 185).

A few years later, G.G. Simpson, the paleontologist known as one of the architects of the evolutionary synthesis, openly regarded Osborn's theories as metaphysical by implication (Simpson 1944a: 152), drawing, at least in appearance, a line under most of his theoretical efforts. Simpson famously opposed orthogenesists' progressionism, their exasperation of polyphyletic evolution, as well as the idea that orthogenesis represented the universal rule of macroevolutionary change.[8] Orthogenetic interpretations too often resulted from the biased interpretation of fragmented data fueled by weak analysis (Simpson 1951: 206). Phyletic linearity and polyphyletism were real facts of macroevolution, yet they were just "modes" of its multifarious course. Rectilinear evolution, for instance, was anything but a universal trend and rather characterized those large populations that evolve at moderate rates (Simpson 1944a: 177). Most importantly, rectilinear evolution could be more successfully explained as a product of selection, i.e., orthoselection, thus without calling on superfluous and indemonstrable hypothesis such as "aristogenesis" (Huxley 1942; Simpson 1944a: 163). In its strictly descriptive use, the term orthogenesis designated a phenomenon that modern synthesis could easily account for (Simpson 1949, 1950).

There is certainly something ironic about Simpson's criticism of Osborn's paleontological work. Simpson had indeed worked in the Department of Vertebrate

[7]Osborn's views of human evolution were highly relevant in this regard. Throughout his career, he strenuously opposed the idea of the African descent of modern humans and instead advocated the "Central Asia" hypothesis, according to which early humans had spread from the Asian plateaus. Osborn further criticized any direct comparison between humans and apes (Osborn 1927). Their phylogenetic relation, Osborn claimed, was too remote to justify any apelike interpretation of human physical and mental features. Between the 1920s and the 1930s, a number of paleontologists and anthropologists espoused doubts on Osborn's criticism (Regal 2002). Even W.K. Gregory (1927) distanced himself from his master and critically labeled Osborn's attitude "pithecophobia."

[8]Further analysis of Simpson's empirical and methodological critique of orthogenesis can be found in Delisle (2009).

Paleontology of the AMNH during the last five years of Osborn's presidency. Paradoxically, the empirical material that Simpson used in *Tempo and Mode in Evolution* to counter orthogenesis came from the resources that Osborn had made available (Rainger 1991: 247). Besides, there might even be something slightly misleading in the almost classical narrative of post-synthesis paleontology shelving decades of former research and theoretical efforts. A multilevel analysis of Simpson's composite work easily reveals elements of continuity between the two generations of paleontologists, as well as a certain inhomogeneity between Simpson's and other architects' views. Simpson himself admitted in *Tempo and Mode* that Osborn's orthogenetic theory had to be interpreted with balance, since "the metaphysical nature of his conclusions as to causes" was "quite independent from his purely physical conclusions as to modes" of evolution (Simpson 1944a: 59). Together with W.B. Scott, Osborn had figured how "latent homologies" may lead to parallel branching, a mode of evolution that both Simpson (1961) and Gould (2002) will thoroughly analyze during the twentieth century (Ochoa 2021). Just as important is that Simpson's theoretical assumptions were not completely aligned with those of other architects (Sepkoski 2019: 693–694). Serrelli (2015: 130) pointed out that Simpson was less radical than Dobzhansky in considering adaptive events as central aspects of evolution, these being rather time-limited patterns. The idea that natural selection was "the only really essential factor in evolution" was indefensible; indeed, "in the present synthesis adaptation, preadaptation, and nonadaptation all are involved, and all can be assigned immediate, if not ultimate, causes" (Simpson 1944a: 74, 77). Although Simpson considered macroevolution in light of the processes of population differentiation that operate at the species level, especially in *Tempo and Mode*, he treated higher taxa evolution as a process that entailed different modes and rates of change, which could provide an explanation to the frequent absence of intermediate fossil forms. Sharp breaks in fossil register were more than the result of imperfect fossilization but rather the *in*visible by-product of quantum evolution: small populations evolving rapidly when isolated in different selection landscapes, this causing a relatively rapid shift to a new equilibrium (Schwartz 2021). According to Gould (1980), Simpson's original pluralistic understanding of quantum evolution, "the dominant and most essential process in the origin of taxonomic units" (Simpson 1944a: 206), was dangerous for the selectionist orthodoxy, which presumably motivated its later minimization in Simpson's *The Major Features of Evolution* (1953).[9]

Even more interestingly, Simpson's theoretical efforts seemed to respond to an epistemological necessity that Osborn had already exposed back in the early twentieth century. In the introduction of *Tempo and Mode in Evolution*, Simpson famously declared: "Synthesis has become both more necessary and more difficult as evolutionary studies have become more diffuse and more specialized" (Simpson 1944a: XV). This statement appears significantly consistent with Osborn's claims for methodological eclecticism in evolutionary studies, which led him to piece

[9] A further examination of this issue is provided in this volume by Ochoa (2021).

together a theoretical system that combined mechanisms that evolutionists had long addressed as conflicting.

Simpson's recognition of Osborn's contribution lies between the lines of *Tempo and Mode*. A few words from Osborn's memorial written by Simpson on the *Dictionary of American Biography* (1944) shed deeper light on the scope of Osborn's scientific inquiry. Here Simpson described Osborn as a scientist who, despite being undisciplined and inattentive to factual data, was far "ahead of his contemporaries." His career overlapped the rise of experimental biology and genetics, whose early practitioners were overly enthusiastic about the explanatory power of their methods of analysis. Osborn opposed to epistemological reductionism and "only toward or after the end of his life did others begin to see the desirability of coordination of paleontology, experimental biology and genetics." "Osborn" Simpson highlighted "had energetically and, it must me judged, so prematurely attempted a synthesis" (Simpson 1944b: 586).

In many respects, the way Osborn tried to reach his premature evolutionary synthesis did not comply with the standards of the scientific community he was still part of, and his plural explanation of evolution turned out to be cast aside very soon. In spite of this, Osborn's attempt to integrate approaches in evolutionary biology outlined a twilight zone between old and new biology which, in turn, escapes classic historiographical categories. The analysis of Osborn's attitude toward natural selection throughout his career suggests that labels such as "anti-Darwinism" and "Darwinism" deserve thoughtful reconsideration and even more careful utilization. Should we consider the assignment of somewhat minor causal role to natural selection a sufficient condition to call upon anti-Darwinism? To what extent does the interpretation of natural selection as an outer and constantly screening regulator of the interaction among inorganic and organic environment, heredity, and ontogeny eclipse Darwin's theory of evolution? Despite his reputation of anti-Darwinian evolutionist, since his early neo-Lamarckian works, Osborn highlighted the need to consider natural selection as one of the causes of evolutionary change, emphasizing the constant interplay of factors and types of variations. In a way, natural selection had limits as long as neo-Darwinians brought its scope to the extreme, which implicitly meant to break away from Darwin's own understanding of the process.

Why then did scholars consider Osborn's work utterly anti-Darwinian? Were Osborn's concessions to Darwin and natural selection just a facade, a rhetorical strategy to play both sides? Several answers might be provided. In the first place, it is probable that Osborn's activism in the American debate between evolution and creationism during the 1920s contributed to create his fame of anti-Darwinist. In fact, to counter the tendency to present science and religion as antagonistic, Osborn advanced a theistic interpretation of orthogenesis, publishing about 370 papers and journal articles on the subject in few years (Gregory 1938). His pamphlet *The Earth Speaks to Bryan*, published on the occasion of the famous Monkey Trial in Dayton (1925), discussed the state of the art in evolutionary biology depicting Darwinism as a quite old and imperfect scientific standpoint. Though natural selection, "one of the many other causes of evolution," was the only one which had been "discovered and demonstrated" (Osborn 1925: 19), Osborn attributed the discovery of the mode of origin of species not to Darwin's *Origin of Species* but to Waagen's *Die Formenreihe des Ammonites Subradiatus* (1869): "Waagen's observation that species do not originate by chance or by accident, as Darwin at one time supposed, but through a continuous and well-ordered process, has since been confirmed by an overwhelming volume of testimony" (ivi: 27–28). Evolution consisted of "the incessant creation of new forms and combinations of energy" which led to human cognitive and spiritual abilities. However, "the creation of this man of a higher order, with his moral, spiritual, and intelligent powers" was "utterly incomprehensible as purely a process of the survival of the fittest" (ivi: 55). As long as "chance" was "the very essence of the original Darwinian selection hypothesis of evolution" (Osborn 1917a: 7), pure Darwinians reduced evolution to chance, a standpoint that, to Osborn, gave rise to harmful philosophical interpretations. As a matter of fact, whereas Darwin's works were "truer and more wonderful than ever," the "speculative or rationalistic side" of his philosophy had "largely failed" (ivi: 62). To some extent, therefore, Osborn's anti-Darwinism seemed mostly confined to his opinions about the potential ideological and philosophical implications of Darwinism. In light of this concern, Osborn defended and popularized evolution by implicitly condemning Darwinism for decades, which arguably turned out to be counterproductive for a comprehensive understanding of his complex evolutionary theory.

There is however a second possible explanation for Osborn's reputation of anti-Darwinist, which straight concerns the historiography of evolutionary studies. One of the most striking outcomes of the "mechanism-centered" historiography (Delisle 2017) is the emphasis on single evolutionary mechanisms rather than on theoretical systems. Multilayered research agendas in evolutionary biology, especially those emerged between 1880s and 1930s, have too often been minimized and reduced to single-mechanism accounts of evolution. In doing so, the narratives of Osborn's work have frequently focused exclusively on the theory of aristogenesis (Mayr 1982; Provine 1986: 232), emphasizing its alleged incompatibility with the concepts of random mutations, natural selection, and evolutionary contingency. Not only did this affect the understanding of important contributions to the debate on evolution but also favored an oversimplified view of progress in science. The transition from orthogenetic paleontology to the post-synthetic one appears profoundly nuanced. The idea of a somewhat abrupt break, besides assuming a theoretical homogeneity within the synthesis (Cain 2009; Delisle 2017; Huneman 2019), loses sight of the multilayered nature of the confrontation that took place between the two generations of scientists.

References

Adams MB (2021) Little evolution, big evolution. Rethinking the history of Darwinism, population genetics, and the "synthesis". In: Delisle RG (ed) Natural selection: revisiting its explanatory role in evolutionary biology. Springer, Cham, pp 195–230

Alexander EP (1997) The museum in America. Innovators and pioneers. AltaMira, Walnut Creek
Allen GE (1978) Life science in the twentieth century. Cambridge University Press, Cambridge
Bateson W (1894) Materials for the study of variation. Macmillan, London
Bowler PJ (1979) Theodor Eimer and orthogenesis. Evolution by "definitely directed variation". J Hist Med Allied Sci 34:40–73
Bowler PJ (1983) The eclipse of Darwinism. Anti-Darwinian evolution theories in the decades around 1900. The Johns Hopkins University Press, Baltimore
Cain J (2009) Rethinking the synthesis period in evolutionary studies. J Hist Biol 42:621–648
Ceccarelli D (2018) Between social and biological heredity: Cope and Baldwin on evolution, inheritance, and mind. J Hist Biol 52:161–194
Clark CA (2008) God or gorilla: images of evolution in the jazz age (Medicine, science, and religion in historical context). Johns Hopkins University Press, Baltimore
Cope ED (1896) The primary factors of organic evolution. Open Court, Chicago
Davidson J (2008) A history of paleontology illustration. Indiana University Press, Bloomington
Delisle RG (2009) Les philosophies du Néo-Darwinisme. PUF, Paris
Delisle RG (2017) Introduction: Darwinism or a kaleidoscope of research programs and ideas? In: Delisle RG (ed) The Darwinian tradition in context: research programs in evolutionary biology. Springer, Cham, pp 1–8
Depew DJ (2017) Darwinism in the twentieth century: productive encounters with saltation, acquired characteristics, and development. In: Delisle RG (ed) The Darwinian tradition in context: research programs in evolutionary biology. Springer, Cham, pp 61–88
Esposito M (2017) The organismal synthesis: holistic science and developmental evolution in the English-speaking world, 1915-1954. In: Delisle RG (ed) The Darwinian tradition in context: research programs in evolutionary biology. Springer, Cham, pp 219–241
Fowler DD (2003) A natural history of man: reflections on anthropology, museum, and science. In: Nash ES, Feinman GM (eds) Fieldiana. Anthropology. Curators, collections, and contexts: anthropology at the field museum, 1893-2002. No. 36. Field Museum of Natural History, pp 11–22
Gould SJ (1980) Simpson, paleontology, and the Modern synthesis. In: Mayr E, Provine EB (eds) The evolutionary synthesis; perspectives on the unification of biology. Harvard University Press, Cambridge, pp 153–172
Gould SJ (2002) The structure of evolutionary theory. Belknap Press of Harvard University Press, Cambridge
Gregory WK (1927) Two views of the origin of man. Science 65:601–605
Gregory WK (1934) A half century of trituberculy. The Cope-Osborn theory of dental evolution with a revised summary of molar evolution from fish to man. Proc Am Philos Soc 73:169–317
Gregory WK (1938) Biographical memoir of Henry Fairfield Osborn. 1857-1935. In: National Academy of Sciences Biographical Memoir, XIX. National Academy of Sciences, Washington, pp 53–119
Grehan JR, Ainsworth R (1985) Orthogenesis and evolution. Syst Zool 34:174–192
Huneman P (2019) Special issue editor's introduction: "revisiting the modern synthesis". J Hist Biol 52:509–518
Huxley J (1942) Evolution. The modern synthesis. Allen and Unwin, London
Johannsen W (1911) The genotype conception of heredity. Am Nat 45:129–159
Levit GS, Olsson L (2006) "Evolution on rails": mechanisms and levels of orthogenesis. Ann Hist Philos Biol 11:97–136
Loeb J (1912) The mechanistic conception of life. University of Chicago Press, Chicago
Loeb J, Morgan TH, Osterhout WJV (1919) Editors' announcement. In: Morgan TH (ed) The physical basis of heredity. Lippincott, Philadelphia
Mayr E (1982) The growth of biological thought. Diversity, evolution, and inheritance. Belknap Press of Harvard University, Cambridge
Mill JS (1843) A system of logic, ratiocinative and inductive, I. John Parker, West Strand, London
Morgan TH (1916) A critique of the theory of evolution. Princeton University Press, Princeton

Ochoa C (2021) Inertia, trend, and momentum reconsidered: G. G. Simpson, an orthogeneticist? In: Delisle RG (ed) Natural selection: revisiting its explanatory role in evolutionary biology. Springer, Cham, pp 261–290
Osborn HF (1887) The origin of tritubercolar type of mammalian dentition. Science 10:300
Osborn HF (1888) Evolution of mammalian molars to and from the tritubercolar type. Am Nat 22:1067–1079
Osborn HF (1889) The paleontological evidence for the transmission of acquired characters. Am Nat 23:561–566
Osborn HF (1890) The paleontological evidence for the transmission of acquired characters. Science 15:110–111
Osborn HF (1891) The present problem of heredity. Atl Mon 67:353–364
Osborn HF (1895) The hereditary mechanism and the search for the unknown factors of evolution. Am Nat 29:418–439
Osborn HF (1897a) Organic selection. Science 6:584–587
Osborn HF (1897b) The limits of organic selection. Am Nat 31:944–951
Osborn HF (1900) Correlation between Tertiary mammal horizons of Europe and America. Ann N Y Acad Sci XIII:1–72
Osborn HF (1902a) Homoplasy as a law of latent or potential homology. Am Nat 36:259–271
Osborn HF (1902b) The law of adaptive radiation. Am Nat 36:353–363
Osborn HF (1905a) The ideas and terms of modern philosophical anatomy. J Philos Psychol Sci Methods 2:455–458
Osborn HF (1905b) Tyrannosaurus and others cretaceous carnivorous dinosaurs. Bull Am Museum Nat Hist 21:259–265
Osborn HF (1907a) Evolution of mammalian molar teeth to and from the triangular type. Macmillan, New York
Osborn HF (1907b) Evolution as it appears to the paleontologist. Science 26:744–749
Osborn HF (1908) The four inseparable factors of evolution. Science 27:148–150
Osborn HF (1909) Darwin and paleontology. In: Fifty years of Darwinism. Modern aspects in evolution. Henry Holt, New York, pp 209–250
Osborn HF (1911) "Mutations" of Waagen, "mutations" of De Vries or "rectigradations" of Osborn. Science 33:328
Osborn HF (1912a) Tetraplasy: the law of the four inseparable factors of evolution. J Acad Nat Sci Phila 15:275–309
Osborn HF (1912b) The continuous origin of certain unit characters as observed by a paleontologist—I. Am Nat 46:185–206
Osborn HF (1912c) The continuous origin of certain unit characters as observed by a paleontologist—part II. Am Nat 46:249–278
Osborn HF (1917a) Application of the laws of action, reaction and interaction in life evolution. Proc Natl Acad Sci U S A 3:7–9
Osborn HF (1917b) The origin and evolution of life. Charles Scribner's Sons, New York
Osborn HF (1925) The Earth speaks to Bryan. Charles Scribner's Sons, New York
Osborn HF (1927) Recent discoveries relating to the origin and antiquity of man. Proc Am Philos Soc 66:373–389
Osborn HF (1930) Fifty-two years of research, observation and publication 1877-1929. C. Scribner's Sons, New York
Osborn HF (1931) Cope: master naturalist. Life and letters of Edward Drinker Cope, with a bibliography of his writings classified by subject. Princeton University Press, Princeton
Osborn HF (1932) The nine principles of evolution revealed by paleontology. Am Nat 66:52–60
Osborn HF (1934) Aristogenesis, the creative principle in the origin of species. Am Nat 68:193–235
Pauly P (1987) Controlling life: Jacques Loeb & the engineering ideal in biology. Oxford University Press, Oxford
Popov I (2018) Orthogenesis versus Darwinism. Springer, Cham
Provine WB (1986) Sewall Wright and evolutionary biology. University of Chicago Press, Chicago

Rainger R (1985) Paleontology and philosophy: a critique. J Hist Biol 18:267–287
Rainger R (1988) Vertebrate paleontology as biology: Henry Fairfield Osborn and the American Museum of Natural History. In: Rainger R, Benson KR, Maienschein J (eds) The American development of biology. University of Pennsylvania Press, Philadelphia, pp 219–256
Rainger R (1991) An agenda for antiquity: Henry Fairfield Osborn and vertebrate paleontology at the American Museum of Natural History, 1890-1935. University Alabama Press, Tuscaloosa
Regal B (2002) Henry Fairfield Osborn. Race and the search for the origins of man. Ashgate, Aldershot
Reingold N (1962) Jacques Loeb, the scientist: his papers and his era. Q J Curr Acquis 9:119–130
Resetarits CR (2012) An anthology of nineteen-century American science writing. Anthem, London
Ruse M (1996) Monad to man: the concept of progress in evolutionary biology. Harvard University Press, Cambridge
Schwartz JH (2021) What's natural about natural selection? In: Delisle RG (ed) Natural selection: revisiting its explanatory role in evolutionary biology, Springer, Cham, pp 329–390
Scott WB (1892) The evolution of the premolar teeth in the mammals. Proc Acad Nat Sci Phila 44:405–444
Scott WB (1894) On variations and mutations. Am J Sci 48:355–374
Sepkoski D (2019) The unfinished synthesis?: paleontology and evolutionary biology in the 20th century. J Hist Biol 52:687–703
Serrelli E (2015) Visualizing macroevolution: from adaptive landscapes to compositions of multiple spaces. In: Serrelli E, Gontier N (eds) Macroevolution: explanation, interpretation and evidence. Springer, Cham, pp 113–162
Simpson GG (1944a) Tempo and mode in evolution. Columba University Press, New York
Simpson GG (1944b) Osborn, Henry Fairfield. In: Starr HE (ed) Dictionary of American biography, XXI. Oxford University Press, London, pp 584–587
Simpson GG (1949) The meaning of evolution. Yale University Press, Yale
Simpson GG (1950) L'orthogenèse et la théorie synthétique de l'évolution. In: Paléontologie et transformisme. Albin Michel, Paris, pp 123–168
Simpson GG (1951) Horses. The story of the horse family in the modern world and through sixty million years of history. Oxford University Press, New York
Simpson GG (1961) Principles of animal taxonomy. Columbia University Press, New York
Spiro JP (2009) Defending the master race: conservation, eugenics, and the legacy of Madison Grant. UPNE, Lebanon
Waagen WH (1869) Die formenreihe des ammonites subradiatus: versuch einer paläontologisehe monographie. Munich, Oldenbourg

Part IV
Other Evolutionary Syntheses

Chapter 8
Little Evolution, BIG Evolution: Rethinking the History of Darwinism, Population Genetics, and the "Synthesis"

Mark B. Adams

Science commits suicide when she adopts a creed.
—T. H. Huxley, quoted by Leo Berg (1926)

He bet on the wrong horse.
—Theodosius Dobzhansky (1973)

Abstract This chapter explores the history of the "micro-/macroevolution" distinction and its significance for the history of Darwinism, its "eclipse," and the "evolutionary synthesis." By noting how Darwin, Galton, DeVries, Fisher, Wright, Haldane, Chetverikov, Severtsov, Filipchenko, Dobzhansky, Goldschmidt, J. Huxley, Mayr, Simpson, Guyénot, Rensch, and others understood the distinction, and the language and arguments they used, the article chronicles the problems it posed for "synthesizing" diverse biological disciplines, and documents how the fundamental disagreements among them were successfully finessed by synthesis advocates. These findings force us to rethink the history of Darwinism, population genetics, and the evolutionary synthesis and their political, methodological, and national dimensions.

Keywords Charles Darwin · Darwinism · "Creative Darwinism" · "Eclipse" of Darwinism · Mendelism · Biometrics · Microevolution · Macroevolution · Megaevolution · Evolutionary synthesis · Population genetics · Paleontology · Systematics · Molecular biology · Theodosius Dobzhansky · Iurii Filipchenko ·

This chapter was written 30 years ago, in 1990; only the footnotes, the section on Julian Huxley (on the editor's prompting), and the "Apologia" at the end were written in 2020. Readers are advised to read that "Apologia" first.

M. B. Adams (✉)
Department of History & Sociology of Science, University of Pennsylvania, Philadelphia, PA, USA
e-mail: madams@sas.upenn.edu

R. G. Delisle (ed.), *Natural Selection*, Evolutionary Biology – New Perspectives on Its Development 3, https://doi.org/10.1007/978-3-030-65536-5_8

Julian Huxley · George Gaylord Simpson · Ernst Mayr · Richard Goldschmidt · Bernhard Rensch · A.N. Severtsov · Émile Guyénot · Leo Berg · Stephen Jay Gould

8.1 Did Population Genetics Save Darwinism?

The scientific field of population genetics arose early in the last century, sometime between 1905 and 1955. Historians quibble about exactly when: accounts date its inception variously from the Hardy-Weinberg law of 1908; a seminal note by R. A. Fisher in 1918; the mathematical papers of J. B. S. Haldane, Sewall Wright, and Ronald Fisher in the 1920s; the *Drosophila* field studies by Chetverikov's group, including Timoféeff-Ressovsky, 1925–1927; Theodosius Dobzhansky's classic 1937 book, *Genetics and the Origin of Species*; or the establishment of the distinct disciplinary status, legitimacy, and language of the field in the decade following World War II.[1]

If historians have disagreed about exactly when population genetics began, they have been remarkably unanimous about the historic importance of its achievement: according to well-established consensus, population genetics "saved" Darwinism and became the linchpin of modern evolutionary biology by making possible the so-called evolutionary synthesis, "synthetic theory of evolution," or "modern synthesis." Indeed, the overwhelming majority of works on the history of population genetics treat it almost entirely as part of the history of Darwinism and evolutionary theory. The "eclipse" of Darwinism during the period 1890–1930, including the "warfare" between Darwinism and Mendelism, has been thoroughly discussed in various books and articles by Bowler, Allen, and a host of British scholars. The rebirth of Darwinism in the evolutionary synthesis of the 1930s, 1940s, and 1950s has been explored in detail by Mayr, Provine, and many others. The role of population genetics in that rebirth has been the subject of important biographies of R. A. Fisher, J. B. S. Haldane, and Sewall Wright (Box 1978; Clark 1968; Provine 1971, 1986). The significance of Chetverikov and his students in initiating studies of the genetics of natural populations has also been the subject of important work (Adams 1968a, 1970, 1980a; Babkov 1985; Chetverikov 2002).

Alas, my own work has convinced me that much of this traditional view of the history of population genetics and its role in the evolutionary synthesis is fundamentally mistaken. Many of the details that I will deal with, of course, are well known to historians of the subject who work with original sources, and I think that

[1]In my search for the origin of the term "population genetics," I asked a host of biologists and historians (including Ernst Mayr, N. W. Timoféeff-Ressovsky, G. G. Simpson, Will Provine, E. O. Wilson, and others) when, where, and by whom they thought the term was first used. All told me they did not know exactly, but guessed that it originated in the 1920s, and that Fisher, Wright, Haldane, and Chetverikov were the likeliest sources. Skimming all their works I could locate, however, the only thing I found was the phrase "the genetics of populations," along with other similar references to "the genetics of wheat," "the genetics of chickens," and so forth. There was nowhere the suggestion that it was a separate field, simply the genetic study of one kind of group or another. I was unable to find a single source using the term "population genetics" prior to Dobzhansky's 1937 book.

the factual details in my account will not be especially controversial. But the standard interpretation has all too often ignored these facts and their implications or else interpreted them in a Whiggish, wishful, or willful way. Properly understood, those facts demand a new interpretation of the history of Darwinism, population genetics, and the evolutionary synthesis.

My strategy will be, first, to present the standard historical view in brief. Then, I will set forth my alternate reading of that same history, redefining the central problem of Darwinism and the reasons for its "eclipse." Next, I will dwell on the founders of population genetics and the evolutionary synthesis in order to see how they dealt with that problem. Finally, I will suggest some alternate explanations for the triumph of the evolutionary synthesis.

8.2 Historiography: The Traditional View

The traditional history of population genetics and the evolutionary synthesis begins with Darwin and the difficulty in his theory uncovered by Fleeming Jenkin: the contemporary conception of heredity employed by Darwin, founded on continuous "blending" inheritance, renders impossible the accumulation of favorable, miniscule variations on which natural selection depends. In this view, Darwin's theory of natural selection was sound, but since he did not know genetics, his theory of heredity and variation was unsound; thus, the problem of the nature of heritable variation is often taken to be the central problem for Darwin's theory (Fisher 1930; Dobzhansky 1937, 1951a, b; Huxley 1942; Dunn 1965; Provine 1971; Allen 1975).

This traditional account sees the "eclipse of Darwinism" at the turn of the century as a consequence of this problem. According to an aphorism of the period, the trouble with Darwin's theory was that it explained the *survival* of the fittest, but not the *arrival* of the fittest. Hence, most biologists accepted the fact of evolution but contested its mechanism. New "experimental" theories attempting to explain the origin of favorable variations attracted scientific attention, notably the "mutation theory" of Hugo DeVries (1906). In Britain, this period witnessed the biometrician-Mendelian debate, traditionally interpreted as a disagreement over the kind of variation that could serve as a basis for evolution, with the biometricians emphasizing continuous variation, and the Mendelians discontinuous.

The usual historical account chronicles the resolution of this dispute through the integration of biometry, Mendelism, and Darwinism in the creation of a new field called "population genetics." Its creation is usually attributed to R. A. Fisher, J. B. S. Haldane, Sewall Wright, and Sergei Chetverikov, who demonstrated, beginning in 1918, that Mendelian particulate heredity destroys the objection to classic Darwinism of Fleeming Jenkin. In effect, so the story goes, these works reversed the theoretical situation. Before, biometricians and Mendelians considered Darwinism and Mendelism to be incompatible. Afterwards, Mendelism appeared to be almost a precondition for Darwinism and natural selection. The new perspective of the four founders was presented in their classic works: Chetverikov (1926), Fisher (1930), Wright (1931), and Haldane (1932).

At the same time, inspired by the works of Fisher and others, Sergei Chetverikov—a Russian butterfly specialist—initiated studies of the genetics of natural populations of *Drosophila* in 1925. In the 1930s, studies of the genetics of natural populations (sometimes called "experimental" or "ecological" population genetics) were developed by N. W. Timoféeff-Ressovsky (Berlin), N. P. Dubinin (Moscow), E. B. Ford (Oxford), and Th. Dobzhansky (Pasadena). Perhaps the most influential work in this vein was Dobzhansky's book *Genetics and the Origin of Species* (1937), which presented the new perspective in a general form accessible to biologists who, as a rule, did not read or understand complicated mathematics. Subsequent works by Julian Huxley (1940, 1942), Ernst Mayr (1944), G. G. Simpson (1944), Bernhard Rensch (1947, 1960), Ivan Schmalhausen (1949), G. Ledyard Stebbins (1950), and other biologists extended the "modern synthesis" to systematics, paleontology, botany, and other disciplines. As Mayr has characterized it, this synthesis involved the gradual acceptance of two conclusions: "gradual evolution can be explained in terms of small genetic changes ('mutations') and recombination, and the ordering of this genetic variation by natural selection; and the observed evolutionary phenomena, particularly macroevolutionary processes and speciation, can be explained in a manner that is consistent with the known genetic mechanisms" (Mayr 1980: 1).

This, in brief, is the classic view that has dominated the historiography of population genetics and evolutionary theory during the last century. Understandably, then, historians have occupied themselves with this unfolding logic, biographies of central figures who foresaw the synthesis, and with the explication of their successes. At times, the internal scientific logic has seemed so inexorable that historians have wondered how the synthesis could have taken so long to come about, concluding that personal and social factors must have played some role. For example, in his pioneering study *The Origins of Theoretical Population Genetics*, Provine noted that "the intense antagonisms generated by Bateson's dislike of Pearson and Weldon and vice versa contributed to a delay of more than a decade in the understanding that Mendelism and Darwinism were complementary," and concluded that "If Bateson and Pearson had collaborated instead of fought, population genetics would have gained a significantly earlier start" (Provine 1971: 177). Likewise, in comparative studies of the evolutionary synthesis in various countries, treatments of America and Britain emphasize that their biologists "got it right" because of their scientific insight and the lack of impeding factors, but authors dealing with Germany and France have looked to institutional, disciplinary, and social conditions to explain why their biologists seem to have lagged behind (Mayr and Provine 1980: 279–384).

This traditional history of population genetics is the one I learned and the one I taught for more than two decades. During the same period, however, my own research on Russian and European genetics and evolutionary theory began to raise some unsettling questions. How could it explain, for example, the many biologists in the 1920s and 1930s who understood and accepted genetics, population genetics, and Darwinism but who saw the relationship between them as problematic? Upon inspection, this view began to look suspiciously Whiggish. Considering the many ways history can be used to legitimate current approaches and agendas, I wondered about the origins of this triumphalist history. A preliminary survey revealed that its source may well be the founders of population genetics themselves: Chetverikov (1926), Fisher (1930), Wright (1931), and Haldane (1932) each introduced his classic work with a historical section outlining a variant of the traditional view stated above—namely, that the major problem with Darwin's theory centered on the nature of heritable variation and that this problem had now been put to rest. Did the founders frame their own contributions by defining the central problem of Darwinism to be the one they had already solved?

Upon investigation, it now appears that this traditional history is fundamentally misleading. In particular, I believe, it misstates a set of related issues: the central problem that Darwin addressed, the fundamental reason for the "eclipse" of his natural selection theory, the adequacy of the "evolutionary synthesis" in dealing with this central problem, and, ultimately, the reasons for its triumph. That central problem concerned the relationship between "little" evolution and "big" evolution, that is, between the origin of varieties, the origin of species, and the origin of higher taxa, or, more generally, between what came to be called "microevolution" and "macroevolution."

8.3 Darwin's Central Problem

In order to get from genetics (and population genetics) to evolution, one must see a link between the origin and character of intraspecific varieties and the origin and character of species and higher systematic categories (genus, family, order, etc.); and to do this, it is necessary to consider the variation within species (intraspecific) and the variation between species (interspecific) as qualitatively the same. If intraspecific variation and interspecific variation differ not in kind, but only in degree, then it is possible, by extension, to envision selection as the creator of new species. But if varieties are fundamentally different from species—if the fundamental character of intraspecific and interspecific variation is essentially different—then the effect of selection on a population cannot explain evolution. The species is the only link between the phenomena of the "selectionist" (breeder) and those of the "evolutionist," who traditionally deals with such larger problems as the sudden origin of radically new forms (vertebrates, or terrestrial vertebrates, for example), "progressive" evolution, and the appearance and elaboration in the fossil record of new structural plans and new systems of respiration, circulation, and mentality.

Darwin clearly recognized the importance of this link for his theory of natural selection—indeed, it explains the structure of his book (Darwin 1964). To establish his argument, he felt it necessary to demonstrate that a process analogous to the selection of domesticated plants and animals occurred under natural conditions and could create new species and higher taxa. Thus, Darwin begins the *Origin* with the chapter "Variation under Domestication," where he tries to establish the effects of conscious and unconscious selection on the creation of new varieties, and the probability of the original derivation of these domesticated varieties from natural species. Only after this does he discuss "variation under nature," the subject of his second chapter, where he tries to demonstrate that the traits used in systematic classification are arbitrary and that they are the same sort of traits whose differences are used to distinguish varieties. Only after these two chapters does he discuss the natural mechanism capable of replacing the breeder: that is, the struggle for existence and the resulting natural selection. But this relationship between varieties and species is so central to his argument that Darwin continually returns to it—in Chapters 5 ("Laws of Variation"), 6 ("Difficulties on Theory"), 7 ("Instinct"), 8 ("Hybridism"), 13 ("Mutual Affinities of Organic Beings: Morphology: Embryology: Rudimentary Organs"), and the final 14th chapter ("Recapitulation and Conclusion"). The book's only visual is in the "Natural Selection" chapter, analogous to a branching tree, which embodies Darwin's theory visually. There, he discusses it as a phylogeny; but an attentive reader will surely note that it could equally represent the origin of breeds and varieties within a single species.

Not only did Darwin understand this to be the central problem of his theory: so too did many of his critics. Indeed, although we have chosen to remember Fleeming Jenkin for his suggestion that blending inheritance would "swamp" new variations, his first critique centered on the implausibility of the view that varieties and species were comparable in their nature and origin and, in particular, that variations within a species could move outside of the range of variation delimiting that species (Jenkin 1867).

We must remember that Darwin advanced two distinct theories: evolution (or, as he called it, "descent with modification") and its mechanism ("natural selection").[2] Ironically, thanks largely to Darwin, by the time of the "eclipse" or "death" of Darwinism, the vast majority of biologists had come to accept his first theory (evolution), but not his second—not the validity of the theory of natural selection. Many of those who accepted the fact of evolution considered its cause or mechanism to be simply unknown. Others preferred orthogenetic theories, according to which the cause of evolution was entirely internal to the organism, because of the apparent directionality in long-term evolutionary trends. For still others, Lamarckian theories seemed more probable because of the perceived universal precision of adaptations in nature. For these scientists, there was little doubt that Darwin's theory could explain the success of the breeder in creating varieties; alas, the problem was rather the origin of species and higher taxa.[3]

[2]What Darwin scholars know, but others seem not to, is that the word "evolution" is absent from the first edition of the *Origin*. (For the reason, see Richards, *The Meaning of Evolution*, 1992.) In addition, it expressed support for the inheritance of acquired characteristics and relied on it much more heavily in later editions.

[3]A popular anti-Darwinian book embracing Lamarckism during the "eclipse" was Peter Kropotkin's *Mutual Aid* (1902), which argued against intraspecific competition in animals, noting that in Siberia, herding and flocks helped protect species from both predation and environmental threats. However, during the "eclipse," undoubtedly the most influential and widely read anti-Darwinian book on evolution was French philosopher Henri Bergson's *L'Évolution créatrice* (1907), translated into English in 1911 as *Creative Evolution*. In place of natural selection, he

Nor did the so-called rediscovery of Mendel's laws change this situation. Even if there was considerable Mendelian variability within a species, what did that mean? It was still possible to conceive of the traits defining the species itself (and thus distinguishing it from other species) as the product of some Lamarckian mechanism, or of some internal "principle," or of some other factor or process as yet unknown—and this held as well for genera, families, and higher taxa. The relevance of Mendelism to the theory of natural selection depended on affirmative answers to two questions: (1) Are all the differences between species reducible to Mendelian genes? (2) Can the natural selection of Mendelian genes create new species and higher systematic categories?

To both of these questions, Hugo DeVries answered in the negative. Like Darwin, he regarded the relationship between varieties and species as the central problem and devoted one of his most influential books (*Species and Varieties: Their Origin by Mutation*) to the subject (DeVries 1906). For him, only "degressive" and "retrogressive" mutations could "Mendelize," but the mutations that established new species—the so-called "progressive" mutations—could not. DeVries's mutation theory was subsequently refuted by Lutz, Renner, and other researchers, but their work did not clarify the larger issue: as it turned out, these "progressive" mutations were not only not Mendelian, they were also not "progressive" in DeVries's sense—that is, they did not produce real, new species.

Similarly, although we have chosen to remember the work of Wilhelm Johannsen on "pure lines" for coining the term "gene," and his distinction between the so-called genotype and phenotype, we should remember that, for his contemporaries, it was most important as proof of the limited, purely intraspecific power of selection. Acting on a population, selection could alter the population to the limits of the genes it contained, but could not go beyond the species or populations limits: initially, selecting for the largest beans would increase their size to a point, but beyond that, further selection had no effect (Johannsen 1903, 1969). Even Francis Galton, the founder of both biometrics and eugenics, would have answered both of the key questions discussed above in the negative: Although his work on humans and peas amply demonstrated the impressive possibilities of selection within a species, he also demonstrated its limits to a species range of variability. Both discovering and emphasizing the so-called reversion to the mean (or, as he phrased it in 1886, "regression toward mediocrity"), he did not regard his own work as a clarification of the problem of the origin of species: to the contrary, he regarded their

explained evolution by embracing both orthogenesis and Lamarckism, attributing its driving force to the *élan vital*, a "vital impetus" analogous to humanity's "natural creative impulse." The book was widely translated, read by millions, and one of the most popular books in the world in the early twentieth century. Both Dobzhansky and Mayr told me that they were much taken with it in their youth, and it inspired their choice of profession. Probably most (if not all) those who cultivated or opposed the "synthesis" of the same age cohort had read it. Of special interest for this paper was his notion of the "emergent properties" that appear in higher and higher forms of life (such as consciousness, intelligence, and language) that then govern and accelerate subsequent evolutionary progress.

origin as a result of some fundamentally different (as yet unknown) process (Galton 1930). Similarly, the biometrician-Mendelian debate actually centered not on the nature of heritable variation so much as on the question of whether intraspecific selection could produce a new species: the biometricians said yes, the Mendelians no.

One of the principal reasons for the eclipse of Darwinism, then, was the widespread conviction that Darwin's argument, based as it was on the role of selection in the origin of varieties, could not account for the origin of species or higher taxa. Many biologists simply did not believe that what delineated species and higher taxa was reducible to Mendelian genes, much less that the natural selection of such genes could create new species and higher taxa.

Darwin's book title had emphasized the origin of *species*, and he apparently believed that if natural selection could explain that, it could also explain the origin of genera, families, orders, and other higher taxa. As we shall see, from Darwin's day to the present, the "species" (and the now 26 different ways of defining them) has remained contested territory, and biologists have debated over exactly where the dividing line between "microevolution" and "macroevolution" should be drawn. Consistent throughout, however, has been the core problem of Darwin's "natural selection" theory, namely, its questionable relevance to what most scientists and the lay public have meant by "evolution."

8.4 The Perspective of the "Founders"

By 1930, however, four crucial biologists did seem to accept Darwin's views: Fisher, Wright, Haldane, and Chetverikov, the founders of population genetics. But how did these founders handle the central problem? How did they demonstrate that the differences between species were reducible to Mendelian genes? How did they demonstrate that the selection of Mendelian genes (which could create varieties) could also create species and higher taxa? How did they "save" Darwinism?

In his classic book, *The Genetical Theory of Natural Selection* (1930), Fisher did not treat this problem directly. Rather, his first chapter begins with a historical introduction, which is one of the earliest expositions of the "traditional view" of the history of population genetics that I have discussed above. Then Fisher uses that history to frame the following logic: there are a number of proposed theories of evolution, including the theory of natural selection; its only difficulty is the concept of blending heredity; the development of genetics invalidates the concept of blending heredity; all the other theories require blending heredity; thus, we must reject the other theories and accept the theory of natural selection. Let us quote from his book:

> The whole group of theories which ascribe to hypothetical physiological mechanisms, controlling the occurrence of mutations, a power of directing the course of evolution, must be set aside, once the blending theory of inheritance is abandoned. The sole surviving theory is that of Natural Selection, and it would appear impossible to avoid the conclusion that if any evolutionary phenomenon appears to be inexplicable on this theory, it must be accepted

> at present merely as one of the facts which in the present state of knowledge does seem inexplicable. (Fisher 1930: 21)

Note what he says here: either natural selection or we just don't yet know. Later, in his sixth chapter, there is a section entitled "Fission of Species," but there he simply assumes, without argument, that the morphological differences between species are the result of the accumulation of Mendelian mutations.

So much for Fisher—what of Sewall Wright? In his classic work "Evolution in Mendelian Populations" (1931), Wright addresses the question only in his introduction (which is also historical), concluding as follows: "As to gene mutation, observation of those which have occurred naturally . . . reveals characteristics which seem as far as possible from those required for a directly adaptive evolutionary process. The conclusion nevertheless seems warranted by the present status of genetics that any theory of evolution must be based on the properties of Mendelian factors, and beyond this, must be concerned largely with the statistical situation in the species" (Wright 1931: 100–101). In his famous paper presented in 1932 at the Sixth International Congress of Genetics at Cornell, Wright notes that "It is only at the subfamily and family levels that clear-cut adaptive differences become the rule" and that "The principal evolutionary mechanism in the origin of species must thus be an essentially nonadaptive one" (Wright 1932). However, he cites this evidence not to show that varieties and species are essentially different (which was one plausible contemporary reading of that data), nor even to address the central question of their relationship, but simply to show the compatibility of such findings with his concept of "genetic drift."

J. B. S. Haldane raised the problem that Wright and Fisher evaded. In his book *The Causes of Evolution* (1932), which also begins with a historical introduction, Haldane defined evolution as "the descent from living beings in the past of other widely different living beings." Haldane acknowledges that selection within a population is not evolution proper: "Many would refuse to dignify the changes which man has effected in the dog as evolution," he notes (Haldane 1932: 4). Furthermore, he describes the purpose of his book as providing answers to a series of questions, among them "What is the nature of heritable differences within a species?" and "Are the differences between species of the same or of a different character?"

So far, so good—but what are his responses to these questions? After having described the character of variation within the species, Haldane reproduces a table that lists the color genes common to the mouse, the Norway rat, the black rat, the deer mouse, the cavy, the rabbit, the dog, the cat, and the ferret, noting that the Russian geneticist Nikolai Vavilov had obtained "similar results . . . in cereals and other plants," as reported in *Nomogenesis* by the Russian ichthyologist, evolutionary theorist, and biogeographer Leo Berg (Haldane 1932: 65–66; Berg 1926). Then he makes the following observation: "Even the ferret and polecat, which have, perhaps erroneously, been placed in different genera, only differ as regards colour by a single gene. Of course the species and subspecies considered must differ by many other genes." His conclusion: "To sum up, interspecific differences are of the same nature as intervarietal" (Haldane 1932: 71, 82).

But hold a minute: the fact that certain species have certain genes in common, and differ in other genes, does not prove that the assortment of a species's genes constitutes the essence of its specificity. In fact, Leo Berg used Vavilov's data to demonstrate just the opposite! Berg considered the fact that very different species have Mendelian genes in common as evidence that species specificity was *not* determined by Mendelian genes, and there is contemporary evidence that Vavilov agreed (Berg 1926; Haldane 1932: 66, 79; Adams 1978b).[4] Thus, Haldane recognized the problem but did not resolve it. To be sure, if one accepted the proposition that varieties and species were qualitatively the same and that the only difference between species laid in their complement of Mendelian genes, all is in order; but if one didn't accept it, Haldane gives no convincing reasons to change one's mind.

The only one of the four founders of population genetics who directly confronted this central problem was Sergei Chetverikov. It was not difficult for him to do so: he had worked for 20 years as one of Russia's leading butterfly taxonomists. An entomologist and systematist, he actually first taught genetics in a course at Moscow University entitled "Theoretical Systematics." In Chetverikov's classic work of 1926, "On Certain Aspects of Evolution From the Viewpoint of Modern Genetics," the entire first section addressed the central problem. There, he argued that the traits determined by Mendelian genes in *Drosophila* (e.g., venation and wing structure) are the very same traits which are "fundamental in modern systematics for distinguishing higher systematic categories" (Chetverikov 1926: 9–10). Elsewhere in this same section, he mobilized his considerable knowledge of the systematics and classification of Diptera and Lepidoptera to make the same point.

Chetverikov's singularity derived not only from his specialty but also from his continental and specifically Russian setting. In this context, for both geneticists and evolutionists alike, it was far from self-evident that "microevolution" and "macroevolution" were the same sort of thing.

8.5 Macroevolution and Microevolution

This fact is well illustrated by the approach to the problem of the relation between genetics and Darwinism of two of Chetverikov's most brilliant contemporaries: Russia's most important evolutionist, Aleksei Severtsov, and its most important geneticist, Iurii Filipchenko.

---

[4] When I was visiting Stephen Jay Gould one evening at his Cambridge home, he told me that he was much taken by this realization after reading my article on Vavilov, and used it as the basis (unattributed) for one of his popular science columns. (His personal thanks was the best apology for failing to mention my article as his source!) As for Haldane, he referred to Berg's work as "in my judgment the best anti-Darwinian book of this century" (p. 12).

Aleksei N. Severtsov (1866–1936) was the most distinguished Russian morphologist of his generation. Throughout his life, he regarded himself as a "pure" Darwinist. He embraced natural selection as sufficient, in principle, to explain all evolutionary phenomena. He rejected both Lamarckism and "autogenesis," a term he coined to refer to all theories based on some hypothetical "internal principle." In addition, during the 1920s and 1930s, he accepted the results of genetics and knew of the work of Chetverikov and his students. But in his final book, written 1935–1936 and published only after his death, he lamented that "Despite the brilliant achievements of hereditary theory, the results of genetic research have contributed almost nothing to the resolution of evolutionary questions" (Severtsov 1939: 80). Of course, for Severtsov (and most of his contemporaries),[5] "evolutionary questions" referred to the fundamental patterns in the history of life on earth: the appearance of new types, "retrogression," "progressive evolution," the origin of vertebrates, the conquest of land, the development of new and more complex forms of respiration, circulation, mentality, and, ultimately, of the psyche (Adams 1980b).

Although he was a geneticist rather than a morphologist, Iurii A. Filipchenko (1882–1930) was of exactly the same opinion. The leading advocate of the new "experimental zoology" in the Russian capital, and a biologist of broad profile and great influence, Filipchenko read many languages and had an encyclopedic command of world scientific literature, a knowledge he shared with his countrymen in many popular and technical books on diverse subjects—including biographies of Mendel and Galton; a history of evolutionary theories; textbooks on plant, animal, and human genetics; and books and pamphlets on wheat genetics, cattle breeding, and eugenics (Adams 1990c).

Like Severtsov, Filipchenko accepted evolution; but, unlike Severtsov, Filipchenko did not accept the theory of natural selection and did not believe that the specificity of categories above the species level was determined by Mendelian genes. In 1923 he argued that intraspecific variation was different in kind from traits characterizing genera and higher taxa, which exhibit "less variation" and "appear significantly earlier during individual development." Unlike "traits characterizing the species," for which "it is known exactly that their carriers are genes localized in the chromosomes of the sex cells ... for the traits of a generic character, not only has no one proved this, but an entirely different proposition is considerably more likely, namely that they are present in entirely special carriers located, not in the nucleus, but in the plasm of the sex cells" (Filipchenko 1923: 213).

In 1927, in a German publication that he dedicated to the 70th birthday of Wilhelm Johannsen, Filipchenko codified this difference in language, distinguishing between two kinds of evolution: "microevolution" (*mikroevoliutsiia*—the evolution of biotypes, jordanons, and Linneons) and "macroevolution" (*makroevoliutsiia*—the

---

[5]Indeed, as Nikolai Krementsov noted, almost a decade earlier, Russia's foremost ornithologist M. A. Menzbir published the same sentiment in almost exactly the same words in two introductory articles in his Russian edition of *The Complete Collected Works of Charles Darwin* (Menzbir 1928, 1929).

evolution of higher systematic groups) (Filipchenko [Philiptschenko] 1927: 93–95). Elaborating this important distinction 2 years later, he declared:

> Thus, we consider that modern genetics, having established the nature of mutations and combinations, will undoubtedly throw light on the evolution of biotypes, jordanons, and Linneons. However, in addition to this, let us say, microevolution, there is the evolution of more major systematic groups, that is, macroevolution, and it undoubtedly lies beyond the purview of genetics, although it is the most interesting for evolutionary theory.
>
> Given this state of affairs we must acknowledge that the resolution of the question of the factors of evolution of higher systematic groups, or our macroevolution, must proceed without reference to the facts of genetics. However nice it would be to be able to lean on them, it seems to us that they are useless for this purpose, because the question of the origin of higher systematic groups lies entirely outside of the field of investigation where genetics works. . . .
>
> As a consequence, it is obvious that current and future evolutionary theory will undoubtedly resolve questions of the "origin of species" (and all lower subdivisions of species) differently from questions of the "origin of genera" (and all higher systematic units). As regards the former, it is impossible not to reckon with the conclusions of genetics, and given the current state of things there can hardly be any basis for disagreement. As regards the latter, to the contrary, at the present time we know very little that is precise, and this opens the field for the broadest speculation. (Filipchenko 1929: 260–261)

The authority of Filipchenko's words derived from his unique status: not only was he Russia's leading geneticist—the author of its most authoritative Russian textbooks, and an expert on the genetics of cattle, soft and hard wheats, and humans, its leading biometrician and its leading "Mendelist" and "Morganist"—he was also one of its leading authorities on evolutionary theory (Adams 1980a: 249).

Filipchenko's views strongly influenced those of his most prominent protégé, Theodosius Dobzhansky. Dobzhansky worked closely with Filipchenko beginning in 1924 and came to idolize him. In 1927, thanks to Filipchenko's nomination, Dobzhansky travelled to the United States to work in the laboratory of T. H. Morgan and, after the beginnings of Stalinism in the Soviet Union, he stayed for the rest of his life. The perspective of his teacher, Filipchenko, is evident in Dobzhansky's evolutionary classic, *Genetics and the Origin of Species* (1937). In it, he co-opts his mentor's neologisms to argue against Filipchenko's reason for coining them.

How does Dobzhansky's classic handle the central problem? In its introduction, Dobzhansky writes:

> Some writers have contended that evolution involves more than species formation, that macro- and micro-evolutionary changes may be distinguished. This may or may not be true; such a duality of the evolutionary process is by no means established. In any case, a geneticist has no choice but to confine himself to the micro-evolutionary phenomena that lie within reach of his method, and to see how much of evolution in general can be adequately understood on this basis. (Dobzhansky 1937: xi)

In the first chapter, he returns to the same subject:

> Experience seems to show, however, that there is no way toward an understanding of the mechanisms of macro-evolutionary changes, which require time on a geological scale, other than through a full comprehension of the micro-evolutionary processes observable within the span of a human lifetime and often controlled by man's will. For this reason we are compelled at the present level of knowledge reluctantly to put a sign of equality between the

mechanisms of macro- and micro-evolution, and, proceeding on this assumption, to push our investigations as far ahead as this working hypothesis will permit. (Dobzhansky 1937: 12)

And yet again, later in the book, he emphasizes that "No major evolutionary change is noticeable in most species of organisms within a human lifetime, hence the supposition that species have become what they are now through evolution by natural selection can be at best no more than a very probable inference" (Dobzhansky 1937: 151). Perhaps this is why Dobzhansky's own comparison of *Drosophila* species emphasized not their different or similar composition of Mendelian genes, but rather the differences in the number and configuration of their chromosomes.[6] In any case, similar language persists in the subsequent editions (Dobzhansky 1941, 1951a).

I was interested in the fact that, although Dobzhansky revered Filipchenko, and indeed got the words "microevolution" and "macroevolution" from him, he took an opposite stance in his book. During a weeklong interview of Dobzhansky at his camp in Yosemite shortly before he died, I raised the matter with "Doby" (a name he insisted on) and asked about his fundamental difference from Filipchenko over the "micro/macro" question. His response was surprising: he shrugged and said, almost indifferently, "He bet on the wrong horse."

Despite Dobzhansky's cautious language, however, his book undoubtedly played an important role in establishing "population genetics" as the core evolutionary discipline. One reason for its importance may be the way he used it to transform evolutionary discourse. In his first chapter, he declares categorically: "Since evolution is a change in the genetic composition of populations, the mechanisms of evolution constitute problems of population genetics" (Dobzhansky 1937: 11). Thus, in one great rhetorical finesse, he defined "evolution" as microevolutionary change, specifically a change in the genetic composition of populations; and then he made a daring claim of territoriality for a new field he called "population genetics" by declaring it to have primacy in dealing with the mechanism of evolution.[7] On this basis, the book subsequently asserted that the problem of the origin of species is resolved. In retrospect, it proved to be a brilliant and consequential move. Consider: When it is uncertain whether population genetics can illuminate what most people mean by "evolution," and evolution itself is under fundamentalist attack for being just a "theory," what to do? Dobzhansky's answer: kill two birds with one stone—*redefine* "evolution," and do so in a way that makes it a *proven fact*!

But certain European biologists were far from convinced. One in particular was Richard Goldschmidt (1878–1958), a very influential German biologist who was the director of the genetics department of the Kaiser Wilhelm Institute of Biology from

[6]For information about Dobzhansky's increasingly cautious approach to the micro/macro issue, see Chap. 11.

[7]I was unable to find a single primary source using the term "population genetics" (as opposed to "the genetics of populations") prior to Dobzhansky's 1937 book, although he himself may well have used it in his own prior studies in the 1930s. Note that "the genetics of populations" is not the name of a field or discipline, whereas "population genetics" is.

1913 to 1935. A Jew, Goldschmidt emigrated to the United States in 1936 and became a professor at the University of California at Berkeley. Like many German zoologists, he regarded macroevolution as the central problem needing explanation. Confronted shortly after his arrival with Dobzhansky's book and the developing Anglo-American "synthesis," in 1940 Goldschmidt published a refutation entitled *The Material Basis of Evolution*. It begins with a scornful attack on the pretensions of this new "Darwinism":

> ... I may challenge the adherents of the strictly Darwinian view, which we are discussing here, to try to explain the evolution of the following features by accumulation and selection of small mutants: hair in mammals, feathers in birds, segmentation of arthropods and vertebrates, the transformation of the gill arches in phylogeny including the aortic arches, muscles, nerves, etc.; further, teeth, shells of mollusks, ectoskeletons, compound eyes, blood circulation, alternation of generations, statocysts ... poison apparatus of snakes, whalebone... Corresponding examples from plants could be given. (Goldschmidt 1940: 6–7)

The body of the book has two large divisions entitled "Microevolution" (pp. 8–183) and "Macroevolution" (pp. 184–395), terms he derived, he states, from Dobzhansky's 1937 book. The first part of his book concludes with the following words (printed in italics): "*Subspecies are actually, therefore, neither incipient species nor models for the origin of species. They are more or less diversified blind alleys within the species. The decisive step in evolution, the first step toward macroevolution, the step from one species to another, requires another evolutionary method than that of sheer accumulation of micromutations*" (Goldschmidt 1940: 183). At the time, Goldschmidt's book made him something of a pariah among those who were advancing the new "synthesis" (notably, Dobzhansky). Ironically, the Goldschmidt passages cited above could well have been penned by Dobzhansky's mentor, Filipchenko.

A second striking example of an unconvinced European biologist was the prominent French evolutionist Émile Guyénot. In his 1944 book, *L'Origine des Espèces*, and repeated in its updated 1964 American edition, Guyénot writes that "We know nothing positive about the genesis of groups larger than species." He continues: "As for the production of new types of organization, such as those characteristic of sub-kingdoms and the majority of classes, neither genetics nor embryology allows us a suspicion of the mechanism" (Guyénot 1964: 123). This leads him to distinguish between the "superficial evolution" treated by population genetics and the "evolution in depth" of the basic structural forms and plans of life, about which we know nothing. The book's final sentence declares, categorically, that "the great stages of evolution escape us entirely" (Guyénot 1964: 139). Thus, like Goldschmidt, Guyénot found population genetics and natural selection wholly inadequate to account for macroevolution, or, as he renamed it, "evolution in depth."

These examples suffice to demonstrate two interesting conclusions. First, for many biologists, population genetics did not prove the validity of Darwinism as a general theory of macroevolution. Second, whether as the study of intraspecific variation and selection or as the chronicling of genetic similarities and differences between species and/or higher taxa, population genetics was widely regarded, even by its strongest advocates, as being incapable in principle of dealing with the macroevolutionary problem.

8.6 Julian Huxley's "Modern Synthesis"

No treatment of this subject would be complete without including the works of the grandson of "Darwin's bulldog," T. H. Huxley, who gave the evolutionary "synthesis" its name. Historians are quite familiar with Julian S. Huxley's seminal contributions, notably *The New Systematics* (1940), which he edited, and *Evolution: The Modern Synthesis* (1942), which he authored. But I believe his story should begin more than a decade earlier (Adams 2000, 2004).

It was triggered by Huxley's younger friend and fellow Etonian, J. B. S. Haldane—not by his *The Causes of Evolution* (1932), but by his very well-known and controversial *Daedalus* (1924), in which he tweaks H. G. Wells for being out of touch. Admitting that "the very mention of the future suggests him," Haldane went on to declare him "a generation behind the time": "When his scientific ideas were formed, flying and radiotelegraphy, for example, were scientific problems, and the centre of scientific interest still lay in physics and chemistry. Now these are commercial problems, and I believe the center of scientific interest lies in biology" (Haldane 1924: 9–10).

Almost immediately thereafter, Wells set about rectifying this failing by drafting his son ("Gip," then a young zoologist at University College, London), and Julian Huxley (the grandson of his own teacher in the 1880s), to produce an encyclopedic popular book to be titled *The Science of Life*. Impressed by Wells's 1921 *Outline of History*, Huxley resigned from his academic job to undertake the project and devoted a whole chapter of his memoirs recalling the experience. There, he confirmed Haldane's criticism: Wells "had forgotten much of his biology, and what he remembered was by now old-fashioned—pre-Mendelian, with little study of animal behaviour or ecology. Thus the bulk of the scientific work would fall on my shoulders" (Huxley 1970: 155–6). As the multiple letters of the time quoted by Huxley demonstrate, almost all of the writing was done by Huxley, Gip, and two secretaries they hired, with Wells serving only as an active editor and handling the commercial side (Adams 2004).

After some 3 years of frenzied work, *The Science of Life* first appeared in three parts, each a long volume. One later edition of the work, published as a complete, single volume, was a very heavy tome comprised of 1510 pages of extremely tiny type (Wells et al. 1934)! In his memoirs, Huxley admitted that the experience had been frustrating and frenetic, but also admitted that it taught him a lot of biology.[8] Within its pages are views that would later fill Huxley's essays, popular writings, and

[8]Will Provine once told me he regarded the work to be the "earliest complete presentation" of the evolutionary synthesis.

subsequent books, including his ideas on religion and "evolutionary humanism." Nowhere do the words "microevolution" or "macroevolution" appear on its pages (although the massive volume covers both, in detail, and much more).

More than a decade later, in Huxley's *Evolution: The Modern Synthesis* (1942), there is only one reference to those two terms in a page-long, not especially negative discussion of Goldschmidt's 1940 book. Nor is there any other reference in the second edition (Huxley 1963, 1964). Nor does Huxley ever explicitly take a position on the "micro/macro" dispute, or discuss it.

Yet his book does reveal where he stands. On page 389 of both editions (1942, 1963/4), in a chapter on "Speciation, Evolution, and Taxonomy," appears the following text[9]:

> The formation of many geographically isolated and most genetically isolated species is thus without any bearing upon the main processes of evolution. These latter, as we shall see in later chapters, consist in the development of new types endowed with mechanisms of higher all-round biological efficiency; in the adaptive radiation of these types to take advantage of all available types of environment and modes of life; in the colonization of new regions of the globe's surface; in the tapping of new resources for exploitation; and in a more rapid turnover of the resources tapped.
>
> These major processes in evolution thus consist essentially in . . . a progressive increase of life's control over and independence of the environment. Superimposed upon these processes, and having little or no bearing upon them, are the processes of species-formation we have just described which are the consequences of accidents in the environment or in the genetic machinery of life. Much of the minor systematic diversity to be observed in nature is irrelevant to the main course of evolution, a mere frill of variety superimposed upon its broad pattern. We may thus say that, while it is inevitable that life should be divided up into species, and that the broad processes of evolution should operate with species as units of organization, the number thus necessitated is far less than the number which actually exist. Species-formation constitutes one aspect of evolution; but a large fraction of it is in a sense an accident, a biological luxury, without bearing upon the major and continuing trends of the evolutionary process.

In other words, microevolution indeed can explain the origin of species, but that explanation has no bearing on macroevolution—evolution writ large, which follows different patterns and processes, and is progressive. Careful readers may note that these words closely resembled Goldschmidt's.

This sustained position raises an interesting question: If, in Huxley's views, microevolution (intraspecific variation and species formation) is different from and has no bearing on "evolution writ large," then what exactly did he mean by "the modern synthesis"? To answer this question, we must look to the volume's preface. Here he mentions that the book derived from an address he gave in 1936 titled *Natural Selection and Evolutionary Progress*, and explains:

> Even among professional zoologists the modern conception of natural selection and its mode of operation is quite different from that in Darwin's day, but much of the research on which the changed outlook is based is so recent that the new ideas have not spread far. The idea of evolutionary progress, on the other hand, has been undeservedly neglected. Thus it seemed

[9]My thanks to Richard Delisle for directing me to this page.

> to me valuable to attempt to give a broad account of the two concepts and their relation to each other. . . . The result is the present volume. (Huxley 1942: 7)

Later in the preface, he highlights his debt to Haldane's *The Causes of Evolution* (1932); praises the books by Dobzhansky (1937) and Goldschmidt (1940), claiming that, although his own book was already in proof when they appeared, "I have tried to take advantage of them where possible"; and notes his obvious debt to Fisher. He then concludes:

> The time is ripe for a rapid advance in our understanding of evolution. Genetics, developmental physiology, ecology, systematics, paleontology, cytology, mathematical analysis, have all provided new facts or new tools of research; the need to-day is for concerted attack and synthesis. If this book contributes to such a synthetic point of view, I shall be well content. (Huxley 1942: 8)

Note that the "synthetic viewpoint" here is *aspirational*, not declarative, and privileges no single discipline over any other. His "synthesis" is not a doctrine but an ongoing process.

At the very opening of Chapter 1, "The Theory of Natural Selection," Huxley sets forth similar ideas, detailing the aspiration he is celebrating:

> Evolution may lay claim to be considered the most central and the most important of the problems of biology. . . .
>
> Biology at the present time is embarking upon a phase of synthesis after a period in which new disciplines were taken up in turn and worked out in comparative isolation. Nowhere is this movement towards unification more likely to be valuable than in this many-sided topic of evolution; and already we are seeing the first-fruits in the re-animation of Darwinism.
>
> By Darwinism I imply that blend of induction and deduction which Darwin was the first to apply to the study of evolution. (Huxley 1942: 13)

After discussing historical disputes over the nature of variation and the so-called eclipse of Darwinism, he returns to his theme:

> Biology in the last twenty years, after a period in which new disciplines were taken up in turn and worked out in comparative isolation, has become a more unified science. It has embarked upon a period of synthesis, until to-day it no longer presents the spectacle of a number of semi-independent and largely contradictory sub-sciences, but is coming to rival the unity of older sciences like physics, in which advance in any one branch leads almost at once to advance in all other fields, and theory and experiment march hand-in-hand. As one chief result, there has been a rebirth of Darwinism. . . .
>
> The Darwinism thus reborn is a modified Darwinism, since it must operate with facts unknown to Darwin; but it is still Darwinism in the sense that it aims at giving a naturalistic interpretation of evolution, and that its upholders, while constantly striving for more facts and more experimental results, do not, like some cautious spirits, reject the method of deduction. . .
>
> It is with this reborn Darwinism, this mutated phoenix risen from the ashes of the pyre kindled by men so unlike as Bateson and Bergson, that I propose to deal in succeeding chapters of this book. (Huxley 1942: 26–28)

What, then, is the synthesis that he is aspiring to? After a period when isolated subfields of biology, whose specialists were largely ignorant of one another's work, had caused an "eclipse" of Darwinism, he is seeing—and encouraging—that they all share their knowledge and contribute what they can to understanding evolution, and noting, indeed celebrating, that this was causing the rebirth of Darwinism. T. H. Huxley had famously served as "Darwin's bulldog"; his grandson now served as Darwinism's.

Dobzhansky had claimed "population genetics" to be the central evolutionary discipline; Severtsov had claimed that status for "morphology"; Mayr had claimed it for "systematics"; Simpson (as we shall see) claimed it for "paleontology." Unlike them, Huxley favored no single "evolutionary" science, but welcomed the diverse contributions of all who could contribute. Two decades later, he would preface the second edition of *Evolution: The Modern Synthesis* (1963, 1964) with a 39-page, unpaginated introduction that began: "In the twenty years since this book was first published, there has been an enormous volume of new work and new ideas on the subject of evolution. . . . The main fact to note is that the neo-Darwinian, synthetic, or integrative theory of evolution that I maintained in 1942 has gained many new adherents and may now be regarded as the established view." He then quickly surveyed and updated the new ideas and approaches to evolution coming from all fields (from geology to DNA). In that survey of 20 years of new work and ideas, he maintained the approach that he had taken throughout: that every field that could, should contribute equally; that all evolutionists should be in active communication and interaction; and that Darwinism in his sense had indeed been reborn.

Assemble all the relevant new information and tools from all fields, let each inform the others, celebrate the rebirth of Darwinism, and extend it to humans, society, and the future: that was the mission and approach he had undertaken for H. G. Wells in *The Science of Life*, and that continued to unfold throughout his popular essays and books, as Darwin's legatee, and as the first director of UNESCO. He might agree or disagree with certain theoretical assertions, but routinely sought to praise all contributors, functioning as a "welcoming evolutionary moderator." But Huxley saw no coupling between microevolution and macroevolution and sustained his broader vision of "evolutionary humanism" and evolutionary "progress" throughout. Given his communal role, his beliefs, and his goals, it may well have seemed to him counterproductive to get overtly involved in the "microevolution/macroevolution" spat, or even to use those words. Better to avoid that "issue" entirely.

8.7 Paleontology and "Macroevolution"

When, then, did the problem of macroevolution get resolved by the synthetic theory? The work usually credited with this is *Tempo and Mode in Evolution* (1944) by the American paleontologist George Gaylord Simpson. In light of this, we might find surprising what Simpson's book actually said. In his introduction, Simpson takes up the issue as a way of demonstrating the primacy of the paleontologist in handling it:

> ...Experimental biology in general and genetics in particular have the grave defect that they cannot reproduce the vast and complex horizontal extent of the natural environment and, particularly, the immense span of time in which population changes really occur. They may reveal what happens to a hundred rats in the course of ten years under fixed and simple conditions, but not what happened to a billion rats in the course of ten million years under the fluctuating conditions of earth history. Obviously, the latter problem is much more important. The work of geneticists on phenogenetics and still more on population genetics is almost meaningless unless it does have a bearing in this broader scene. Some students, not particularly paleontologists, conclude that it does not, that the phenomena revealed by experimental studies are relatively insignificant in evolution as a whole, that major problems cannot now be studied at all in the laboratory, and that macro-evolution differs qualitatively as well as quantitatively from the micro-evolution of the experimentalist. Here the geneticist must turn to the paleontologist, for only the paleontologist can hope to learn whether the principles determined in the laboratory are indeed valid in the larger field, whether additional principles must be invoked and, if so, what they are. (Simpson 1944: xvii)

In his third chapter ("Micro-evolution, Macro-evolution, and Mega-evolution"), Simpson takes up the "old but still vital problem of micro-evolution as opposed to macro-evolution," claiming in a footnote that "the terms are Goldschmidt's" (but failing to note that Goldschmidt credited Dobzhansky, who in turn got them from Filipchenko). This is how Simpson defines the terms:

> Micro-evolution involves mainly changes within potentially continuous populations, and there is little doubt that its materials are those revealed by genetic experimentation. Macro-evolution involves the rise and divergence of discontinuous groups, and it is still debatable whether it differs in kind or only in degree from micro-evolution. If the two proved to be basically different, the innumerable studies of micro-evolution would become relatively unimportant and would have minor value in the study of evolution as a whole. (Simpson 1944: 97)

He then claims, without evidence, that "the great majority of geneticists and zoologists believe that the distinction is only in degree and combination" (Simpson 1944: 97).

Simpson then handles the problem of "macroevolution" by redefining it. How surprising it is to a paleontologist, he notes, to find that the "higher categories" discussed in a monograph on macroevolution "are subgenera, at the highest possible evaluation, and indeed would be called 'species' by most paleontologists." He continues:

> If the term "macro-evolution" is applied to the rise of taxonomic groups that are at or near the minimum level of genetic discontinuity (species and genera), the large-scale evolution studied by the paleontologist might be called "mega-evolution" (a hybrid word, but so is "macro-evolution"). The assumption, as in Goldschmidt's work, that mega-evolution and macro-evolution are the same in all respects is no more justified than the assumption, so violently attacked by Goldschmidt and others, that micro-evolution and macro-evolution differ only in degree. (Simpson 1944: 98)

He then adds: "As will be shown, the paleontologist has more reason to believe in a qualitative distinction between macro-evolution and mega-evolution than in one between micro-evolution and macro-evolution." By substituting "megaevolution" for what many had meant by "macroevolution," was Simpson attempting a rhetorical finesse of his own?

Simpson's book, then, does not argue that what he calls "megaevolution" is explained by population genetics. Gould has rightly called the book a "consistency argument" (Gould 1980: 153–172). Even so, Simpson notes that "the facts are that many species and genera, indeed the majority, do appear suddenly in the record, differing sharply and in many ways from any earlier group, and that this appearance of discontinuity becomes more common the higher the level, until it is virtually universal as regards orders and all higher steps in the taxonomic hierarchy," and he adds: "The essentially paleontological problem is also of crucial interest for all other biologists, and, since there is such a conflict of opinion, nonpaleontologists may choose either to believe the authority who agrees with their prejudices or to discard the evidence as worthless" (Simpson 1944: 99). Indeed, Simpson warns the reader in his preface that "for almost every topic discussed in the following pages the data are insufficient" (Simpson 1944: xviii).

By the 1953 version of this book, published under the title *The Major Features of Evolution*, Simpson had added no new evidence for his dismissal of the macroevolutionary problem, but his tone had become more assured. Gould sees the changes from the 1944 version as evidence of "a much harder, much less generous, much more uncompromising line" which he terms the "hardening" of the synthesis (Gould 1980: 166). In his introduction, Simpson characterizes the "new and exciting idea" of his 1944 book as its attempt "to apply population genetics to interpretation of the fossil record, and conversely to check the broader validity of genetical theory and to extend its field by means of the fossil record," and then adds, with apparent satisfaction, that "This idea is now a commonplace" (Simpson 1953: ix).

He does not address the problem of macroevolution until the beginning of the 11th chapter, "Higher Categories." There, recalling his 1944 discussion about micro-, macro-, and megaevolution, Simpson remarks: "At present I am inclined to think that all three of these somewhat monstrous terminological innovations have served whatever purpose they may have had and that clarity might now be improved by abandoning them. . . ." Even so, he admits that "The study of evolution of higher categories, say for present purposes from families upward, does involve some of the most important problems of evolution," that there are "proper doubts as to whether the factors found experimentally in low-level evolution are those effective at high levels," and that "the absence of experimental control and the incompleteness of the fossil record make it difficult, although eventually not impossible, to draw final and decisive conclusions regarding widely variant, basic interpretations" (Simpson 1953: 339–340).

There is no doubt that Simpson's stand was important for the success of the synthesis. But what did Simpson offer by way of arguments or evidence to convince those otherwise predisposed? Clearly, for those continental biologists who, like Filipchenko, Goldschmidt, and Guyénot, believed that "big" (macro or mega) evolution was qualitatively different from microevolution, even Simpson's work gave them no reasons to change their opinion. In this connection, it is instructive to compare Simpson's treatment with that written by Severtsov only a few years earlier. Both held that the really important evolutionary issues are macroevolutionary ones; that studying the historical evolution of life on earth is the primary evolutionary science; that its data provide the ultimate test for theories concerning the evolutionary mechanism; that "autogenesis," "orthogenesis," "imminent principles," and neo-Lamarckism can be ruled out; and that Darwin's theory was essentially correct. Furthermore, both were aware of the results of population genetics and agreed that recent developments in genetics were brilliant and sound. But there agreement ended: Severtsov lamented the irrelevance of population genetics to evolutionary science, Simpson toasted its relevance. In the absence of evidence, Severtsov was not satisfied that population genetics provided all that was needed to explain macro- (or "mega-") evolution, whereas, in the absence of evidence, Simpson was.

A second figure credited with extending the evolutionary synthesis to the problem of macroevolution is the German morphologist Bernhard Rensch. His central contribution was a book written in Germany during the latter years of the war (Rensch 1947). In the introduction to the second edition, arranged by Dobzhansky and published in 1960 in America, Rensch addresses the central problem:

> It has become customary to distinguish between problems of 'macro-evolution' and 'micro-evolution'. As these two terms merely designate 'larger' and 'smaller' events without any clear borderline, and as they are linguistic mixtures of Greek and Latin roots, I prefer to use the terms 'infraspecific' and 'transspecific' evolution. Thus I hope to indicate a little more clearly the difference between those phylogenetic processes that occur within a species or lead to a new species and those that occur beyond the species limit and lead to new genera, families, and lesser divisions, and thus to new constructional types.
>
> At present, transspecific evolution is one of the central problems of evolutionary thinking. A number of biologists and paleontologists say that undirected mutation, selection, and isolation are not sufficient for a workable explanation or the causation of major phylogenetic phenomena and their regularity.... There can be no doubt that [the view that transspecific evolution is accounted for by infraspecific evolution] is somewhat 'unsatisfactory' to those who otherwise look entirely to the cause-and-effect principle in their research work. (Rensch 1960: 1–2)

Thus wrote one of the figures credited with extending the evolutionary synthesis to macroevolutionary problems—in 1960. Note also that, like Simpson, he chose to rename the dichotomy, in his case to "infraspecific" and "transspecific."[10]

When, then, was the problem of "big evolution" (whether called "macroevolution," "megaevolution," "transspecific evolution," or "evolution in depth") finally settled? The answer is, quite simply: It never has been. The appearance of the theory of punctuated equilibrium by Eldredge, Gould, and others, the attacks on the "hardening of the synthesis" by some punctuationalists, the republication of Goldschmidt's 1940 book with a highly flattering introduction—all are testimony

[10]For a fuller explication of Simpson's evolving views and the philosophical underpinnings of Rensch's stance, see the chapters by Ochoa (Chap. 10) and Levit and Hossfield (Chap. 9). One paleontologist I spoke with remarked, "For my generation, Simpson was our god!" Others respected his work, but regarded him as something of an outlier, saw no possible relevance of population genetics to their work, didn't much care about or accept the "synthetic theory," and tried to ignore it. I asked one well-known paleontologist (who shall remain nameless) what the contributors to the synthesis (e.g., Dobzhansky, Mayr, Simpson, et al.) may have had in common, as he knew most of them. "I can tell you one thing they all had in common," he replied: "They were all sons-of-bitches."

that at least some of the persistent reservations concerning macroevolution remained (Goldschmidt 1982).[11]

8.8 Sociopolitical Dimensions of the Synthesis

Let us recapitulate by returning to Mayr's 1980 characterization of the evolutionary synthesis, in his introduction to the ultimate statement of its nature and history, which he thoughtfully orchestrated, and published when almost all of its other creators were dead (Mayr and Provine 1980). Claiming it was what Huxley had meant by the "modern synthesis," Mayr defined it as "the gradual acceptance of two conclusions: gradual evolution can be explained in terms of small genetic changes ("mutations") and recombination, and the ordering of this genetic variation by natural selection; and the observed evolutionary phenomena, particularly macroevolutionary processes and speciation, can be explained in a manner that is consistent with the known genetic mechanisms" (Mayr 1980: 1).

Parsing the rhetoric, the phrase "explained in a manner that is consistent with" seems carefully chosen. One cannot help but wonder precisely what it means. That no macroevolutionary phenomenon contradicts the validity of genetic findings? (No one claimed they did!) That macroevolutionary phenomena can be explained by the findings of population genetics? (Most professionals did not believe they could!) If it means the first, it is indisputable; if the second, it is more aspirational than descriptive. But a third possibility comes to mind: that the findings of population genetics and the "macroevolutionary" sciences are both valid and "consistent with" each other because they are unrelated—two entirely different subjects! (As we have seen, that is close to what Huxley actually meant!)

The traditional view is that population genetics, as developed by Fisher, Haldane, Wright, Dobzhansky, and others, played a central role in this synthesis. As to the first conclusion alluded to by Mayr—the one concerning "microevolution"—the role of population genetics was central. But, as we have seen, population genetics did not, did not claim to, and could not prove or even address the second conclusion concerning macroevolution—the reservation largely responsible for the "eclipse of Darwinism" in the first place. This simple but remarkable fact casts the evolutionary synthesis in a much different historical light. Mayr was, I think, correct in his observation that, at least in some disciplines and countries, many scientists gradually accepted that the old problem of macroevolution had been sufficiently resolved by the "evolutionary synthesis." In light of the foregoing analysis, we must ask where, how, and why it was gradually accepted.

[11] At a conference at Penn in 1990, I sat in the second row behind Ernst Mayr when he was being chided by an advocate of punctuated equilibrium at the podium immediately in front of him, who contrasted his theory (looking directly at Mayr) with "evolution by creeps," eliciting audience laughter. Instantly, Mayr interrupted in a loud voice, shouting: "It's much better than evolution by jerks!" Note in the next paragraph Mayr's specification of "gradual evolution."

Geographically, in the 1920s and 1930s, the evolutionary synthesis was largely rejected in Germany and France and largely pursued and accepted in three countries: Britain, the United States, and the Soviet Union. At the height of popularity of the evolutionary synthesis, a new historical literature began to emerge, seeking to explain its international history, and account for its rejection in France and Germany. Interestingly, the conviction that microevolution and macroevolution were qualitatively different was one of the informing principles of that rejection. In a path-breaking book, Jan Sapp painstakingly documented the development of ideas concerning cytoplasmic inheritance and made clear that, especially in France, many looked to it for the determinants that establish species and higher taxa (Sapp 1987). In a beautifully crafted article, Jonathon Harwood explored interwar Germany, demonstrating that the Grundstock hypothesis, the Plasmon theory, and research on Dauermodifikationen were aimed at this same macroevolutionary problem (Harwood 1985). In light of the inability of the synthesis to address macroevolution in any logically or evidentially compelling way, this rejection makes sense.

How, then, to explain its acceptance in Britain, America, and Russia? Let me offer some preliminary thoughts about the last two. Perhaps social and political factors played a role. In both countries, there were interesting, and in some respects opposite, "selection" pressures operating on both genetics and Darwinism that are worth sketching out.

In a book review (of *The Evolutionary Synthesis*) many years ago, Bentley Glass asked a pregnant question: Could it really be the case that creationism had no effect whatever on the development of the evolutionary synthesis in America? After the infamous Scopes Trial in 1926, the religious "creationist" opposition to Darwin denied the reality of evolution altogether, and the scientific arguments that they employed were based on the proposition that species and higher systematic categories were qualitatively different from intraspecific varieties and variations. (This, incidentally, remains a central tenet of so-called scientific creationism.) In this context, we should consider the utility of the definition of the word "evolution" deployed by Dobzhansky—"a change in the genetic composition of populations" (Dobzhansky 1937: 11). The self-evident conclusion is clear: evolution is not a "theory" but a demonstrable (indeed a demonstrated!) fact.

Perhaps in the United States, the battle to establish the validity of evolution in the face of creationist religious opposition solidified the identification between the general theory and those experimental researches that could demonstrate, and thereby "prove," that theory "as a fact." This situation may well have encouraged individual scientists to downplay their reservations about the macroevolutionary question in order to protect evolutionary biology as a whole: uncertainties about mechanism may have seemed less pressing when the validity of evolution itself was under attack, especially when those uncertainties were being deployed by the creationists to disprove evolution. Dobzhansky's extensive correspondence with one scientific creationist suggests that this criticism weighed, at least on this central figure in forging the evolutionary synthesis in America.[12] In Britain, Haldane had a similarly extensive correspondence of the same sort (Lunn and Haldane 1935).

In the Soviet Union, of course, there operated other forms of sociopolitical selection. While Darwinism was under religious attack in Coolidge's America, official Soviet ideology after the Bolshevik revolution was pro-Darwinist; indeed, Berg's *Nomogenesis* ran into trouble with the censor because of its explicit anti-Darwinian character. In Roosevelt's America, Darwinism continued under fundamentalist attack, but genetics was enthroned; in Stalin's Russia, just the opposite was true. After 1932, it was almost impossible to publish anything in Russia openly opposing Darwinism. In fact, in the late 1930s, "Darwinism" actually became a discipline with its own university departments. The nature of that "Darwinism," however, was hotly debated. Severtsov's student, Ivan Schmalhausen, assumed the chair of Darwinism at Moscow University, but I. I. Prezent, a philosophical ally of Lysenko, assumed the chair at Leningrad University. Lysenko and Prezent, of course, denied the validity of genetics and the reality of the gene altogether, and their so-called Creative Darwinism denied that intraspecific competition played any evolutionary role whatsoever. In 1948 and subsequent years, Lysenko gained Stalin's (and then Khrushchev's) support, and genetics was officially banned, going underground, until the partial rebirth of Soviet genetics following Khrushchev's ouster in 1964 (Adams 1978a). In this context, with Darwinism ideologically enthroned but genetics under attack as a "bourgeois science," it became important for Russian geneticists to be identified as "pure" Darwinists, and useful for them to demonstrate that their researches were not only consistent with Darwinism, but helped to demonstrate and reinforce it. Once again, sociopolitical pressures, albeit of a rather different character and intensity, may have encouraged scientific solidarity (Adams 1987, 1989a, 1991).

This brings us to the "triumph" of the synthesis in the late 1940s and the 1950s—when, as Stephen Jay Gould put it, the synthesis "hardened," the tone of its advocates changed from cautious and tentative to proud and assertive, and they came to treat doubters like Goldschmidt not as legitimate skeptics, but as heretical crackpots. What happened that might account for this remarkable difference between pre- and postwar attitudes and views?

One searches in vain for any scientific breakthrough concerning macroevolution that might account for this. The cause may reside again, at least in part, in sociopolitical factors—in the world war and its outcome. The victorious allies were Russia, Britain, and the United States—which just happened to be the countries where, for local reasons, the evolutionary synthesis had flourished; the "vanquished" countries were France and Germany, where that synthesis had been rejected by those who spent the war under Nazi rule. The war's end—and the revelation of Nazi atrocities in which German biologists participated—hardly predisposed Western biologists to

[12] I have not read this unpublished correspondence, but Doby mentioned it in our 1973 interview, and Ron Numbers, a specialist in "science and religion" and the then editor of ISIS, had read the letters and emphasized their importance in a 1990 conversation.

take instructions from German biologists who resided in Germany throughout the Nazi period on matters of "race" (whether intraspecific varieties or subspecies) and its relationship to macroevolutionary progress. Later, of course, German biology was selectively relegitimated when particular works, such as that by Rensch (which seemed to more or less accept the new line), were published in English.

Furthermore, as John Beatty and subsequent researchers have demonstrated, the study of the Hiroshima victims and the fear of fallout led to the AEC's massive funding of population genetics. And this newly prestigious discipline, well financed because of its expertise on microevolution within human populations, was linked, as it happened, to a view that macroevolution was unproblematic.

I suspect, though, that in the United States (and possibly Britain), academic politics played a bigger role. So-called Big Science, which had developed earlier following the revolution in the USSR, took hold in the United States during World War II, and postwar America saw massive financial government support to science ... especially to the right *kind* of science, useful science that was experimental, often conducted in laboratories, and addressed perceived immediate needs. Universities and faculties throughout the United States had to react swiftly to the new, powerful, emergent "molecular biology," and university administrators became addicted to government money. The situation in biology became especially tense, as my own professors explained to me (informally). For many years, in most universities, "molecular biology" was the winning, front-cutting-edge subject, and when "traditional" biologists retired, their replacements were almost always "molecular." The laboratory's promises and prestige overwhelmed field biologists, and all the more so after the Watson-Crick publication of the double helix structure of DNA in 1953.[13]

How could traditional biology survive? The "synthesis," it seems to me, served as a unifying, rallying call to save traditional biology under a new label parallel to its threatening rival: "evolutionary biology."[14] With "population genetics" at the core of

[13]When I was in college at Harvard in the early 1960s, the only lights visible in the middle of the night were the molecular biology labs, working around the clock, trying to "break" the "genetic code." I remember sitting in a sparsely attended seminar taught by E. O. Wilson which was interrupted by dozens of students seeking the location of Watson's seminar on molecular biology, which, to the seven of us, seemed to clearly annoy him. Once the seminar began, Wilson announced: "Now, from the molecules of today, to the *biology* of tomorrow!"

[14]In this circumstance, "Darwinism" was increasingly a word to be avoided. (There is no "Einsteinism" or "Watson-Crickism.") The comparable "Darwinism" analogues (Aristotelianism, Platonism, Lamarckism, Freudianism, Marxism, Leninism, Mendelism) are terms referring to particular movements or traditions based on the thinking, philosophy, worldview, ideology, or theory of a single individual, not a field of inquiry or discipline identifying a generic subject. In the Soviet Union—the home of Marxism-Leninism—the field of "Darwinism" had not only scientific but also philosophical and ideological import. Although Huxley celebrated a "rebirth of Darwinism," he wisely characterized the scientific study of evolution as a "modern synthesis" created by many scientists from many fields. Given creationist opposition to evolution in the United States, any field called "Darwinism" was unlikely to be acceptable as a university department and could not successfully compete with "molecular biology" or attract government funding—hence the wisely crafted term "evolutionary biology." Throughout its history, the word "Darwinism" has been used in radically different ways, in different countries, by different people, for different purposes, to

this new incarnation, traditional biology was no longer a smorgasbord of zoological or botanical specialties focusing on different species groups or higher taxa (ichthyology, ornithology, herpetology, systematics, etc.). It was reborn as a "consensus" experimental, mathematical, laboratory discipline explaining and mastering evolution on every level (from flies, to crops, to people, and beyond).[15]

Soviet scientists had long since mastered such self-rebranding to satisfy the government that controlled their funding. Darwinism was part of official materialist ideological orthodoxy, so all scientists, regardless of specialty, became "Darwinists." When "eugenics" was officially condemned in 1930, all eugenics labs and journals were closed; but the network that supported it managed to create the new Soviet field of "medical genetics," and a large institute to study it, only 4 years later (Adams 1990a). When Stalin empowered Lysenko in 1948, condemning genetics as "idealist, capitalist science," and closing its labs and institutes, powerful physical scientists came to their rescue, and geneticists survived and continued their work in physics, chemistry, and mathematics institutes under different labels ("radiation biophysics," "cybernetics," "the chemistry of high molecular weight compounds," and so forth) (Adams 1978a, 2001). When Khrushchev was ousted in 1964, genetics was quickly re-established, only then publishing work that had been conducted during the interim (Adams 1978a).

American scientists also found similar ways of adapting to the new honeypot. Indeed, even before the government, the Carnegie and Rockefeller foundations had been supporting science selectively since the turn of the twentieth century. The establishment of "genetics" as an institutionalized field depended in these earlier decades on its ties with the popular, progressive "eugenics" promises of the day (and some of the earliest college textbooks in the field included both "eugenics" and "genetics" in their titles). Only after genetics had become firmly established and institutionalized did it distance itself from eugenics. After World War II, when "molecular biology" became the unifying new home for virologists, bacteriologists, organic chemists, biochemists, and so forth, "evolutionary biology" emerged as the consensus label for traditional biologists to move forward with "experimental

mean different things. Most "Darwinists" of Darwin's day had all the same qualms as the "anti-Darwinians" of the early twentieth century. The last edition of Darwin's *Origin* (1872) embraced the inheritance of acquired characteristics, but three decades later, those who agreed with Darwin were branded "anti-Darwinian." The "Darwinism" of the Soviets ("Creative," and a part of their ideology!) was almost the complete inverse of the contemporary "Darwinism" of the "synthesis." Like other named "isms," the term's use is always problematic, because its meaning and utility are always granularly context dependent.

[15]When preparing this chapter for publication, I sent the final draft to several colleagues. On July 7, 2020, I received an email from Nikolai Krementsov, which I quote: "When I was working in APS, I saw a letter from Doby (from Brasil) to, if I remember correctly, Dunn, in which Doby complained that biology writ large . . . is going to be overshadowed by the molecular stuff . . . and that they should be doing something about it." Krementsov added that Doby made "a direct reference to AEC funding." "I can't remember, of course, the date or the exact wording . . . but the meaning of the letter supports very strongly your suggestion about coining "evolutionary biology" as a counterweight to "molecular [biology]."

evolution." Even so, those critical of the new consensus had a trying time.[16] Others managed to survive by rebranding their fields. When traditional "geology" began to seem an old-fashioned, field-based science most useful to energy companies, some departments managed to survive (and even prosper) by rebranding themselves as departments of "Environmental Studies" or "Climate Science."

Historians should distrust narratives that explain the acceptance of a theory in certain times and places "because it is right," while seeking social or political reasons for its rejection elsewhere. Explanatory symmetry must be demanded: acceptance and rejection both require explanation. No matter what we think is true right now, we have no way of knowing whether the future may see things differently. After all, alas, historically, what we now regard as truth has not always triumphed, and what were once dismissed as "mystical," "unscientific," "crackpot" ideas have turned out to be the basis for new sciences. Recall the times of "scientific consensus" that the earth could not move (the Copernican revolution!); that continents cannot move (plate tectonics!); that humans have 48 chromosomes (only in the late 1940s did we learn we only have 46); that water is scarce in the cosmos (it is almost everywhere!); and that all bodies far from the sun are frozen and dead (quite the contrary!)

I have tried to suggest some factors that may have contributed to the acceptance of the "evolutionary synthesis" in postwar America, Russia, and Britain, and against its acceptance elsewhere. Indeed, given the above, the temptation may be the opposite: Why did so many scientists come to accept (apparently) what could not be demonstrated or proved? Like many things in science that cannot be proved, but are accepted and believed, perhaps "normal science" requires a large dose of (dare I say it?) faith—or perhaps hope, dogged persistence, intuition, ideology, vision, self-interest, or pragmatism. Or, in Dobzhansky's words, perhaps scientists like to bet on the horse that seems to be winning.

8.9 Broader Implications

This new perspective on the history of evolutionary theory over the last century has historiographic implications. For one thing, it suggests that we need a new history of population genetics and raises questions about its origins. Traditionally, the rise of population genetics is viewed principally as the necessary and sufficient condition for the evolutionary synthesis. But if population genetics did not, did not claim to, and could not illuminate macroevolution, how are we to explain its development? Evidently, by looking at what, only later, came to be called microevolution. I would suggest that population genetics almost certainly developed in order to clarify, not

[16]A renowned US embryologist told me he had spent his sabbatical doing research in Estonia, because the kind of embryology he was pursuing no longer existed in the United States, where all the newly hired embryologists for decades had made their departments entirely "molecular."

"natural selection," but "selection" in the original Darwinian sense: that is, plant breeding, animal breeding, and human breeding—eugenics.

Such a perspective helps to illuminate many features of its history that are sometimes ignored. For example, in his classic book of 12 chapters, Fisher devoted no less than five to eugenics (Fisher 1930: chapters 8–12). Sewall Wright has repeatedly emphasized the importance of animal breeding in the development of his own ideas and in the development of the discipline of population genetics generally (Wright 1967). Even Haldane, as the final chapter of his 1932 classic reveals, was motivated by his visionary hopes for human evolution in the distant future. (Haldane 1932: 144–170; Adams 2000). Finally, consider the origin of the term "gene pool," so central to population genetics. The term was coined by Dobzhansky in a popular paperback in 1946 against racism, and deployed in an evolutionary professional context in 1950, where it was used to define a "Mendelian population" (and thereby "evolution") (Dunn and Dobzhansky 1946; Dobzhansky1950, 1951b). But, as I have argued elsewhere, Dobzhansky probably translated the term from the Russian word *genofond* or "gene fund," used in the 1920s by his colleague Aleksandr Serebrovskii (Adams 1979). A eugenicist and a poultry breeder, Serebrovskii coined "gene fund" to dramatize his call for the creation of a "socialist" eugenic society in the Soviet Union and to analyze the movement of human tribes in Dagestan by studying their domesticated chickens (Adams 1990b).

These four examples suffice to show that the agricultural and eugenic dimensions of population genetics in the 1920s and the early 1930s were far from incidental. Literature on the history of eugenics has burgeoned, and work on the history of Mendelism and genetics in the United States, Britain, Russia, Sweden, and elsewhere has been making its agricultural and eugenic ties evident (Kimmelman 1983; Kimmelman and Paul 1988; Kevles 1985; Müller-Hill 1984, 1988; Weingart 1989; Adams 1989b).

I believe that this analysis may also have a broader historiographic implication for the discipline of the history of science itself. For many years, some historians and philosophers of science and technology have explored evolutionary approaches to their history, and some major historical books and articles have set forth the history of science and technology as a kind of Darwinian evolutionary process (Kuhn 1962; Adams 1968b, 1979; Nash 1967: 254–295; Fleming 1967; Gillespie 1968; Richards 1987; Hull 1988; Basalla 1988; Sapp 2003; many works by Donald Campbell and Stephen Toulmin). The utility of this approach—which owes more to Charles Darwin, Ernst Mach, and Alphonse de Candolle than to the much-maligned Herbert Spencer—is still far from generally accepted in our field.[17] But one need not follow

[17]It will come as no surprise to historians of science that there are many dozens of metaphors, images, and narratives that have been used to model, characterize, or describe both the evolution of science writ large and of particular sciences. Those less impressed by the "Darwinian evolutionary model" than I am are invited to explore three other metaphors for the history of evolutionary biology in Esposito's interesting piece in this volume.

this approach to see suggestive parallels between the recent history of science and the earlier history of evolutionary biology.

We know that science has evolved, but how it has evolved remains open to dispute. Like orthogenesists of an earlier age, some historians have been sufficiently impressed with the diachronic continuities and long-term progressive trends in science to posit a largely autogenetic development of science, dominated by internal principles and factors. Like neo-Lamarckians, other historians have seen the manifold adaptations of science to its cultural, social, political, and economic milieux as proof that all science is socially constructed. In the 1960s and 1970s, like Darwinists, many historians saw the problem of the relation of the "internal" and "external" in the evolution of science as the central problem and challenge of our field. Some have sought to formulate "evolutionary epistemology." Others have attempted to understand the "ecology of knowledge" in supportive institutions and environments, the emergence of scientific "varieties" and "species" in the form of fields and disciplines, the role played by resources and institutional niches in their emergence, and the competition between fields for limited resources. There are those who emphasize the "macromutations" produced by minds of genius and others who emphasize the effects of intellectual and social selection on contemporary scientific variants.

Like biologists over the last century, then, we too speak of the evolution of science but dispute the mechanism. Microevolutionary studies of science of extreme granularity have proved remarkably fruitful in recent decades. Whether such studies can be used to address the "macroevolutionary" trends in the history of science and history more generally—among them the classic issues of scientific and technical progress and the origin and adaptive radiation of new kinds of natural knowledge—remains to be seen. There would seem to be a pervasive assumption, even in the absence of compelling demonstration, that they can. If history is anything to go by, the strong assertion of that certainty and its gradual acceptance—even without evidence—may, for a time, suffice.

Speaking of which, it has not escaped my attention that in the 1920s and 1930s, the tensions between the "little" and the "big" were not unique to evolutionary biology. In economics, for instance, as the field was transitioning from a social, philosophical subject into its modern, mathematical incarnation, issues regarding the relationship between "microeconomics" (and its "indifference curves") and "macroeconomics" (the overall economy, GDPs, and the like) were at the forefront: were they qualitatively different, or could the latter be explained or accounted for by the former? Likewise, in the new physics, there were the tensions between the almost infinitely "little" (governed by quantum mechanics—microphysics if you will) and the almost infinitely "big" (governed by relativity—macro- or mega-physics). These tensions vexed theoretical physicists, and do to this day, since the two are incompatible, and seemingly contradict one another. I suspect these are not the only examples. (What of the philosophers of science disputing "reductionism"? Or political "scientists" disputing the inherent tensions in federalism?)

Why, or how, would the same fundamental issue arise, apparently simultaneously and independently, in at least three such seemingly disparate fields? Could this striking parallelism be the result of selection? Or is it just a coincidence, resulting from the "microevolutionary" processes within each? Or, perhaps, are there broader "macroevolutionary" patterns, regularities, emergent properties, or trends in play? I do not know. But, in light of this history of the evolutionary synthesis, it seems a question well worth pondering.

Acknowledgments I wish to express my appreciation to Ronald Numbers, Richard Burian, John Beatty, Robert Richards, and Scott Gilbert, and most especially to Donald Forsdyke (Queens University), Nikolai Krementsov (University of Toronto), and Daniel Todes (The Johns Hopkins University), for their encouragement and support, as well as their advice and suggestions over the years for improving this chapter.

Apologia

This paper originated with a weeklong interview of Theodosius Dobzhansky in Mather Camp, Yosemite, during his field trip there in 1973, 2 years before his death. I had studied with Ernst Mayr, G. G. Simpson, and E. O. Wilson; I knew Russian, had just become an assistant professor, and was researching Doby's firsthand experiences in Russia during the 1920s. Each day, after he had finished his field work, I would join him on his porch, chatting (and recording) one day in English, the next in Russian. His respect and admiration for his Leningrad mentor, Iurii Filipchenko, was almost worshipful: he pronounced his mentor the finest geneticist, biometrician, breeder, and evolutionary theorist in pre-revolutionary Russia and throughout the 1920s.

It was from his mentor that he adopted the microevolution, macroevolution terms that his 1937 book made standard. He also noted that his mentor had created those neologisms in 1927 (when Doby was his assistant and protégé) to distinguish varieties and species from genera and higher taxa, to make the case that, however useful in understanding the former, genetics could *not* illuminate the latter, namely, evolution. Yet, a decade later, now at Columbia, Dobzhansky deployed his mentor's neologisms to argue just the opposite! I was puzzled, and I asked him, given his enormous respect for his mentor, how he accounted for the difference. I was expecting a technical answer. (Was it the influence of the Morgan School, his work with lady beetles, or some experience he had had?) Instead, he shrugged and said, almost indifferently, "He bet on the wrong horse." (Doby was an inveterate horseback rider and was riding in Central Park when he was injured; he convalesced for weeks in hospital, where, at the urging of L. C. Dunn, he used the time to draft his 1937 classic book from memory, only adding bibliography when he could return to his office.)

"Bet"? "On the wrong horse?" I was startled, having never (in my innocence) thought of science as a "horse race" or a "betting" matter. That comment changed my perception. This was not a scientist who had been certain of his own approach, but rather someone who realized it might have gone either way, and chose the option that, if it turned out to be right, would both justify and empower his newly coined specialty, "population genetics." In subsequent years, I began to explore the evolutionary synthesis in greater detail, losing patience with triumphalist accounts, and gradually developing the views and evidence for this paper. Aside from teaching an annual lecture course on the history of evolutionary thought at the University of Pennsylvania, I broadened and deepened my understanding in preparation for various invited talks at universities, meetings, and other settings (at the University of Wisconsin, Ohio University, the University of Minnesota, Washington University in St. Louis, the University of Chicago, Virginia Tech, and Woods Hole, as well as in New York, Montreal, Paris, Geneva, Annecy, Uppsala, Oslo, Bergen, Moscow, Leningrad, Ischia, and the Naples Station).

I drafted the earliest variant of this article in the late 1980s. I had given a paper at a meeting in Paris and had to deliver it by a date certain in order to have it translated into French for publication—a deadline I missed. Calling on my high school French AP course, and consulting with various friends who were native speakers, I set about trying to translate my English ramblings but found that what seemed good academic English did not read well in French. So I set forth to create a new, short paper, in a more Cartesian, French style. It was subsequently published as "La génétique des populations était-elle une génétique évolutive?" in *Histoire de la génétique*, ed. Jean-Louis Fischer and W. H. Schneider (Paris: A.R.P.E.M., 1990), pp. 153–171.

Fearing that many of my colleagues could not (or would not) read French—and engrossed in the subject—I wrote a more extended English version of my thoughts. That was the initial draft of this paper. As I shared my findings with colleagues, I was strongly encouraged to prepare it for publication in ISIS by its editor and others. I came up with an almost final version, but ISIS changed management and it remained in the drawer. Over the years, various colleagues (including biologists, historians, and philosophers) encouraged me to find some way to make it publicly available, so they could cite it. After several such requests (a few quite insistent), I received an inquiry from Richard Delisle (whom, unfortunately, I have never met), sent him a copy of that paper, and he suggested publication in this volume. I have updated it here and there, and am grateful for the feedback and criticism from him and other colleagues, new and longstanding. Only the discussion of Julian Huxley is newly added, at the prompting of the editor. The standard references are from 1990; the footnotes provide new, 2020, relevant information, perspectives, comments, reflections, memories, and personal experiences.

I am neither a biologist nor a philosopher, but an historian. At one point, with the passage of time, I thought that by now, surely, many would have already realized the failings of the traditional view. But, alas, the "traditional" narrative still holds sway, as I learned when co-teaching a lecture course on the history of evolution just 4 years ago with a biologist who taught the introductory evolution course for the biology department. His take on "evolution" was all population genetics, equations, fitnesses, and claims of "evolution" (all intraspecific!), sprinkled with molecular biological updating, emphasizing Dobzhansky's claim that Darwinian evolution was not a theory, but a proven fact—and no mention whatsoever of higher taxa, fossils, extinctions, or the evolving history of life and nature on our planet. I am not so foolish as to think that my 1990 paper's publication will rectify the prevailing narrative, but I am very pleased to finally be able to make it available to interested colleagues and a broader readership.

References

Adams MB (1968a) The Founding of Population Genetics: contributions of the Chetverikov School, 1924–1934. J Hist Biol 1(1):23–39
Adams MB (1968b) Towards an evolutionary model of the history of science. 67 pp, manuscript
Adams MB (1970) Towards a synthesis: population concepts in Russian evolutionary thought, 1925–1935. J Hist Biol 3(1):107–129
Adams MB (1978a) Genetics and the Soviet Scientific Community, 1948–1965. University Microfilms, Ann Arbor
Adams MB (1978b) Nikolai Ivanovich Vavilov. In: Dictionary of scientific biography, XV (supplement I). Charles Scribner's Sons, New York, pp 505–513
Adams MB (1979) From 'gene fund' to 'gene pool': on the evolution of evolutionary language. Stud Hist Biol 3:241–285
Adams MB (1980a) Sergei Chetverikov, the Kol'tsov Institute, and the evolutionary synthesis. In: Mayr E, Provine W (eds) The evolutionary synthesis: perspectives on the unification of biology. Harvard University Press, Cambridge, MA, pp 242–278
Adams MB (1980b) Severtsov and Schmalhausen: Russian morphology and the evolutionary synthesis. In: Mayr E, Provine W (eds) The evolutionary synthesis, pp 193–225
Adams MB (1987) Darwinism as science and ideology. In: Burke JG, Adams MB et al (eds) Science and culture the western tradition. Gorsuch, Scarisbrick, Scottsdale, AZ, pp 199–234
Adams MB (1988) A missing link in the evolutionary synthesis (essay review of I. I. Schmalhausen, *Factors of evolution*). Isis 79:281–284
Adams MB (1989a) The politics of human heredity in the USSR, 1920–1940. Genome 31 (2):879–884
Adams MB (ed) (1989b) The wellborn science: eugenics in Germany, France, Brazil, and Russia. Oxford University Press, New York
Adams MB (1990a) La génétique des populations était-elle une génétique évolutive? In: Fischer J-L, Schneider WH (eds) Histoire de la génétique. A.R.P.E.M., Paris, pp 153–171
Adams MB (1990b) Aleksandr Sergeevich Serebrovskii. In: Dictionary of scientific biography, supplement II. Charles Scribner's Sons, New York
Adams MB (1990c) Iurii Aleksandrovich Filipchenko. In: Dictionary of scientific biography, supplement II. Charles Scribner's Sons, New York
Adams MB (1990d) Georgii Dmitrievich Karpechenko. In: Dictionary of scientific biography, vol 17, supplement II. Charles Scribner's Sons, New York, pp 460–464
Adams MB (1991) Through the looking glass: the evolution of Soviet Darwinism. In: Warren L and Koprowski H (eds) New perspectives on evolution. Alan R. Liss, New York, pp 37–63
Adams MB (1994a) Theodosius Dobzhansky in Russia and America. In: Adams MB (ed) The evolution of Theodosius Dobzhansky. Princeton University Press, Princeton, NJ, pp 3–11
Adams MB (ed) (1994b) The evolution of Theodosius Dobzhansky: essays on his life and thought in Russia and America. Princeton University Press, Princeton, NJ
Adams MB (2000) Last judgment: the visionary biology of J. B. S. Haldane. J Hist Biol 33 (3):457–491
Adams MB (2001) Networks in action. In: Allen GE, MacLeod RM (eds) Science, history and social activism: a tribute to Everett Mendelsohn (Boston studies in the philosophy of science), vol 228. Kluwer, Dordrecht, pp 255–276

Adams MB (2004) The quest for immortality: visions and presentiments in science and literature. In: Post SG, Binstock RH (eds) The fountain of youth: cultural, scientific, and ethical perspectives on a biomedical goal. Oxford University Press, New York, pp 38–71

Adams MB (2008) Eugenics. In: Ravitsky V, Fiester A, Caplan A (eds) The Penn Center guide to bioethics. Springer, pp 371–382

Allen GE (1975) Life science in the twentieth century. Wiley, New York

Babkov V (1985) Moskovskaia shkola evoliutsionnoi genetiki. Nauka, Moscow

Basalla G (1988) The evolution of technology. Cambridge University Press, Cambridge

Berg LS (1926) Nomogenesis, or evolution determined by law. Constable, London; reprint, MIT Press, Cambridge, MA, 1969

Bergson H (1911) Creative evolution (trans Arthur Mitchell). Henry Holt, New York

Box JF (1978) R. A. Fisher: the life of a scientist. Wiley, New York

Brink RA (ed) (1967) Heritage from Mendel. University of Wisconsin Press, Madison

Chetverikov SS (1926) O nekotorykh momentakh evoliutsionnogo protsessa s tochki zreniia sovremennoi genetiki. Zhurnal eksperimental'noi biologii ser A 2(1):3–54

Chetverikov SS (2002) Sergei Sergeevich Chetverikov: dokumenty k biografii, neizdannye raboty, perepiska i vospominaniia. Nauchnoe nasledstvo, vol 28. Nauka, Moscow

Clark R (1968) J.B.S.: the life and work of J.B.S. Haldane. Hodder and Stoughton, London

Coleman W, Limoges C (eds) (1979) Studies in the history of biology, vol 3. Johns Hopkins University Press, Baltimore, MD

Darwin C (1964) On the origin of species: a facsimile of the first edition. Harvard University Press, Cambridge, MA

DeVries H (1906) Species and varieties: their origin by mutation. In: MacDougal DT (ed) Lectures delivered at the University of California, 2nd edn. Open Court, Chicago, IL; Kegan Paul, London

Dobzhansky Th (1937) Genetics and the origin of species. Columbia University Press, New York

Dobzhansky Th (1941) Genetics and the origin of species, 2nd revised edn. Columbia University Press, New York

Dobzhansky Th (1950) Mendelian populations and their evolution. Am Nat 84:401–418; reprinted in Dunn 1951

Dobzhansky Th (1951a) Genetics and the origin of species, 3rd revised edn. Columbia University Press, New York

Dobzhansky Th (1951b) Mendelian populations and their evolution. In: Dunn LC (ed) Genetics in the 20th century: essays on the progress of genetics during its first fifty years. Macmillan, New York, pp 573–589

Dunn LC (ed) (1951) Genetics in the 20th century: essays on the progress of genetics during its first fifty years. Macmillan, New York

Dunn LC (1965) A short history of genetics. McGraw-Hill, New York

Dunn LC, Dobzhansky Th (1946) Heredity, race and society. Mentor (New American Library), New York

Fischer JL, Schneider WH (eds) (1990) Histoire de la génétique. A.R.P.E.M, Paris

Fisher RA (1930) The genetical theory of natural selection. Oxford University Press, Oxford

Fleming D (1967) Attitude: the history of a concept. Perspect Am Hist 1:285–365

Galton F (1930) The life, letters and labours of Francis Galton, 4 vols. Pearson K (ed). Cambridge University Press, Cambridge

Gillespie CC (1968) Remarks on social selection as a factor in the progressivism of science. Am Sci 56(4):439–450 (first presented at the XIIth international congress of the history of science held in Paris, 25 August–1 September 1968)

Goldschmidt R (1940) The material basis of evolution. Yale University Press, New Haven, CT; reprint 1960 Pageant, New Jersey

Goldschmidt R (1982) The material basis of evolution, introduction by Stephen Jay Gould. Yale University Press, London

Gould SJ (1980) G. G. Simpson, paleontology, and the modern synthesis. In: Mayr E, Provine W (eds) The evolutionary synthesis, pp 153–172
Guyénot É (1944) L'Origine des espèces. Presses Universitaires de France, Paris
Guyénot É (1964) The origin of species. Walker, New York
Haldane JBS (1924) Daedalus, or science and the future. Kegan Paul, Trench, Trubner, London
Haldane JBS (1932) The causes of evolution. Longmans, Green, London; facsimile reprint 1966, Cornell University Press, Ithaca, NY
Harwood J (1985) Geneticists and the evolutionary synthesis in interwar Germany. Ann Sci 42:279–301
Hull DL (1973) Darwin and his critics: the reception of Darwin's theory of evolution by the scientific community. Harvard University Press, Cambridge, MA
Hull DL (1988) Science as a process: an evolutionary account of the social and conceptual development of science. Chicago University Press, Chicago, IL
Huxley JS (ed) (1940) The new systematics. Clarendon, Oxford
Huxley JS (1942) Evolution: the modern synthesis. George Allen & Unwin, London
Huxley JS (1963) Evolution: the modern synthesis, 2nd edn. George Allen & Unwin, London
Huxley JS (1964) Evolution: the modern synthesis, 2nd edn. Wiley, New York
Huxley JS (1970) Memories, vol 1. Harper & Row, New York
Jenkin F (1867) The origin of species. North Br Rev 46:277–318; reprinted in Hull DL (1973) Darwin and his critics: the reception of Darwin's theory of evolution by the scientific community. Harvard University Press, Cambridge, MA, pp 302–344
Johannsen W (1903) Ueber Erblichkeit in Populationen und in reinen Linien. Gustav Fischer, Jena
Johannsen W (1969) Heredity in populations and pure lines: a contribution to the solution of the outstanding questions in selection. In: Peters JA (ed) Classic papers in genetics. Prentice-Hall, Englewood Cliffs, NJ, pp 20–26
Kevles D (1985) In the name of eugenics. Knopf, New York
Kimmelman B (1983) The American Breeders' Association: genetics and eugenics in an agricultural context, 1903–1913. Soc Stud Sci 13:163–204
Kimmelman B, Paul D (1988) Mendel in America: theory and practice, 1900–1919. In: Rainger R, Benson KR, Maienschein J (eds) The American development of biology. University of Pennsylvania Press, Philadelphia, PA, pp 281–310
Kropotkin P (1902) Mutual aid: a factor of evolution. McClure Phillips, New York
Kuhn TJ (1962) The structure of scientific revolutions. Chicago University Press, Chicago, IL
Lunn A, Haldane JBS (1935) Science and the supernatural: a correspondence between Arnold Lunn and J. B. S. Haldane. Eyre and Spottiswoode, London
Mayr E (1944) Systematics and the origin of species. Columbia University Press, New York
Mayr E (1980) Prologue: some thoughts on the history of the evolutionary synthesis. In: Mayr E, Provine W (eds) The evolutionary synthesis. pp 1–48
Mayr E, Provine W (eds) (1980) The evolutionary synthesis: perspectives on the unification of biology. Harvard University Press, Cambridge, MA
Menzbir MA (1928) Mendelizm i ego otnoshenie k darvinizmu. In Menzbir MA (ed) Polnoe sobranie sochinenii Charlza Darvina (1928–9), vol 3, pt 1, pp 3–19
Menzbir MA (1929) Teorii nasledstvennosti. In: Menzbir MA (ed) Polnoe sobranie sochinenii Charlza Darvina (1928–9), vol 4, pt 2, pp 424–469
Müller-Hill B (1984) Tödliche Wissenschaft. Rowohlt, Reinbeck bei Hamburg
Müller-Hill B (1988) Murderous science. Oxford University Press, New York
Nash LK (1967) The nature of the natural sciences. Harvard University Press, Cambridge, MA, especially "The Evolution of Scientific Theories", pp 254–297
Peters JA (ed) (1959) Classic papers in genetics. Prentice-Hall, Englewood Cliffs, NJ
Philiptschenko J [Filipchenko IA] (1923) Izmenchivost' i metody ee izucheniia. Gosizdat, Petrograd
Philiptschenko J [Filipchenko IA] (1927) Variabilität und Variation. Verlag von Gebrüder Borntraeger, Berlin

Philiptschenko J [Filipchenko IA] (1929) Izmenchivost' i metody ee izucheniia, 4th edn. Gosizdat, Leningrad
Provine WB (1971) The origins of theoretical population genetics. Chicago University Press, Chicago, IL
Provine WB (1986) Sewall Wright and evolutionary biology. Chicago University Press, Chicago, IL
Rainger R, Benson KR, Maienschein J (eds) (1988) The American development of biology. University of Pennsylvania Press, Philadelphia, PA
Rensch B (1947) Neuer Probleme der Abstammungslehre. Ferdinand Enke, Stuttgart, 2d edn, 1954
Rensch B (1960) Evolution above the species level. Columbia University Press, New York
Richards RJ (1987) Darwin and the emergence of evolutionary theories of mind and behavior. University of Chicago Press, Chicago, IL (see the conclusion, entitled "Darwinism is Evolutionary")
Richards RJ (1992) The meaning of evolution. Chicago University Press, Chicago, IL
Sapp J (1987) Beyond the gene: cytoplasmic inheritance and the struggle for authority in genetics. Oxford University Press, New York
Sapp J (2003) Genesis: the evolution of biology. Oxford University Press, New York
Schmalhausen II (1949) The factors of evolution: the theory of stabilizing selection. Blakiston, Philadelphia, PA; reprinted 1986 University of Chicago Press, Chicago, IL
Severtsov AN (1939) Morfologicheskie zakonomernosti evoliutsii. Izdatel'stvo Akademii Nauk SSSR, Moscow
Simpson GG (1944) Tempo and mode in evolution. Columbia University Press, New York
Simpson GG (1953) The major features of evolution. Columbia University Press, New York
Stebbins GL (1950) Variation and evolution in plants. Columbia University Press, New York
Warren L, Koprowski H (eds) (1991) New perspectives on evolution. Alan R. Liss, New York
Weingart P (1989) Health, race, and German politics between national unification and Nazism, 1870–1945. Cambridge University Press, Cambridge
Weingart P, Kroll J, Bayertz K (1988) Rasse, Blut, und Gene. Frankfurt-am-Main, Suhrkamp
Wells HG, Huxley JS, Wells GP (1934) The science of life. Literary Guild, New York (previous eds 1929, 1930, 1931: Doubleday, New York)
Wright S (1931) Evolution in Mendelian populations. Genetics 16(2):97–159
Wright S (1932) Mutation, inbreeding, crossbreeding and selection in evolution. Proc Sixth Int Congr Genet 1:355–366
Wright S (1967) The foundations of population genetics. In: Brink RA (ed) Heritage from Mendel. University of Wisconsin Press, Madison, WI, pp 245–263

Chapter 9
When Panpsychism Met Monism: Why Did the Philosopher Theodor Ziehen Become a Crucial Figure for the Evolutionary Biologist Bernhard Rensch?

Georgy S. Levit and Uwe Hossfeld

Abstract Theodor Ziehen was a well-known German psychiatrist and experimental psychologist of the first half of the twentieth century. But he was also an obscure philosopher who developed a very sophisticated and radical form of panpsychism. While Ziehen's work left few traces within the history of philosophy itself, his epistemology enjoyed significant influence within German evolutionary biology. Most prominently, Ziehen had a great impact on the "co-architect" of the German evolutionary synthesis, the zoologist Bernhard Rensch. Our paper has two major objectives: first, to sketch Ziehen's distinctive contribution to philosophy and, second, to explain his importance for Rensch. Our hypothesis is that Ziehen's monism and nomotheism constituted the philosophical foundation of Rensch's evolutionary universalism. Monism was a prominent philosophical position within the German tradition of evolutionary biology beginning with Ernst Haeckel and remained influential thereafter due to Rensch and some other of his contemporaries. Nomotheism, the idea of elevating biological regularities to the level of universal laws also became prominent in biology due to Ernst Haeckel and, in a modified form, was promoted by Rensch as well. For Rensch, universal selectionism best satisfied the philosophical requirements of monism and nomotheism. Furthermore, Rensch's monism and his version of determinism (polynomic determination) turned out to be a selectionist interpretation of the idea of directionality in evolution.

Keywords Theodor Ziehen · Bernhard Rensch · Ernst Haeckel · Determinism · Panpsychism · Nomotheism · Evolutionary synthesis · Monism

G. S. Levit (✉) · U. Hossfeld
Biology Education Research Group, Friedrich-Schiller-University Jena, Jena, Germany

R. G. Delisle (ed.), *Natural Selection*, Evolutionary Biology – New Perspectives on Its Development 3, https://doi.org/10.1007/978-3-030-65536-5_9

9.1 Introduction

Theodor Ziehen was a well-established German psychiatrist and psychologist and a marginal philosopher who developed a very sophisticated and radical form of positivism (Ruschmeier 1999; Gerhard and Blanz 2002; Gerhard 2004; Levit and Hoßfeld 2020). His contemporary, the founder of bolshevism, Vladimir Lenin, considered Ziehen a proponent of empirio-criticism, as developed by Ernst Mach and Richard Avenarius. Yet Ziehen never achieved their prominence in philosophy (Fig. 9.1).

Despite his small profile in the history of philosophy, Ziehen became a prominent figure in German evolutionary biology. This was primarily due to the fact that the key "co-architect" of the German evolutionary synthesis, Bernhard Rensch, regarded Ziehen as his major philosophical inspiration. Rensch's own philosophy laid the methodological foundations for his evolutionary biology and as a result Ziehen became central to Rensch's entire theoretical system (Ruschmeier 1999; Levit et al. 2008). Ziehen's work also heavily influenced another towering figure of modern biology, the founder of biological systematics, Willi Hennig (Rieppel 2007, 2016). Ziehen even made an impact on the "German Darwin", Ernst Haeckel, who in his last years became aware of Ziehen's work, cited him, and exchanged some letters with him. Drawing on his psychological work, Ziehen also published a joint paper with the German geneticist and developmental biologist Valentin Haecker (Hossfeld et al. 2017, 2019) (Fig. 9.2).

Ziehen's obscurity as a philosopher is remarkable. For example, the extensive *Stanford Encyclopedia of Philosophy* provides only three references to Ziehen in the context of discussing logic and psychology, but without mentioning his major philosophical publications.[1] *The Oxford Companion to Philosophy*, which mentions both Ernst Mach and Richard Avenarius (with quite similar philosophical profiles), completely neglects Ziehen (Honderich 2005). *The Cambridge History of Philosophy (1870–1945)* includes only one reference to one Ziehen's psychological publication (Baldwin 2003, p. 104) completely ignoring his fundamental philosophical works. Perhaps even more startling, this is also true of German-language textbooks and encyclopaedias. The dictionary *Philosophie und Naturwissenschaften* (Philosophy and Natural Sciences), specifically aimed to elucidate the connection between science and philosophy, does not mention Ziehen at all (Hörz et al. 1997) (Fig. 9.3).

Given this obscurity, the question stands why he was of such interest to German biologists and, most importantly, to Rensch. Our hypothesis is that Ziehen's panpsychism was a good conceptual fit with a specific version of monism prominent in German evolutionary biology during both the first and the second Darwinian revolutions (Levit and Hossfeld 2017). Furthermore, the overemphasis Ziehen placed on laws and lawfulness nourished Rensch's determinism. These two cornerstones of Rensch's metaphysics (monism and determinism) allowed him to establish an idea of evolutionary progessionism. After turning from neo-Lamarckism to

[1]https://plato.stanford.edu/search/searcher.py?query=Ziehen

Fig. 9.1 Portrait of Theodor Ziehen (UAHW, Rep. 40-VI, Nr. 2, Bild 43)

Fig. 9.2 Valentin Haecker (second from the left) handing his rector's office over to Theodor Ziehen in 1927 at the University of Halle (UAHW, Rep. 42, N1338)

panselectionism in the mid-1930s, Rensch instrumentalized the theory of natural selection to serve these philosophical objectives (Levit et al. 2008).

In this paper, we will first contextualize Ziehen's philosophy by introducing the key figures of the Darwinian revolutions in Germany. Then we will outline the main themes of Ziehen's epistemological work within the tradition of monism, before returning to a discussion of Bernhard Rensch and his interest in Ziehen. Finally, we will demonstrate the role of natural selection in Rensch's all-embracing theoretical system.

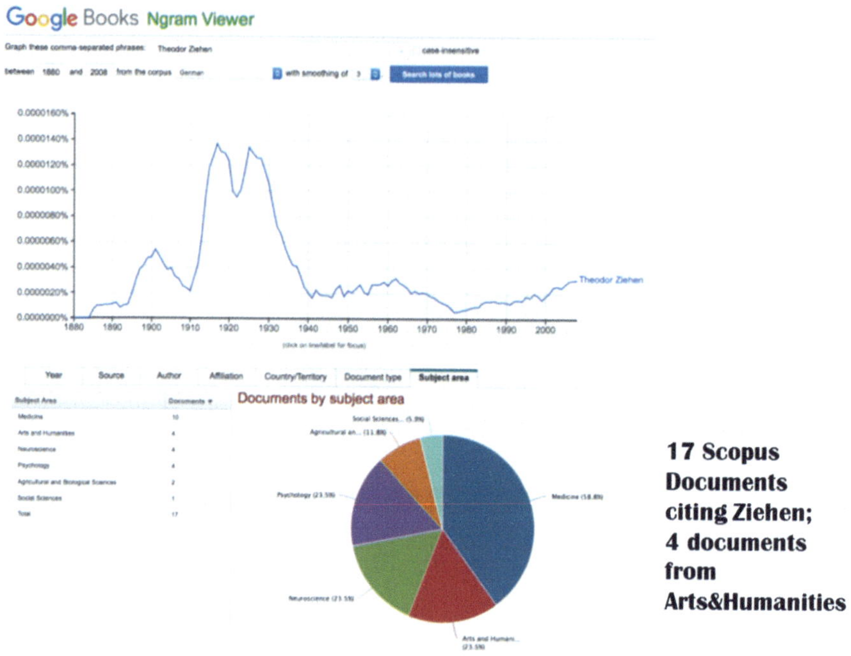

Fig. 9.3 Diagram showing Ziehen almost unknown in the history of humanities (established by Dr. Dmitry Prokudin, St. Petersburg)

9.2 Who Is Who in German Evolutionary Biology: The First and the Second Darwinian Revolutions in Germany

Most historians of biology agree that there were three major historical forms of Darwinism (Reif et al. 2000; Bowler 2003; Kutschera and Niklas 2004; Junker 2004; Levit and Hoßfeld 2011; Kolchinsky 2014; Delisle 2017). *Classical Darwinism* is Darwin's own theory, which advanced the very idea of organic evolution and common descent and introduced the principle of natural selection within a broad theoretical context.

Classical Darwinism was to be followed by a split between *neo-Darwinism* and *old-school-Darwinism* (Levit and Hoßfeld 2006). At the end of the nineteenth century, Canadian-born English psychologist George John Romanes recognized the crucial importance of the question "whether natural selection has been the sole, or but the main cause of organic evolution" (Romanes 1895, p. 1). Answering this question, Romanes opposed Darwin, who admitted that natural selection has been assisted by the "subordinate principles", and Alfred Russel Wallace along with August Weismann, maintaining that natural selection should be regarded as the only cause of evolution. To denote the "the pure theory of natural selection to the exclusion of any supplementary theory", Romanes coined the term neo-Darwinism (Romanes 1895, p. 12). Under "supplementary theories", Romanes understood "Lamarckian factors" (use-inheritance) and the theory of sexual selection. The original Darwinian line of thinking—which preserved the priority of natural selection but added both Lamarckian and selectionist factors along with moderate orthogenesis and some mutationism—was continued by the "old-Darwinian" school represented, first of all, by Haeckel and his successor at Jena University, Ludwig Plate (Levit and Hoßfeld 2006; Gliboff 2012).

Finally, the third form, the *Synthetic Theory of Evolution* (STE) or the *Modern Synthesis* or the *Evolutionary Synthesis* originated in the early 1930s, after a period characterized by the "eclipse" of Darwinism (Bowler 1983) and the associated dominance of alternative (non-Darwinian) theories of evolution. According to Ernst Mayr, the Synthesis was completed in 1947, and the so-called period of "post-synthesis" then began (Mayr 1999, p. 20). The synthetic theory of evolution was "cosmopolitan" insofar as it proposed a logically coherent and empirically well-substantiated theoretical system which incorporated several branches of biology including classical genetics, population genetics, molecular genetics, systematics, evolutionary morphology, developmental biology, paleontology, etc. Within the STE, "non-selectionist factors of evolution, especially isolation, chance events, and population size are emphasized. Selection is regarded as important, but only as one of several evolutionary factors" (Reif et al. 2000). With all these factors taken into account, the STE succeeded in proposing a coherent theory of macroevolution, although its approach may be seen as reductionist, i.e. as reducing macroevolutionary processes to microevolutionary ones.

In Germany, two major figures responsible for the growth of evolutionary theory in the first and the second period were Ernst Haeckel and Bernhard Rensch, respectively. Haeckel was a younger contemporary of Darwin and one of his most influential (although controversial) proponents on the continent (Hossfeld 2010; Hopwood 2015; Levit and Hossfeld 2019). Haeckel belonged to the old-Darwinian current—his goal was to exactly follow Darwin in his description of evolutionary mechanisms. Rensch was, arguably, the most influential figure on the international scene of what is known as the Modern Synthesis or the Second Darwinian revolution in Germany (Levit and Hossfeld 2017). Simply put, Haeckel and Rensch were the two most important advocates of Darwinism in German lands in both the "classical" and "synthetic" periods of evolutionary biology.

Rensch was not the only German advocate of the evolutionary synthesis, however. Before and during the Second World War, he was a member of a movement involving dozens of German scientists including Gerhard Heberer, a zoologist and anthropologist from Jena who was, during the Third Reich, one of the crucial figures of the German evolutionary synthesis. Heberer influenced the Synthesis in Germany in two ways. First, he organized and edited the multi-author book *Die Evolution der Organismen* (1943). This collection is the most comprehensive statement of the Synthesis published during its formation phase in Germany. It resembled Julian Huxley's *The New Systematics* (Huxley 1940; see also Hoßfeld 1997). The nineteen contributors to *Die Evolution* predominantly advocated for selectionism and oppose neo-Lamarckism, saltationism and orthogenesis. All contributors may be considered Darwinians, but Heberer, Rensch, Nikolai Timoféeff-Ressovsky and Walter Zimmermann arguably contributed most to the Synthesis. The topics in the book range from a philosophical analysis of the theory of descent and the refutation of typology to biological proofs of the theory of descent; methods of phylogenetics; ethology; phylogeny of plants, animals and humans; as well as population genetics, selection theory and macroevolution.

The only philosopher who contributed to Heberer's volume was the Darmstadtian theorist Hugo Dingler, who can be considered the "official" philosopher of the "synthetic" movement in Germany. Dingler wrote the first chapter of the book with the ambitious title, "Die philosophische Begründung der Deszendenztheorie" (The Philosophical Foundation of Evolutionary Biology). Although Dingler worked very closely with the German "synthetic" Darwinian movement, it was not he who determined the theoretical pathways of the major German synthetic evolutionist Rensch; it was Ziehen.

9.3 Theodor Ziehen's Life Path

It is difficult to attach Ziehen to any particular scientific discipline. He was a psychologist, neurologist, psychiatrist and philosopher who enjoyed great fame during his lifetime, but who has since been almost completely neglected by the history of science and philosophy: "Theodor Ziehen belongs among the great universal thinkers of the end of the nineteenth century and the twentieth century. Unfortunately, his accomplishments have largely been forgotten, although contemporaries compared him to Einstein and Leibniz" (Gerhard and Blanz 2004).

Ziehen was born on 12 November 1862 in Frankfurt am Main as the son of Protestant theologian, philologist and writer, Eduard Ziehen (Ziehen 1923, p. 211). His brother Ludwig Ziehen is known as a pedagogue and historian. Ziehen attended a so-called humanist high school (*humanistisches Gymnasium*), where he became proficient in classical languages (Greek and Latin) and showed philosophical interest especially in the philosophy of Plato, Kant and Schopenhauer and even in Indian philosophy. Ziehen himself wrote about this period of his life: "Already at that time I decided that philosophy is the ultimate objective of my life" (Ziehen 1923, p. 220).

Despite his love for philosophy, Ziehen decided to study medicine because it was the only way to receive a stipend which he urgently needed. In 1881, he enrolled at a Würzburg university, where he attended classes in the history of philosophy by Georg Neudecker (born 1850) who, in his turn, was significantly influenced by Fichte. Following Neudecker's advice, Ziehen studied modern philosophers, paying special attention to Spinoza, Hume, Hegel and George Berkeley. From the side of the natural sciences, he was deeply impressed by the famous botanist Julius Sachs (1832–1897) (Ziehen 1923, p. 221). Two years later (1883), he moved to Berlin to continue his medical education and received his doctorate in 1885 with a PhD thesis entitled, *Über die Krämpfe infolge elektrischer Reizung der Großhirnrinde* (On the Spasms of the Cerebral Cortex as a Consequence of Electrical Stimulation). His decision to specialize in psychiatry was connected with its proximity and relevance to psychology and philosophy. Besides medicine and philosophy, Ziehen studied mathematics and theoretical physics.

In 1885, he began to work as an assistant volunteer at the famous mental hospital in Görlitz under the guidance of Karl Ludwig Kahlbaum, and in 1886, he moved to Jena to become a "senior doctor" (Oberarzt) in Otto Binswanger's psychiatric clinic (Castell 2003, p. 422). Ziehen remained in Jena for 14 years. At that time, Jena was one of the major centres of Darwinism due to the activities of Ernst Haeckel with whom Ziehen was in contact (Fig. 9.4). In Jena, Ziehen also came into contact with Friedrich Nietzsche, who was one of Ziehen's patients (Nenadić 2011).

In 1887, Ziehen completed his *Habilitation* (Dr. sc. thesis) with the topic *Sphygmographische Untersuchungen an Geisteskranken* (Sphygmographic Studies on Mentally Ill Patients) and became a "Privatdozent" (lecturer) in psychiatry (Castell 2003, p. 423).

In 1892, Ziehen became an "extraordinary" (associate) professor in Jena, and in 1896, he left the clinic to open a private neurological practice (Castell 2003, p. 424). In 1900, he was offered the position of professor of psychiatry in Utrecht (Holland), and 3 years later (1903), he became the chair in psychiatry at the University of Halle. After spending only half of a year in Halle, Ziehen moved to Berlin to become the director of a newly established clinic for psychiatry and neurology at the famous Charité Hospital (Nenadić 2011), a post he maintained until his retirement in 1912, during which time he received (in 1910) a doctorate (*honoris causa*) from the Philosophical Faculty at Berlin University for his important contributions to philosophy (Castell 2003, p. 425). However, his position as clinic director left him little time for philosophical occupations, and in 1912, he moved with his family to a small villa in Wiesbaden, where he spent a few years devoted to psychology and philosophy as an independent scientist (Nenadić 2011). During the First World War, Ziehen helped establish the Flemish University in Ghent, but in 1917, he came back to Halle as a professor of philosophy, co-director of philosophical seminars and keeper of a neurophysiological collection. In this period, he shared his time between philosophy, child psychology and pedagogy. It is also in this period when Ziehen and Rensch first crossed paths as Rensch began studying biology, chemistry and philosophy at the University of Halle in 1920 (Dücker 2000, p. 3). Rensch was deeply impressed by Ziehen's philosophy and personality, and as he put it in his autobiography: "The great event was for me the lectures of Professor Theodor Ziehen on the history of philosophy, epistemology and logic as well as on *Naturphilosophie*. I was fascinated by this paramount personality right from the start" (Rensch 1979, p. 35). In Rensch's autobiography, published in 1979, Ziehen is the most cited person second only to Darwin (Fig. 9.5).

In 1922, Ziehen became dean of the philosophical faculty, and in 1927 he replaced Valentin Haecker as the rector of the university in Halle (Nenadić 2011). When Rensch completed his PhD in 1922, it was Ziehen who signed his certificate. In 1930, Ziehen retired and moved to Wiesbaden, where he died on 29 December

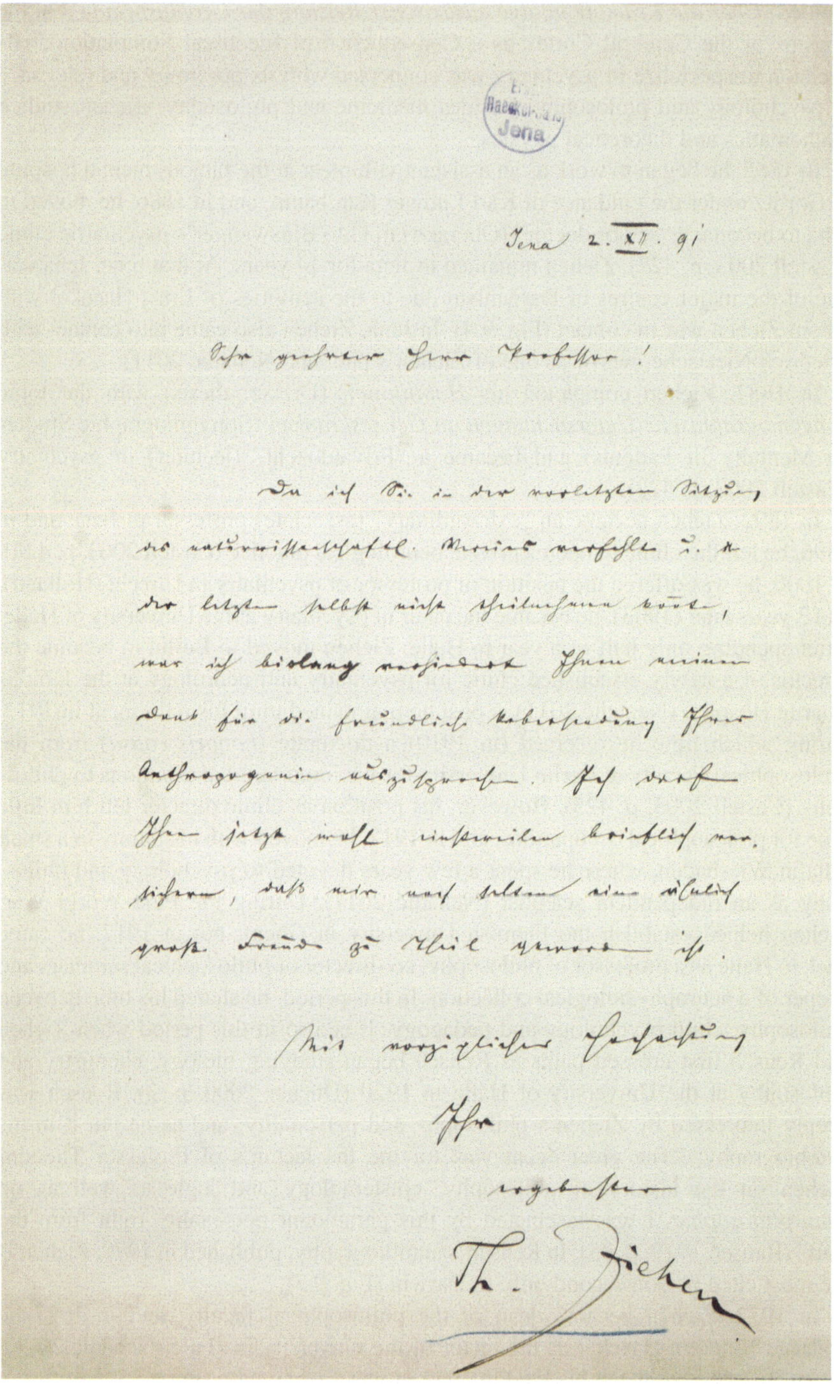

Jena 2. XII. 91

Sehr geehrter Herr Professor!

Da ich Sie in der vorletzten Sitzung der naturwissenschaftl. Verein verfehlte u. an der letzten selbst nicht theilnehmen konnte, war ich bislang verhindert Ihnen meinen Dank für die freundliche Uebersendung Ihrer Anthropogenie auszusprechen. Ich darf Ihnen jetzt wohl einstweilen brieflich versichern, dass mir noch selten eine ähnlich grosse Freude zu Theil geworden ist.

Mit vorzüglicher Hochachtung

Ihr

ergebenster

Th. Ziehen

Fig. 9.4 One of four letters of Theodor Ziehen to Ernst Haeckel (EHH Archiv der FSU Jena). From their letter exchange (1891–1908), it follows that they took part in joint meetings and exchanged publications

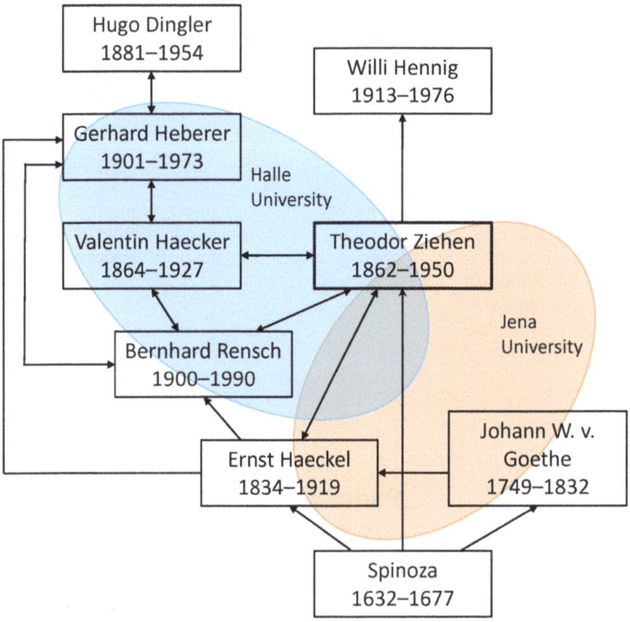

Fig. 9.5 Theodor Ziehen's personal network

1950. He avoided membership in the Nazi Party, and never supported National Socialism (Castell 2003, p. 426), which was unusual among German physicians and especially psychiatrists; 45% of German physicians belonged to the Nazi party, about 7 times the mean rate for the employed male population of Germany (Seeman 2005). Psychiatrists were among most enthusiastic participants in the regime (Robertson et al. 2017).

Ziehen's entire oeuvre amounts to more than 400 works on various subjects including two dozen monographs. A significant portion of his later works were devoted to philosophy (Fig. 9.6).

9.4 Theodor Ziehen's Philosophy

Ziehen published the second edition of his *Erkenntnistheorie* in two parts, the first in 1934 and the second in 1939 (Ziehen 1934, 1939). These two volumes include the most mature version of his philosophy. The first volume has the subtitle, *Allgemeine Grundlegung der Erkenntnistheorie. Spezielle Erkenntnistheorie der Empfindungstatsachen einschließlich Raumtheorie* [General Foundation of Epistemology. Special Epistemology of Sensations Including the Theory of Space]. The second volume is devoted to several related topics and is subtitled, *Zeittheorie. Wirklichkeitsproblem. Erkenntnistheorie der anorganischen Natur*

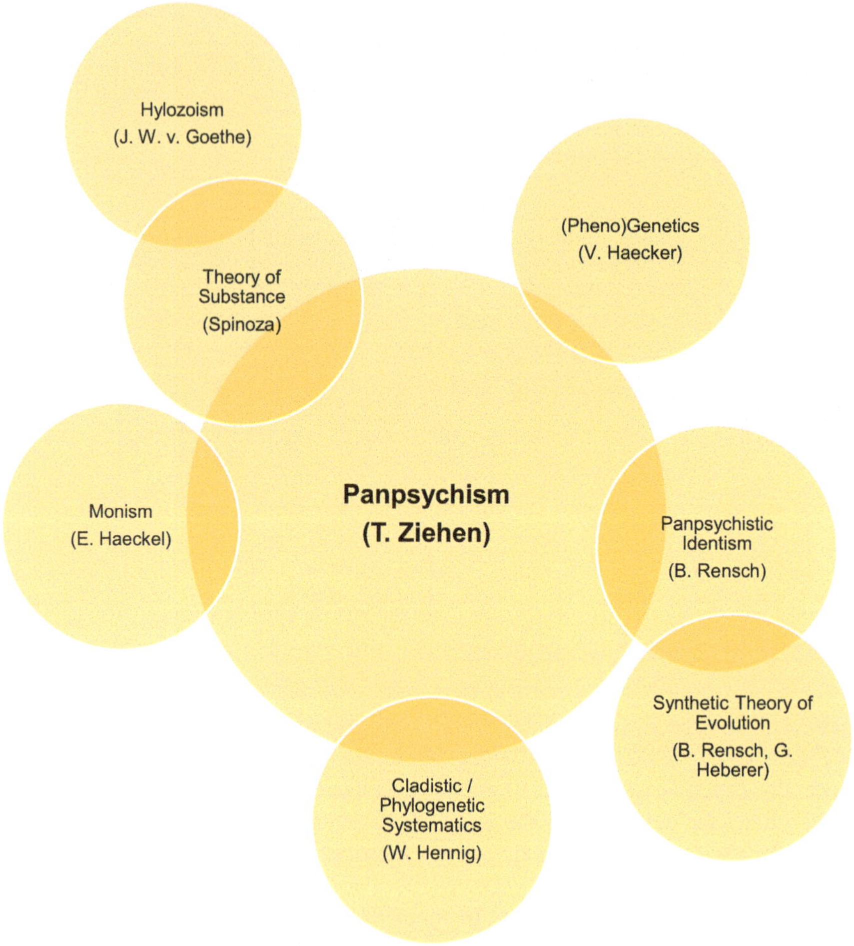

Fig. 9.6 Theodor Ziehen's conceptional context

(erkenntnistheoretische Grundlagen der Physik). Kausalität [Theory of Time. The Problem of Reality. Epistemology of the Inorganic Nature (Epistemological Foundations of Physics). Causality].

Right in the first sentence of the first volume, Ziehen formulates the very basic idea of his panpsychism, the so-called Gignomenal-principle: "The Given is the raw material of our entire knowledge" (Ziehen 1934, p. 1). The word "gignomenal" comes from the term "Gignomene", invented by Ziehen to describe "the Given" [das Gegebene]. In the first approximation, "the Given" is that which we experience, it is our sensations and perceptions: "Everything that is given is either sensation or perception. Initially, we get sensations; then they are followed by recollections and perceptions" (Ziehen 1907, p. 4).

Elsewhere in the text, Ziehen offers a clear and short definition of what he calls "the basic principle of idealism" [*idealistisches Grundprinzip*]: "The Given is only psychic, in an ordinary sense; the so-called material things are not given, they will be extrapolated [erschloßen]" (Ziehen 1922, p. 2). This means that "the given" does not simply reflect the material things out there; they will be constructed by our psyche.

For Ziehen, the "given" exists in past, present and future forms (Ziehen 1934, p. 7), and knowledge is, in the first approximation, "the Given" as processed by our mind. The "knowledge complexes" combined within a certain system (i.e. the system of knowledge) establish "science". Epistemology is the most fundamental of all sciences and is, in fact, the "gignomenology", i.e. the science of the given (Ibid., p. 8).

Another central notion of Ziehen's epistemology is "the principle of immanence". In 1915, in the *Grundlagen der Psychologie* (Foundations of Psychology), Ziehen briefly explained that "the principle of immanence" includes three logically interconnected claims: (1) It is impossible to establish a meaningful concept of type or kind [*Gattung*] covering all the given [*Gignomene—plur.*]; (2) it is impossible to distinguish gignomena[2] from non-gignomena; (3) it is impossible to imagine something that would be totally different from gignomena. In other words, the "immanent philosophy" is beyond the opposition between "physical" and "psychical". The "immanent philosophy" rejects both the "metaphysical" and the "metapsychical" (Ziehen 1915, pp. 11–12). In 1934, Ziehen emphasized that "transcendent" knowledge is impossible since all knowledge derives from the "given", i.e. we cannot trespass the boundaries of our own mind into the material world. Knowledge is therefore always immanent, and this is the core of the "immanence principle" (Ziehen 1934, p. 12). In Ziehen's vocabulary, the word "immanent" describes that which can be derived from the "given" and does not essentially differ from it, i.e. that which is derived from the individual mind's experiences and in its essence cannot be separated from these experiences. The "immanence principle" gives rise to the "positivist principle", according to which series of sensations and perceptions relate to each other in ways that can be described as "laws"—indeed, the very notion of the "law" is of crucial importance for Ziehen. In this sense, then, epistemology is the science of immanent laws of the "given". The "positivist" principle directs epistemology towards the obligatory search for such laws (Ibid., p. 16).

The "immanence principle" leads to "epistemological relativism", which Ziehen articulates in the formula of cognition (*Erkenntnis*): $Erk = F(O)$, where O stands for object and F stands for "cognitive functions" [*Erkenntnisfunktionen*], i.e. the functions of a cognizing mind. The formula symbolizes the inaccessibility of the object for the cognizing mind, which is always dealing with "functions" processing information on objects out there. Knowledge is always based on available gignomena and

[2]One of Ziehen's reasons for introducing the term *Gignomene* was to encompass the possibility of using plural and single forms of the "given". In the following, we use *gignomena* for plural and *gignomenon* for a single form.

correspondingly is relative not only in relation to *F* but also in relation to *O* (Ziehen 1934, p. 19). It is important to stress that *F* only operates on objects of the mind (objects already existing in the mind); Ziehen identified this aspect of his thought with Berkeley's subjective idealism (Ziehen 1907, p. 7).

All gignomena are classified into sensation-, perception-, thought-, feeling- and will-gignomena (Ziehen 1934, pp. 25–26). Accordingly, gignomenology (epistemology) is subdivided into the epistemologies of sensation, perception, thought, feeling and the will. This is the counter-intuitive part of Ziehen's philosophy. If sensations and perceptions are direct conductors of the "given", then free will comes from within the individual; it is not given by the outer world. But for Ziehen, the "processes of willing" [*Willensvorgänge*] either can be traced back to all other forms of gignomena or are dependent on other forms of gignomena. This free-will paradox[3] will be crucial also for Rensch's philosophy, but Rensch made his argument more straightforward by arguing that humans live in a lawful, deterministic world and that, in all probability, there is no such thing as free will (Rensch 1991, p. 151).

Another of Ziehen's ideas which became important for Rensch is the concept of reduction. Ziehen distinguishes all "given" (in sensations, perceptions, etc.) into a "reduced (causal) component" and "parallel component" (Ziehen 1934, p. 44). For example, our daily observations of a yellow disk (the sun) flying through the heavens do not itself suggest what sort of law lies behind its motion. This pure observation is the "parallel component" (so-called *N*-component) of a gignomenon.

Sensations are "parallel", because our subjective perceptions (e.g. yellow, green) run in parallel to, e.g. visual cortex. Ziehen illustrates this idea of causal and parallel lawfulness with two examples. The falling of a stone in proximity to a person causes a cascade of physiological reactions including the activation of the cerebral cortex. This cascade of reactions is repeatable and independent of individual sensations. This is an example of a causally lawful ("objective") process, which is reflected by "parallel" ("subjective") processes in my brain. The perception of a "blue sky" becoming "red" illustrates parallel lawfulness even better, because there is no "blue" nor "red" really out there, and the sky may appear blue or red for a variety of reasons. In any case, there is no causal relation connecting "blue" or "red" sensations; they just run in parallel to something extramental (in our terms). At the same time, my perception of "blue" or "red" cannot be explained in terms of physics, and cannot be equated with extramental processes. The way to "objective" knowledge Ziehen denoted by the term "reduction".

The "reduced or reduction component" of the "given"—the *R*-component—is akin to what we might think of as the ability to build abstract categories and to draw regularities based on observations. Ziehen considered the *R*-component "an objective component" of the "given", as opposed to the subjective *N*-component, which depends on the nervous system (Ziehen 1922, p. 6). The process of reduction is, in fact, the process of eliminating individual (subjective) sensations to arrive at a more

[3]Ziehen and Rensch were under Spinoza's influence; compare, e.g. with Spinoza, *Ethics*, Part 1, Th. 32, esp. Cor. 1, 2 (Spinoza 1996).

generalized picture of the world. The *R*-component in its ultimate form is science. Commenting on the above example of the sun, Ziehen wrote: "I decompound the sun experience in an *R*-component, which essentially corresponds to my scientific knowledge, and an *N*-component remaining completely outside of the scientific knowledge" (Ziehen 1934, p. 30). Blue sky, yellow sun, feelings of warmth and observations of sunsets belong to the *N*-components. The very name "reduction" appears because the discovery of regularities is the result of removing certain features of the "given". For example, to come to the laws of optics, one should "eliminate" the colours (all these subjectively experienced as "blue", "yellow", etc.). Every gignomenon (everything "given") can therefore be characterized by the formula, G = R + N (Ziehen 1922, p. 6), where G means gignomena.

The above is, roughly, Ziehen's epistemology. What kind of ontology did it presuppose? Ziehen was an explicit monist. But how does Ziehen's epistemology relate to monism, which was so popular among German scientists at the time (Ziche 2000; Nöthlich et al. 2006; Hoßfeld 2013)? In attempting to bridge his epistemology and ontology, Ziehen asked whether *R*-components can actually exist without *N*-components. In other words, he wondered whether there are isolated *R*-components and *N*-components (Ziehen 1934, p. 49). Ziehen's monism led him to regard as misguided the efforts of philosophers who divide everything existing into "psychic" and "physical" parts. Correspondingly, these philosophers are also obliged to ponder whether the "psychic" and the "physical" exist independently, i.e. whether both possess the status of "substance", or whether they somehow depend on each other. The champions of the first view are "psychophysical dualists", while the advocates of the second are "materialists" or "parallelists". Ziehen rejects both dualism and materialism. Yet, he also rejects what he calls faked "identism" or "contrition-systems". Examples of such seeming "identists" are, according to Ziehen, Spinoza and Fechner who initially accepted dualism, but later regretted it and proclaimed the identity of physical and psychical (Ibid. p. 47). Ziehen opposed to these views his own claim that all the "given" contains *R*- and *N*- components. They differ only and exclusively due to "the lawful relations" they are involved in. Ziehen expressed this aspect of his metaphysic in the formula: $\mathrm{G} = \mathrm{R} \sqcap\!\!\!\!\perp \mathrm{N}$. The "given" is not split or separated into two components but changes according to *R*-laws or *N*-laws. Distinguishing between two components is thus a "nomistic" procedure. As to the hypothetical independent existence of the psychic and the physical, Ziehen emphasized that the "given" is all that exists and that *R*- and *N*-components exist only as far they are *in* the "given". To exist means to be given. In colloquial language, Ziehen's view can be expressed by saying that only the "psychic" exists, but this way of putting it is imprecise and misleading (Ibid., p. 48). In fact, *R*- and *N*-components are not even components; they are more accurately understood as *directions* in which the "given" changes. Correspondingly, the question can be reformulated in the following way: Which are primary, *R*-changes or *N*-changes? Do isolated *R*-changes or *N*-changes exist? It is with this question that Ziehen's philosophy meets his psychology and psychiatry, because he insists that it can be answered only with empirical studies and observations (Ibid., p. 50). Based on his own scientific work, Ziehen concluded that neither *R*-changes nor *N*-changes exist independently. Although one can speak of molecules and atoms, which cannot be immediately given in sensations, atoms and molecules always act to *induce* sensation, and thus they are never completely absent therein. The existence of isolated *R*-changes is therefore improbable, although logically not prohibited. The experiments and observations of neuropathology, Ziehen claimed, disprove the independent existence of *N*-changes.

Ziehen called the champions of the philosophical view that *R*-changes cannot in principle exist without *N*-changes "hylozoists" or "hylopsychists", and counts among them Spinoza, Fechner, Spencer and Haeckel. The major mistake of hylozoists is in thinking there is a clear difference between psychic and material "things", whereas according to Ziehen, "the bearer of all changes is completely neutral, i.e., it is neither psychic nor physical and only the laws of changes in two directions can be distinguished" (Ziehen 1934, p. 51). Conscious of drawing on previous forms of monism, Ziehen used the term "neutral monism" to describe the idea that the "ultimate something" [*letztes Etwas, zugrunde Liegende*] is "neutral" in this sense, although he expressed reservations about its explanatory usefulness: "nothing is achieved by bald proclamations of such neutrality or identity" (Ziehen 1939, p. 124). Ziehen claimed that Spencer spoke of "internal" and "external" as corresponding to "soul" and "matter". Other "identists" argued likewise. So, instead of true identism, they actually proposed the existence of three components: internal, external and a hypothetical observer. Ziehen believed he could replace this dualist (actually trialist) view with his own, according to which the "bearer" of the changes is a genuine unity (Ziehen 1934, p. 51). The term "identism" reflects the idea that the "bearer" remains self-identical despite changes. At the same time, Ziehen was against the radical view that every smallest "single unity" (an *R-unity*) is connected to a corresponding "soul" (*N-unity*); there are no "atom-souls", "electron-souls", etc. as Ernst Haeckel would suggest (Ziehen 1939, p. 113). Rather, there are complex entities (complexes of atoms) connected to a "parallel component" (very roughly a "soul" in colloquial language).

The most difficult aspect of Ziehen's philosophy is the transition from his epistemology to his ontology. The idea of the "given" alone and the idea of ensouled "complex entities" alone seem to be comprehensible. However, the idea of "true" identism—his way of avoiding dualism—is rather obscure. The problem is that Ziehen's ontology was not developed by him explicitly, on the same evidentiary basis as empirical epistemology: rather, Ziehen's ontology followed logically from his epistemology.

In a paper published in 1924, he tried to make the link between his epistemology and monistic ontology more explicit (Ziehen 1924). Ziehen pointed out that the so-called idealism is misguided in that it claims that only the psyche ("spirit" in Berkley's terms) is the bearer of the psychic [*Psychisches als Träger des Psychischen*], the same view known as "psychomonism". According to Ziehen, this view is wrong because it presupposes a "material-psychic" dichotomy as if "psychic" referred to a separate kind of reality. Ziehen appealed to the example of the sun: in the sensation of a yellow sun disk, one can distinguish "causal" (same as *R*-components) and "parallel" (same as *N*-components) elements. Here, the word "causal" refers to a chain of events known from physics and physiology, in this case the chemico-physical composition of the sun, the light waves issuing therefrom, the physiological reaction of the visual cortex and so on. "Parallel" refers to the subjective appearance of the sun as yellow disk of a certain size. This is the point where Ziehen makes an ontological transition: "Finally our concept becomes of special significance because, along with many great philosophers (e.g., Spinoza, Fechner, Paulsen), we ascribe parallel processes (in ordinary terminology, 'ensoulment' [Beseelung – auth.]) to all the given. Only with this step to hylopsychism is the unity of the worldview rendered complete".[4] The body-soul problem appears then, Ziehen argued, in a different way. If there are no two distinct substances composing the world, then both materialism and spiritualism are wrong. Both brain physiology and brain pathology demonstrated that there are consistent relationships between the cerebral cortex and psychic processes, but sensations and thoughts do not mysteriously dwell within or outside of cortex cells: "We claim that these cells or, better to say, their reductions remain in dichotomous lawful relations to the rest of the word and to each other" (Ziehen 1924). The watch is not connected to the sense of time by a special mechanism; time and watch also do not merely run in parallel, either. There is only one watch and one dial-plate, Ziehen argued, but the clockwork runs in accord with two different laws, whereas the dial-plate (our mind) displays the whole effect as one [*Gesamteffekt*].

In summary, Ziehen's "true" ontological identism follows from his epistemology, which in turn is connected to his work in psychology and psychiatry. All these disciplines (epistemology, ontology, psychology and psychiatry as part of experimental medicine) worked together for Ziehen to establish a monist and identist worldview. Ziehen's obsession with laws (parallel laws, causal laws, etc.) led him to a view known as nomotheism, i.e. an identification of all-embracing lawfulness with divinity. The latter elements of his philosophy are strongly reminiscent of Haeckel's Gott-Natur (God-Nature) as the idea that the totality of natural laws will be identified with "God". Ziehen's monism, identism, his emphasis on laws and the aura of experimentally provable philosophy attracted German natural scientists, and especially Rensch (Fig. 9.7).

9.5 Bernhard Rensch's Panpsychistic Identism as a Philosophy of Universal Evolutionism

Bernhard Rensch was one of the best-known German "architects" of the evolutionary synthesis, and he crucially contributed to the growth of Darwinism in Germany and worldwide. At the same time, Rensch created an exotic and sophisticated

[4]German original: "Eine bedeutsame Erweiterung erfährt unsere Auffassung schließlich noch dadurch, daß wir mit vielen großen Philosophen Parallelprozesse—in der üblichen Terminologie 'Beseelung'—allem Gegebenen zuschreiben. Erst mit diesem Schritt zum Hylopsychismus wird die Einheitlichkeit des Weltbildes vollständig" (Ziehen 1924).

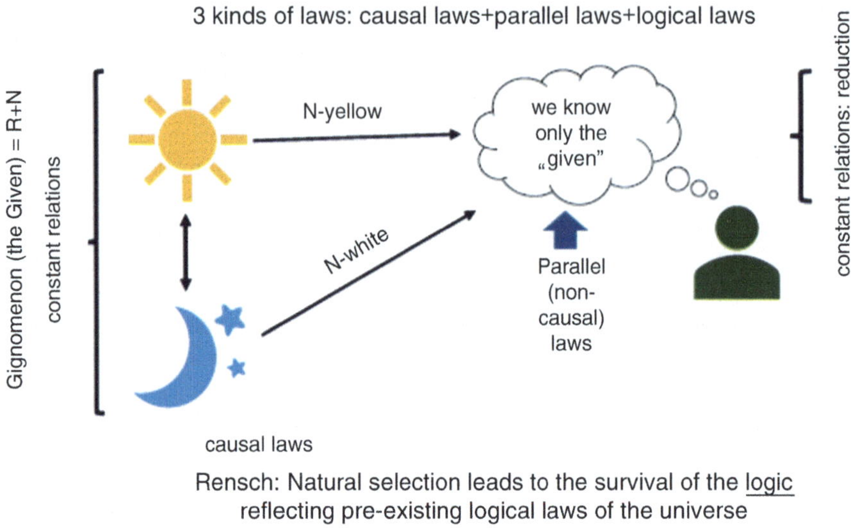

Fig. 9.7 The schematic representation of Ziehen's philosophy

evolutionary metaphysics, which became an integral part of his universal evolutionism. Rensch's philosophy, which he called *panpsychistic identism* (Rensch 1988, p. 36), was a version of monism; thus, Rensch followed in Ernst Haeckel's footsteps not only in establishing Darwinian doctrine but also in promoting the monist worldview. This made Rensch a philosophical opponent of Ernst Mayr as, according to the latter, Rensch was "forced to adapt pan-psychic or hylozoic theories of matter" by denying Mayr's emergentism (Mayr 1982, p. 64).

Rensch was not a selectionist throughout his whole career. Until the end of the 1930s, Rensch was a neo-Lamarckian—he maintained that major macroevolutionary transitions can be explained by the direct adaptation of organisms to their environment (Rensch 1933, pp. 48, 58). At that time, Rensch saw natural selection as an auxiliary mechanism of evolution, but on no account the only or major directing force thereof (Ibid., p. 54). The Lamarckian movement in Germany was from the very beginning tightly united with orthogenesis and the idea of progressive evolution. Orthogenesis was an issue for Rensch already in his early neo-Lamarckian publications and remained central after his "selectionist turn". Yet, if in the "pre-synthetic" period Rensch was ready to accept the concept of directed evolution, in the "synthetic" period, he strongly rejected it. Rensch, along with other champions of the Modern Synthesis including, for example, the American paleontologist George G. Simpson, was convinced that macroevolution can be explained without appealing to saltationism and orthogenesis and that neo-Lamarckian mechanisms are thus redundant to such explanations. Instead, they claimed that paleontological data should be connected with the new explanatory paradigms appearing in the fields of genetics and microsystematics (Mayr 1982, p. 607). However, Rensch never entirely abandoned the idea of progressive and directed evolution. His philosophy became an asylum for the elements of orthogenesis which he banned from the evolutionary theory, in particular the idea of compulsory evolutionary ascent: "The entire *Höherentwicklung* (anagenesis) or biological ascent [*biologischer Aufstieg*] of organisms is therefore for Rensch inevitable [*zwangsläufig*]" (Overhage 1959, p. 75).

The first voluminous paper in which Rensch considers selectionism as a serious alternative to neo-Lamarckism was the *Typen der Artbildung* (The Types of Speciation) (Rensch 1939). His arguments in this article appeared to be quite "synthetic" since he claimed that random mutations and selection can be generally seen as sufficient to explain major transitions in evolution and that by elucidating the "higher categories" and "special regularities of paleontology", there is no need for other explanatory patterns. There is no reason, Rensch argued, for postulating another set of totally hypothetical laws other than those used to explain microevolution.

Rensch's writings on evolutionary biology between 1929 and 1947 demonstrate the astonishing continuity in topics, methodology and empirical generalizations despite the shift in his understanding of evolutionary mechanisms. The continuity of Rensch's theoretical system can to some extent be explained by the guiding role played by the general philosophical principles explicitly and implicitly underlying the entire system (Levit et al. 2008).

Rensch made his philosophy explicit to the international scientific community in his best known "synthetic" book, *Neuere Probleme der Abstammungslehre*: *Die Transspezifische Evolution* (Rensch 1947). The book, which became known in the English-speaking world as *Evolution Above the Species Level*, was written in Prague during the Second World War. The draft of the contents of the book found in the *Archives of the Academy of Sciences in Prague* shows that Rensch conceived his work from the very beginning as a deep theoretical investigation of evolution with explicit methodological and philosophical reflections (Levit et al. 2008). As RG Delisle has pointed out, Rensch's book "is often taken to propound the view that macroevolutionary events are explained by microevolutionary processes. It has perhaps not been sufficiently realized, however, that the microevolution/macroevolution equation in his conception only constitutes the intermediate link of a gigantic and universal causal chain binding together all the cosmic entities, from microphysical phenomena all the way up to the most complex life forms" (Delisle 2009). Already in the first edition of the book, Rensch abundantly cited Theodor Ziehen, his major philosophical inspiration. In this text, Ziehen is mentioned more often than Darwin. In his autobiography, Rensch recalls that he thoroughly—"chapter for chapter"—studied Ziehen's double-volume *Erkenntnistheorie* (Ziehen 1934, 1939) in 1942, right before putting together the *Abstammungslehre* (Rensch 1979, p. 106).

From the standpoint of evolutionary theory, the objective of Rensch's text was to substantiate the Darwinian theory of macroevolution. Truly original also was the line of argumentation in favour of universal selectionism he developed as a contribution to epistemology, philosophy of science and metaphysics, and which he wrote under the influence of Ziehen. In subsequent books, Rensch developed this initial concept into a full-blown philosophical system, which laid the foundation for his evolutionary views.

In one of his latest works, *Probleme genereller Determiniertheit allen Geschehens* (The Problems of the general Determinacy of all Occurrences), Rensch presented his *pantheistic* metaphysics as a holistic and scientifically based worldview (Rensch 1988, p. 11). In constructing his philosophy, Rensch proceeded from the general epistemological assumption that "the only entirely reliable foundation for a philosophical worldview is the indisputable reality *[Wirklichkeit]* of the phenomena [. . .] of consciousness" (Rensch 1988, p. 11). This sounds like a repetition of Ziehen's "immanence principle", but Rensch goes further and provides his epistemology with an evolutionary explanation. The human ability to analyse our experience is an inherited feature acquired over the course of evolutionary history. The essential human ability to "draw conclusions" can be explained by the ways in which the human mental apparatus is adapted to the regularities of the external world. Our advanced degree of adaptation to the lawfulness of the "extra-mental" reality forms the foundation of the correlation between the mental and extra-mental worlds. Rensch makes Ziehen's philosophy less exotic by claiming that there is, indeed, only one single reality, but one with two fundamentally different aspects: the mental and the material.

One of Rensch's central philosophical claims is that both the organic and the inorganic worlds are causally determined—there is no such thing as an acausal process (Rensch 1988, pp. 15–16). Insofar as natural selection is a deterministic mechanism, Rensch thus sees it as the major factor determining organismic evolution. Indeed, he claims that biological progress can be fully explained in causal terms of Darwinian selectionism, although he admits the influence of stochastic events such as random mutations within the selection process.

The evolution of human cultures proceeds mostly on the level of non-heritable characters, Rensch continues, and it is important to distinguish between psychic and neurophysiological phenomena. Rensch refers to Karl Popper's concept of the relationship between psychic and neurophysiological as an example of a *dualistic concept*. Popper assumed that there are two principally different essences [*Seinswesen*]: psychic phenomena, on the one hand, and the neurophysiological processes on the other. Rensch formulates his own position in contrast to Popper's dualism. If purely psychic phenomena, for example, volition, could influence muscle contractions, Rensch argues, it would violate the law of energy conservation making the purely biochemical explanation of muscle contractions impossible (Rensch 1988, p. 34). Another possibility would be to assume that mental [*geistige*] processes run in parallel to events in the material world. This position Rensch labels *psychological parallelism*. Yet, psychological parallelism cannot explain why physiologically identical brain processes can cause various mental effects.

Following Ziehen, Rensch argued that for a human being, the only indisputable objects are his/her own psychic phenomena resulting from immediate experiences: perceptions, imaginations, feelings and thoughts. It is only through analysis of these experiences that we come to develop concepts of an extra-mental world, which appears as a visible and testable reality. Matter appears as "the ultimate something", which will perhaps in the future be described only in terms of interactions of various forces, causal chains, and fundamental constants. Rensch appealed to Ziehen's psycho-physiological epistemology to introduce a "monistic principle" (Rensch 1971, p. 29). As with any kind of philosophical monism, the "monistic principle" constitutes an ultimate, ontologically definable reality which cannot be multiplied or decomposed into further elements. Rensch argued that the reduction of elementary mental features (sensations and perceptions) to their foundations will inevitably lead to the concept of "the ultimate something" that underlies the world and cannot be decomposed into finer elements (Rensch 1988, p. 35). According to Rensch, one can appeal to Spinoza's concept of substance in order to avoid dualism and to give a name to this "ultimate something".

Along these lines, Rensch arrived at his concept of *psychophysical identism* (Rensch 1988, p. 36). He presented this concept for the first time in the *Abstammungslehre* (1947), although at that time he employed another term (with the same meaning) *hylopsychism.* Quite in accord with Ziehen, Rensch claimed: "We would like to point out here once more that this worldview is an idealistic one, since what is primarily given to us is the 'psychic'; there is definitely no opposition between subject and object, matter and soul; even the abstract reductionist world [*Reduktwelt*] of natural scientists should not be searched for outside of the 'conscious' [*Bewußten*]" (Rensch 1947, p. 372).

In the mature philosophical text, *Biophilosophie* (Rensch 1968; English: 1971), Rensch converted Ziehen's identism into the so-called "identistic" foundation of the philosophy of biology and coined the term "panpsychistic-identical or polynomistic worldview". Rensch formulated two basic "facts" constituting panpsychistic identism: "(1) The only reality of which we can be absolutely certain relates to experienced phenomena, which include sensations, mental images, feelings, and volitional processes as a whole. (2) Man does not consist of two separate components—matter and mind, or body and soul, but represents an indivisible psychophysical unity" (Rensch 1971, p. 299). Both claims reveal the strong influence of Ziehen, but Rensch approached the second claim in a somewhat different manner than did his predecessor. Rensch even acknowledged that on this issue, his view was at odds with that of Ziehen (Rensch 1988, p. 36). As Rensch commented in his autobiography, at the time of the third edition of the *Abstammungslehre* (Rensch 1970), he had departed from Ziehen's idea of "proto-psyche" as an immanent property of matter in favour of the view that the "psychic" is the very essence of everything "material" (Rensch 1979, p. 184). Rensch sought to distance himself from the slightest shadow of dualism, and to this end he discovered the need of a "neutral observer" within his concept impossible in Ziehen's world. Despite this and several other discrepancies, Rensch's philosophical thought always remained within Ziehen's paradigm. One may wonder whether Rensch's major claims are self-contradictory in as far as they conflict with Ziehen's epistemology and ontology, but such issues lie outside the scope of this paper. Our objective is just to demonstrate the great impact of Ziehen's philosophy on the central figure of the second Darwinian revolution, and not to go into detailed analysis of Rensch's epistemological views.

Rensch's own panpsychism, which grew out of Ziehen's panpsychism, led him to conclusions which proved crucial for the further development of his version of evolutionism. First of all, he formulated the hypothesis of psycho-phylogeny proving that all psychic abilities develop gradually in the course of biological evolution. Rensch noticed that even protists react to [electrical, physical, etc.] impulses similarly to how organisms with a nervous system react to the same impulses. But if we admit that organisms developed psychic abilities continuously over the entire course of evolutionary history, why should we ascribe "the Psychic" only to the first stages of *biological* evolution without looking for its roots in the geological and astronomic pre-history of evolution? Rensch claimed that we can go down to the level of proto-phenomena (preceding any kind of material evolution, whether abiotic or biological) which underlie the phenomenological nature of the material world: "the proto-phenomena precede even the inanimate pre-stages of phenomena, and respectively matter is of a *proto-phenomenal nature*" (Rensch 1972, p. 406). In other words, protopsychic properties are immanent to matter. *Nihil est in intellectu, quod non fuerit in sensu* can therefore be substantiated in phylogenetic terms, Rensch concluded. Rensch's identism is thus a kind of monism, which was inspired by Ziehen.

Rensch's monist worldview led him to regard nature as deterministic. This type of monism brought about Rensch's determinism. Postulating the pre-phenomenal nature of matter had the corollary that every particle of perceivable reality became supplied with a tiny unit of "intelligence". Since this intelligence is an essential attribute of the Universe, the evolution of the Universe implies a sort of pre-programmed movement in the direction of human-like intelligence. This aspect of Rensch's view is strongly reminiscent of Teilhard de Chardin, who saw the Universe as a "closed quantum", where nothing can appear which did not already exist. Teilhard also saw biological evolution as a continuation of pre-biological evolution, but he went further than Rensch would never go by creating a theology of evolution. To make his universal determinism compatible with the natural scientific worldview, Rensch concealed his obviously teleological concept within the paradigm of universal selectionism, since selectionism was widely accepted and respected and was considered "free" of any teleology.

Rensch's identism-based selectionism is universal, because it penetrates all aspects of evolution: "On the foundation of such panpsychic identism not only the evolution of the Solar System, the Earth, the plants and animals as well as humans can be presented as a continuous process but also the evolution of human and animal mental abilities" (Rensch 1991, p. 258).[5] Even cultural evolution, Rensch asserted, is directed by natural selection, with the only difference being that *genetically non-heritable* features play the major role therein. For Rensch, the primary factor responsible for advancing human civilization over the course of history is science, which is driven by the selection process as well. The evolution of human cultures has

[5]German original: "Auf der Grundlage eines solchen panpsychistischen Identismus stellt sich also nicht nur die Evolution des Sonnensystems, der Erde, der Pflanzen und Tiere sowie des Menschen, sondern auch der tierischen und menschlichen geistigen Fähigkeiten als ein einheitlicher kontinuierlicher Ablauf dar [. . .]" (Rensch 1991, p. 258).

been determined, first of all, "by *growing scientific knowledge*[6] bringing about new techniques and social institutions", while scientific knowledge itself makes progress due to the positive selection of concepts (Rensch 1988, p. 116).

The same is true for the religious evolution, Rensch claimed. Religion is a kind of explanation for external world events developed by humans, who progressively exclude "improbable" explanations in favour of more consistent and "probable" ones: "In fact, the selection of various ways of thinking [*Denkmöglichkeiten*] took place", Rensch concludes (1988, p. 61). The evolution of religious beliefs is a process determined psychically and physiologically and proceeding in a way analogous to biological evolution. Again, religious beliefs are not genetically heritable—instead, the mechanism of biological heredity is replaced by the transmission of intellectual heritages. The law-governed [*Regelhaftigkeit*] progress[7] of religious beliefs manifests itself in the adaptation of beliefs to the general growth of knowledge and an increase of spirituality [*Vergeistigung*] in religious systems. For example, the majority of polytheistic religions evolved towards monotheism. The great religious systems, once appeared, began splitting into sub-systems, sects and so on, showing also in this respect a clear parallelism to organic evolution (Rensch 1988, p. 116).

In sum, Rensch advocated for a kind of all-embracing evolutionism and selectionism. Natural selection is the major source of lawfulness in evolution, and although it differs from the lawfulness of physics, "it is nevertheless possible to characterize evolutionary regularities [*Regelhaftigkeiten*] as laws [*Gesetzlichkeiten*]" (Rensch, 1991, p. 107). Indeed, Rensch insists that the origin of humans from their ape-like ancestors "was presumably a lawfully determined [*gesetzmäßig bedingter*] process" (Rensch 1991, p. 225). Rensch's universal selectionism is thus packed into his universal determinism. He does not reduce his concept of "lawfulness" to the vulgar understanding of determinism and coins the term, "polynomic determination", to describe his own view. Polynomic determination implies that the whole range of biological, physical, chemical, social and other natural laws control the process of evolution, and although the interactions of these laws with one another bring about seemingly stochastic events, in fact, all such events can be explicated in terms of the intersection of lawful processes. Correspondingly, there is a significant chance that organic and cultural evolution would occur on other planets with comparable chemico-physical conditions to those on Earth, and the organisms on such planets would and evolve in a comparable way (Rensch 1991, p. 108). Thus, Rensch's anthropocentric determinism goes hand in hand with universal selectionism (Fig. 9.8).

[6] Rensch's italics.

[7] Here, Rensch employs the same term, "Höherenentwicklung", which was central for discussion about evolutionary progress in German lands in the first half of the twentieth century.

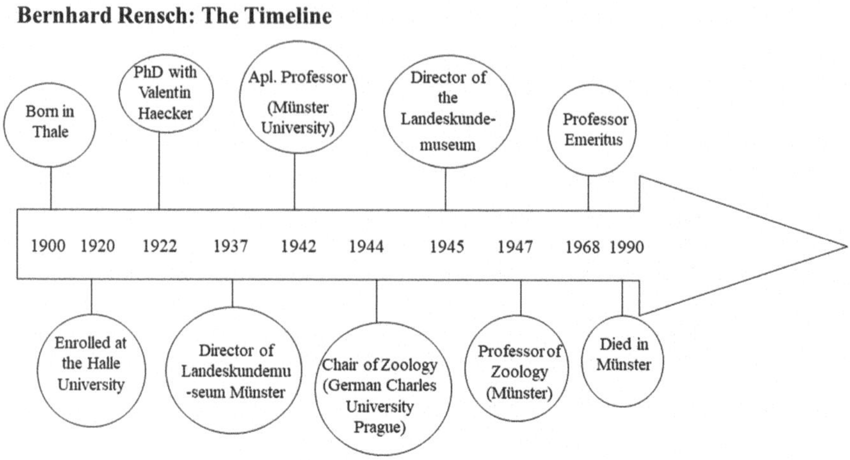

Fig. 9.8 The timeline of Bernhard Rensch

9.6 Once Again: Why Was Bernhard Rensch So Fascinated by Theodor Ziehen?

Rensch was not the only German biologist influenced by Ziehen. As O. Rieppel has shown, the philosophical foundation of Willi Hennig's systematics is based on the work of Ziehen, Rudolf Carnap, Ludwig von Bertalanffy and Hugo Dingler (Rieppel 2006, 2007). Ziehen's cooperation with one of the forerunners of experimental biology, genetics and developmental physiology and a former rector of the University of Halle, Valentin Haecker, is also well-known. Yet, among the biologists who made use of Ziehen's philosophical work, none did so more deeply than Rensch. Why did Rensch choose Ziehen's epistemology and not other philosophies compatible with synthetic evolutionism? There were several philosophers at that time who were influential among German evolutionists who could have potentially taken Ziehen's place or whose ideas Rensch could have combined with Ziehen's. The most obvious candidate for this role would be the "official" philosopher of German Modern Synthesis, Hugo Dingler. Why did Rensch completely ignore Dingler and concentrate only on Ziehen?

Rensch cited Dingler neither in the first and second edition of the *Abstammungslehre* (Rensch 1947, 1954) nor in the *Biophilosophie* (Rensch 1968, 1971) or autobiography (Rensch 1979). Dingler's views were incompatible with Rensch's for two main reasons. First, Dingler was a supporter of national socialism and in that sense was close to Heberer. Dingler entered the SS as a "promoting member" in 1933, but was unable to become a member of NSDAP until 1940, as he had to wait for Hitler to personally grant him membership since Dingler had been a Freemason in the 1920s (Junker and Hossfeld 2002). Rensch, on the other hand, belonged to the liberal wing of German evolutionists and was not a member of any national socialist organizations. Second, Dingler's ideas were unacceptable to

Rensch because Dingler espoused a strong version of methodological voluntarism which Carnap labelled "radical conventionalism" (Rieppel 2016, p. 266). As Rieppel has put it: "At the core of Dingler's philosophy of science stands instead a deductively structured theoretical system based strictly on universals, what Dingler called the world of ideas of ancient Greek philosophers" (Rieppel 2012). While Dingler "firmly rejected the possibility of a rational justification of inductive inference" (Rieppel 2012), Rensch's reasoning was explicitly based on inductive inferences. Dingler's methodological approach was thus entirely alien to Rensch. As Rensch put it: "It is by induction that science at first usually proceeds to the ordering of statements, in other words to classification" (Rensch 1971, p. 11).

Another potential candidate for the role of Rensch's major philosophical inspiration was Ernst Mach, who, similarly to Ziehen, championed "physiological psychology" and monism.[8] This suffices to explain Rensch's general interest in Mach, but it is worth asking why Rensch still preferred Ziehen over him. The answer lies partly in Rensch's own understanding of philosophy philosophical trajectories. Rensch thought of Ziehen's work as the culmination of a tradition which included Mach, Avenarius and other related philosophers. Thus, Rensch read these philosophers through Ziehen's eyes, as it were. He often counted them among Ziehen's theoretical predecessors, sometimes mentioning that psychism, conscientalism, hylozoism and the empiriocriticism of Avenarius only "represent variations or combinations of the theories mentioned" in Ziehen (Rensch 1971, p. 159). It is noteworthy that Mach himself seemed to hold mixed opinions about Ziehen. For example, in the introduction to the Russian edition of his "Analysis of sensations", first published in 1908 and later mentioned by Lenin, Mach wrote that at the end of the 1880s, after he had come into contact with the works of Avenarius (1843–1996), Wilhelm Schuppe (1836–1913) and Ziehen, he concluded that they were each following very close—if not identical—paths (Mach 2005, p. 43). Yet, in a private letter to the Austrian philosopher Wilhelm Jerusalem (1854–1923) on 1 July 1915, Mach was rather sceptical about Ziehen's philosophical achievements: "I have partly read Ziehen[9] and am of your opinion about him. For the branding together [Umstempelung] of Plato and Goethe, if I may be allowed to give an opinion, strikes me as rather comical" (Blackmore et al. 2001, p. 225). However, it was exactly this "synthesis" of Plato and Goethe that made Ziehen so popular among German evolutionists and especially with Rensch.

Rensch's belief that Mach was Ziehen's philosophical predecessor is visible already in the first edition of *Abstammungslehre* (Rensch 1947). Rensch emphasized that there are a few philosophers open to scientific questions and able to combine natural science, psychology and epistemology. The tradition he is explicitly leaning on here includes philosophers like Spinoza, Descartes, Locke, Berkley, Hume, Kant, Spencer and Mach, but "especially Theodor Ziehen" (Rensch 1947, p. 332). Accordingly, he read those philosophers with Ziehen in mind. For example, when writing

[8]https://plato.stanford.edu/entries/ernst-mach/

[9]With all probability, he read Ziehen's *Die Grundlagen der Psychologie* (1915).

about Kant's "appearances" [*Erscheinungen*], Rensch immediately comments that they are the same as Ziehen's gignomena (Ibid., p. 333). But to return to the question, why "especially Ziehen"?

The "special" status Rensch granted Ziehen is illustrated by the fact that the whole tenth chapter of the *Abstammungslehre*, "Evolution of the Phenomena of Consciousness" [*Evolution der Bewusstseinserscheinungen*] (Rensch 1947, pp. 331–373), is written under the immediate and very strong influence of Ziehen. In the second sentence of the initial section of the chapter (following the introduction) labelled "B. Epistemological foundations", Rensch introduced Ziehen's most basic notion of "the given", and later in the text, the whole section C. bears the title, "Evolutions of the Gignomena". Here, Rensch borrowed the term most central to Ziehen's thought, a term which nobody else used in the entire history of philosophy. Rensch also often cited the most detailed and obscure parts of Ziehen's epistemology (Ziehen 1934, 1939). Furthermore, Rensch shared Ziehen's major claim that epistemology *must* proceed from "the given" and that there is no other way to build an applicable epistemology. He uses this claim to justify his rejection of psychophysical dualism and to proclaim the "animatedness" [*Allbeseelung*] of the all matter—with identism being the more sophisticated version of this view. Rensch articulated this latter component of his view even more strongly than Ziehen did his, but the whole argumentative structure comes from Ziehen, although Rensch presents it more clearly.

In fact, Rensch borrowed all the crucial notions of Ziehen's epistemology. He speaks not only of "gignomena" but also of "parallel components" (Rensch 1947, pp. 370–371) and "processes of reduction" (Ibid. p. 334) following in the steps of Ziehen's panpsychism. A new dimension Rensch adds to Ziehen's philosophy is evolution. Rensch spent more than 40 pages of the *Abstammungslehre* analysing the "evolution of the gignomena" and "parallel processes and somatic evolution" (Ibid., pp. 340, 370). The compatibility of universal evolutionism with Ziehen's universal panpsychism is the major reason why Ziehen's thought was "especially" well-suited to serve as the foundation of Rensch's methodology. Ziehen's panpsychism was fully compatible with Rensch's universal evolutionism because Ziehen developed a terminology which facilitated the detachment of the "psychic" or "conscious" from the "human" and the conversion of the former into an abstract and universal category similar to energy or matter. In that sense, Rensch completed Ernst Haeckel's mission by looking for a *Weltseele* (a World's Soul). We should emphasize here that in the subsequent two editions of the *Abstammungslehre* (1954, 1972), Rensch only strengthened the "philosophical" part of his evolutionary theory by constantly keeping Ziehen in the forefront of philosophical chapters. This constitutes more strong evidence in favour of seeing Ziehen's epistemology as a necessary integrative component of Rensch's theory of evolution.

There is also a more subtle argument explaining Rensch's interest in Ziehen. As we have argued elsewhere (Levit et al. 2008), the most characteristic concept in Rensch's monism, which he acquired from the German *Naturphilosophie* tradition and which he shared with Haeckel, was anthropocentrism. Yet, the strict synthetic selectionism adapted by Rensch in the mid-1930s presupposed that evolution has no orthogenetic characteristics, i.e. no pre-programmed directionality, and is a purely stochastic process. In order to preserve the anthropocentricity of the tradition Rensch belonged to, he transplanted the idea of the directionality of evolution from the world of phenomena into the pre-phenomenal nature of things. The evolution of the whole universe became a directed process, but *biological* evolution *sensu strictu* remained stochastic and directed only by natural selection. Ziehen's version of "identism" or "neutralist monism" was most compatible with this philosophical undertaking.

A logical consequence of Ziehen's panpsychism and identism for Rensch was "the idea that the extramental world is structured according to the laws of logic" so that the very ability to think logically becomes an evolutionary adaptation to that extra-mental world (Rieppel 2007). Another crucial consequence of Ziehen's influence on Rensch was the latter's tendency to see "laws" everywhere in biological and social evolution. To develop a completely law-based worldview was one of the declared objectives and cornerstones of Ziehen's philosophy. His "nomotheism" as an identification of God with lawfulness (Gabriel 2004) brings Ziehen, again, close to Haeckel. We can therefore characterize Ziehen as a philosophical "catalyser" of monist evolutionary biology.

9.7 Conclusions

Ziehen was a well-known psychiatrist and an obscure philosopher who left almost no traces in the history of philosophy itself, but whose epistemology enjoyed a significant influence within German evolutionary biology. Our hypothesis is that Ziehen became a visible figure in evolutionary theory mostly because German biology was fundamentally guided by the philosophy of monism. Monism was a characteristic feature of the German tradition of evolutionary biology, and was not as prominent in other scientific cultures (Levit and Hossfeld 2017). Both of the most important figures of the first and the second Darwinian revolutions in Germany (Haeckel and Rensch) were explicit monists. Monism and evolutionary theory were, for Haeckel, parts of the same research program—the "monistic doctrine of evolution" (*monistischen Entwicklungslehre)*—which ultimately aimed at unifying science and religion on a biological foundation (Hossfeld 2010; see also Levit & Hossfeld, first contribution in this volume). Rensch, being a major German "co-architect" of the evolutionary synthesis, developed his own version of synthetic Darwinism into an all-embracing metaphysical system based on a kind of Spinozism situated within the same tradition as Haeckel's monism. Ziehen became Rensch's major philosophical inspiration because Ziehen's specific version of monism was convertible into Rensch's panpsychistic identism, which in its turn served as the foundation of his universal selectionism and evolutionism.

Ziehen's obsession with "laws" also heavily influenced Rensch's thought. Indeed, the very idea of "the law" was ubiquitous in Ziehen's work and remerged prominently in Rensch's philosophy as well. However, the majority of Rensch's "laws" were not seen as such by other evolutionary biologists (e.g. Mayr 1982,

Fig. 9.9 Portrait of Bernhard Rensch, 1954, presented to Gerhard Heberer (Archive of Gerhard Heberer in private archive of Uwe Hossfeld)

p. 37). Rensch's tendency to elevate rules and regularities to the level of laws was a consequence of his philosophical worldview and brings his biological universalism close to that of Ernst Haeckel, the crucial figure of the first Darwinian Revolution in Germany (Fig. 9.9).

Rensch's philosophy was not the "philosophy" of a retired biologist but was instead the core of his entire theoretical system. As our archival research and textual analysis demonstrates, Rensch was from the very beginning convinced that his major "synthetic" book, *Evolution Above the species level*, must include a "philosophical" part. Rensch's all-penetrating nomological selectionism was for him an instrument for establishing a monist agenda and solving the mind-body issue. Rensch's theoretical system therefore gave natural selection a role both crucial and subordinate. In the realm of empirically explicable phenomena, it was irreplaceable, but on a more fundamental metaphysical (metatheoretical) level, it served merely as a tool for drawing the determinist picture of the universe. Rensch's monism and the idea of "polynomic determination" turned out to be a selectionist version of the fundamental idea of directionality in evolution which, on the surface, was formulated in terms of the Modern Synthesis. In fact, however, "Rensch's determinism was a logical consequence of his panprotopsychistic identism" (Ruschmeier 1999, p. 171). In other words, Rensch established a fundamental and logically coherent metaphysics which subordinated to itself all "lower-level" (i.e. more empirical) theories, including selectionism.

Rensch's metaphysics allowed him to preserve the directedness of evolution while simultaneously advocating for all the basic postulates of the Modern Synthesis, which explicitly rule out orthogenesis. Accepting random variation means that evolution loses its orthogenetic characteristics and becomes a stochastic process. In order to preserve the anthropocentricity of the tradition he belonged to, Rensch removed orthogenesis from the world of phenomena and placed it into the pre-phenomenal nature of things. It is not just by chance that he compared his version of identism to Averroes' concept of *natura naturans*, nor that he appealed to Spinoza already in the first edition (1947) of the *Neuere Probleme*. As Rensch puts it: "The evolving forms are substantiated already in the essence of the 'matter' and the lawfulness of the world" (Rensch 1991, p. 528).

Acknowledgements We appreciate suggestions made by Prof. Ian G. Stewart (Halifax, Canada) during his stay at Jena University as an Erasmus-Plus Professor. We are thankful to Dr. Thomas Bach from Ernst-Haeckel-House (Jena, Germany) for providing us with letters of Theodor Ziehen to Ernst Haeckel; Prof. Dmitry Prokudin (St. Petersburg State University, Russia) for establishing citation diagrams; Dr. Alexander Lvov (St. Petersburg State University, Russia) for providing insightful philosophical comments; and Dr. Luise Knoblich (Jena, Germany) for designing visual images.

Cameron Yetman (Canada) contributed a lot to establishing the final version of this paper.

References

Baldwin T (ed) (2003) The Cambridge history of philosophy. Cambridge University Press, Cambridge

Blackmore J, Itagaki R, Tanaka S (eds) (2001) Ernst Mach's Vienna 1895–1930. Boston studies in the philosophy of science, vol 218. Kluwer Academic Publishers, Dordrecht

Bowler PJ (1983) The eclipse of Darwinism. The John Hopkins University Press, Baltimore & London

Bowler PJ (2003) Evolution: the history of idea, 3rd edn. University of California Press, Berkeley, London

Castell R (2003) Geschichte der Kinder- und Jugendpsychiatrie in Deutschland in den Jahren 1937 bis 1961. Vandehoek & Ruprecht, Göttingen

Delisle RG (2009) The uncertain foundation of neo-Darwinism: metaphysical and epistemological pluralism in the evolutionary synthesis. Stud Hist Phil Biol Biomed Sci 40(2):119–132

Delisle RG (2017) Darwinism or a kaleidoscope of research programs and ideas? In: Delisle R (ed) The Darwinian tradition in context. Springer, Cham

Dücker G (ed) (2000) 100 Jahre Bernhard Rensch: Biologe - Philosoph – Künstler. LIT Verlag, Münster

Gabriel G (2004) Ziehen, Theodor. In: Enzykopädie Philosophie und Wissenschaftstheorie. J.B. Metzler, Stuttgart

Gerhard U-J (2004) Theodor Ziehen, M.D., Ph.D., 1862–1950. Am J Psychiatr 161:1369

Gerhard U-J, Blanz B (2004) Theodor Ziehen. Am J Psychiatr 161(8):1369

Gerhard U-J, Blanz B (2002) Theodor Ziehen (1862–1950). Nervenarzt 73:905–907

Gliboff S (2012) Monism and morphology at the turn of the twentieth century. In: Weir TH (ed) Monism. Palgrave Macmillan, N.Y., pp 135–158

Heberer G (ed) (1943) Die Evolution der Organismen. Gustav Fischer, Jena

Honderich T (ed) (2005) The Oxford companion to philosophy, 2nd edn. Oxford University Press, Oxford

Hopwood N (2015) Haeckel's embryos: Images, evolution, and fraud. The University of Chicago Press, Chicago & London

Hörz H, Liebscher H, Löther R, Schmutzer E, Wollgast S (eds) (1997) Philosophie und Naturwissenschaften. Pahl-Rugenstein, Bonn

Hoßfeld U (1997) Gerhard Heberer (1901–1973) Sein Beitrag zur Biologie im 20. Jahrhundert. Berlin, VWB

Hossfeld U (2010) Absolute Ernst Haeckel. Orange Press, Freiburg

Hoßfeld U (2013) Protestantismus und Monismus - Das Beispiel Ernst Haeckel. In: Spurenlese. Kulturelle Wirkungen der Reformation. Hg. von der Reformgeschichtlichen Sozietät der Martin-Luther-Universität Halle-Wittenberg, Evangelische Verlagsanstalt, Leipzig, pp 219–241

Hossfeld U, Watts E, Levit GS (2017) Valentin Haecker (1864-1927) as a pioneer of phenogenetics: building bridge between genotype and phenotype. Epigenetics 10(4):247–253
Hossfeld U, Levit GS, Watts E (2019) 100 years of phenogenetics: Valentin Haecker and his examination of the phenotype. Mol Genet Genom 294(2):445–456
Huxley J (ed) (1940) The new systematics. At the Clarendon Press, Oxford
Junker T (2004) Die zweite Darwinsche Revolution: Geschichte des Synthetischen Darwinismus in Deutschland 1924 bis 1950. Acta Biohistorica, Bd. 8. Marburg
Junker T, Hossfeld U (2002) The architects of the evolutionary synthesis in national socialist Germany: science and politics. Biol Philos 17(2):223–249
Kolchinsky EI (2014) The unity of evolutionary theory in the 20th century divided world. Nestor-Historia, St. Petersburg
Kutschera U, Niklas KJ (2004) The modern theory of biological evolution: an expanded synthesis. Naturwissenschaften 91:255–276
Levit GS, Hoßfeld U (2006) The forgotten "old Darwinian" synthesis: the evolutionary theory of Ludwig H. plate (1862-1937). NTM Int J Hist Ethics Nat Sci Technol Med 14:9–25
Levit GS, Hoßfeld U (2011) Darwin without borders? Looking at "generalised Darwinism" through the prism of the "hourglass model". Theory Biosci 130:299–312
Levit GS, Hossfeld U (2017) Major research traditions in twentieth-century evolutionary biology: the relations of Germany's Darwinism with them. In: Delisle R (ed) The Darwinian tradition in context. Cham, Springer Nature, pp 169–193
Levit GS, Hossfeld U (2019) Ernst Haeckel in the history of biology. Curr Biol 29:R1269–R1300
Levit GS, Hoßfeld U (2020) Biology and Panpsychism: German evolutionists and a philosopher Theodor Ziehen (1862–1950). Vestnik of the Saint-Petersburg University: philosophy and conflict studies, 36(2): 240–253
Levit GS, Simunek M, Hoßfeld U (2008) Psychoontogeny and psychophylogeny: the selectionist turn of Bernhard Rensch (1900-1990) through the prism of panpsychistic identism. Theory Biosci 127:297–322
Mach E (2005) Analysis of sensations [translated into Russian]. Territoria Budustchego, Moscow
Mayr E (1982) The growth of biological thought. The Belknap Press of Harvard University Press, Cambridge, Mass
Mayr E (1999) Thoughts on the evolutionary synthesis in Germany. In: Junker T, Engels E-M (eds) Die Entstehung der Synthetischen Theorie: Beiträge zur Geschichte der Evolutionsbiologie in Deutschland. Verlag für Wissenschaft und Bildung, Berlin, pp 19–30
Nenadić I (2011) Der übermenschliche Patient: Nitzsches Behandlung in der Jenaer Psychiatrie. Weimar-Jena die Große Stadt 4(1):31–45
Nöthlich R, Weber H, Hoßfeld U, Breidbach O, Krauße E (2006) "Substanzmonismus" und/oder "Energetik": Der Briefwechsel von Ernst Haeckel und Wilhelm Ostwald (1910–1918). Zum 100. Jahrestag der Gründung des Deutschen Monistenbundes. VWB, Berlin
Overhage O (1959) Um Die Ursachliche Erklarung Der Hominisation Biologisch Hypothesen über den Ursprung des Menschen. Acta Biotheoretica, Supplementum Primum, vol XII. E.J. Brill, Leiden
Reif W-E, Junker T, Hoßfeld U (2000) The synthetic theory of evolution: general problems and the German contribution to the synthesis. Theory Biosci 119:41–91
Rensch B (1933) Zoologische Systematik und Artbildungsproblem. Akademische Verlagsgesellschaft, Leipzig
Rensch B (1939) Typen der Artbildung. Biol Rev 14:180–222
Rensch B (1947) Neuere Probleme der Abstammungslehre. Die transspezifische Evolution. Ferdinand Enke Verlag, Stuttgart
Rensch B (1954) Neuere Probleme der Abstammungslehre. Die transspezifische Evolution. Zweite, stark veränderte Auflage. Ferdinand Enke Verlag, Stuttgart
Rensch B (1968) Biophilosophie auf erkenntnistheoretischer Grundlage (Panpsychistischer Identismus). Gustav Fischer Verlag, Stuttgart
Rensch B (1970) Neuere Probleme der Abstammungslehre. Ferdinand Enke Verlag, Stuttgart

Rensch B (1971) Biophilosophy. Columbia University Press, New York
Rensch B (1972) Neuere Probleme der Abstammungslehre. Die transspezifische Evolution. Dritte, durch einen Anhang erweiterte Auflage. Ferdinand Enke Verlag, Stuttgart
Rensch B (1979) Lebensweg eines Biologen in einer turbulenten Zeit. Gustav Fischer, Stuttgart, New York
Rensch B (1988) Probleme genereller Determiniertheit allen Geschehens. Parey, Berlin/Hamburg
Rensch B (1991) Das Universale Weltbild. Wissenschaftliche Buchgesellschaft, Darmstadt
Rieppel O (2006) On concept formation in systematics. Cladistics 22:474–492
Rieppel O (2007) The metaphysics of Hennig's, phylogenetic systematics: substance, events and laws of nature. Syst Biodivers 5(4):345–360
Rieppel O (2012) Hugo Dingler (1881–1954) and the philosophical foundation of the German evolutionary synthesis. Biol Theory 6:162–168
Rieppel O (2016) Willi Hennig as a philosopher. In: Williams D, Schmitt M, Wheeler Q (eds) The future of phylogenetic systematics: the legacy of Willi Hennig. Cambridge University Press, Cambridge, pp 356–376
Robertson M, Light E, Lipworth W, Walter G (2017) Psychiatry, genocide and the National Socialist State: lessons learnt, ignored and forgotten. In: Marczac N, Shields K (eds) Genocide Perspectives V. UTS ePRESS, Sydney, pp 69–89
Romanes GJ (1895) Darwin and after Darwin: post-Darwinian questions: heredity and utility. Open Court, Chicago
Ruschmeier I (1999) Das philosophische Werk Bernhard Renschs. Eine kritische Auseinandersetzung mit Renschs wissenschaftlichem Weltbild anhand klassischer Probleme der Philosophie. (= Naturwissenschaft – Philosophie – Geschichte (Hg. v. Hucklenbroich P) Band 9). Lit-Verlag. Münster, Hamburg, London
Seeman MV (2005) Psychiatry in the Nazi Era. Can J Psychiatr 50:218–225
Spinoza B (1996) Ethics. Penguin Classics, London
Ziche P (2000) [Hg.] Monismus um 1900. Wissenschaftskultur und Weltanschauung. VWB, Berlin
Ziehen T (1907) Psychophysiologische Erkenntnistheorie. Zweite Auflage. Gustav Fischer, Jena
Ziehen T (1915) Die Grundlagen der Psychologie. B.G. Teubner, Leipzig und Berlin
Ziehen T (1922) Grundlagen der Naturphilosophie. Verlag Quelle & Meyer, Leipzig
Ziehen T (1923) Theodor Ziehen. In: Schmidt R (ed) Die Philosophie der Gegenwart in Selbstdarstellungen, vol Bd. 4. Meiner, Leipzig, pp 219–236
Ziehen T (1924) Das Leib-Seele-Problem. Dtsch Med Wochenschr 38:1–7
Ziehen T (1934) Erkenntnistheorie. Teil I. Gustav Fischer, Jena
Ziehen T (1939) Erkenntnistheorie. Teil II. Gustav Fischer, Jena

Chapter 10
Inertia, Trend, and Momentum Reconsidered: G. G. Simpson—An Orthogeneticist?

Carlos Ochoa

> *The structure of an ancestral group inevitably restricts the lines of possible evolutionary change. That simple fact greatly increases the probability that among the number of descendant lineages several or all will follow one line.*
> *George Gaylord Simpson (1961)*

Abstract George Gaylord Simpson, one of the architects of Modern Synthesis, was one of the main figures of paleontology who discredited and rejected the theory of orthogenesis in his discipline. Following the neo-Darwinian agenda, he thought that this theory had little basis to be proven. Since then, orthogenesis has been defined in textbooks as a "metaphysical," "vitalistic," or "theological" theory. However, in the present analysis, I demonstrate that Simpson indirectly advocated for an explanation of orthogenesis through his explanation of the concept of "parallelism." In other words, Simpson did not end orthogenesis but rather ended up defending the phenomenon of orthogenesis through the concept of parallelism. I argue that Simpson maintained pluralistic ideas upon including constraints into his evolutionary system as a complementary factor to the argument of natural selection.

Keywords George Gaylord Simpson · Modern synthesis · Paleontology · Orthogenesis · Parallelism · Constraints · Macroevolution · Otto Schindewolf · William Berryman Scott · Natural selection

10.1 Introduction

Once upon a time, more than a hundred years ago, orthogenesis (or directed evolution) was one of the most popular and accepted theories in evolutionary biology, particularly in paleontology. But with the rise of the Modern Synthesis movement

C. Ochoa (✉)
Faculty of Science, Department of Evolutionary Biology, Universidad Nacional Autónoma de México, Coyoacán, Mexico City, Mexico
e-mail: carlos008a@ciencias.unam.mx

R. G. Delisle (ed.), *Natural Selection*, Evolutionary Biology – New Perspectives on Its Development 3, https://doi.org/10.1007/978-3-030-65536-5_10

during the 1930s and early 1940s, geneticists and taxonomists perceived that orthogenesis was inconsistent with the assumptions of population genetics, but little could be done to disprove it. Until one day, in 1944, an American paleontologist named George Gaylord Simpson published a book whereby he reconciled population genetics with paleontological data. Following the neo-Darwinian agenda, he thought that orthogenesis had little basis to be proven. And that was how Simpson ended all theoretical consideration of orthogenesis in paleontology, and neo-Darwinians lived happily ever after.

The previous paragraph, written with a little bit of sarcasm, is a model of the type of narrative that has been used to discredit the theory of orthogenesis while highlighting the achievements of the Modern Synthesis. In any case, the narrative of the history of the Modern Synthesis, which treats this movement as a scientific revolution, has been challenged by some modern historians who attempt to reframe the history of evolutionary thought (e.g., Amundson 2005; Delisle 2009, 2011, 2017; Ochoa 2017, 2021; Stoltzfus 2017; Adams 2021; Schwartz 2021; van der Meer 2021). On that basis, this chapter aims to contribute to the historiographic development of the Modern Synthesis, but under the "no-traditional" approach. This study attempts to reconsider not only how the history of evolutionary biology has been written but also the basis on which some current evolutionary positions have developed.

In the early years of the twentieth century, anti-Darwinian theories such as saltational evolution and orthogenesis dominated in the disciplines that studied macroevolution, e.g., morphology and paleontology. In consequence, during the development of the Modern Synthesis, the first architects entered the scene slightly skeptical about supporting one of the most important assumptions of this movement, that is, that the larger-scale evolutionary phenomena (macroevolution) could be explained by observable phenomena and occurrences on a smaller evolutionary scale (microevolution). Shortly thereafter, paleontologist George Gaylord Simpson, in his book *Tempo and Mode in Evolution* (1944), stated an argument whereby paleontological data was congruent with population genetics research and, therefore, with a microevolutionary approach. When the architects took notice of this work, they were more confident in their conclusions about the premise of extrapolation. As a result, alternative theories to Darwinism were subsequently marginalized. Without a doubt, that is why we can say that Simpson's contributions to paleontology were vital for the development of the Modern Synthesis.

Despite this narrative, we have been noticing that Simpson's works stayed aloof from the central assumptions of Darwinism (Ochoa 2017; Popov 2018), and although he attempted to interpret the observations of paleontology with the known mechanisms of population genetics, many of his central assumptions are far from supporting a reductionist point of view. In particular, I shall demonstrate that although Simpson discredited and rejected the theory of orthogenesis in his writings, he indirectly advocated for an explanation of orthogenesis through his explanation of the concept of "parallelism" (a type of homoplasy). In other words, we shall see that Simpson did not end orthogenesis, but rather ended up defending the phenomenon of orthogenesis through the concept of parallelism. Simpson as a character is quite complex, so we cannot say that he just reduced all paleontological observations to microevolutionary facts. Hence, this chapter endorses the idea that Simpson maintained pluralistic ideas upon including constraints into his evolutionary system as a complementary factor to the argument of natural selection.

To bring this about, I first explore in section 10.2 how the figure of Simpson was constructed in the Modern Synthesis narrative. I stress that Simpson's works stayed away from a reductionist view. In the following section 10.3, I briefly discuss the meaning of orthogenesis. Specifically, I review the ideas of two anti-Darwinian paleontologists of the early and mid-twentieth century, respectively, William Berryman Scott and Otto H. Schindewolf, who related the phenomena of orthogenesis with the modern term of "parallelism." In the last section 10.4, I analyze Simpson's ideas concerning orthogenesis, and I expose how even though he rejected the theory in his writings, I have discovered that he defended the idea of orthogenesis and even gave support and explanation for its occurrence indirectly through the term of "parallelism."

10.2 Creating the Hero of the Modern Synthesis

George Gaylord Simpson was a well-known American paleontologist for his role as an architect of the Modern Synthesis. Simpson began his work by studying the Mesozoic and Cenozoic mammals of North America as well as those of the Tertiary in South America. He was appointed to an Alexander Agassiz professorship at the Museum of Comparative Zoology, Harvard University. Later, he joined the University of Arizona and became professor emeritus of geosciences. Simpson inaugurated a new period in the study of vertebrate paleontology, and, along with Edward Drinking Cope, Henry Fairfield Osborn, and William Berryman Scott, he figured as one of the most outstanding American paleontologists in the history of evolutionary thought. He founded the Society of Vertebrate Paleontology, belonged to distinguished scientific associations, and published many papers and books, whose subjects included paleontology and evolution (Wittington 1986).

But perhaps the most prominent influence is the publication of Simpson's book, *Tempo and Mode in Evolution* (1944), which contributed to the Modern Synthesis by making compatible the views of population genetics with paleontological data. But, why was Simpson's book so special? According to Gould (1980, pp. 157–160), firstly, *Tempo and Mode* was a unique book in the sense that it was outside from the paleontological tradition of its time. Paleontologists wrote about evolution, but most of them made fossil descriptions, or they only established their phylogeny. Few devoted considerable attention to the study of the processes and mechanisms of evolution. Secondly, the book contains graphs, frequency distributions, and representative models: "No paleontological innovation could have been more stunning than this." Simpson's book introduced the novelty of quantified information, unique in comparison with traditional paleontological works. Thirdly, and more importantly, he displayed a good argument supporting that Darwinism might be consistent with the fossil record; therefore, no special theories of macroevolution were needed (such as saltational evolution and orthogenesis). And in the opinion of Cain (1992), *Tempo and Mode* was the bridge that linked population genetics with paleontology. This allowed greater confidence in the extrapolationist premise, that is, that micro-evolutionary processes could also explain the macroevolutionary realm, although the book itself gave a somewhat exclusive role to paleontology when dealing with a macroevolutionary approach.

In this regard, it is important to stress here that during the beginning of the twentieth century, most paleontologists supported ideas contrary to traditional Darwinism, for example, by considering that evolution might also occur by leaps and that variations do not occur randomly, therefore, paleontologists maintained that their discipline had autonomy in the study of macroevolution and the history of life (Bowler 1983, 2017; Gould 2002; Levit et al. 2008; Ochoa 2017). Later, the Modern Synthesis did not gain general acceptance until paleontological data were consistent with Darwinism. By regarding the meaning of the Modern Synthesis, for example, Mayr's rhetoric (1980a) tells us that these advances facilitated the removal of those "communication gaps" that did not allow understanding among paleontologists, geneticists, and naturalists. Thus, the Modern Synthesis was presented as an agreement between different research traditions, excluding anti-Darwinian theories that "had wrong ideas on the nature of inheritance and variation."

This narrative was not accidental, since, in the writings of the architects, we can observe that Simpson's work provided some confidence with respect to extrapolationist premise that macroevolution is nothing more than accumulated microevolution. This was a crucial point for the development of the Modern Synthesis.[1] Theodosius Dobzhansky, for example, in his monumental book *Genetics and the Origin of Species* (1937), which was published 7 years before Simpson's book, argued about the difficulties of assuming the extrapolationist premise. He stated that geneticists had been limited to studying only the phenomena of micro-evolution since it is not possible to experiment for millions of years; at that point, the best advice was to conceive that microevolution and macroevolution were the same but at a different timescale: "For this reason we are compelled at the present level of knowledge reluctantly to put a sign of equality between the mechanisms of macro- and micro-evolution, and, proceeding on this assumption, to push our investigations as far ahead as this working hypothesis will permit" (Dobzhansky 1937, p. 12). However, in a later edition of his book, Dobzhansky is more confident in his hypothesis[2] because he found support in the works of paleontologists and morphologists who were in favor of his research:

[1]Whether microevolution is different from macroevolution has been one of the central issues in evolutionary thought; see Adams (2021) in this volume.

[2]However, it is important to note that Dobzhansky had doubts about the central role of natural selection in macroevolution through his life; for this discussion, see van de Meer (2021) in this volume.

> All that is possible is to examine the evidence bearing on macroevolution which has been accumulated by paleontologists and morphologists, and to attempt to decide whether it agrees with the hypothesis that all evolutionary changes are compounded of microevolutionary ones... The three authors [Simpson, Schmalhausen, and Rensch] find nothing in the known macroevolutionary phenomena that would require other than the known genetic principle for causal explanation. (Dobzhansky 1951, p. 17)

Likewise, Ernst Mayr, in his book *Systematics and the Origin of Species* (1942), published 2 years before Simpson's *Tempo and Mode*, is equally distrustful but at the same time enthusiastic to reach an agreement to solve the discrepancies between the disciplines that study microevolutionary principles and the disciplines that focus on a macroevolutionary approach:

> Geneticists and most taxonomists have devoted most of their attention to microevolution, and the field of macroevolution was left more or less to the paleontologist and the anatomist. This has led to difficulties and misunderstandings, since paleontologists, taxonomists, and geneticists talk three different languages, and all three of them have certain mistaken ideas about the basic facts and axioms of their sister disciplines. To state that orthogenesis proves that evolution proceeds without selection would be just as erroneous as to state that orthogenetic series do not exist. (Mayr 1942, pp. 291–292)

However, in *Animal Species and Evolution* (1963), Mayr offered his vote of confidence in Simpson's works as follows: "It is not the task of this volume... to refute these theories [saltational evolution and orthogenesis] and to cover in detail the entire area of transpecific evolution [i.e., macroevolution]. This has been done superbly by Simpson ... with emphasis on the paleontological evidence" (p. 586). Similarly, many years later, Mayr stressed the fact that Simpson "was one of the most important architects of the synthesis. He engineered the marriage of paleontology with genetics and more broadly with the rest of evolutionary biology" (Mayr in Mayr and Provine 1980, p. 153). Also, that "Simpson was responsible for bringing paleontology and macroevolution into the synthesis... He acquired this understanding by studying the genetic literature and was able, by integrating it with his fine knowledge of paleontological evidence, to arrive at an interpretation of macroevolutionary events in *Tempo and Mode of Evolution* (1944) that was fully consistent with the findings of the new genetics" (Mayr 1980a, p. 37). And finally, he states: "The gap between the findings of genetics laboratories and the findings of paleontologists seemed to be unbridgeable because of the difficulties of a genetic interpretation of macroevolution. This gap, in turn, delayed the synthesis until the division was closed by zoologists and paleontologists (Rensch and Simpson)" (Mayr 1980b, p. 134).

In short, we can ascertain that in the first writings of the founders of the Modern Synthesis, they were doubtful concerning the conclusion on macroevolution. Up until that time, disciplines such as embryology, morphology, and paleontology stayed away from the achievements accomplished by population genetics, experimental genetics, and the "new systematics." For example, the greatest adversary of the Modern Synthesis, Richard Goldschmidt, argued in *The Material Basis of Evolution* (1940) that: "The general picture of evolution resulting from such deliberations is in harmony with the facts of taxonomy, morphology, embryology, paleontology, and the new developments of genetics. The neo-Darwinian theory of the geneticists is no longer tenable" (p. 397). But once Simpson's *Tempo and Mode* book appeared, the architects felt more confident with respect to the extrapolationist premise. With this, I do not mean that Simpson's work alone facilitated this endeavor, for example, surely other contributions such as those of Julian Huxley and Bernhard Rensch strengthened this point of view as well. In any case, Simpson's writings were essential for creating this narrative of the Modern Synthesis: the microevolutionary mechanisms provided by genetics and taxonomists are congruent with what we know about paleontology and macroevolution. For example, Laporte (2000, p. 1), in his biographical work on Simpson, mentions that: "Simpson's book applied the concepts and conclusions of the new discoveries in genetics to the large body of fossil evidence of life's long history, and claimed that the 'microevolution' of the geneticist could indeed be extrapolated to explain adequately the 'macroevolution' of the paleontologist."

It is interesting to note that this rhetorical assumption, characteristic and essential to the argument of the Modern Synthesis, still resonates in some recent historical works dealing with the topic of macroevolution and orthogenesis. Bowler (2017, p. 209), for example, claims that: "The work of George Gaylord Simpson would cement the palaeontologists' rejection of the non-Darwinian stance adopted by the previous generation." Correspondingly, Pigliucci (2017, p. 93; his italics) argues that: "It was George Gaylord Simpson's (1944) magistral role within the Modern Synthesis to clear away any remnants of orthogenesis from paleontology … he convincingly argued that the sort of so-called 'micro'-evolutionary processes accounted for by Darwinism could be extrapolated to geological timescales, thus yielding the *appearance* of macro-evolutionary changes of a qualitatively different nature. In reality, Simpson argued, the second is simply a scaled up version of the former." Similarly, Turner (2017, p. 337) comments that: "Simpson's *Tempo and Mode in Evolution* (1944), which is often cited as an effort to bring paleontology into the modern synthesis, represented a major theoretical turn away from earlier ideas such as orthogenesis… and the neo-Lamarckian theorizing about macroevolution that one finds in the work of Edward Drinker Cope." Nevertheless, Turner wonders whether this was really the only merit that Simpson had for the Modern Synthesis: "Still one big issue that Simpson's work had not entirely resolved was whether paleontology would have much more to contribute to neo-Darwinian evolutionary theory."

In this regard, Sepkoski (2019) emphasizes the fact that paleontology played a very important role in the development of the Modern Synthesis, particularly due to Simpson's influence both institutionally and theoretically. However, he argues that it seems odd to realize that many paleontologists of the 1970s and 1980s such as Stephen Jay Gould, Niles Eldredge, and Steven Stanley, among others were dissatisfied with the achievements of paleontology in the Modern Synthesis. They even argued that this discipline was relegated by the movement. For example, Gould (1980, p. 170) considered that "Simpson's synthesis unified paleontology with evolutionary theory, but at a high price indeed– at the price of admitting that no fundamental theory can arise from the study of major events and patterns in the history of life."

But an important point of this discussion is that Sepkoski (2019, p. 693) noticed that *Tempo and Mode* differs considerably from the opinion of the other architects, for example, by considering the assumption that the fossil record is incomplete. On the contrary, Simpson points out that the absences in the fossil record might be real, and part of his argument in *Tempo and Modes* relies on the claim that higher taxa have evolutionary rates and modes different than those of species. One of his central contributions to this discourse is with respect to the development of the concept of "quantum evolution,"[3] for example. And although it is not precisely a mode that moves away from the mechanisms provided by population genetics, it is a manifesto for the no-extrapolation. However, according to Gould (2002, pp. 528–531), Simpson changes his mind regarding quantum evolution. While in *Tempo and Mode* he highlighted that quantum evolution was the most important concept of his research, in *The Major Features of Evolution* (1953) (which is considered as the second edition of *Tempo and Mode*), he doesn't consider it as an exclusive mode, but rather as a type of phyletic evolution. This interpretation brings us to the assumption that Simpson yielded to extrapolation in later works.

In fact, Sepkoski (2019, pp. 694–695) follows Gould's narrative, and he laments that Simpson changed his mind since the arguments in *Tempo and Mode* could be outlined as a work that would bestow theoretical autonomy to paleontology regarding evolution; and the concept of quantum evolution, specifically, was one of the most central contributions toward this endeavor. Finally, Sepkoski (2019) concluded that paleontology has played an important role in the development of the Modern Synthesis; however, this contribution was not entirely revolutionary since other perspectives came to light in subsequent years which called for a better contribution of paleontology to evolution.

In any case, I suspect that Simpson's contribution to paleontology was eclipsed, on the one hand, by this historiographic narrative of the achievements in paleontology developed by the other architects (particularly Mayr) to corroborate the extrapolationist premise and, on the other hand, by the same paleontologists of the 1970s and 1980s who used this narrative to promote and defend their theories on macroevolution, particularly by Stephen J. Gould and Niles Eldredge during the development of punctuated equilibrium. An example of this may be seen in Gould and Eldredge's (1977) discourse when arguing in favor of "speciation" instead of "phyletic gradualism" as the dominant mode of biological evolution: "speciation is the raw material of macroevolution, and genetic substitution within populations cannot be simply extrapolated to encompass all events in the history of life. We therefore challenged the central assumption that secured the admission of

[3] To account for the origin of the species and higher taxa, Simpson argued that there were different modes of evolution. Speciation (splitting up of a population) pertains to species and subspecies. Phyletic evolution (directional shift of average characters of the entire population) was related to the genus level. Quantum evolution (a rapid change of the population from a state of instability to stability) corresponds to higher categories such as families, classes, and orders.

paleontology into the modern synthesis of evolutionary theory (Simpson 1944, 1953): change in gene frequency within populations is the building block of major evolutionary events" (p. 139).

As we can see, they quote *Tempo and Mode* and *The Major Features of Evolution* as the source of the perspective they seek to challenge. However, as we saw above, *Tempo and Mode* is beyond the reductionist approach for not simply based on the notion of "change in gene frequency within populations." Moreover, unlike Eldredge and Gould, Simpson thought that higher taxa are real and, therefore, genuine units of evolution. Even despite what most biologists may think, he still maintained this non-reductionist point of view in *The Major Features of Evolution*:

> The study of the evolution of higher categories, say for present purposes from families upward, does involve some of the most important problems of evolution. For one thing, evolution on this scale cannot be directly studied by experimental method ... The main themes of this book relate to evolution on a broad scale, hence largely to categories above the species. The time has now come to undertake explicit consideration of what higher categories are and how they arise. (Simpson 1953, pp. 339–340)

In fact, a very different idea of Simpson as a character has also been noted by Popov (2018, p. 54), who writes far removed from historiographic orthodoxy about the subject of orthogenesis. For example, Simpson, he says, "remained aloof from his fellow 'architects', and seemed to enjoy his aloofness, too. 'Orthodox' Darwinists doubted that Simpson's views on evolution were quite right." Even the editors of the Russian edition of *Tempo and Mode* realized that Simpson "came to Darwinism 'by the backdoor' of doubts and exceptions." Therefore: "Deviations from the traditional scheme can be easily found in his work."

Perhaps many questions arise from all of this, like how "anti-Darwinist" could Simpson's works end up being? If *Tempo and Mode* stays away from the assumptions of the other architects of the Modern Synthesis, why did these architects use this famous text to foster and strengthen their views? Was *Tempo and Mode* just a rhetorical means to promote the Modern Synthesis movement? And, if there was a radical change in his stance, did Simpson feel pressured by the other architects and correspondingly had to sacrifice his macroevolutionary ideas? Attempting to answer all these questions would obviously require many more historiographical studies that may exceed the objectives of this humble essay. However, I shall demonstrate that one of the most emblematic ideas of the anti-Darwinian movement and macroevolution in early-twentieth-century paleontology is present in Simpson's works: orthogenesis. But before looking into this topic in detail, it is pertinent to know the historical context of the theory of orthogenesis, and we will see how this theory fostered the development of the modern term of "parallelism" as well.

10.3 On the Origin of the Term "Parallelism": Orthogenesis and Paleontologists

One of the evolutionary controversies that have been debated is the meaning of concepts from comparative anatomy such as homology and homoplasy. Homology refers to the identity of structures in different lineages regardless of form and function, while homoplasy pertains to those similarities of form which evolved independently in two separate lineages. Regarding homoplasy, biologists recognize two kinds: parallelism and convergence. Parallelism generally refers to independent evolution in closely related groups, mainly due to internal constraints that limit and channel evolutionary change, while convergence is described as functionally similar structures which arise independently in distantly related groups but caused mainly due to the action of natural selection (Ochoa and Barahona 2014).

With this in mind, it is interesting to note that Ochoa and Barahona (2014) have found that the terms "parallelism" and "convergence" have an underlying origin in anti-Darwinian theories (see also Gould 2002, pp. 1081–1086). To clarify, in their study, they demonstrate that the term "parallelism" emerges from the theory of orthogenesis, but they also gathered that the meaning of the theory of orthogenesis is not precisely the one shown in textbooks, as we shall see hereafter.

Currently, the theory of orthogenesis has been misunderstood, and it has generally been associated with metaphysical, theological, or divine conceptions. Nevertheless, some new interpretations of orthogenesis have come to light (e.g., Grehan and Ainsworth 1985; Gould 2002; Levit and Olsson 2006; Popov 2009, 2018; De Renzi 2014; Ochoa and Barahona 2014; Ulett 2014, 2016; Ochoa 2017), and far from any progressive or teleological view, they propose that this theory had well-founded empirical and epistemological bases and whose phenomenology probably relies on developmental constraints.

In general, orthogenesis was a theory which indicated that some characters of related groups follow the same evolutionary trend and whose causes are attributed to internal factors. For example, species of different genera but belonging to the same family possess the ability to produce the same characters independently; this change, however, is not necessarily due to adaptation. Some naturalists claimed that there might be some internal factor that controlled the evolutionary pathway, e.g., genetic processes or developmental constraints, and although natural selection may also act to produce these transformations, it was limited to act with few alternatives of change (Ochoa and Barahona 2014).

Orthogenesis arose as a theory opposing Darwinism in the sense that Darwinists claimed that variation occurs spontaneously in multiple directions. On the contrary, the followers of orthogenesis argued that variations were directed and limited to a few possibilities of change, that is, that lineages were predisposed to vary toward certain directions and not others (Levit and Olsson, 2006). In any case, orthogenesis was conceptualized via three main phenomena: (1) the observation of the independent origin of characters in closely related groups; (2) the observation of characters that evolve beyond the permitted functional limits, meaning without any selective control due to environmental pressures; and (3) the observation that there are constraints that restrict a wide range of morphological possibilities (Kellogg 1907; Ochoa and Barahona 2014; Ochoa 2017). So, having made this clear, let us take a closer look at how orthogenesis is related to the concept of "parallelism" by William Berryman Scott and Otto Schindewolf, two of the most influential paleontologists involved in evolutionary debates in the early and mid-twentieth century, respectively.

10.3.1 Scott's Parallelism

William Berryman Scott was one of the most distinguished paleontologists of American paleontology in the late nineteenth and early twentieth centuries. He began his career along with the paleontologist Henry Fairfield Osborn[4] who knew him since their youth. Scott worked all his academic life in Princeton, where he initially worked as a professor in geology and paleontology, but later held the chair of the Department of Geology until his retirement. In Princeton, he founded a paleontological school turning the university into one of the largest research and teaching centers of vertebrate paleontology in North America. For a long time, Scott played an important role in the national and international affairs of the scientific community in the area of paleontology and evolution, and like Osborn, he received several awards and belonged to different scientific associations (Simpson 1948). One important aspect of Scott's contributions is that he coined the terms "parallelism" and "convergence" in their modern sense; however, these terms were initially proposed under anti-Darwinian theories, as Ochoa and Barahona (2014) have pointed out. For example, convergence, or independent acquisition of similar characters in distant lineages, was explained by the theory of inheritance of acquired characters also known as Neo-Lamarckism, while parallelism, or independent acquisition of similar characters in closely related lineages, was explained by the phenomenon of orthogenesis. But for the moment, we shall devote more attention to this last term.

In a paper, Scott argues in favor of orthogenesis as a process that dominates the evolutionary trends of ancient mammalian lineages: "So far as the series of fossil mammals which we have been considering are concerned, the developmental history appears to be very direct, and subject to comparatively little fluctuation, advancing steadily in a definite direction, though with slight deviations" (Scott 1891, pp. 370–371). According to Scott, this constant march of change can be observed in horse evolution, for example, where premolars acquired the shape of molars one by one in the entire group. Additionally, their faces and limbs elongated, their digits

[4]Osborn was a champion of the orthogenesis theory as well. He integrated natural selection as one of the several causes which explains evolutionary trends. For more information, see in this volume Ceccarelli 2021.

reduced slowly (except the middle digit which became bigger and more pronounced), and the overall stature increased. Likewise, the different genera such as *Palaeosyops*, *Diplacodon*, and *Titanotherium* from the Bridger, Uinta, and White River formations, respectively, display repeated characters that originated independently in each of their respective formations.

Scott comments that these similarities are observed in the *Oreodon* from the White River and in *Eporeodon* from John Day, and although these cases could be explained as multiple origins of the genus, some species of an ancient genus could have acquired similar characters simultaneously. From this evidence, we might conclude that "In many genera the cycle of variation appears to be a singularly small one" or that "the limited plasticity of the mammals, except along certain definite lines, is very marked" (Scott 1891, p. 371). That is, the production of variation is highly constrained by the structure of the lineage, and, therefore, the possible descending lineages would have few alternatives for change.

According to Scott, these constraints have severe consequences for lineages, since by having limited variation, if a sudden change of environment occurs, then inevitably some groups might be in danger of extinction because their bodies would not be "sufficiently plastic to adapt themselves readily to new conditions" (Scott 1891, p. 371). Scott states that if we consider the channeled change of their parts, this does not mean that the directions must be completely constant perpetually. On the contrary, these trajectories could be steady for a certain period, and later they might change course toward a completely different direction. For example, regarding the evolution of the Pecora, the hornless groups from the lower Miocene, such as *Amphitragulus*, *Paleomeryx*, etc., display a continuous increase in the size of the upper canines which are still conserved in hornless deer-like *Moschus*, *Hydropotes*, etc., whereas in typical cervids and bovids, the antlers and horns developed gradually, while the upper canines gradually decreased until they became rudiments or even disappear.

These facts allow us to postulate that change in evolutionary direction is steady and that the transformation path can change slightly; although once the trend is directed, the probability of taking a different trajectory is minimal. Of course, what drives this evolutionary trend are the constraints that are provided by the internal structure of the lineage. This is something very similar to what might be called today "developmental constraints"[5]: "just as the power of regeneration of lost parts diminishes as we ascend in the scale of animal life, so plasticity of organization and capacity for differentiation of structure in widely different directions diminishes also" (Scott 1891, p. 372).

However, Scott thought that variation was so constrained that the body only permits change toward a few alternative paths (or even only one). If the lineage does not have to change, the structure will remain in morphological stability, meaning that although many groups have displayed great variability across time, these alterations might be minimal; in consequence, the lineages will maintain their fundamental

---

[5]Or we should use the term "developmental bias" to be more precise.

structures for a long time.[6] As we can see, this phenomenon seems to be a hard version of orthogenesis. But here the meaning of orthogenesis has to do with the constraints generated by the structure itself. Scott argues that the options for change are so narrow that there is only one avenue to go forward. In his sense, the evolutionary march advances in one direction, and that is why change is uniform.

For Scott, this constrained mechanism does not necessarily work alone, since the march of change toward specific adaptations might also be controlled by an auxiliary mechanism, natural selection: "It may, perhaps, be the outcome of future investigations, that while variations are generally due to the union of changing hereditary tendencies, mutations are the effect of dynamical agencies operating long in a uniform way, and the results controlled by natural selection" (Scott 1891, p. 388). That is to say, since Scott's orthogenesis describes the reduction alternatives for change and although those changes proceed by themselves in a single evolutionary pathway, natural selection acts here as auxiliary mechanism pushing evolution in a single direction.

From these assumptions arises the concept of parallelism; since the evolution of characters in mammalian lineages is constrained to a few evolutionary pathways, more closely related lineages would tend to produce the same characters more frequently: "the various species of the ancestral genus may acquire the new character independently of each other (parallelism)" (Scott 1891, p. 362). Later, in another paper, he defined this concept more clearly: "By parallelism is meant the independent acquisition of similar structure in forms which are themselves nearly related . . . and thus in one or more respects come to be more nearly alike than were their ancestors" (Scott 1896, p. 185).

Another important point is that in these writings, Scott also coins the term convergence, but he separates parallelism from convergence by the mechanisms which act upon them, making it possible to distinguish them theoretically: parallelism is explained by structural constraints, as we saw above, while convergence is explained by inheritance of acquired characters. However, in terms of independent evolution, Scott stated that parallelism must be more common than convergence: "even though the resemblances have been independently acquired, because parallelism is a more frequently observed phenomenon than convergence," and given that structural constraints induce themselves to follow channeled pathways during evolutionary change, "the more nearly related any two organisms are, the more likely are they to undergo similar modifications" (Scott 1896, p. 186). In short, the relationship of ancestry, or the shared structure among related individuals, constrains the possibilities of change, restricting the form to a few avenues whose variations are unique to each group (more closely related lineages will more likely produce similar characters in each of their descendants).

In conclusion, Scott coined the term "parallelism" based on the theory of orthogenesis. For this paleontologist, orthogenesis describes the constant evolutionary trends which occur in the evolution of closely related lineages. Therefore,

[6] This phenomenon is now recognized as stasis (see Gould and Eldredge 1977).

orthogenesis could be explained by either structural constraints or by natural selection as an auxiliary mechanism acting upon a few pathways of change. With this in mind, it is important to note that Scott's orthogenesis has no mention of anything about "metaphysical" or "divine" processes. On the contrary, his orthogenesis is based on detailed observations of the fossil record which displays the evolutionary trend. We also do not find any "theological" conception in his view, since the trends are due to the structural constraints and not due to metaphysical internal programming that drives lineages toward a specific point. Likewise, these constraints might be aided by natural selection that would act by pushing the trend.

10.3.2 Schindewolf's Parallelism

Otto Heinrich Schindewolf was a German paleontologist of the mid-twentieth century, famous for his work on fossil ammonites, but, above all, he stands out for having challenged the orthodoxy of the Modern Synthesis during his time. He worked at the University of Tübingen, serving as director from 1956 to 1957. In general, Schindewolf was one of the most important and influential paleontologists of German paleontology during the period after the Second World War and beyond. One of his most notable achievements was the publication of the book *Grundfragen der Paläontologie* [*Basic Questions in Paleontology*] in 1950, which became a classic in German paleontology literature for two decades. In fact, this book and his ideas had remained hidden from American literature until an English translation appeared in 1993 (Reif 1993).

First of all, it is pertinent to comment that Schindewolf was always intrigued by the phenomenon of independent evolution. However, like many paleontologists of the time, he thought orthogenesis accounted for part of these phenomena and particularly associated this theory to the concept of parallelism. For example, Schindewolf defines parallelism in the following way: "The unfolding of the individual stocks does not take place in a single lineage but rather in a varying number of parallel lineages at the same time" (Schindewolf 1993, p. 274). Yet, to recognize this kind of independent evolution, it was necessary to observe that the lineages "differ from one another in certain characters, which are indicators of their [evolutionary] independence, but agree in other … those that demonstrate their relatedness" (Schindewolf 1993, p. 274).

In any case, "parallel lineages exhibit similar remodeling even though they are related only at their roots and, moreover, evolve independently" (Schindewolf 1993, p. 274). As for the above definition, we can note that parallelism is explained as a phenomenon of internal channeling in lineages, a process which inevitably leads structures toward the same evolutionary conclusions, that is, orthogenesis and parallelism are governed by the same causes: "Other informative glimpses into the nature and causes of orthogenesis are provided by parallel evolution …, an extremely common, probably even universal phenomenon of the typostatic phase of evolution" (Schindewolf 1993, p. 274). Furthermore, this channeling depends on the common ancestry and, thus, upon a degree of ancestry: "the common ancestor must have transmitted a specific factor" whereby lineages "evolved along a directed, undeviating course" (Schindewolf 1993, p. 275).

Some examples of the parallelism shown by Schindewolf (1993, pp. 274–275) are the following: With respect to the evolution of different ammonite lineages, we can perceive that although they all stem from the same common ancestor, they independently acquire the same characters such as septa, siphon, sutures, etc. For example, in the genus *Clymenia*, there was an evolutionary trend to go from a round shell to a triangular shell. According to Schindewolf, this occurred in three different lineages that evolved independently and in parallel from the same common ancestor. The palaeodictyopterans (ancestors of all flying insects), on the other hand, display ancient characters, for example, all four wings are identical and cannot be flexed. During the Carboniferous, this group gave rise to the protoblattoids, whose wings can be folded back horizontally over the abdomen. During the course of evolution, the delicate forewings became gradually chitinous and rigid, until elytra were formed. Subsequently, the forewings took on the function of protecting the more delicate hind wings and the abdomen. Now, during the Permian, five or six lineages (e.g., beetles, cockroaches, etc.) arose from protoblattoids; it is interesting to note however that all of them underwent independent evolution shifting from forewings to elytra. Therefore, this character "was acquired independently as the result of similar latent evolutionary potential."

In fish, changes in the caudal fin, the scales, and the skull evolved independently from a common ancestor. For example, the different groups of actinopterygians (a subclass of bony fishes) show a similar evolutionary trend: going from a heterocercal caudal fin to homocercal one; from rhomboidal ganoid scales to thin, round cycloid or ctenoid scales; trends in the shape of the cranium and the modification of the endocranium; increased ossification of the vertebral column; reduction of the clavicles; etc. In Schindewolf's words: "these groups [Chondrostei, Holostei, and Teleostei] are polyphyletic, which means that they consist of a cluster of separate, independent lineages, which [...] developed [their structures] in parallel in the same direction and reached and passed through the various evolutionary stages more or less at the same time" (Schindewolf 1993, p. 275).

With these and other examples, Schindewolf postulated that parallelism, being a phenomenon of constraint which limits variation and drives lineages inside the same channeled evolutionary pathway, is a universal phenomenon and applies to all levels of taxonomic categories; that is, the structure constrains the evolutionary course at different levels steering the closely related lineages toward different modes of structural conclusion:

> As our examples show, parallelism in evolution expresses itself in quite different categories and orders of magnitude. It is found in lineages within genera and families as well as in categories of higher taxonomic and evolutionary rank, where the phenomena are the same as in the smallest unit, the species. And in species, too, there is parallel evolution in numerous separate reproductive lines, as we see in individual clans, races, and subspecies. (Schindewolf 1993, p. 276)

But how can we explain this phenomenon of parallelism or orthogenesis? One type of explanation refers to structural constraints; these are due to the fact that the structure of the ancestor restricts and drives the subsequent modifications of the descendant group in question: "the set of rudiments in the first representatives of each lineage largely determines later evolution." Therefore, every time an organ is modified, the alternatives for change become more and more reduced, so that the structure in itself would be the constraint of phenotypic possibilities: "that subsequent differentiational steps entail a progressive narrowing of evolutionary creative potential" (Schindewolf 1993, p. 273; his italics).

Part of this explanation lies in the constraints produced by a developmental system, something not very different from what we know today as developmental constraints considering that, for Schindewolf, any phenotypic flexibility exists only in the early stages of ontogeny. Once this period has passed, there is a defined and constrained avenue toward a specific point. This brings us to the conception that the individual constrains himself, meaning that the body does not easily allow possibilities for change. Therefore, in the development of a specific organ, once the change begins (although it can happen in different proportions), the evolutionary trajectories are constrained bit by bit until the potential of change is reduced: "In this manner, the compulsory course of events leading to the final stages of one ontogeny is transmitted through evolution to the following one, which then carries it further" (Schindewolf 1993, p. 273; his italics). Based on this, Schindewolf claims that, although there should be variations without any favored direction, variations with directed trajectories would have greater opportunities to be retained, at least those large-scale variations originated during the early stages of ontogeny.

Moreover, Schindewolf argues that such parallel evolution also occurs because closely related lineages have related genes and a potential to develop similar mutations; this points out that the channeling of the structure is ruled by these common genes: "We may assume, then, *that the orientation and parallelism of individual lineages is essentially guided by a common genetic base, which reacts the same way in each line*" (Schindewolf 1993, p. 276; his italics). Or differently stated, "[the] internal reasons for parallelism … *reside in the matching genotypes linking the lineages in question, which allow only a limited number of possible directions*" (Schindewolf 1993, p. 278; his italics).

Now, Schindewolf points out that parallel evolution is by no means guided by external factors (i.e., natural selection). He lists three main arguments to rule out this possibility:

1. *Independent evolution has occurred in different places and environments*: "The transformation of individual lineages has taken place many times under extremely different environmental conditions in widely disparate areas, and despite this, the result has been parallelism."
2. *These similar changes have happened at different times and geological periods*: "the points at which transformation takes place in the individual parallel lineages are by no means always simultaneous, which means that the same external influences cannot be inferred. To take an example from the amphibians… the

phyllospondyls attain as early as the Permian the evolutionary stage of skull transformation that does not appear in the labyrinthodonts until Triassic."

3. *Independent evolution of similar characters is always associated with closely related groups*, since non-related lineages that undergo the same environmental conditions do not develop the same structural shape in a similar way: "it is always only a certain cluster of closely related lineages that brings forth these parallel transformations, whereas other series of forms existing at the same time, which are subject to the same external influences, behave in a completely different way and show a different evolution of characters."

Taking into account these three arguments, Schindewolf concludes that: "the critical, deciding factors are always internal and depend on the potential for evolutionary creativity of the organism itself" (Schindewolf 1993, p. 277).

With this, I do not want to portray that Schindewolf did not take into consideration the mechanism of natural selection. The issue here is that he believed that this mechanism did not play a relevant role in the evolution of higher taxa, but rather its role was microevolutionary. Natural selection would work by introducing the final and adaptive variations of already established types: "selection concerns and affects only the most superficial layer of characters" (Schindewolf 1993, p. 341; his italics).

So, by regarding Schindewolf's ideas, we can conclude that his concept of parallelism was strongly linked to the theoretical conception of orthogenesis which entails directional changes in defined avenues. It is important to emphasize that Schindewolf's orthogenesis resides neither in mysticisms nor purposes provided by a divine guiding force; rather, it is a phenomenon that lies on ontogenetic and genetic processes that constrain evolution to a few alternatives, without assigning these directional changes to external factors.

10.4 Simpson and Orthogenesis

Of course, no paleontologists of the time could let the phenomenon of orthogenesis slide, and Simpson dealt with the topic in detail. But it is important to note that Simpson, as one of the main figures of paleontology, apparently attempted to reject the theory of orthogenesis in his writings. According to Delisle (2009), he denied orthogenetic approach in three basic points: (1) empirically, because orthogenetic trends do not preside in all cases in the evolutionary history of lineages; (2) methodologically, since there must be no other kinds of mechanisms than those seen in genetic research; and (3) epistemologically, given that he rejected a domain of physicalism and determinism in the biological realm. However, we shall see that he indirectly advocated for an explanation of orthogenesis through his explanation of the term "parallelism," since as we have seen above, the phenomenon of parallelism was a description of the phenomenon of orthogenesis.

The history of Simpson's opinion on orthogenesis begins in *Tempo and Mode*. In chapter 5, titled "Inertia, Trend, and Momentum," he states that orthogenesis is a phenomenon by which lineages evolve continuously in a certain direction, even though it might result in extinction. Then, he lists a series of theories which attempt to explain this principle, among which are (1) direct interaction of organism and environment (the inheritance of acquired characters); (2) the effect of natural selection on the survival and distribution of spontaneous mutations (orthoselection); and (3) the occurrence of definite modifications of direction without reference to the environment (orthogenesis per se). On the latter, he discusses that there has been an extremely wide range of theories which are contrary to the others, but they have in common only the idea of either inherent trend or metaphysical conception which describes that evolution goes forward in a straight line:

> Most theories of this school, however, involve an element of predestination, of a goal, a perfecting principle, whether as a vitalistic urge, or a metaphysical necessity, or a frankly theological explanation of evolution according to which it is under divine or otherwise spiritual guidance. (Simpson 1944, p. 152)

Later, he quotes some followers of this view as Henry Fairfield Osborn, Louis Vialleton, Teilhard de Chardin, and Robert Broom in spite of the fact that not all of them maintained this metaphysical or vitalistic notion.[7] Whatever the means which accounts for orthogenesis, we can see that he recognized it as a fact; in other words, Simpson admitted orthogenesis in a descriptive sense without a clear explanation. He mentions: "There is no possible doubt but that some degree of rectilinearity is common in evolution. It can seldom or never be maintained that the evolution . . . is exactly rectilinear or literally undeviating, but the best part of the paleontological record is made up of lines that evolve approximately in one direction over long periods of time" (Simpson 1944, p. 152). He also points out that a "tendency for phyla to continue to evolve in much the same direction for considerable periods of time, rectilinear evolution or orthogenesis in a purely descriptive sense, is a common evolutionary phenomenon" (Simpson 1944, p. 177).

With this, Simpson was not at odds with orthogenetic trends in its descriptive sense, but he was careful by interpreting all evolutionary tendencies in this way (Delisle 2009). For Simpson, orthogenesis could be understood as the evolutionary mode of "phyletic evolution," that is, in large populations with moderate (horotelic) rates presented at lower levels within the evolutionary constancy of the adaptive zone. This, of course, shows that orthogenesis was not exclusive for all evolutionary levels, but only for levels that displayed continuous evolution in a period of environmental stability:

> The probability that a tendency toward rectilinearity is not characteristic of evolution as a whole, but only of certain levels of change under certain common but far from universal conditions, is in itself a potent argument against the third school of orthogenesis, that

[7]For instance, the orthogenesis theories of Osborn and Vialleton are far from being metaphysical or vitalistic. Simpson only recognized this fact in Osborn's theory. On the other hand, Teilhard de Chardin developed a thought which had nothing to do with orthogenesis, while Robert Broom openly recognized a divine plan.

> involving an inherently directional process or a metaphysical perfecting principle. (Simpson 1944, p. 153)

What mechanism could account for orthogenesis? It has been highlighted that Simpson precluded those theoretical proposals which possess deterministic or vitalistic notions because they lacked real mechanisms; as a consequence, he focused on neo-Darwinian agencies which he thought were more adequate than the others from the current science point of view (Delisle 2009). And Simpson attempted to associate those mechanisms to the paleontological data as far as possible. But how far could Simpson go with the genetics of the neo-Darwinian account?

According to Simpson, there are at least two genetic factors that might explain this phenomenon: first, that inheritance is primarily conservative: "In groups that do show rectilinearity ... every animal is nearly like its parents in thousands of characters and may differ significantly only in a few characters." Second, those mutations are not so random: "It is not only improbable but also inconceivable that mutations in every imaginable direction occur with equal frequency" (Simpson 1944, p. 154).

Given these considerations, it is easy to realize that Simpson admits there are certain structural and genetic constraints when analyzing the evolutionary trends of lineages:

> The conservative factor of heredity greatly limits the possible avenues of evolution for any given type of organism. If an animal is like its ancestors in all but a few respects, the differences must necessarily exist in connection and in harmony with the more extensive, more complex inherited elements of structure. (Simpson 1944, p. 154)

On this, it should be noted that, in 1944, Simpson's description of orthogenesis was very similar to Schindewolf and Scott's explanation of orthogenesis and parallelism (see Table 10.1). Still, for Simpson, the conservation of inheritance and the limits in the trajectory of variation given by the genotype were not the only factors that provided constraint; he also considered natural selection as a mechanism that limited the change which hinges essentially on a constant environment:

> The fact that species are not constructed *de novo*, but on the basis of genotypes already existing strongly limits the possible avenues of change, and the possibilities are still further limited by natural selection, the restricting influence of which is obvious and is admitted by all theorists to some extent. Most often it would happen that no sort of available modification would be definitely advantageous and next most frequently that only one would be of selective value. In this situation a character would usually tend not to change or to change only in one direction. (Simpson 1944, p. 154)

So, which of both forces, environment versus internal control, would be the predominant in constraining evolutionary change toward a definite trend? Simpson proposes that mutations, even though restricted, are still random; any advance would surely be a non-adaptive trend, and if the trajectory was frequent in a single direction, these would probably provide a degenerative and non-progressive development. Therefore, natural selection would have the final decision on any evolutionary advance, and orthogenesis might be explained rather as a kind of orthoselection (an expression coined by Ludwig Plate):

Table 10.1 Comparison between explanations of orthogenesis and parallelism: Simpson's orthogenesis (1944), Simpson's parallelism (1945, 1961), Schindewolf's orthogenesis (1950), and Scott's orthogenesis (1891). Explanation 1 is based on structural or developmental constraints. Explanation 2 is based on genetic constraints. Explanation 3 is based on natural selection as an auxiliary mechanism. Note the great similarity between the explanations

	Explanation 1 Based on structural or developmental constraints	Explanation 2 Based on genetic constraints	Explanation 3 Based on natural selection as an auxiliary mechanism
Simpson's orthogenesis (1944)	The conservative factor of heredity limits the possible avenues of evolutionary change	Mutations do not occur in every direction with equal frequency	Natural selection limits and steers the changes in one direction
Simpson's parallelism (1945)	Descendant lineages evolve in the same way because they inherited genetic factors which control development	Homologous genes tend to mutate in the same way	Natural selection acts upon homologous characters exposed to the same environment
Simpson's parallelism (1961)	The structure of an ancestral group inevitably limits the lines of possible evolutionary change	Homologous mutations and relative mutation rates produce change in the same direction	The probability that descendant lineage follows one line will be further reinforced by natural selection
Schindewolf's orthogenesis (1950)	The structure of the ancestor (its developmental system) limits and drives the subsequent modifications of the descendant group	Common genetic base reacts in the same way and limits the number of possible directions	
Scott's orthogenesis (1891)	The internal structure of the ancestral lineage constrains the possible avenues of evolutionary change		Natural selection acts on a few options and then pushes change in a single direction

> Theories of orthogenesis, strictly speaking, by direct environmental action are consistent with many of the same observational data as theories of orthoselection. If it is true, as it seems to be, that rectilinearity is most common at certain levels of evolution and in certain types of populations, orthoselection offers an acceptable explanation for this limitation, and orthogenesis due to direct influence of the environment does not. (Simpson 1944, p. 157)

Simpson's next strategy was to refute paleontological evidence that evolution was advancing on a strictly steady course. It has been pointed out that Simpson attempted to demonstrate that not all trends followed a narrow path; rather some divergence is perceived in some cases (Delisle 2009). He even asked if those tendencies were not more than a product of the paleontologists' imagination. In any case, Simpson interpreted some orthogenetic examples by means of their adaptive account. The first case Simpson described was the evolution of the horse because, in his opinion, it had been a recurring example to account for orthogenesis. According to him, some of the horse's traits that have been described as progressive during their evolution such as the proportion of the body, skull, the brain, the limbs, the reduction of the foot, and the dental cusps evolved by some selective value. For example, intelligence was related to brain size, efficiency and speed in locomotion is linked to the evolution of limbs, and the ability to chew is associated with the modification of the molar cusps.

In his book *The Meaning of Evolution* (1949, pp. 131–159), Simpson analyzed the most cited cases of orthogenesis related with non-adaptive trends, among which were the tooth growth increase in saber-toothed tigers, the development of horns in Irish elk, the coiled shell of *Gryphaea*, and the horn of the titanotheres. With regard to saber-toothed tigers, he said that the size of the canines was very variable in different groups, and the size of the sabertooth has a strictly adaptive purpose for each of the species; then, it should not be interpreted simply as growth increase without adaptive value due to internal factors. Besides, the development of the sabertooth doesn't display any evolutionary trend by itself; rather, this argument had spread as a myth of paleontology based on a misinterpretation of the data. The same is true in the case of Irish elk since the growth increase of their horns should not be interpreted as said by the popular orthogenesis myths, that is, as disproportionate development which inevitably led them to extinction; rather the horn size was proportionate to their body, and they surely had an adaptive value as well, for instance, they used them for battle or defense.

Regarding the coiled shell of *Gryphaea*, other orthogenesis myths stated that this lineage went extinct because the exaggerated coiled shell hindered the ability to open their shells. Simpson explained that this peculiarity rather was an advantage so these oysters could settle in the soft mud of the seafloor: "There is really no good reason to believe that the change was carried to an inadaptive degree by the trend" (Simpson 1949, p. 154). Furthermore, about the horn of the titanotheres, he wondered whether its incipient stage had an adaptive value. In *Tempo and Mode*, Simpson admitted that the evolution of the horns was the only evidence in favor of orthogenesis, but later he mentioned that slight variations could have driven their development along with natural selection. He emphasized this view in his books.[8]

Finally, it is important to point out that Simpson changed his mind significantly concerning orthogenesis. In the beginning, he admitted orthogenesis was a description of evolutionary trends and proposed constraints as a possible mechanism. However, later, he suggested that the cause of these trends would have to do with adaptation. For example, he wrote:

> Thus the whole trend is adaptive from beginning to end [...]. Adaptation has a known mechanism: natural selection acting on the genetics of populations [...] In seeking the orienting factor in evolution we have seen that in some cases this must, by all reasonable inferences, be adaptation and in all, even the most doubtful, it could be adaptation. (Simpson 1949, pp. 158–159)

But what happened with those internal factors of non-random mutations and conservation of inheritance? In *The Major Features of Evolution*, we see that

[8]See Simpson (1949, p. 155, 1953, pp. 270–271).

Simpson (1953, p. 272) still argues that natural selection is limited; however, now, these constraints are interpreted as the ability to produce mutations randomly and not necessarily by structural constraints as he said previously. In any case, the argument remains strictly adaptationist. Indeed, he used the phenomenon of convergence to support this fact and mentions examples of how natural selection can produce structures independent of common ancestry (see below). Therefore, Simpson is convinced that "the direction of evolution is largely determined by adaptively oriented selection acting on adaptively unoriented materials [mutations] that limit possible avenues of change" (Simpson 1953, p. 273).

Likewise, constraints also are understood as the absence of some characters, for example, many rhinos which do not have horns could have had a selective advantage by having horns, but the mutation [genetic change] that gives rise to horns never appeared. Here, the mutation is a constraint, just because without it, the rhinos would not have had horns. Subsequently, we see that he resorts to Fisher's argument[9] to reject orthogenesis because he claims that the mutations alone cannot channel change toward any particular direction:

> Within the firm framework of modern population genetics, it seems that mutation pressure, alone, could not orient evolution unless mutations were strongly directional and occurred with such high frequencies as to overcome omnipresent selection [. . .] The possibility is by no means excluded that in some instances and phases of evolution a predominant direction of mutation might have been an effective factor. It does, however, seem warranted to conclude from evidence already available that this is not the usual or even a common way in which orientation occurs. If direction of mutation and of selection did somehow and sometimes happen to coincide, this would unquestionably accelerate change. If very high mutation against selection occurred, this would probably be inadaptive and presumably might cause extinction. There seems to be no convincing evidence that it has ever done so, in fact. (Simpson 1953, pp. 273–274)

Interestingly, today we can write this chapter in the history of evolutionary thought under the title "that was how Simpson ended all theoretical consideration of orthogenesis in paleontology," given that no genetic factor could account for these evolutionary trends, except for natural selection working upon randomly occurring mutations. However, we should not forget that Simpson lost the thread of the real meaning of orthogenesis. He refuted orthogenesis by solely considering that the direction is not guided by mutations alone and not that the steering would have been constrained by the internal structure of the ancestral lineage. These evolutionary trends, for Simpson, are due to natural selection acting upon mutations. In any case, where he didn't lose the thread was in his defense by the term "parallelism," which is precisely where his belief in orthogenesis dwells, as we will see below.

---

[9]Ronal Aylmer Fisher, well-known for founding the population genetics, in his famous book *The Genetical Theory of Natural Selection*, argued that orthogenesis was false because experimental experience does not support the assumption that the mutations can be directed toward certain points. Most mutations, according to him, occur randomly and without any adaptive value. However, if the mutations had a steady direction, he reasons, they would have to possess a constancy of rate greater than the required selection to fix the new mutation into the population (see Ochoa 2017).

10.4.1 Simpson's Parallelism

In *Principles of Animal Taxonomy* (1961), Simpson defined the term parallelism as follows: "Parallelism is the development of similar characters separately in two or more lineages of common ancestry and on the basis of, or channeled by, characteristics of that ancestry" (Simpson 1961, p. 78; his italics). Meaning that although the characters are "absent in the common ancestry, they may arise in some or all descendant taxa as *parallel* developments channeled by characteristics, genetical or other, of the ancestry" (Simpson 1961, pp. 77–78; his italics). Later he highlights that: "parallelism is the independent occurrence of similar changes in groups from a common ancestry and *because* they had a common ancestry" (Simpson 1961, p. 103). Interestingly, this definition perfectly describes the phenomenon of orthogenesis, as we saw with Scott and Schindewolf. Remember, Simpson never denied orthogenesis in its descriptive sense (see above). Therefore, I dare say that he recognized orthogenesis as a phenomenon through the concept of "parallelism." This fact can help us to find the true beliefs of our hero.

As we saw, Scott and Schindewolf proposed explanations of orthogenesis through constraints (see Table 10.1). Yet, Gould (2002 p. 1086) mentions that in the early years of the Modern Synthesis, Darwinians paid little attention to constraints because they felt it was a process contrary to natural selection and not part of the evolutionary process. So, when Simpson recognized and advocated the concept of parallelism, he found himself to be in a theoretical dead end, because if natural selection is the mechanism that steers evolutionary trend, and if convergence is due to natural selection,[10] then why should we assume that parallelism is different from convergence?

In his 1945 monograph on the classification of mammals, Simpson had his first conflict with the concept of parallelism; therein he recognized that this term might be explained in the same way as convergence. This fact creates difficulties when one attempts to differentiate parallelism from convergence, and even from homology. If homology represents common ancestry, while convergence is the similarity by non-common ancestry, then parallelism looks like an additional term that contains some amount of homology and some amount of convergence (see Gould 2002, pp. 1088–1089). In any case, parallelism would be based on a phenomenon that has to do more with common ancestry:

> It is a complication that a third sort of process also produces similarities: parallelism. The term is descriptive rather than explanatory and refers to the fact that distinct groups of

[10]In *Principles of Animal Taxonomy* (1961), Simpson defines the term of convergence as follows: *Convergence is the development of similar characters separately in two or more lineages without a common ancestry pertinent to the similarity but involving adaptation to similar ecological status* (Simpson 1961, pp. 78–79; his italics). That is, the characters "may and frequently do also arise as independent *convergent* adaptations to similar ways of life in taxa of quite different ancestries" (Simpson 1961, p. 78; his italics). For Simpson, and unlike the case of parallelism, similarities instigated under convergence is the product of the action of natural selection on organisms living in similar environments.

> common origin frequently evolve in much the same direction after the discontinuity between them has arisen, so that at a later stage the phyla may have characters in common that were not visible in the common ancestry but that tend, nevertheless, to be more or less in proportion to the nearness of that ancestry. This proportional tendency distinguishes parallelism from convergence, but the distinction is far from absolute. The two phenomena intergrade continuously and are often indistinguishable in practice. (Simpson 1945, p. 9)

Later, Simpson attempted to explain parallelism under three different situations:

1. *Homologous genes tend to mutate in the same way.* Once the two lineages split from their common ancestor, they end up with homologous genes which produce the same mutations: "such groups may and, in all probability, frequently do develop the same characters, not typical of their ancestry or directly inherited but nevertheless due to inheritance: the inheritance of genes prone to mutate in the same way" (Simpson 1945, p. 9).
2. *Proportions and some other characters are commonly the result of relative growth rates that are genetically controlled.* Once the two lineages split from their common ancestor, they tend to evolve in the same way, for example, toward a large size, because they inherited genetic factors which determinate that growth: "they may develop characters of proportion that are the same in both and that are quite different from those of the common ancestry but that were, nevertheless, inherited from that ancestry" (Simpson 1945, p. 10).
3. *Groups of common origin inherit many homologous characters, and they frequently also retain or develop similar habits and environmental preferences.* Lineages that share a common ancestor will have similar structures that will be exposed to the same environmental situations; therefore, evolution will probably take the same pathway: "Then convergence that is strictly such and not attributable to an ancestral genetic factor is likely to occur in them, and if it does, it acts in the same direction as the conservative element of homology" (Simpson 1945, p. 10).

In this light, we see that Simpson's explanation for orthogenesis provided in 1944 is not much different from these explanations on parallelism we have just seen. They are even alike Schindewolf's explanation for orthogenesis (see Table 10.1). Simpson had fought up until then to maintain the term parallelism, which phenomenologically can be perceived in the fossil record by paleontologists, regardless of the theory used to explain it. However, things were not working out as he had hoped.

In 1946, Otto Haas and Simpson wrote one of the best historiographical works on the concepts of comparative anatomy. In their conclusions, they defined homoplasy as "similarities between organisms or their parts, organs or structures, due not to common ancestry but to independent acquisition of the similar characters" (Haas and Simpson 1946, p. 344). In this definition, they include convergence and parallelism as special cases within homoplasy by which the increase or decrease in similarity may be assessed during evolution.

It is interesting to note that Haas and Simpson discussed plenty to validate Scott's parallelism since they were aware that their description was based on "such vague or metaphysical ideas" as phylogenetic "predetermination" or "latent" characters. They emphasized the importance of the geometric sense of parallelism, believing that it was the most appropriate sense for phylogenetic descriptions. Interestingly, Simpson's disagreement with the conclusions of parallelism is reflected in this paper. In fact, he supported maintaining Scott's original meaning: "one of us (Simpson 1945, pp. 9–10) has emphasized that there are at least two known genetic processes that would lead to these results without recourse to such vague or metaphysical ideas as phylogenetic 'predetermination' or 'latent' characters... Thus there is a real validity in the idea of parallelism in the sense of Scott and of those who have followed his definition, although the basis for this validity has usually been obscured" (Haas and Simpson 1946, p. 336). According to Haas and Simpson, there is great difficulty in accepting these phenomena because Scott did not have a knowledge of genetics which would have helped him to recognize that convergence and parallelism are the same processes. They stressed:

> The intended distinction would have been validated if it had earlier been expressed in genetic terms. Redefined in such terms, parallelism would be similarity in structure due to a common genetic basis (and so far resembling homology) but not reaching morphological expression until after the separation of the two or more lines involved (and in this differing from homology). This seems really to be the concept toward which earlier students were progressing, gropingly in the absence of clearer genetic knowledge, from two different directions: when they spoke of parallelism in the sense of Scott as similarity arising in nearly related lines (i.e., with a considerable common genetic basis), and when they spoke of latent or potential homology and the like (i.e., of a common genetic basis for characters not morphologically evident until a later stage in phylogeny). (Haas and Simpson 1946, pp. 336–337)

Therefore, the phenomenon could be recognized as a special case of homoplasy (independent similarity) that reached a certain level of homology due to close ancestral relationships. Still, the question rests on whether it was appropriate to maintain the word "parallelism" for the description of this phenomenon:

> It is possible to gather together broadly diverse expressions and ideas of the past and to relate them to a definite and modern restatement of a phylogenetic principle. It remains, however, to decide whether to define or redefine parallelism as the term applicable to this principle. On historical grounds this would be justified and perhaps preferable [keep the word parallelism] because this genetic definition is closer to the usage of Scott and of most later authors than is the strictly geometric definition. (Haas and Simpson 1946, p. 337)

Yet, the main issue had been deciding whether this type of parallelism was not a kind of convergence acting on different degrees with respect to common ancestry. Hence, the authors faced one of the greatest responsibilities for the future consequences of the term "parallelism": whether to opt for a term that entailed structural constraint, which relied on a few theoretical resources to be properly explained by genetics (i.e., parallelism still had only the descriptive sense without any adequate explanation by means of hereditary factors) or to opt for the geometric sense that could be more useful for a description of the phylogeny. Despite all endeavors, they ended up choosing the geometric sense of parallelism:

> Nevertheless, we have decided against such a proposal (very reluctantly, as concerns one of us [Simpson]) because of the inappropriateness of the term [parallelism]. Parallelism does

have a vernacular, or at least a non-phylogenetic sense, which is applicable to the phenomenon in question only in a descriptive way, whereas the essence of genetic definition would be to supply a theory or opinion as to the cause underlying the descriptive situation. This genetic principle is not the only one that can produce a sequence descriptively parallel. Moreover, the result of this genetic factor is not necessarily parallelism in a descriptive sense. It can, indeed, be convergence in this sense although convergence has almost always been considered as essentially distinct from parallelism. One of the reasons for this view, and probably the decisive one, has been that convergence may be found between groups far distant from each other in the zoological system. (Haas and Simpson 1946, p. 337)

So, given that the original term for parallelism could not be adequately defined because there was no genetic theory that properly accounted for this phenomenon, Haas and Simpson opted to include parallelism within a general phenomenon called homoplasy. This had to be so until the boundaries between convergence and parallelism were well-defined: "However, it seems advisable to retain the last term [homoplasy] as the more comprehensive one all the same, all the more so since the boundaries between parallelism and convergence are quite indefinite and it is frequently difficult to decide which of these special terms applies" (Haas and Simpson 1946, p. 325).

Interestingly, despite these conclusions, Simpson still maintained confidence in the term parallelism as distinguishable from convergence in *Principles of Animal Taxonomy* (1961). He was not satisfied with the decision of the abovementioned paper: "Some students (for example, Haas in Haas and Simpson 1946) have preferred a more purely descriptive definition, especially by the geometrical model of parallel lines, symbolizing two lineages both changing but not becoming significantly either more or less similar" (Simpson 1961, p. 103). Simpson then mentions that parallelism has been considered useful by many taxonomists when bearing in mind common ancestry. In other words, parallelism must have something special: "The distinction of parallelism from convergence is vital" (Simpson 1961, p. 106).

Given Simpson's continued defense in favor of the term parallelism (as a term that arose from an anti-Darwinian theory with an emphasis on constraints and channelization of change), some fundamental questions inevitably come about: Did Simpson advocate for the theory of orthogenesis through the concept of parallelism? If this is so, did he propose a mechanism based on structural constraints that supported it? Any answer to these questions could be controversial. However, from my point of view, he could have postulated an explanation for orthogenesis that linked structural constraints with the action of natural selection (just as Scott did) as well as genetic constraints as reinforcement (as Schindewolf also did at the time) (see Table 10.1):

Parallelism has several theoretical bases that help one to understand and also to recognize it. *The structure of an ancestral group inevitably restricts the lines of possible evolutionary change. That simple fact greatly increases the probability that among the number of descendant lineages several or all will follow one line.* That probability will be further reinforced by natural selection in a geographically expanding and actively speciating group if the ecologies of diverse lineages remain similar in respect to the adaptations involved in the parallelism. The degree of dependence on similar ecology resembles that of convergence, but the retention of homologous characters from the relatively near common ancestry usually

> distinguishes parallelism. The parallel lineages (unlike those only convergent) furthermore start out with closely similar coadapted genetic systems, and similar changes are more likely to keep the system adequately coadapted. Tendency toward genetic parallelism is also strongly reinforced by recurrent "homologues" mutations and similar relative mutation rates. (Simpson 1961, p. 106; my italics)

To summarize, we can highlight some interesting statements about this topic. In his well-knowing book *Tempo and Mode*, Simpson recognized the phenomenon of orthogenesis, even though he discredits the theories that supported it by saying that they were based on theological, metaphysical, and vitalistic conceptions. Of course, Simpson does not agree that orthogenesis occurs without adaptation; nonetheless, he provides some explanations such as structural constraints and the mutations that do not occur in all directions (see Table 10.1). In later publications, Simpson proposed that all evolutionary trends were adaptative, and he rejected orthogenesis because mutations rates in a single direction were simply not possible, so the only guide of those evolutionary trends was natural selection. However, when he faced the issue of why parallelism should be different from convergence, he ended up defending the concept of parallelism and finally argued for structural and genetic constraints as its possible causes. So, if parallelism was the description of orthogenesis in anti-Darwinian terms, and if orthogenesis was originally based on arguments of structural constraints, I dare say that Simpson did not end orthogenesis, but rather ended up defending the phenomenon. In other words, Simpson was an orthogeneticist.

10.5 Conclusion: Orthogenesis Arises from Ashes Like a Phoenix

Without a doubt, Simpson's role in paleontology and his views on macroevolution were vital for the construction of the Modern Synthesis. But the traditional narrative tells us that Simpson ended macroevolutionary anti-Darwinian theories such as saltational evolution and orthogenesis—theories that were defended by most paleontologists during the first third of the twentieth century—given that paleontological data was congruent with population genetics and microevolution after all. This narrative may have been established by the other architects who used Simpson's portrait of macroevolution to favor their microevolutionary perspectives. In consequence, Simpson's works only served to support the premise that macroevolutionary phenomena could be explained through the sum of microevolutionary processes. The paleontologists of the 1970s and 1980s inherited this narrative and took it as a paradigm that had to be challenged, that is, paleontology should have a greater contribution other than only corroborate the extrapolationist premise. They probably also benefited from this narrative to promote and defend their macroevolutionary theories (e.g., punctuated equilibrium).

Still, when we read Simpson's works, we find something very different from what the traditional narrative tells us. As I have pointed out throughout this work, while Simpson is guilty of discrediting the theory of orthogenesis in paleontology, he defends the theory indirectly through the concept of parallelism. In fact, like anti-Darwinian paleontologists, such as Scott and Schindewolf, Simpson attempted to explain orthogenesis though structural and genetic constraints (see Table 10.1). This corroborates the fact that Simpson maintained pluralistic ideas by including constraints into his evolutionary system as a complementary factor to the argument of natural selection.

If this is so, I think that Simpson's contribution to paleontology and evolution has not been appreciated correctly because of the traditional narrative. Simpson also blamed himself, during the establishment of the Modern Synthesis, since he showed that his ideas were congruent with the ideas of the other architects. Perhaps this was due to the social pressure generated by the same Modern Synthesis movement that had the desire to achieve the unification of biological science by the agreement between the different disciplines (see Smocovitis 1996). Also, as said by Delisle (2011, p. 57): "For several of its architects, the evolutionary synthesis became a theoretical trap: forced to explain all sorts of evolutionary phenomena within a narrow theoretical corpus, these scholars had to recognize, usually only implicitly, that a number of such phenomena could not properly be explained by this corpus."

On the other hand, the fact that Simpson had ideas contrary to the central assumptions of the Modern Synthesis not only rests on a reinterpretation of history but also on some current theoretical proposals about evolution (e.g., Extended Synthesis). Because if the Modern Synthesis is incongruent with macroevolution (see Adams 2021 in this volume), and if this Modern Synthesis, theoretically challenged, builds a theoretical framework that serves as a basis for any current theory of evolution, those theories that are cemented under the Modern Synthesis must, therefore, be likewise fragile and incoherent (see also Esposito 2021 in this volume). And as Delisle (2011, p. 58) claims: "current calls for an expanded and extended Darwinism need to avoid erecting a straw man when considering its history." But not everything is lost; recently, there is a theoretical proposal developed under a pluralistic approach to evolution (see Diogo 2017).

Finally, this chapter also attempts to do historical justice to all those assumptions stated under the name of "orthogenesis," considering that it has been demonstrated by this and other works that orthogenesis was not a "metaphysical," "vitalistic," or "theological" theory with "progressive" connotations (Grehan and Ainsworth 1985; Gould 2002; Levit and Olsson 2006; Popov 2009, 2018; De Renzi 2014; Ochoa and Barahona 2014; Ulett 2014, 2016; Ochoa 2017; Ceccarelli 2021). Rather, it was a theory that describes the evolutionary trends that exist in the evolution of lineages which today we recognize phenomenologically under the term of "parallelism" (Wake et al. 2011; Ochoa and Rasskin-Gutman 2015; Monnet et al. 2015), and whose causes are probably due to the developmental bias (Arthur 2004; Jablonski 2020; Moczek 2020), sometimes along with a little help from natural selection.

Acknowledgments I thank Sandra Naitzé Martínez Byer and Aaron Thompson for the English review. I am very grateful to Richard Delisle for the trust that he has placed in me, his support, and his suggestions that improved this work. I also want to thank the reviewers for their suggestions.

References

Adams MB (2021) Little evolution, BIG evolution: rethinking the history of Darwinism, population genetics, and the "synthesis". In: Delisle RG (ed) Natural selection: revisiting its explanatory role in evolutionary biology. Springer, Cham, pp 195–230

Amundson R (2005) The changing role of the embryo in evolutionary thought: roots of evo-devo. Cambridge University Press, Cambridge

Arthur W (2004) The effect of development on the direction of evolution: toward a twenty-first century consensus. Evol Dev 6:282–288

Bowler PJ (1983) The eclipse of Darwinism: anti-Darwinian evolution theories in the decades around 1900. Johns Hopkins University Press, Baltimore

Bowler PJ (2017) Alternatives to Darwinism in the early twentieth century. In: Delisle RG (ed) The Darwinian tradition in context: research programs in evolutionary biology. Springer, Cham, pp 195–218

Cain JA (1992) Building a temporal biology: Simpson's program for paleontology during an American expansion of biology. Earth Sci Hist 11:30–36

Ceccarelli D (2021) Recasting natural selection: Osborn and the pluralistic view of life. In: Delisle RG (ed) Natural selection: revisiting its explanatory role in evolutionary biology. Springer, Cham, pp 171–194

De Renzi M (2014) Evolutionary trends promoted by internal causes versus orthogenesis-a discussion. In: Royo-Torres R, Verdú FJ, Alcalá L (eds) XXX Jornadas de Paleontología de la Sociedad Española de Paleontología. ¡Fundamental! vol 24, pp 57–60

Delisle RG (2009) Les philosophies du néo-darwinisme: conceptions divergentes sur l'homme et le sens de l'évolution. PUF, Paris

Delisle RG (2011) What was really synthesized during the evolutionary synthesis? A historiographic proposal. Stud Hist Phil Biol Biomed Sci 42:50–59

Delisle RG (2017) From Charles Darwin to the evolutionary synthesis: weak and diffused connections only. In: Delisle RG (ed) The Darwinian tradition in context: research programs in evolutionary biology. Springer, Cham, pp 133–167

Diogo R (2017) Evolution driven by organismal behavior: a unifying view of life, function, form, mismatches, and trends. Springer, Cham

Dobzhansky T (1937) Genetics and the origin of species. Columbia University Press, New York

Dobzhansky T (1951) Genetics and the origin of species, 3rd edn. Columbia University Press, New York

Esposito M (2021) Cathedrals, corals and mycelia: three analogies for the history of evolutionary biology. In: Delisle RG (ed) Natural selection: revisiting its explanatory role in evolutionary biology. Springer, Cham, pp 11–38

Goldschmidt R (1940) The material basis of evolution. Yale University Press, New Haven

Gould SJ (1980) G. G. Simpson, paleontology, and the modern synthesis. In: Mayr E, Provine WB (eds) The evolutionary synthesis: perspectives on the unification of biology. Harvard University Press, Cambridge, pp 153–172

Gould SJ (2002) The structure of evolutionary theory. Harvard University Press, Cambridge

Gould SJ, Eldredge N (1977) Punctuated equilibria: the tempo and mode of evolution reconsidered. Paleobiology 3:115–151

Grehan JR, Ainsworth R (1985) Orthogenesis and evolution. Sys Zool 34:174–192

Haas O, Simpson GG (1946) Analysis of some phylogenetic terms, with attempts at redefinition. Proc Am Phil Soc 90:319–349

Jablonski D (2020) Developmental bias, macroevolution, and the fossil record. Evol Dev 22:103–125

Kellogg VL (1907) Darwinism to-day: a discussion of present-day scientific criticism of the Darwinian selection theories, together with a brief account of the principal other proposed auxiliary and alternative theories of species-forming. G. Bell and Sun, London

Laporte LF (2000) George Gaylord Simpson: paleontologist and evolutionist. Columbia University Press, New York

Levit GS, Olsson L (2006) "Evolution on rails": mechanisms and levels of orthogenesis. Ann Hist Philos Biol 11:99–138

Levit GS, Meister K, Hoßfeld U (2008) Alternative evolutionary theories: a historical survey. J Bioecon 10:71–96

Mayr E (1942) Systematics and the origin of species. Columbia University Press, New York

Mayr E (1963) Animals species and evolution. Harvard University Press, Cambridge

Mayr E (1980a) Prologue: some thoughts on the history of the evolutionary synthesis. In: Mayr E, Provine WB (eds) The evolutionary synthesis: perspectives on the unification of biology. Harvard University Press, Cambridge, pp 1–48

Mayr E (1980b) The role of systematics in the evolutionary synthesis. In: Mayr E, Provine WB (eds) The evolutionary synthesis: perspectives on the unification of biology. Harvard University Press, Cambridge, pp 123–136

Mayr E, Provine WB (eds) (1980) The evolutionary synthesis: perspectives on the unification of biology. Harvard University Press, Cambridge

Moczek AP (2020) Biases in the study of developmental bias. Evol Dev 22:3–6

Monnet C, Klug C, De Baets K (2015) Evolutionary patterns of ammonoids: phenotypic trends, convergence, and parallel evolution. In: Klug C, Korn D, De Baets K, Kruta I, Mapes RH (eds) Ammonoid paleobiology: from macroevolution to paleogeography. Springer, Dordrecht, pp 95–142

Ochoa C (2017) El eclipse del antidarwinismo: la historia detrás de la síntesis moderna. Centro de Estudios Filosóficos, Políticos y Sociales Vicente Lombardo Toledano, México

Ochoa C (2021) The origins of theoretical developmental genetics: reinterpreting William Bateson's role in the history of evolutionary thought. In: Delisle RG (ed) Natural selection: revisiting its explanatory role in evolutionary biology. Springer, Cham, pp 261–290

Ochoa C, Barahona A (2014) El Jano de la morfología: de la homología a la homoplasia, historia, debates y evolución. UNAM; Centro de Estudios Filosóficos, Políticos y Sociales Vicente Lombardo Toledano, México

Ochoa C, Rasskin-Gutman D (2015) Evo-devo mechanisms underlying the continuum between homology and homoplasy. J Exp Zool B Mol Dev Evol 324:91–103

Pigliucci M (2017) Darwinism after the modern synthesis. In: Delisle RG (ed) The Darwinian tradition in context: research programs in evolutionary biology. Springer, Cham, pp 89–103

Popov I (2009) The problem of constraints on variation, from Darwin to the present. Ludus Vitalis 17:201–220

Popov I (2018) Orthogenesis versus Darwinism. Translated by Natalia Lentsman. Springer, Cham

Reif W-E (1993) Afterword of OH Schindewolf, basic questions in paleontology. University of Chicago Press, Chicago, pp 435–453

Schindewolf OH (1993) Basic questions in paleontology: geologic time, organic evolution, and biological systematics. Translated by Judith Schaefer. Original (1950). University of Chicago Press, Chicago

Schwartz JH (2021) What's natural about natural selection? In: Delisle RG (ed) Natural selection: revisiting its explanatory role in evolutionary biology. Springer, Cham, pp 329–390

Scott WB (1891) On the osteology of *Mesohippus* and *Leptomeryx*, with observations on the modes and factors of evolution in the Mammalia. J Morphol 5:301–406

Scott WB (1896) Paleontology as a morphological discipline. Science 4:177–188

Sepkoski D (2019) The unfinished synthesis? Paleontology and evolutionary biology in the 20th century. J Hist Biol 52:687–703

Simpson GG (1944) Tempo and mode in evolution. Columbia University Press, New York

Simpson GG (1945) The principles of classification and a classification of mammals. Bull Am Mus Nat Hist 85:1–350

Simpson GG (1948) Biographical memoir of William Berryman Scott, 1858-1947. BMNAS 25:175–203

Simpson GG (1949) The meaning of evolution: a study of the history of life and its significance for man. Yale University Press, New Haven

Simpson GG (1953) The major features of evolution. Columbia University Press, New York
Simpson GG (1961) Principles of animal taxonomy. Columbia University Press, New York
Smocovitis VB (1996) Unifying biology: the evolutionary synthesis and evolutionary biology. Princeton University Press, Princeton
Stoltzfus A (2017) Why we don't want another "Synthesis". Biol Direct 12. https://doi.org/10.1186/s13062-017-0194-1
Turner DD (2017) Paleobiology's uneasy relationship with the Darwinian tradition: stasis as data. In: Delisle RG (ed) The Darwinian tradition in context: research programs in evolutionary biology. Springer, Cham, pp 333–352
Ulett MA (2014) Making the case for orthogenesis: the popularization of definitely directed evolution (1890-1926). Stud Hist Phil Biol Biomed Sci 45:124–132
Ulett MA (2016) Directed evolution, history of. In: Kliman RM (ed) The encyclopedia of evolutionary biology. Academic Press, Oxford, pp 441–443
van der Meer JM (2021) The concept of natural selection in Theodosius Dobzhansky. Its development and interpretation. In: Delisle RG (ed) Natural selection: revisiting its explanatory role in evolutionary biology. Springer, Cham, pp 291–328
Wake DB, Wake MH, Specht CD (2011) Homoplasy: from detecting pattern to determining process and mechanism of evolution. Science 331:1032–1035
Wittington HB (1986) George Gaylord Simpson: 16 June 1902 – 6 October 1984. Biogr Mems Fell R Soc 32:526–539

Chapter 11
The Concept of Natural Selection in Theodosius Dobzhansky: Its Development and Interpretation

Jitse M. van der Meer

Abstract Throughout his career, Dobzhansky (1900–1975) doubted whether natural selection could explain evolution. My first thesis is that his doubt was qualified and declined during his life, but never disappeared. I offer five reasons for his hesitation. First, evolution was possible without natural selection. He included drift, cooperation, and polyploidy with natural selection as causes of evolution. During his career, he gained confidence in natural selection but kept an open mind about other causes of evolution. Second, the selective neutrality of alleles required for drift was not established, and this affected the assessment of selection. Third, Dobzhansky wanted a causal account of evolution. But initially no causal connection between short-term natural selection and adaptation and long-term speciation could be established by experiment. Later, empirical support for a role of natural selection in evolution became available, but was insufficient. Fourth, it was unlikely that the timing of reproductive isolation would make adaptive evolution possible. Fifth, the selection-balance controversy exposed a variety of additional reasons for doubts about the role of selection. This skepticism about the role of selection in evolution makes Dobzhansky an unlikely participant in the so-called hardening of the synthetic theory of evolution. My second thesis is that Dobzhansky interpreted natural selection as natural evil and that this interpretation can be explained in the context of his religion.

Keywords Theodosius Dobzhansky · Selection · Drift · Evolution · Adaptation · Shifting balance · Religion · Hardening of the synthesis · Stephen Jay Gould

J. M. van der Meer (✉)
The Pascal Centre, Redeemer University, ON, Canada
e-mail: jmvdm@redeemer.ca

R. G. Delisle (ed.), *Natural Selection*, Evolutionary Biology – New Perspectives on Its Development 3, https://doi.org/10.1007/978-3-030-65536-5_11

11.1 Introduction

Dobzhansky (1937: 149, 1941: 186, 1951: 76, 1970a: 2) defined natural selection as the process in which organisms with better adapted gene complexes have more offspring than organisms with lesser adapted gene complexes. It was a key concept for Dobzhansky throughout his life. Yet, from 1937 to 1970, he expressed a qualified skepticism about its role in evolution. Natural selection was primarily the cause of adaptation, only secondarily of evolution (Dobzhansky 1937: 150–151, 1941: 187–188, 1951: 77–79; Dobzhansky to J. L. King, June 8, 1970b, Dobzhansky Papers). In this same period, he also routinely mentioned extinction, stasis, and diversity as characteristics of evolution. They also could be explained by natural selection. Perhaps that was its secondary role as Dobzhansky saw it. So, whence his skepticism?

My first thesis is that his skepticism was qualified and declined during his life, but never disappeared. I offer five reasons for his skepticism (Sect. 11.5.1). To explain these reasons will describe how natural selection acquired its explanatory role in Dobzhansky's contribution to the synthetic theory of evolution and how this affected his theory choice. This description will cover the Russian period, 1915–1927 (Sect. 11.2); the three editions of *Genetics and the Origin of Species*, 1927–1951 (Sect. 11.3); and the debate about the classical and balance theories, 1951–1975 (Sect. 11.4). I describe his philosophy of science, the empirical support for natural selection, and the relative roles of natural selection and drift in adaptation. This description suggests five reasons for his skepticism about selection (Sect. 11.5.1) which make him an unlikely participant in the view that selection is the sole cause of evolution, i.e., in the hardening of the synthesis (Sect. 11.5.2).

If natural selection causes evolution and evolution is progressive, how could Dobzhansky interpret natural selection as natural evil even if he left progress undefined? Progress and natural evil are neither contraries nor contradictories. My second thesis is that Dobzhansky's religious vision of evolutionary Progress explains why he interpreted natural selection as natural evil. Specifically, natural evil referred to extinction and evolutionary stasis—the other two products of natural selection which Dobzhansky interpreted as the price of progress. This requires a description of his worldview and of how Dobzhansky interpreted natural selection in terms of his religious vision of Progressive evolution (Sect. 11.6). Please note that "Progress" capitalized refers to the religious vision of cosmic evolution. Lower case "progress" refers to increase in the complexity of organisms in the course of biological evolution.

11.2 Theory Development: 1915–1927

11.2.1 Dobzhansky vs Filipchenko on Natural Selection

Early in his career, Dobzhansky attributed an important but hypothetical role to natural selection. From December 1923 to December 1927, he worked in the Department of Genetics headed by Juri Filipchenko at Leningrad University. Contrasting the two will outline important features of Dobzhansky's early theory of evolution. Filipchenko started his career comparing among others the germ layers of insects, millipedes, and centipedes. Germ layers differ between higher categories of classification and are, therefore, used to characterize them. He concluded that traits varying between species of different higher taxa such as germ layers and organ systems are fundamentally different from traits that vary within a species. Filipchenko proposed that above-species traits mark stages in biological evolution. He also suggested that a non-corpuscular cytoplasmic substance (plasmon) determines above-species traits, whereas chromosomal genes determine species traits. This suggestion originated in his work on polyploid wheat which is larger than diploid wheat. Since they do not differ in genotype, but only in the number of chromosomes, Filipchenko (1929: 13–14) reasoned that the size difference must be due to non-chromosomal causes (Filipchenko to Dobzhansky, Jan. 22, 1930 In: Kolchinskii et al. 2002: 399–400). Taking size, germ layers, and organ systems as above-species traits, he proposed that they are determined by this non-corpuscular cytoplasmic substance. Hence, the above-species traits are inherited in a non-Mendelian way. Given that Mendelian genes determine species traits, variation within a species or microevolution is to be explained in terms of Mendelian genetics. Here, Filipchenko follows Johannsen's concept of biological type (genotype) or pure line, and he introduced Dobzhansky to it (based on Filipchenko 1912; Dobzhansky 1962–1963: 150–151; Adams 2008a: 248; Alexandrov 1994: 51). Filipchenko's distinction between cytoplasmic (non-Mendelian) inheritance of above-species traits and chromosomal (Mendelian) inheritance of species traits implied that Darwinian evolution and Mendelian genetics are mutually independent. He believed that macroevolution or the origin of taxa above the species level involves the sudden appearance of new forms and is not affected by external influences such as natural selection. In the thesis for his master's dissertation, written in late 1912, Filipchenko emphasized that "The process of the evolution of organisms can be explained neither by so-called Lamarckian factors, nor by selection—it is one of the basic features of living beings" (Aleksandrov 1994: 50; see also Burian 1994: 129–140).

Unlike Filipchenko, Dobzhansky's published views included that Mendelian genetics is part of the mechanism of evolution, that variation within and between species is of the same kind, that both must be explained by Mendelian genetics, and thus that evolution below the species level (microevolution) is of the same kind as evolution above the species level (macroevolution) (Dobrzhanskii 1924: 417–426 quote tr. by Todes In: Alexandrov 1994: 58). The particulate character of Mendelian inheritance ensures the transmission of hereditary variants between generations even in the case of intra-specific blending inheritance. Dobzhansky distinguished this sharply from the mixing of non-corpuscular cytoplasmic substances according to Filipchenko: "If germ plasm can combine as a dye combines with water, then the amount of variation present in a sexually reproducing random breeding population must be halved in every generation." Mutation would be too slow to replace variation lost at this rate (Dobzhansky 1937: 122). Importantly, the loss of variation would render natural selection irrelevant. This would have been a major reason for Dobzhansky's rejection. Finally, Dobzhansky rejected the sudden appearance of new species. He argued that closely related species differ in many genes which are unlikely to mutate simultaneously to produce a new species (Dobrzhanskii 1926a: 43–44).

These are the features of Dobzhansky's theory of evolution before he moved to the United States. During his stay in Filipchenko's lab, Dobzhansky was already committed to a role for natural selection among other mechanical causes of evolution including mutation and recombination governed by Mendelian laws (Dobrzhansky 1924: 426; Dobrzhanskii 1926a). He proposed natural selection as the cause of the geographical differentiation of a species (*Harmonia axyridis*) into races. At this point, the role of natural selection clearly was hypothetical. Dobzhansky reported that the differentiation of races was correlated with their geographic distribution. But he noted that there was no empirical evidence to support a causal relation with regional differences in climate or plant distribution (Dobrzhansky 1924). Nevertheless, natural selection was clearly important to Dobzhansky because he took the risk of disagreeing with his mentor. In fact, their public disagreement extended beyond natural selection to the meaning of genetics for evolution (Konashev 1994: 79). In the end, Filipchenko and Dobzhansky agreed not to "debate the gene-versus-cytoplasm problem, because it was obvious that we were not getting anywhere, with this" (Dobzhansky 1962–1963: 152. Background in Adams (2008a: 297–303, 2008b: 249; Alexandrov 1994). In sum, the young Dobzhansky considered natural selection as a crucial but hypothetical component in evolutionary theory.

11.2.2 Early Hypothetical Role of Natural Selection in Speciation

The hypothetical role Dobzhansky assigned to natural selection was to integrate adaptation and speciation in the evolutionary process. This becomes clear in his response to Chetverikov. During 1924–1926, Chetverikov developed the concepts of hidden variation, gene pool, genotypic milieu, genetic drift, and selection and incorporated them in the first balance theory of population genetics. This theory stated that mutations are mostly recessive, random with respect to the environment, and preserved in a population in the absence of selection. Further, selection removes recessive mutations more slowly than dominants and acts not on single alleles, but on alleles and their genetic background (Chetverikov 1926; Adams 1968). In the absence of selection from a randomly mating population, all genes are maintained at a constant frequency. The balance theory is named after these balanced frequencies. It received support from hidden genetic variation in heterozygotes, mutation frequencies in natural populations, and non-adaptive differentiation between different populations of the same species of *Drosophila melanogaster* (Timoféeff-Ressovsky and Timoféeff-Ressovsky 1927; Chetverikov 1927a, b).

While Chetverikov developed his theory, Dobzhansky studied geographical variation in ladybird beetles and pleiotropy in *Drosophila* (Konashev 1994: 77). They had met in 1922 in Moscow. Later, Chetverikov gave him *Drosophila* mutants which he used in Kiev to study the anatomy of the genitalia. Dobzhansky kept contact when he worked in Filipchenko's department from 1924 to 1927 and comments on his generosity with time and ideas (Dobzhansky 1962–1963: 124–25, 1980: 233–35, 237; Konashev 1994: 70). Hidden genetic information was discovered by Chetverikov's group in 1925 and 1926 (Chetverikov 1926, tr. Barker, 178; Adams 1980: 263; Dobzhansky 1980: 235). If Dobzhansky did not learn about it from Chetverikov, he would have from Timoféeff-Ressovsky, a member of Chetverikov's group. In 1925, Timoféeff-Ressovsky moved to the Kaiser Wilhelm Institute in Berlin where he identified mutations hidden in a natural *Drosophila* population. Dobzhansky visited him in December 1927 in transit to the United States (Adams 1980: 261, 2008b: 160; Dobzhansky to Filipchenko, Dec. 7, 1927 in: Kolchinskii et al. 2002: 59). It is, therefore, beyond reasonable doubt that he knew about genetic variation hidden in natural populations before it was published.

Yet Dobzhansky ignored hidden variation in three publications that appeared during that time (Dobrzhanskii 1926a, b; Dobzhansky 1927). Why? Starting between 1924 and 1926, Dobzhansky held that evolution comprises *one single process* including geographic variation, selection, adaptation, and speciation and that selection causes geographic variation (Dobrzhansky 1924: 416, 418, Dobzhansky 1924: 26–27, 1926a). The first explicit statement that speciation is a stage in evolution follows soon after (Dobzhansky 1933: 100, 1935: 354). In contrast, Chetverikov had proposed that,

> in the evolutionary development of the organic world, *two processes* proceed side by side, their paths sometimes crossing, but they are still strictly separate in their causes as well as in the consequences resulting from them: one is the process of differentiation, of splitting-up, leading in the end to *speciation*— (isolation is its basis); the other leads to *adaptation*, to the progressive evolution of organic life, and its cause lies in ... *natural selection* (Chetverikov 1926, tr. Barker, 188–89. This distinction was first made by Chetverikov 1905).

The two parallel causal chains in Chetverikov are:

1st Isolation → differentiation → speciation
2nd Natural selection → adaptation → progressive evolution

Chetverikov disconnected speciation from selection. Selection causes adaptation, but not isolation and speciation. If speciation was adaptive, it was a coincidence because it happened alongside or was followed by evolution (Chetverikov 1926: tr. Barker, 184, 188; Adams 1970: 111). Dobzhansky rejected it because he held that selection is a cause of speciation which is a causal step in evolution. Since evolution involves speciation, evolution was disconnected from selection. In my opinion, Chetverikov intended isolation and speciation to be followed by selection and adaptation, but this was not necessary and largely implicit. He cited Jordan (1905):

> At first glance, it may appear that the very fact of isolation taken by itself cannot play any role in the process of evolution (Jordan 1905) (Chetverikov 1926: tr. Barker, 179).

But he connects selection with speciation when he sees it "leading ceaselessly to modification of the species" (Chetverikov 1926: tr. Barker, 181), when he writes about the transformation of the species as the replacement of the unadapted by the adapted form (Chetverikov 1926: tr. Barker, 182), and when he explains the existence of adaptive differences between closely related forms (species) as the result of selection following differentiation (Chetverikov 1926: tr. Barker, 188, 192). Only once he seems to ignore this connection writing about "adaptive evolution" (Chetverikov 1926: tr. Barker, 184). By emphasizing the causal independence of these two processes, he invited an interpretation in which speciation did not cause evolution.

The second reason why Dobzhansky might have rejected Chetverikov's theory is non-adaptive speciation by random statistical variation (drift). Drift is the statistical phenomenon that the frequency of an allele in a large population may not be faithfully represented when its size is drastically reduced.[1] Assuming that mutation frequency does not depend on population size, Chetverikov argued that in a small population, heterozygous carriers of the same allele are more likely to have offspring homozygous for that allele than in a large population. This produces variation in a direction unrelated to the environment. Since selection is not involved, there would be no adaptation. Thus, Chetverikov concluded that there was "no basis for denying the possibility of *nonadaptive* evolution" (Chetverikov 1926: tr. Barker, 192, conclusion #17). This would have made it difficult for Dobzhansky to see evolution as adaptive. In sum, the disconnect between selection and speciation and thc possibility of non-adaptive speciation suffice to explain Dobzhansky's rejection of Chetverikov's balance theory. The best explanation for his insistence on a role for natural selection is its promise as an explanation for adaptation and speciation. But at this time, this role of natural selection remains hypothetical. Moreover, Dobzhansky had no knowledge of the possibility that selection could follow drift and act on differences between the populations of a species as suggested by Wright (1931, 1932). His acquaintance with Wright's work started in 1932 (Provine 1986: 332).

[1]To avoid anachronism, I am not using the contemporary definition: Drift is an indiscriminate sampling process in which any heritable physical differences between entities (organisms, gametes, etc.) are causally irrelevant to differences in reproductive success. By contrast, natural selection is a discriminate sampling process in which any heritable physical differences between entities (organisms, gametes, etc.) are causally relevant to differences in reproductive success. Millstein (2002), quoted from: Millstein (2017)

11.2.3 Natural Selection in Dobzhansky's Research Program

The causal role of natural selection is clearly articulated in the research program discernable in the work of the young Dobzhansky. At 15, Dobzhansky read the *Origin* of Darwin and accepted Darwinian evolution as fact (Dobzhansky 1962–1963:19, 1980: 235. 237). Nine years later (1924), he decided that an understanding of evolution must satisfy the demands of method (explanation in terms of mechanical causes) and observation (Mendelian genetics). For instance, his Cartesian method demanded that the causes of speciation are seen as mechanical causes (Dobrzhanskii 1926a). I assume that they were seen as efficient causes (Dobzhansky 1969). He explained the laws of Mendel by attributing causal agency to hereditary factors in 1924 (Dobrzhanskii 1924). The evolution of genetic races is the causal product (Folgeerscheinung) of natural selection, and races are a stage in the evolution of species (Dobrzhansky 1924: 418, 420; see also Dobrzhanskii 1926a). The latter is causally connected with macroevolution (Sect. 11.2.1). Other than that, he used a nontechnical notion of cause. It simply means that genes and natural selection produce observable effects. We may interpret that as an ontological connection, i.e., the lawful production of an effect.

But his commitment to mechanical accounts cannot explain why he rejected the theories of Filipchenko and Chetverikov because they too used mechanical explanation. Filipchenko's plasmon was a material substance (Adams 1980: 249, 2008a). During 1921 and 1922, Dobzhansky read reviews of the work of Morgan and co-workers (Dobzhansky 1962–1963: 118–19, 1980: 235; Konashev 1994: 67). He accepted Morgan's theory of evolution by mutation and selection which included consistency with Mendelian genetics (Filipchenko 1919, 1922; Provine 1971: 125–126). Natural selection as the cause of evolution and his commitment to mechanical explanation explain why he insisted on a causal link between microevolution and macroevolution against Filipchenko. Building on Morgan, he asked how one can connect taxonomic categories with those revealed by the study of genetic variation in natural populations (Dobrzhansky 1924: 401–21, Dobzhandsky 1924: 245, 248; Dobrzhanskii 1924, 1926a).[2] His working hypothesis in 1924: speciation begins with the gradual concentration of a trait from a wide distribution throughout a species to a geographically localized distribution described as a subspecies which is subsequently transformed into a species. Speciation is driven by mutation and selection (Dobrzhansky 1924: 418, 420, Dobrzhanskii 1926a). Echoing Morgan (Provine 1971: 120–21, 125–26), Dobzhansky proposed in 1924 that mutation transforms old into new genes which are combined with other genes by meiotic recombination. Natural selection removes non-adaptive combinations, does not affect neutral combinations, and consolidates harmonious combinations. Evolution is the creation of such new and harmonious combinations of genes (Dobrzhanskii 1924: see quote tr. by Todes in: Alexandrov 1994: 58). This theory of evolution was

[2]For the relationship between the Mendelians (the experimentalists) and the Darwinians (the naturalists), see Mayr (1980).

mechanical and Mendelian. Filipchenko's plasmon theory was rejected because its associated blending inheritance destroyed variation, and without variation, there is no natural selection. The plasmon theory was also inconsistent with Mendelian genetics, but that was not the reason why Dobzhansky rejected it. At that time, Dobzhansky's theory had no empirical support except for Morgan's discovery of small heritable variations (Provine 1971: 121). This qualifies his program as a metaphysical commitment to using Mendelian genetics in explaining evolution (van der Meer 2007). Dobzhansky began to replace the metaphysics with empirical support. In that same year (1924), he did crosses to determine whether differences in the reproductive structures of *Drosophila* were pleiotropic or polygenic.

This direction reveals a program that was both visionary and rooted in local circumstances. The specific interest was in how genetic differences are related to reproductive isolation (Dobzhandsky 1924). But his focus on pleiotropy and polygeny also suggests a wider interest in how natural selection acts on complexes of mutually related genes. This was aimed at understanding how speciation and gradual evolution are possible if selection acts on groups of related genes integrated in what Dobzhansky called a population genotype. It also could provide him with empirical support for his probabilistic argument that the many coordinated mutations he believed were required by Filipchenko's theory of the sudden appearance of new species are unlikely to occur. Further, if Dobzhansky could show that the inheritance of an *organ system* (the reproductive system) followed the laws of Mendel, he would have disproved Filipchenko's plasmon theory as well as the associated distinction between variation within and between species. Correspondence suggests that his choice of an organ *system* was inspired by his desire to disprove Filipchenko's plasmon theory (Dobzhansky to Filipchenko, July 23, 1928, February 7, 1929 and Filipchenko to Dobzhansky, Jan. 22, 1930 In: Kolchinskii et al. 2002: 82–83, 158, 399–400; Dobzhansky and Bridges 1928). This was indeed the conclusion he arrived at 3 years later (Dobzhansky 1927: 384). Dobzhansky also started to break down this distinction on a second front by stressing that mutations within one particular species often are a species trait in other species and added that evolution was gradual (Dobrzhanskii 1926a, from Alexandrov 1994: 59).

Finally, Filipchenko had introduced Dobzhansky to Johannsen's concept of biological type (genotype) or pure line (based on Filipchenko 1912; Dobzhansky 1962–1963: 150–51; Adams 1980: 248; Alexandrov 1994: 51). Dobzhansky published two more studies exploring variation, selection, and the formation of races in ladybeetles (Dobrzhansky 1924: 401–21) and pond snails (Dobrzhanskii and Kossakovskii 1925: 481–482; Rumiantsev from Alexandrov 1994: 56). In these and other studies as well as in a review, he replaces the essentialistic interpretation of the Mendelian concept of a wild-type or "biotype" introduced by Johanssen with a statistical interpretation which refers to a set of traits that varies with the environment and is not considered essential to a species (Dobrzhanskii 1924; Dobrzhansky 1924; Dobzhandsky 1924; Dobrzhanskii 1926a: 40; Dobrzhanskii and Kossakovskii 1925; Rumiantsev 1928). By late 1926, Chetverikov's group had much evidence for hidden variation. Dobzhansky ignored it because evolution was causally uncoupled from selection. In sum, since 1924, Dobzhansky had a research program in population genetics. It had a method (mechanical explanation), a problem (production of hereditary variation and adaptation, relation of taxonomic to biogeographic species), and a working hypothesis (natural selection causes evolution). But while Chetverikov's balance theory included drift, Dobzhansky does not mention it.

11.3 Genetics and the Origin of Species (1927–1951)

11.3.1 Introduction

During this period, Dobzhansky refined his research program by introducing stages in the evolutionary process as levels of analysis and explanation. Phenomena at different levels have categorically different descriptions. For instance, Mendelian genetics applies to individuals and complements Darwinian evolution which applies to groups. Different rules of change apply at the individual and group levels. The way Dobzhansky (1937: 11, 13, 119–20, 1941: 11, 13–14, 152–53, 1951: 15–18, 76) describes the relations between the rules for different levels satisfies the definition of "complementation" even though he does not use that term. (More on complementation in Sect. 11.3.3.) This difference means that the rules are not logically related. For levels to be logically independent means that the regularities governing each level cannot be logically inferred from the level below or deduced from the level above, while processes at a given level require processes at the next level down as their causal substructure. Further, processes at the group level are divided into short-term reversible changes and long-term irreversible ones. This yields three levels of analysis which are interpreted as causal steps in evolution (Dobzhansky 1937: 13, 119–120, 180–91, 1980: 235. 234–35).

Level 1 phenomena involve individual organisms. They provide the initial conditions (mutation, recombination) for level 2 phenomena which involve reversible changes in populations (drift, natural selection, adaptation). Level 2 phenomena—the physiology of populations—are determined by the physiology of individuals at level 1.

> The former are determined by the latter only in the same sense in which the structure of a human state may be said to depend upon the physiology of its members (Dobzhansky 1937: 120; see also 11, 1941: 11, 1951: 15).

Level 2 also provides the initial conditions for long-term irreversible evolution at level 3. Natural selection is postulated to operate at the individual and at the group level[3] as the cause of long-term irreversible changes, i.e., macroevolution (level 3).

[3]Dobzhansky (1937) appears to treat a population as an "individual" by attributing a physiology to it (120, 185). Thus, selection between populations could be construed as selection between "individuals." On the other hand, he refers to selection between "colonies," i.e., groups within a population (190). This would seem to suggest that he acknowledges selection between individuals

Level 2 further provides boundary conditions (environmental factors) for level 1 phenomena.

This level system is the most important addition to his program. Epistemically, it implies that theories about phenomena at different levels cannot be considered as alternatives because the phenomena are categorically different. For instance, "The theory of mutation relates to a different level of the evolutionary process than that on which selection operates, and therefore the mutation and selection theories are not alternatives" (Dobzhansky 1937: 150, 1941: 187, 1951: 76). Rather, they complement each other in the sense in which mutation and selection complement each other in explanations of adaptive evolution. Ontically, Dobzhansky clarified that the relations between the levels are causal, not logical. This prescribes that natural selection is the cause of evolution—a notion that would become a challenge to interpret.

11.3.2 Level 1: Individual Organisms

Level 1 phenomena are the raw materials for natural selection. They consist of heritable changes—mutation and recombination—and their effects on genetic networks and associated physiological processes in individual organisms. Effects include incompatibility of egg and sperm, incompatibility of external genitalia, different courtship behaviors, and different reproductive seasons. They enter into complex interactions with the environment resulting in differential reproduction (Dobzhansky 1937: 254–256, 1941: 280–88, 1951: 207–208). Therefore, Dobzhansky considers them as the component causes of natural selection at level 2. Together these interactions are the level 2 "mechanisms" that produce evolution.

All three editions present solid support for mutation and recombination as the causes of random variation between individuals within populations both in the laboratory and in the field (Dobzhansky 1937: 41–47, 118–119, 127, 1951: 50, 63, 74). Natural selection continues its function as theory. Although the particulate nature of hereditary material was established, Dobzhansky continued to argue against forms of blending inheritance because they were used into the 1930s in support of cytoplasmic inheritance. His continued rejection of blending testifies to the role of variation in his research program and ultimately to the continued importance of natural selection. This brings us to the next level.

(intra-group) and between colonies (inter-group). In both cases, natural selection operates at the population level.

11.3.3 Level 2: Groups of Organisms: Short Time Scale

Early Importance of Selection

The second level encompasses short-term reversible changes in allelic frequencies in groups which are the condition for adaptation and progressive evolution. This is the level at which natural selection acts (Dobzhansky 1937: 13, 150, 1941: 187, 1951: 76). Questions include the following: What causes variation and adaptation of groups? What are the physiological rules for groups as distinct from individuals? Dobzhansky envisions that a population has "a definite genetic constitution, which is evidently a function of the constitution of the individuals composing the group." It has its own "physiology of populations" as distinct from the "physiology of individuals" suggesting that he considered populations as analogous to individuals (Dobzhansky 1937: 11, 120). Evidence for the production of variation by group level mechanisms emerged. Dubinin and coworkers showed that the concentration and kind of mutations vary geographically and within one population from year to year. In groups, allelic frequencies could drift randomly between localities or seasons. They concluded that natural selection can produce evolution by acting on genetic differences not only between individuals in a population but also between populations in a species (Chetverikov 1927a, b; Dubinin et al. 1934; Adams 1970, 1980: 262).

This conclusion required an important change in the balance theory of Chetverikov. As we saw, Dobzhansky first ignored this theory because it did not have selection as the cause of evolution, and this prevented evolutionary progress. Six years later, Dubinin and Romaschoff (1932) connected selection and evolution. They agreed with Chetverikov (1926) that isolation and speciation could follow or accompany selection and adaptation and that the frequency of manifestation of mutations is inversely proportional to population size. However, they added that selection causes adaptive mutations to spread faster in small populations than in large ones because in small populations, hidden mutations are more likely to be exposed to natural selection. This connected the random variation between small isolated populations within a species underlying Chetverikov's model of speciation with selection and adaptation leading to evolution. As a result, progressive evolution was conceivable though unsupported by evidence. Dobzhansky called it "a novel result—and a very startling one" (Dobzhansky 1955: 4). He doesn't say why he was startled. It couldn't have been the introduction of natural selection because he had already included it in his own account of evolution. Van der Meer (2007) suggests it is the link between selection and evolution. By 1937, Dobzhansky had accepted the balance theory of Chetverikov (1926) as modified by Dubinin and Romaschoff (1932) with selection as the cause of speciation (Dobzhansky 1937: 128–129).

Continuing Importance of Selection

The concept of natural selection developed between 1937 and 1951. In 1937, Dobzhansky emphasized that:

> The theory of natural selection is primarily an attempt to account for adaptation of organisms to their environment. Whether it explains not only adaptation but evolution as well is another matter. (Dobzhansky (1937: 150–51, 1941: 187–88)

He then acknowledged those for whom selection drives evolution and others who feel there is not enough evidence for this view. Only in 1951 does he cautiously point to the increasing support for selection as a cause of evolution:

> The development of population genetics in the last two decades has considerably strengthened the theory of natural selection. It is fair to say that, among the two opinions just cited (selection as the cause of evolution and its skeptics), the first is believed by a majority of modern evolutionists to be much nearer the truth than the second. (Dobzhansky 1951: 77–79; Letter from Dobzhansky to J. L. King, June 8, 1970, Dobzhansky Papers)

But the growing influence of selection prompts him to stress that cooperation also is adaptive and that this does not contradict the theory of natural selection (Dobzhansky 1951: 79). This possibility is not mentioned in the earlier editions.

Heterozygote Superiority Maintains Adaptability

Dobzhansky (1937) used this modified balance theory to explain how rare and harmful mutations maintain genetic variation and contribute to the adaptability and viability of a species. Most individuals are heterozygous at most loci in their genome. A recessive allele *a* in a heterozygote *Aa* is not exposed to natural selection if it is not expressed in the phenotype. Recessive alleles are exposed to selection if they are expressed in heterozygotes or in homozygotes *aa*. The homozygotes allow a population to adapt to a new environment. In heterozygotes, the same alleles are shielded from selection because either they are not expressed or their expression has become suppressed by modifier genes. Further, most alleles produced by mutation are recessive. Over time, a population adds new untried mutations to recessive alleles which provided adaptation in the past and can do so again in the future. Thus a population maintains its adaptability in the face of mutation and natural selection because natural selection favors heterozygotes—a notion that underwent several improvements (Dobzhansky 1943, 1951: 122–123; Dobzhansky and Levene 1955: 806 [GNP XXIV, 629]; Beatty 1987a: 289–90). As in Chetverikov's theory, the concept of balance in the theory of Dobzhansky refers to Hardy's prediction that in a sexually reproducing and randomly mating population, the relative frequencies of alleles remain constant over the generations. Evolution is a shift of equilibrium frequencies caused by mutation, selection, migration, and inbreeding. Balance also describes a dynamic equilibrium between opposite effects of these causes of evolution. "[E]volution results when one group of them [causes] is temporarily gaining the upper hand over the other group" (Dobzhansky 1937: 123–127). The balance theory of 1937 became the working hypothesis by which to investigate the behavior of group variation produced at the individual and group levels.

Role of Isolation in Maintaining Adaptability

Recombination and mutation, Dobzhansky argued, produce new gene combinations which allow a population to explore new environments. But recombination also disrupts existing adaptive gene combinations. Disruption is reduced when a population divides into small groups, and each group receives fewer genes from other groups. That is, breeding between populations is reduced or blocked so that fewer or no genes enter a population and disrupt its existing adaptive gene combinations by recombination. But reducing population size also promotes *in*breeding which reduces the production of new gene combinations causing too much adaptation to an existing environment and a failure to adapt to new circumstances. Extinction would follow. Thus, we have two opposing demands that must be satisfied to maintain adaptability. Dobzhansky concluded that if isolation comes too early, variation is lost by inbreeding, but if it comes too late, recombination disrupts too many adapted gene complexes. If the timing is right, the conflict does not arise. "Favorable conditions for progressive evolution are created when a certain balance is struck: isolation is necessary but it must not come too soon" (Dobzhansky 1937: 229–230, 1941: 255–256, 1951: 180, compare Dobzhansky 1970a: 312).

Relative Roles of Drift and Selection

Throughout 1937–1951, Dobzhansky relies on Wright's theory. According to this theory, all other things being equal, the effect of drift and selection on variation between populations depends on effective population size and magnitude of selective force. Establishing the relative roles of drift and selection in adaptation and evolution is complex in this theory (Dobzhansky 1937: 127–134, 180–182, 1941: 161–165, 168, 332–335, 1951: 161–63, 171, 205–206, 1970a: 232–234). Effective population size determines which of the two dominates. Drift produces variation between small populations. The magnitude of selective force then determines whether natural selection would remove this variation and produce adaptation or whether drift could produce adaptation and evolution apart from selection. This could only be sorted out empirically. The lack of facts about these roles could have strengthened Dobzhansky's skepticism about selection as an explanation for evolution. Drift and selection are both population-level phenomena and thus could be alternatives in principle. Since drift is random with respect to the environment (Dobzhansky 1937: 128, 131, 1941: 161,164, 1951: 280–281), it could produce adaptation by accident. More importantly, alleles subject to drift are assumed to be neutral with respect to selection. If true, drift would lead to evolution without selection. However, under natural conditions, the size of a population fluctuates and with it the relative importance of drift and selection. As Dobzhansky emphasized:

> [I]n practice the separate colonies [of a species] are exposed to a variety of environments. Although in the abstract the differentiation may be pictured as taking place under the influence of the restriction of the population size alone, or under the influence of selection alone, in nature the process is going on because of both these factors, one or the other gaining the upper hand, probably only temporarily. (Dobzhansky 1937: 183–184, 1941: 335, 1951: 156, 162–163, 1970a: 261)

Thus, in natural populations, drift and selection can either complement each other or be alternatives depending on the circumstances.

What about the theoretical status of the shifting balance theory? Between 1937 and 1970, Dobzhansky used it as his working hypothesis. I just noted that he maintained both drift and selection in his account of evolution. Specifically, he is committed to drift and the balance theory despite a lack of direct empirical demonstration and periods when there were no observations that could be interpreted as drift (Beatty 1987a, b; Dobzhansky and Queal 1938: 249; also in: Lewontin et al. 1981: 130 [GNP I, 130]; Dobzhansky 1939: 410 [GNP IV, 170, 190], 1948: 171, [GNP XVI, 508], 1951: 164–165, 1952: 242 [GNP XX, 568], 1970a: 262, 278–80; Dobzhansky and Levene 1948: 538 [GNP XVII, 516]; Dobzhansky et al. 1963: 101 [GNP XXXI, 738]; Powell et al. 1976: 505 [GNP XLIII, 895]). He also continued to interpret empirical evidence provided by others in terms of drift (Dobzhansky 1951: 173–76, 205–206, 1970a: 238–266).

What about the empirical status of the shifting balance theory? Mutation and migration tend to obscure the effect of drift and selection on allelic frequencies. Direct evidence for the random fixation of allelic frequencies was not available until 1954. As for selection, it had been one among other theoretical causes of evolution between 1924 and 1926 (Dobzhansky 1937: 134, 150, 159, 186). Selection can maintain variation provided heterozygotes are superior to homozygotes because this establishes constant frequencies of the dominant and recessive alleles in the heterozygotes all other things being equal. Drift, migration, and geographical isolation were added as hypothetical causes of group variation in the balance theory of 1937. Soon after, Dobzhansky started his work on inversion polymorphism in chromosome 3 of *Drosophila pseudoobscura*, and empirical evidence for selection began to accumulate. Dobzhansky and Queal (1938) found that different inversions occurred with different frequencies in different locations. They attributed the apparent randomness of this variation to drift. But Dobzhansky (1943) discovered that the frequency of third chromosome inversions in isolated populations of *Drosophila pseudoobscura* varied seasonally. Now he attributed the seasonal variation to natural selection by seasonal environmental changes rather than to drift. Yet he had considerable doubts about the relevance of seasonal variation for evolution.

> The changes observed in *Adalia* and *Drosophila* seem to follow the seasonal climatic cycle. If they are constantly recurring and completely reversible, they represent, from the point of view of long-range evolutionary progress, largely a wasted motion. It may even be that species undergoing changes as rapid as these never actually reach the highest theoretically

> possible adaptive levels: the species may at any one time be adapted best to the conditions which have just lapsed (Dobzhansky 1943).

That is, in any one season, *Drosophila* may be adapted to the preceding season. Thus, there isn't enough time for an advantageous gene to reach 100% fixation. But it was worth to consider it as a working hypothesis (Dobzhansky 1947b). Further research, however, culminated in a convincing demonstration of the effect of natural selection (Dobzhansky 1947a, GNP XVII; 1947b, GNP XIV; Dobzhansky and Levene 1948, GNP XII). For instance, Dobzhansky (1947b, GNP XIV) subjected flies with controlled initial chromosome frequencies to variations in temperature, light-dark cycle, and geographic origin. He observed among others that chromosomes with the same gene arrangement but from different geographic origin behaved differently when exposed to the different experimental conditions. The dependence of survival on geographic location demonstrated not only differences in selection between places but also the causal role of selection in the field. Dobzhansky and Levene (1948, GNP XII) established the same conclusion for the dependence of survival on seasonal changes first observed by Dobzhansky (1943).

In the second edition of *Genetics and the Origin of Species*, Dobzhansky (1941) reviewed his early results in the wider context of selection in plants and animals. He knew of similar seasonal changes in the beetle *Adalia bipunctata* (Timoféeff-Ressovsky 1940) and reviewed other evidence for natural selection singling out microorganisms and cyanide resistance in scale insects as most convincing (Dobzhansky 1941: 188–196). Variation-maintaining selection received a broader basis in 1953–1954 when this mechanism was found to apply to all heterozygotes (Dobzhansky and Spassky 1953; Dobzhansky and Wallace 1953; Dobzhansky and Spassky 1954a, b).

Drift and Selection Equally Important

The growing evidence for selection made little difference for the theoretical status of the shifting balance theory. On the one hand, Dobzhansky (1947b, Dobzhansky and Levene 1948) had shown that selection occurs in natural populations. He noted evidence of the same in other organisms (Dobzhansky 1951). He continued to appreciate selection in 1970 when new theoretical and empirical studies moved Dobzhansky to conclude that populations of all sizes except very small ones have large amounts of genetic variety for natural selection to act on (Dobzhansky 1970a: 258–259). He no longer repeats his mantra that natural selection is primarily the cause of adaptation and only secondarily of evolution at large. On the other hand, the importance of selection was relative to that of drift which must be regarded "as an important, or even as the prevalent evolutionary mode" because it explains "The ubiquity of polymorphisms and the high heterozygosity found in natural populations" (Dobzhansky 1970a: 261). He keeps asking for further studies on the role of drift in evolution in 1951 as well as in 1970 (Dobzhansky 1951: 165, 1970a: 266). The problems with drift remained the same. When drift is not followed by selection, its randomness could lead to non-adaptive evolution or extinction. There was no evidence relative to the selective neutrality of alleles assumed in drift (Dobzhansky 1970a: 259–266). Privately he also continues to express his skepticism (Letter from Dobzhansky to J. L. King, June 8, 1970b, Dobzhansky Papers). Thus, the importance of drift remained an open question, and, therefore, a modified skepticism regarding selection continued. Dobzhansky was the first to acknowledge that the evidence for drift was inadequate for a general theory of progressive evolution. Nevertheless, he continued to maintain the relevance of both selection and drift as complementary mechanisms of evolution until the end of his career.

There are at least five situations which require both drift and selection for their explanation. All are related to population size. First, in populations of intermediate size, drift and selection can act simultaneously. This happens when gene A is neutral with respect to selection and is subject to drift, while gene B in the same population is not (Dobzhansky 1937: 190–191, 1941: 168, 1951: 170–171, 1970a: 278–279). Second, an existing species may consist of small and large populations. Each population is evolving in its own direction. A species-level account of the differences between its populations requires drift and selection. Third, when a large population divides into several much smaller ones, drift replaces selection. Both drift and selection are necessary to explain the differences between the large population that no longer exists and the smaller populations that differentiated from it. Fourth, both drift and selection are required to explain the features of the stages a population passes through before it becomes a species. Once a new species, drift may explain its non-adaptive features, while selection explains its adaptations. Fifth, both drift and selection are needed to explain speciation as such. Drift is a source of variation between the populations of a species. This variation is required for natural selection to be effective in producing different species. The reason why both drift and selection are equally required in these accounts is because these are accounts of what Dobzhansky had considered to be a single evolutionary process since the 1920s (Sect. 11.2.2).

So far, the most important *theoretical* development during this period was that Dubinin and Romaschoff (1932) linked natural selection to evolution. This is what moved Dobzhansky to accept their balance theory. Second, theoretically, selection and drift are equally important mechanisms of evolution. The most important empirical development was the growth of evidence for the existence of genetic variation. But this applied only to one species of fruit fly. Moreover, there was no empirical support for drift. Further, Dobzhansky sees the role of natural selection in evolution as one among other causes such as drift and cooperation. Finally, the low probability of a favorable timing of isolation, the lack of empirical information on several fronts, and the possibility of evolution without selection are possible reasons for skepticism about the role of natural selection in evolution. In sum, from a theoretical perspective, selection and drift remained equally important. Further, the shifting balance theory has various implications that justify skepticism about the role of selection in evolution. From an empirical perspective, convincing evidence for selection in *Drosophila pseudoobscura* raised his confidence in that rôle, but this did not diminish the possible relevance of drift. As I claim in my first thesis, during 1927–1951, his skepticism declined, but it did not disappear.

11.3.4 Level 3: Groups of Organisms: Long Time Scale

The third level concerns the long-term irreversible stabilization (fixation) of variation and adaptation at the group level that result from the interaction of mutations and recombination at the individual level with processes at the group level such as migration, selection, and reproductive isolation (compare Dobzhansky 1937: 11–13, 1941: 11–14, 1951: 14–18). The questions were: What are the causes of evolution? How are adaptations turned into irreversible historical changes? It is in this context that Dobzhansky tried to understand the role of natural selection.

In 1924, Dobzhansky observed that traits characterizing different species in one location could characterize different populations within a species in other locations (Dobrzhansky 1924: 418, 420; Dobrzhanskii 1926a: 31–44). But there was not enough evidence to convince him that such differences between populations could become irreversible. He pictured a species in terms of the flow of genetic information between its members. He then identified blocks to gene flow as a first step toward irreversibility, that is, as the likely causal link between microevolution and macroevolution. The presence or absence of gene flow could explain that traits differentiating populations within a species in one location could differentiate species in another location. Throughout the three editions, evidence grows for the existence of reproductive isolation by the interruption of gene flow between populations (intraspecific hybrid sterility) as well as between species (interspecific hybrid sterility) and by mechanical, sexual, and ecological isolation (Dobzhansky 1937: 233–254, 1941: 258–277, 1951: 182–199). By 1951, Dobzhansky's assessment was that natural selection explains adaptation. Yet, whether it explained evolution remained an open question (Dobzhansky 1951: 77).

This skepticism is understandable. During 1927–1937, Dobzhansky tried but failed to disprove Filipchenko's belief that microevolution is unrelated to macroevolution. This is a preoccupation in the correspondence between Dobzhansky and Filipchenko (Kolchinskii et al. 2002). He wrote to Filipchenko that it is not possible to prove that plasmons do not exist (Dobzhansky to Filipchenko, July 23, 1928 In: Kolchinskii et al. 2002: 82–83). Dobzhansky hoped that the historical process of evolution could be repeated experimentally. For instance, Filipchenko explained intersexes in *Drosophila* as a result of a mixing of cytoplasmic substances at fertilization. Dobzhansky countered that the occurrence of intersexes depends on the balance between the number of autosomes and sex chromosomes (Dobzhansky to Filipchenko, July 23, 1928, and Filipchenko to Dobzhansky, Jan. 22, 1930 In: Kolchinskii et al. 2002: 82–83, 399–400; Dobzhansky and Bridges 1928). This meant that sex is a genetic trait with Mendelian rather than cytoplasmic inheritance (e.g., Baur 1930: 197). Given that Classes are often differentiated by organs of reproduction, Dobzhansky had an empirical argument for the continuity between microevolution and macroevolution by showing that Mendel's laws applied to the genetics of the reproductive organ system.

Unfortunately, this argument applied to the evolution of Species, but not of Classes. A species is a group, the members of which can have fertile offspring. This becomes impossible when the reproductive organs are too different. Since the latter difference is the result of genetic influences, there is a direct causal effect of genes on the structure of reproductive organs, viz., the inability to have fertile offspring and speciation. But, he continued, genes for Class traits could not be identified directly because crosses between representatives of different Classes are not possible. Thus, the issue of the continuity of microevolution and macroevolution could not be resolved experimentally. He also doubted that the results of experiments on variation-producing agents such as gene and chromosome mutation could be applied in nature. Finally, even if they applied, observation and laboratory studies could never span the time periods required by evolution (Dobzhansky 1937: 12–14, 1941: 12–15, 1951: 16–17). No wonder Dobzhansky was reluctant to accept the continuity between microevolution and macroevolution as a working hypothesis in the 1937 and 1941 editions. Seasonal changes in natural populations, however, convinced him that:

> Instances of rapid transformation [. . .] have accumulated to such an extent that this concession (that evolutionary changes are too slow to be observed directly) is no longer necessary. (for a review see Dobzhansky 1941) (Dobzhansky 1941, 1943: 184 [GNP IX, 327])

But this means merely that by 1943, he was convinced that the evidence reviewed in 1941 warranted the conclusion that small evolutionary changes can be observed in live organisms.[4] One year later, his correspondence with the creationist Marsh (1944) prompted him to express confidence that the theoretical continuity between microevolution and macroevolution has a basis in reality.

> The proofs of evolution derived from comparative anatomy, from paleontology, and from geographic distribution are admittedly inferential, but things certainly look so that if one assumes evolution they become quite reasonable and self-consistent. (Dobzhansky to Marsh, Dec. 7, 1944. See also Dobzhansky to Marsh, Dec. 22, 1944, Frank Lewis Marsh Papers)

However, these proofs allowed only an inference of evolution, not a causal explanation. Five years later, Simpson (1949) and Schmalhausen (1949) justified the inference of evolution from population genetics by convincing Dobzhansky that population genetics was consistent with paleontology, classification, embryology, and comparative anatomy (Dobzhansky 1951: 16–17).

Simpson (1949) argued that variation, rate and character of mutation, effective population size, length of generations, and natural selection explain the development not only of races and species but also of fossil phyla. For instance, the average rate of evolutionary change of fossils matched that based on rates of change of allelic

[4]This conclusion is warranted by the definition of evolution as any change of allelic frequencies. The definition is ambiguous because it leaves unspecified whether the change is reversible or irreversible (evolution) which is the point at issue. Hence, the conclusion is ambiguous. Proof of evolution requires evidence of irreversible change.

frequencies in population genetics (Simpson 1944: 45–46, 93–96). Further, population genetics explained trends in the fossil record such as increasing size of body and brain as adaptations. Too much adaptation as in the oversize antlers in the Irish Elk was just what Dobzhansky expected from inbreeding in small populations isolated at an unfavorable time (Simpson 1944: 159–174; Dobzhansky 1937: 230, 1941: 255–256, 1951: 180, compare Dobzhansky 1970a: 312). Finally, fossil populations and species existed, and their evolution could be explained in terms of the adaptation of populations used by Dobzhansky (Simpson 1944: 91–93). Dobzhansky noted that Simpson (1943) had defined the relationships between the species concepts used in taxonomy, population genetics, and morphology (Dobzhansky 1951: 263–264). He had taken the first step himself in 1924 when he reported that traits characterizing different populations within one species in one location could characterize different species in another location. The genetic category of a population matched the systematic category of a race or species (Dobrzhansky 1924: 418, 420; Dobrzhanskii 1926a: 31–44; Dobzhansky 1951: Ch. 9). Simpson's work encouraged taxonomists to create consistency with population genetics by reclassifying many species as races of a single species and higher taxa as species (Dobzhansky 1951: 258–259). In sum, Simpson convinced Dobzhansky that specific features of the fossil record can be inferred from population genetics.

Schmalhausen (1949) proposed a common cause for the evolution of populations and embryos. Since most mutations are harmful, selection would favour other mutations that make harmful mutations recessive. Dominant alleles would control normal embryonic development and protect it from mutation. These dominant alleles would also control physiological buffers of developmental processes which could account for the stability of overall anatomical plan revealed by comparing embryos. As a result, embryonic development could be interpreted in terms of population genetics (Schmalhausen 1949: 150–152, 205–245; Dobzhansky 1951: 106–107). Disruption of embryonic development by mutation could no longer be used as an objection to interpreting anatomical similarities between embryos of different phyla in terms of common ancestry. Also, the balance theory could now explain the differences between embryos from higher categories of classification. The gap created by Filipchenko between microevolution and macroevolution had been bridged by an inference of macroevolution from microevolution. Accordingly, Dobzhansky (1951: 16–17) listed evidence from paleontology (Simpson 1944, 1949) as well as from comparative anatomy and embryology (Rensch 1947; Schmalhausen 1949) that had become available.

The consistency of second-level population phenomena with third-level evolutionary changes became another major step in the conceptual development of the synthetic theory of evolution. In particular, selection as the main cause of adaptation and evolution had now been generalized to include the development of individuals (embryos), groups (populations, species), and higher taxa (up to phyla). As a result, Dobzhansky felt that the extrapolation of the causes of microevolution to macroevolution was justified. This explains an observation by Burian (1994: 137) that Dobzhansky (1951) had removed the separate treatment of evolutionary statics and dynamics in earlier editions and treats all three levels under evolutionary dynamics.

Further, levels 2 and 3 are no longer considered categorically different. He still distinguishes reversible from irreversible changes, but he considers the distinction as quantitative and, therefore, arbitrary (Dobzhansky 1970a: 429). Accordingly, Dobzhansky no longer mentions his three stages of evolution as levels of description and analysis. But he still incorporates them in a more general idea of a level system (Dobzhansky 1970a: 4, 431).

However, level 2 remained the weakest link in what he originally viewed as the causal chain from level 1 to 3. This concerns the production of variation which was in question because drift had to be demonstrated and the superiority of heterozygotes had to be more common. It also concerns the maintenance of variation because isolation had to be timed just right. Finally, the assumption of a causal step from level 2 to 3 had to be replaced with an inference.

In sum, at first, there was no justification for the assumption that evolution from level 2 to 3 was a causal process. Experimentation could not provide this justification. Instead, Dobzhansky (1941) referred to the evidence cited by Darwin that had convinced him in his youth. By 1944, evidence for small-scale changes at level 2 had accumulated to the point that it became the justification for the assumption. By 1949, the consistency between level 2 and level 3 phenomena shown by Simpson and Schmalhausen strengthened this justification. But it remained an inference from the fossil record that could not be supported experimentally (Dobzhansky 1951: 17). His Cartesian mechanicism demanded a causal role for natural selection in evolution. But it was out of the question. This realization may have contributed to his skepticism about the role of natural selection.

11.4 Natural Selection in the Classical and Balance Theories (1951–75)

11.4.1 The Classical Theory

By the 1950s, a competitor for the balance theory of Dobzhansky emerged: the classical theory of Muller also known as the selection or gradualist theory. At issue was the role of natural selection. Underlying the classical theory is Muller's expectation that most heritable traits are unstable because they are the cumulative result of harmful mutations over long periods of time. Most fossil characters, however, are very stable over long periods. He concluded that natural selection must remove most mutations to explain this stability. This includes recessive mutations newly arisen in heterozygotes. Otherwise, after a number of generations have produced enough of such heterozygotes, their inbreeding would create unstable characters in offspring homozygous for the mutation. Natural selection eliminates such heterozygotes because the recessive mutation has a phenotypic effect. Even the slightest deviation from the normal phenotype is selected against; otherwise, recessives in heterozygotes could not be removed. This means that in contrast to the balance theory, there is only one single allele for every locus in a population that can be considered normal in a given environment. Since most mutations are deleterious, this would be the rule:

CLASSICAL:	A1B5C2DE1....Z	A1B5C2DE1....Z
	A2B5C2DE1....Z	A1B5C2DE1....Z
	Individual 1	Individual 2

A rare new mutation (A2 in Individual 1) will be heterozygous when it arises (A2/A1 in Individual 1). If in a given environment A2 is more optimal than all existing alleles at that locus, its frequency will rise because natural selection will remove all other existing alleles both in heterozygous and homozygous form. After some generations, homozygotes for the new optimal allele will appear due to inbreeding. They are selected over heterozygotes because they have the more optimal phenotype. In heterozygotes, the optimal allele is never completely recessive to other alleles at its locus. Thus, homozygotes and heterozygotes for that allele have different phenotypes, and both are subject to selection. As a result, an optimal allele will eventually replace all other alleles at its locus in the entire population. Thus, according to the classical theory, most individuals in a population are homozygous at most loci because only one allele of the set for a given locus is optimal (normal, wild type) in a given environment. Heterozygosity is predicted to be rare because mutation is rare and selection removes heterozygous carriers of mutations.

Under both the classical and balance theories, mutation was the source of conflict (Dobzhansky's term) between organism and environment because it is random with respect to the environment. As noted, in the classical theory, the primary function assigned by Muller to natural selection was to remove alleles from a population. By contrast, the balance theory explains adaptability as the result of natural selection that favors heterozygotes and, thereby, maintains the frequency of recessive alleles at a given level. To maintain genetic reserves is to be adaptable. Moreover, adaptation can also be achieved by cooperation (Dobzhansky 1951: 79). The classical theory postulates that heterozygotes are rare because most new recessive mutations affect the expression of the dominant allele negatively. For a mutation to remain in existence, it would have to be adaptive in the environment in which it arises. The low probability of mutation combined with the even lower probability of an adaptive mutation would leave a population no freedom to adapt to new environments. Dobzhansky reasoned that under the classical theory, virtually all new alleles introduced by mutation would be eliminated irrespective of the environment (Dobzhansky 1962: 296). He rejected the classical theory because a population would be unable to accumulate a stock of potentially adaptive alleles, and natural selection could not produce adaptation and evolution. The lack of clarity about how natural selection operates must have obscured its role in evolution.

11.4.2 Comparing Theories in the Light of Observation

A simple choice between the two theories was not possible (details: Lewontin 1974). Firstly, the available facts did not discriminate between them. This was acknowledged by both Muller and Dobzhansky. For instance, both agreed that recessive genes exist and are subject to selection when they affect the expression of the dominant allele. But they focused on different aspects of the history of life, and this led to different interpretations of the long-term effect of selection. For Muller, the long-term stability of fossil characters warranted the assumption that all new mutations are subject to selection which would reduce variability. In contrast, for Dobzhansky, selection maintained adaptability by favoring the development of complete dominance and heterozygote superiority.

These different views were possible due to a lack of empirical constraints. There was some evidence that complete dominance resulted from suppression of a new mutation by its counterpart at the same locus or by alleles at different loci, but the spread of suppression across the gene pool was unclear because percentages of heterozygosity varied widely. Further, evidence for heterozygote superiority widened from inversion heterozygotes on the third chromosome to most single genes (Vetukhiv 1953; Brncic 1954) but was confined to *D. pseudoobscura*. Dobzhansky tried to generalize inversion polymorphism in *Drosophila* to other species and taxons, but failed. The only other well-supported instance was that of sickle-cell anemia in humans (Allison 1955). During the 1950s, inversion polymorphism was known only in plants, and during the 1960s, it was discovered in other insects as well as in rats and mice (Dobzhansky 1970a: 151). Moreover, between 1951 and 1975, there was as much evidence for the superiority of homozygotes as for that of heterozygotes. This puzzle was not clarified until the late 1960s when heterozygotes were found to be superior only when bred from individuals in the same population so that they have the same genetic background (Dobzhansky and Spassky 1968 [GNP XL], review in Dobzhansky 1970a: 189). In sum, according to the balance theory, selection explains adaptability, but the facts about maintaining adaptability were confusing.

Secondly, the decision was not between mutually exclusive theories, but about their relative importance. Given the confusing empirical picture, it was reasonable for Muller to ask whether the balance theory was applied in taxons other than *Drosophila* (Muller to A. R. Gopal-Ayengar, April 17, 1957, Muller Papers). It might apply in some taxons and the classical theory in others. Further, within a single population, the balance theory might apply to some loci and the selection theory to others, or they might even apply to the same loci at different times and places (Muller 1956: 488–491; Dobzhansky: 1955: 4; Wallace 1957: 58). Another problem was that the theories did not entail definite percentages of heterozygotes. The classical theory postulated 5–10% of loci heterozygous in the average individual, and the balance school assumed 50–90% (Crow 1987). Hence, the theories could not be tested statistically, and one could not discriminate between them. The theories were different in degree, not in kind. Compounding all these difficulties was the need to measure amounts of variation, but this could not be done during the lifetime of Dobzhansky (Lewontin 1974). Finally, the theories did not predict how much variation was needed for adaptability (Lewontin 1974). If a small amount of heterozygosity would suffice, a decision between the theories would be impossible. When Dobzhansky declared victory for the balance theory based on an average of 30 % heterozygosity extrapolated from allozyme surveys, it was premature (Lewontin 1974: 118–119). Suffice it to say that debates about the classical and balance theories must have obscured the role of natural selection in evolution.

11.5 Biological Interpretation of Natural Selection

The biological interpretation of natural selection by Dobzhansky focuses on causal range, empirical support, taxonomic range, and effectiveness. I will use these aspects to suggest five reasons for Dobzhansky's hesitation about the role of natural selection in evolution, not in adaptation. He does not offer reasons himself.

11.5.1 Reasons for Ambivalence About Natural Selection

Introduction

I suggest that the initial reason for his skepticism toward natural selection as the cause of evolution may have been uncertainty about the scope of other possible causes of evolution. There were other possible causes of evolution:

> ... the historical process (cf. Dubinin 1931), the molding of the hereditary variation into racial, specific, generic, and other complexes, is due to action of the environment through natural selection and other channels .... (Dobzhansky 1937: 120)

Other channels included polyploidy in plant and animal speciation (Dobzhansky 1937: 192–227, 1941: 223–253, 1951: 287–297, 1970a: 162–163), drift in small populations in general (Dobzhansky 1937: 133–134, 180–182, 1941: 164–165, 168, 332–335, 1951: 161–163, 171, 205, 1970a: 232–234), and cooperation (Dobzhansky 1951: 79). I suggest five reasons for his skepticism.

Evolution Without Natural Selection

In theory, drift could explain speciation apart from natural selection. But, as noted, Dobzhansky doubted that in evolving populations, under natural conditions, drift and selection would act in isolation. This problem exists only in the abstract, not in nature (Dobzhansky 1937: 183–184, 1941: 335, 1951: 156, 1970a: 261). The shifting balance theory entailed that drift creates the variation between small populations that selection could act on. But empirical evidence for drift was not forthcoming. Therefore, I suggest in particular that uncertainty about the scope of drift in relation to selection contributed to his hesitancy about the role of natural selection in evolution.

Selective Neutrality of Alleles

It was not known whether there are selectively neutral alleles as required by drift. Uncertainty about drift translates into uncertainty about the role of selection in evolution because in natural populations, drift and selection can complement each other or operate together in some other way.

Causal Efficacy of Natural Selection Cannot Be Demonstrated by Experiment

The causal efficacy of natural selection is its power to cause adaptive evolution by changing allelic frequencies in a population. The efficacy of natural selection in causing adaptation, extinction, stasis, and diversity was being demonstrated experimentally and by field observation. For instance, when speciation experiments were possible, Dobzhansky is initially careful to distinguish questionable from probable empirical evidence. The three editions of *Genetics and the Origin of Species* reveal a growing confidence in the role of natural selection in adaptation from questionable (Dobzhansky 1937: 159, 160, 164) to probable or reasonable (Dobzhansky 1937: 151–152, 154–158, 161, 1941: 188–219, 1951: 82–103). At the same time, its causal efficacy in evolution could not be demonstrated experimentally because of the long time span of evolution.

> No major evolutionary change is noticeable in most species of organisms within a human lifetime, hence the supposition that species have become what they are now through evolution by natural selection can be at best no more than a very probable inference. (Dobzhansky 1937: 151, 1941: 12, 1951: 17)

Macroevolution could be inferred from the consistency of the balance theory with paleontology, classification, embryology, and comparative anatomy (Dobzhansky 1951: 17). This was an empirical inference, but not a causal one. Put otherwise, there was empirical evidence for progressive evolution in the fossil record, but no empirical evidence that natural selection could produce it. There was no causal continuity between micro- and macroevolution. But by 1970, this distinction is considered to be "arbitrary" (Dobzhansky 1970a: 429). Empirical support for the causal role of natural selection is taken for granted (Dobzhansky 1970a). Yet he kept up his skepticism in private correspondence (Dobzhansky to J. L. King, June 8, 1970b, Dobzhansky Papers). One contributing reason, I suggest, was the impossibility in principle of demonstrating by experiment or observation that natural selection is the cause of evolution. This source of skepticism is associated with his Cartesian mechanicism. It entailed a commitment to explanation in terms of causal mechanisms. In order to establish a causal mechanism in population genetics, experimentation or field observation is required. This was inherently impossible because evolution surpasses a human lifetime. He had to content himself with an inference which is not as definitive as causal manipulation. This lowered his epistemic confidence in the causal role of selection.

Dobzhansky's stance on this point is puzzling because there was a ready solution at hand. In the same work in which Dobzhansky (1937: 11, 12, 1941: 12, 1951: 16, 17) exchanged causal explanation of macroevolution for inference of macro- from microevolution, he defined evolution as a change in the genetic composition of a population. After he demonstrated in the late 1940s that selection can change allelic frequencies, he could have claimed to have a causal explanation for macroevolution. But he never replaced inference with causal explanation.

The taxonomic range of natural selection is the number of different species in which natural selection operates. This number was slowly growing during Dobzhansky's life which may have ameliorated his skepticism about natural selection.

Critical Timing of Isolation

Dobzhansky envisioned the maintenance of variation in a way that may have made him skeptical about the role of natural selection in evolution. As noted, the timing of reproductive isolation is critical because it must satisfy two opposing demands. On one hand, the timing must be early enough to reduce the influx of alleles from other populations because this would disrupt adaptive gene complexes accumulated in the population. Failure would result in natural selection eliminating unadapted individuals. On the other hand, the timing must not be too early to allow the influx of alleles from other populations because this avoids inbreeding. Failure would reduce variation, i.e., adaptability and natural selection would remove unadapted individuals. The result would be stasis or extinction. This situation suggests two more reasons for skepticism about natural selection. First, the likelihood of striking this balance is low. Second, it is difficult to establish whether in fact variation due to allelic influx balances adaptation protected by blocking allelic influx.

Uncertainties Associated with the Selection: Balance Controversy

Finally, the controversy about the role of selection in evolution that was at issue in the selection-balance debates exposes several more reasons for reservations about selection. Little was known about the extent to which recessive alleles are subject to selection, the relative frequencies of heterozygotes and homozygotes, the taxonomic extent of heterozygote superiority, the conditions under which selection applies to some alleles and drift to others in the same population, and the amounts of variation.

Discussion and Conclusions

Two developments gave Dobzhansky more confidence in the role of selection in adaptive evolution. These are the experimental demonstration of the efficacy of selection in natural populations of *Drosophila pseudoobscura* and the addition of other species to which this applies.

The causal continuity between micro- and macroevolution started as a working hypothesis demanded by his Cartesian mechanicism and ended being replaced with an inference. This shift indicates a deeper methodological problem that could not have been addressed at the time. Today we can recognize differences between experimental natural sciences addressing microevolution and historical natural sciences dealing with macroevolution. They require different methodologies with different standards of evidence (Cleland 2001, 2002, 2011). As far as I know, these differences had not been addressed during Dobzhansky's life. Adams (2021, this volume) concludes that the question of the continuity "was never settled." I would add that the question of causal continuity was unsolvable within the Cartesian program but that the inference Dobzhansky accepted could be justified as an interpretation within the methodology of the historical natural sciences.

Dobzhansky attributes creativity to natural selection (Dobzhansky 1980: 450). He is referring to the ability of natural selection to create new combinations of genes within an integrated population genotype (Dobzhansky 1937: 185, 186, 187, 1941: 215, 339, 343, 1951: 277, 278). That is, the creativity of natural selection is its ability to increase the adaptability of a population by adding new adaptive gene complexes. He attributes creativity also to evolution (Dobzhansky 1951: 77, 1970a: 430–431) and appears to see this as the result of the creativity of natural selection (Dobzhansky 1951: 278–279, 1970a: 431).

But how can macroevolution be creative if natural selection remains one among other causes of evolution even if an important one? Further, how can Dobzhansky think both that natural selection is the source of the creativity of evolution and that its role in evolution is questionable? It appears that he counted on occasions when conditions allowed natural selection to operate and contribute its creativity to evolution. Dobzhansky indicates that the evolution of a species can end in extinction, stasis, diversification, or adaptation. In 1968, Dobzhansky made a point he had not made before: *progress* is not an essential feature of evolution. Natural selection had a role in some, but not all mechanisms of evolution (Dobzhansky 1968: 31). In 1970, he repeated that "evolutionary progress [is seen] in some lines of descent, extinction in many more lines, and evolutionary stasis in the rest" (Dobzhansky 1970a: 391). Dobzhansky (1974) asserted that there is progress in complexity when the entire process of evolution is considered, but on the level of particular details, there is also regress and extinction. Thus, progress is not essential for evolution and happens only occasionally (Dobzhansky 1951: 279, 1970a: 431, 1974: 311–312). "Natural selection has tried out an immense number of possibilities and has discovered many wonderful ones. Among which, to date, the most wonderful is man" (Dobzhansky 1970a: 431).

11.5.2 Did Dobzhansky Participate in the Hardening of the Synthetic Theory of Evolution?

The relative roles of drift and selection in evolution have been central in discussions of the so-called hardening of the synthetic theory of evolution. Whether Dobzhansky participated in it depends on how the hardening is characterized. Its common denominator is a shift in emphasis from drift which is non-adaptive to selection as a mechanism of adaptive evolution (Gould 1982, 1983, 2002). Two main grounds have been offered for such a shift in Dobzhansky. The first is that Dobzhansky gradually gave more weight to selection over drift. The second related ground is that Dobzhansky moved the application of the shifting balance theory from populations to species, that is, from short-term reversible level 2 processes to long-term irreversible level 3 processes.

The First Ground

Gould (1982: xxxv) introduced the first ground with a comparison between the editions of 1941 and 1951. The 1951 edition "clearly reflects this hardening." "He [Dobzhansky] deleted the two chapters that contained most material on nonadaptive or nonselected change (polyploidy and chromosomal changes, though he includes their material, in reduced form, within other chapters)" (Gould 1982: xxxv). Gould is correct in observing that Dobzhansky gradually gave more weight to selection, but he appears to overlook that this reflects growing empirical support confined to selection as opposed to drift. Gould did not consider that theoretically Dobzhansky kept giving equal weight to selection and drift. Gould's argument might apply to the 1970 edition because it mentions polyploidy less often and no longer links it with speciation. But even then, Dobzhansky still declared drift to be essential for evolution (Dobzhansky 1970a: 261–263).

Provine (1986: 349, 388–397) supports the first ground by noting the replacement of drift in GNP I (Dobzhansky and Queal 1938) with selection in GNP IX (Dobzhansky 1943) and GNP XII (Dobzhansky and Levene 1948). Provine (1986: 452) takes this replacement as prompted by the discovery of cyclic seasonal changes of inversion polymorphism and as a mark of hardening. Dobzhansky (1951) interpreted the correlation of seasons and chromosome frequencies as empirical evidence for a causal relation between selection and evolution. Provine (1986: 452) challenged "anyone who compares the second and the third edition to conclude anything other than that Dobzhansky had become more selectionist in the decade 1941–51." I agree with the changes that have occurred in Dobzhansky's publications. But I do not take them as marking a hardening in the sense of diminished importance of drift.

To assess the first ground, I refer to the evidence I have presented showing that throughout the period in which hardening is said to have occurred—1940 to 1955—Dobzhansky entertained multiple mechanisms of evolution. To make this more specific, I distinguish between theory and empirical evidence as Dobzhansky (1970a: 263) did.

On the theoretical side, Dobzhansky's encounter with Filipchenko and Chetverikov shows that Dobzhansky attributed an important but hypothetical role to natural selection in the 1920s. It was the decisive issue in the theoretical choices he made early in his career (Sect. 11.2.1). The theory of natural selection did not need to gain relevance. It needed empirical support, and that is what Dobzhansky (1947a, b, 1948, 1949, 1950) and Dobzhansky and Levene (1948) provided. Further, from 1937 to 1970, his theoretical commitment was to the equal importance of drift and selection. Their relative importance was not a zero sum game (Sect. 11.3.3).

On the empirical side, Dobzhansky produced convincing evidence for selection by the late 1940s (Dobzhansky 1947b; Dobzhansky and Levene 1948, Sect. 11.3.3). He acknowledged others who produced evidence for selection. But this is the ordinary course of events in research, and in Dobzhansky, it is not anything other than just that. Accumulating evidence made Dobzhansky more confident in the selection aspect of the shifting balance theory. But he had to suspend judgment about drift and wait for more testing. Thus, empirically, the relative importance of drift and selection remained undecided. This must have contributed to his hesitancy about the actual role of selection as a cause of evolution.

The Second Ground

The second ground for the hardening relates to a shift in the application of Wright's fitness landscape. Gould (1982: xxxvi) notes that through the 1940s and 1950s Dobzhansky raised the level at which he applied Wright's adaptive landscape from drift in populations (Dobzhansky 1937) to selection at the species level. Gould interprets this as a sign of hardening. Provine (1986: 344) joins Gould when he writes that "there was more selection and less random drift at levels above that of local populations, where drift might still be of central importance . . . ." Gould and Provine appear to take the move from the population to the species level as an epistemic move. That is what it would have to be if it marks a change in the application of a theory. Instead, I argue that the level at which Dobzhansky applies Wright's fitness landscape depends on the stage of evolution from population to species that is being addressed. This ontic interpretation is the one Wright gives to his shifting balance theory. It also makes more sense than an epistemic interpretation simply because a population can evolve into a species. Drift is a source of variation required for natural selection to be effective. Therefore, both drift and selection are equally required in a single evolutionary process. This is consistent with Dobzhansky's vision in the 1920s that evolution comprises one single process (Sect. 11.2.2).

Discussion and Conclusions

On the theoretical side of the shifting balance theory, Dobzhansky remained committed to the equal importance of selection and drift as mechanisms of evolution. Relative changes in emphasis on selection and drift are due to the availability of empirical evidence, to fluctuations in its interpretation by Dobzhansky[5], and to the influence of Wright (Provine 1986: 402–403). As a result, his skepticism about the role of natural selection in adaptive evolution never disappeared. But it diminished when empirical evidence accumulated. This is my first thesis. I conclude that Dobzhansky did not participate in the hardening of the synthesis whether viewed from a theoretical or from an empirical perspective, let alone from a dogmatic or ideological one. The very ambivalence of Dobzhansky regarding natural selection as the most important or even the sole mechanism of adaptive evolution argues against it. So do the uncertainties about the scope of drift. Of course, this does not mean that there was no hardening of the synthesis in other participants in the synthesis.

This conclusion goes one step beyond that of Beatty (1987a: 274). I agree when he writes that "Dobzhansky himself never overestimated the significance of the selectionist successes. In reporting them, he was usually careful to point out that they did not have much bearing on the all-importance of selection ...." But I disagree when he suggests that Dobzhansky attached more weight to selection than drift. He asserts, for instance, that:

> Dobzhansky *always* attributed a major role to selection. He never considered drift an important "alternative" to selection. The most he ever accorded to drift was a role that "complemented" the role of selection. In this sense, Dobzhansky was never *simply* a proponent of the importance of drift. Moreover, Dobzhansky was never *simply* a selectionist." (Beatty 1987a: 275)

I agree that selection was always important to Dobzhansky. But unlike Beatty, I argued that Dobzhansky maintained the theoretical claims of the shifting balance theory and attributed equal importance to drift and selection. I supported this with five reasons (Sect. 11.3.3).

11.6 Religious Interpretation of Natural Selection

My second thesis is that Dobzhansky's religious vision of progressive evolution explains why he interpreted natural selection as natural evil. Dobzhansky believed that evolution is cosmic, progressing through biological, social, and religious phases. From a religious perspective, Dobzhansky envisioned the kingdom of God emerging through stages of natural, cultural, and spiritual progress. The purpose of this history is the redemption of the cosmos by means of divine action in both nature and society.

---

[5]For example, the emphasis of taxonomists on the non-adaptive character of species traits may have caused Dobzhansky to emphasize drift as a mechanism of speciation (Provine 1986: 292–293).

This pervasiveness of divine action is a belief characteristic of both Eastern and Western Christianity. Dobzhansky articulated this belief in the 1960s:

> [E]volution, like everything in the world, is a manifestation of God's activity. [...] I see no escape from thinking that God acts not in fits of miraculous interventions, but in all significant and insignificant spectacular and humdrum events. (Letter to J. Greene, November 23, 1961 in Greene and Ruse 1996: 462; Dobzhansky 1967: 25)

Dobzhansky's conviction that the cosmos needs to be rescued from moral and natural evil is the context for his interpretation of natural selection as natural evil.

Let us first establish that Dobzhansky did interpret natural selection as natural evil. Late in his career, Dobzhansky wrote some popular science works in which he was explicit in his interpretation of the struggle for existence in nature as natural evil and suffering (Dobzhansky 1951, 279, 1967, 120, 1975, 205, 211).

To explain why Dobzhansky interpreted natural selection as natural evil, we need context. Ruse (1996: 400–401) showed that in the debate of the early 1950s over the classical and balance theories, empirical evidence did not discriminate between the two. Yet Dobzhansky preferred the balance theory. Ruse explains this preference as a result of Dobzhansky's commitment to a religious vision of progressive evolution. According to the balance theory, but not the selection theory, populations accumulate adaptive gene complexes over time. As a result, their adaptability increases, and this characterizes evolution as progressive. Thus, the balance theory could be interpreted in terms of his religious vision of cosmic evolutionary progress. Since this could not be done under the classical theory, Dobzhansky preferred the balance theory.

How does this religious vision of progressive evolution explain that Dobzhansky interpreted natural selection as natural evil? His belief that the same God acts in both natural and social evolution creates the expectation of similarities of divine action between the two. One of these analogies is that between social evil and natural evil. Dobzhansky interpreted natural selection as natural evil because he thought about nature in terms of what he knew about society. Specifically, he thought about nature in terms of his experience of violence and moral evil in the Ukrainian War of Independence (1917–1922). As society eliminates individual people, the environment eliminates individual organisms. He explicitly articulates the relation between organism and environment as a conflict relation (Dobzhansky 1937: 126–27, 230, 1941: 159, 255–256, 1951: 73, 77–79, 180, 1938: 445–49) and only rarely as cooperation (Dobzhansky 1951: 78). Sources of conflict include mutation, recombination, and environmental change. Conflict describes a class of relations between organism and environment that can result in extinction, evolutionary stasis, diversification, or evolutionary progress. Their common denominator is that organisms with better adapted gene complexes have more offspring than organisms with lesser adapted gene complexes—a process he referred to as natural selection (Dobzhansky 1937: 149, 1941: 186, 1951: 76, 1970a: 2). His interpretation of natural selection as natural evil applies in particular to extinction and stasis. They are read as the price required for evolutionary progress. This is how his religious vision of progressive evolution explains that Dobzhansky interpreted natural selection as natural evil.

There are no indications that this is how the young Dobzhansky interpreted natural selection.

Now that I have sketched Dobzhansky's worldview, I can explain how the progressive character of evolution combined with its cosmic scope may have supplied another reason for his skepticism about natural selection. In his popular-scientific and philosophical publications, Dobzhansky presents himself as an emergentist:

> Inorganic, organic, and human evolutions occur in different dimensions, or on different levels, of the evolutionary development of the universe. ... [T]he different dimensions are characterized by different laws and regularities, ... (Dobzhansky 1967: 44)

The next level up is superimposed on the previous level with feedback relations between them. These levels are stages in cosmic evolution (Dobzhansky 1967: 42, 44; Delisle 2008, 2009). Earlier we encountered this description in the three levels of biological evolution described in his professional work (Dobzhansky 1937: 11, 13, 119–120, 1941: 11, 13–14, 152–153, 1951: 15, 18, 76) which he incorporates in this more general idea of a cosmic level system (Dobzhansky 1970a: 4, 431). According to Dobzhansky, progress is an increase in complexity that characterizes evolution as it moves up the levels. From the perspective of the cosmic scope of evolution and its progressive character, there is another reason for skepticism about natural selection. Natural selection cannot explain the transition from inorganic to organic patterns of organization. As Dobzhansky points out, the inorganic does not have the self-replication required for natural selection to work (Dobzhansky 1967: 46–48).

11.7 Conclusions

1. My first thesis is that his doubt about the role of natural selection in evolution was qualified and declined during his life, but never disappeared. When Dobzhansky started his career in the 1920s, the main elements of his theory of biological evolution were in place (Sect. 11.2.1). They remained unchanged throughout his career. Included was a pluralist view of the mechanisms of evolution, all hypothetical. From this theoretical perspective, natural selection was one of a number of possible causes of evolution. Other causes included drift, cooperation, and polyploidy. During his career, empirical information accumulated that supported selection as a mechanism of evolution. But he did not gradually exclude all mechanisms of evolution except selection. Rather, from a theoretical perspective, he never questioned the equal importance of selection and drift in the shifting balance theory. On the empirical side, lack of evidence kept the scope of drift in limbo. But accumulating evidence gave him a growing confidence that selection acts in natural populations. Yet, his reservations about selection as the cause of progressive evolution never disappeared for the reasons listed (Sect. 11.5.1).

2. My second thesis is that Dobzhansky interpreted natural selection as natural evil and that this interpretation can be explained in terms of his religious vision of progressive evolution. Dobzhansky believed that God acts similarly in both nature and society. This led him to interpret nature in terms of what he knew about society and vice versa. He saw such a similarity between conflict in nature eliminating organisms by natural selection and conflict in society eliminating individuals by war. Since conflict in society involved moral evil, he thought about conflict in nature—natural selection—in terms of natural evil.

Acknowledgments I thank the Jackman Foundation for support through the Reverend Edward Jackman. My appreciation goes to the late Helen Semenuk for translations from Russian and Ukrainian and to Ronald Numbers for telling me about the Frank Lewis Marsh Papers. I thank both the American Philosophical Society Library, Philadelphia, and the Center for Adventist Research, James White Library, Andrews University, Berrien Springs, Michigan, for granting me permission to quote from the Dobzhansky Papers and the Frank Lewis Marsh papers, respectively. Finally, I thank Richard Delisle for the invitation to contribute this chapter and for challenging me to improve it.

References

Adams MB (1968) The founding of population genetics: contributions of the Chetverikov school, 1924-1934. J Hist Biol 1:23–39

Adams MB (1970) Towards a synthesis: population concepts in Russian evolutionary thought, 1925-1935. J Hist Biol 3:107–129

Adams MB (1980) Sergei Chetverikov, the Kol'tsov Institute, and the evolutionary synthesis. In: Mayr E, Provine WB (eds) The evolutionary synthesis: perspectives on the unification of biology. Harvard University Press, Cambridge/London, pp 242–278

Adams MB (2008a) Filipchenko [Philiptschenko], Iurii Aleksandrovich. Complete Dictionary of Scientific Biography. Charles Scribner's Sons, https://www.encyclopedia.com/science/dictionaries-thesauruses-pictures-and-press-releases/filipchenko-philiptschenko-iurii-aleksandrovich. Accessed 27 May 2020

Adams MB (2008b) Chetverikov, Sergei Sergeevich. Complete dictionary of scientific biography. Charles Scribner's Sons 2008

Adams MB (2021) Little evolution, BIG evolution: rethinking the history of Darwinism, population genetics, and the "synthesis." In: Delisle RG (ed) Natural selection: revisiting its explanatory role in evolutionary biology. Springer, Heidelberg, pp 195–230

Alexandrov DA (1994) Filipchenko and Dobzhansky: issues in evolutionary genetics in the 1920s. In: Adams MB (ed) The evolution of Theodosius Dobzhansky: essays on his life and thought in Russia and America. Princeton University Press, Princeton, pp 49–62

Allison AC (1955) Aspects of polymorphism in man. Cold Spring Harbor Symp Quant Biol 20:239–255

Baur E (1930) Einführung in die Vererbungslehre. Borntraeger, Berlin

Beatty J (1987a) Dobzhansky and Drift: facts, values, and chance in evolutionary biology. In: Krüger L, Gigerenzer G, Morgan MS (eds) The probabilistic revolution. Volume 2: ideas in the life sciences. The MIT Press, Cambridge, MA, pp 271–311

Beatty J (1987b) Weighing the risks: stalemate in the classical/balance controversy. J Hist Biol 20:289–319

Brncic D (1954) Heterosis and the integration of the genotype in geographic populations of *Drosophila pseudoobscura*. Genetics 39:77–88

Burian RM (1994) Dobzhansky on evolutionary dynamics: some questions about his Russian background. In: Adams MB (ed) The evolution of Theodosius Dobzhansky: essays on his life and thought in Russia and America. Princeton University Press, Princeton, pp 129–140
Chetverikov SS (1905) Volny Zhizni [Waves of life]. Dvevnik zootdeleniia, III, no. 6
Chetverikov SS (1926) O nekotorykh momentakh evolutionnogo protsessa s tochki zreniia sovremennoi genetiki. Zhurnal Eksperimental'noi Biologii, ser. A 2(1):3–54 [On certain aspects of the evolutionary process from the standpoint of modern genetics. Proceedings of the American Philosophical Society 105:167–195 (1961). Barker M (tr.), Lerner IM (ed)]
Chetverikov SS (1927a) Eksperimental'noe reshenie odnoi evoliutsionnoi problemy [Experimental solution of one evolutionary problem]. Trudy Tret'ego Vserossiiskogo S''ezda Zoologov, Anatomov i Gistologov v Leningrade 14-20 dekabria 1927 g. [Proceedings of the third all-Russian congress of zoologists, anatomists, and histologists at Leningrad, 14–20 December 1927] Leningrad 1928: 52–54
Chetverikov SS (1927b) Über die genetische Beschaffenheit wilder Populationen. Verhandlungen des Fünften internationalen Kongresses für Vererbungswissenschaft, Berlin, 1927, II. Leipzig, 1928: 1499-1500. Reproduced in: Zeitschrift für induktive Abstammungs- und Vererbungslehre 46:38–39
Cleland CE (2001) Historical science, experimental science, and the scientific method. Geology 29:987–990
Cleland CE (2002) Methodological and epistemic differences between historical science and experimental science. Philos Sci 69:447–451
Cleland CE (2011) Prediction and explanation in historical natural science. Br J Philos Sci 62:551–582
Crow JF (1987) Muller, Dobzhansky, and Overdominance. J Hist Biol 20:351–380
Delisle RG (2008) Expanding the framework of the holism/reductionism debate in Neo-Darwinism: the case of Theodosius Dobzhansky and Bernhard Rensch. Hist Phil Life Sci 30:207–226
Delisle RG (2009) The uncertain foundation of neo-Darwinism: metaphysical and epistemological pluralism in the evolutionary synthesis. Stud Hist Phil Biol Biomed Sci 40:119–132
Dobrzhanskii F (1924) Nasledstvennost' i mutatsiia. [Inheritance and mutation]. Chevolek i Priroda 5(6):417–426
Dobrzhanskii FG (1926a) Mutatsii i vidoobrazovanie [Mutations and speciation]. Priroda 5 (6):31–44
Dobrzhanskii FG (1926b) Otchet. Report of the Bureau of genetics and eugenics. University of Petrograd, pp 225–226
Dobrzhanskii FG, Kossakovskii L (1925) Trudy I Vserossiiskogo Gidrologicheskogo S″ezda [Proceedings of the first all-Russian hydrological congress], pp 481–482
Dobrzhansky T (1924) Die geographische und individuelle Variabilität von Harmonia axyridis Pall. in ihren Wechselbeziehungen. Biologisches Zentralblatt 44:401–426
Dobzhandsky T (1924) Über den Bau des Geschlechtsapparats einiger Mutanten von Drosophila melanogaster Meig. Zeitschrift für Induktive Abstammungs- und Vererbungslehre 34:245–248
Dobzhansky T (1924) Die weiblichen Generationsorgane der Coccinelliden als Artmerkmal betrachtet (Col.). Entomologische Mitteilungen 13:18–27
Dobzhansky T (1927) Studies on the manifold effect of certain genes in *Drosophila melanogaster*. Zeitschrift für induktive Abstammungs- und Vererbungslehre 43:330–388
Dobzhansky T (1933) Geographical variation in lady-beetles. Am Nat 67:97–126
Dobzhansky T (1935) A critique of the species concept in biology. Philos Sci 2:344–355
Dobzhansky T (1937) Genetics and the origin of species. Columbia University Press, New York
Dobzhansky T (1938) The raw materials of evolution. Sci Monthly 46:445–449
Dobzhansky T (1939) Mexican and Guatemalan populations of *Drosophila pseudoobscura*. Genetics 24:391–412. [GNP IV: 171-91]
Dobzhansky T (1941) Genetics and the origin of species, 2nd edn. Columbia University Press, New York

Dobzhansky T (1943) Temporal changes in the composition of *Drosophila pseudoobscura*. Genetics 28:162–186. [GNP IX: 305-328]

Dobzhansky T (1947a) Adaptive changes induced by natural selection in wild populations of *Drosophila*. Evolution 1:1–16. [GNP XVII]

Dobzhansky T (1947b) Genetics of natural populations XIV. A response of certain gene arrangements in the third chromosome of *Drosophila pseudoobscura* to natural selection. Genetics 32:142–160

Dobzhansky T (1948) Altitudinal and seasonal changes produced by natural selection in certain populations of *Drosophila pseudoobscura* and *Drosophila persimilis*. Genetics 33:158–176. [GNP XVI: 495-512]

Dobzhansky T (1949) Observations and experiments on natural selection in *Drosophila*. Hereditas 35:210–224

Dobzhansky T (1950) Genetics of natural populations XIX. Origin of heterosis through natural selection in populations of *Drosophila pseudoobscura*. Genetics 35:288–302

Dobzhansky T (1951) Genetics and the origin of species, 3rd edn. Columbia University Press, New York

Dobzhansky T (1952) Changes induced by drought in *Drosophila pseudoobscura* and *Drosophila persimilis*. Evolution 6:234–243. [GNP XX: 560-569]

Dobzhansky T (1955) A review of some fundamental concepts and problems of population genetics. Cold Spring Harbor Symp Quant Biol 20:1–15

Dobzhansky T (1962) Mankind evolving: the evolution of the human species. Yale University Press, New Haven, CT

Dobzhansky T (1962–1963) The reminiscences of Theodosius Dobzhansky. Oral History Research Office, Columbia University, New York

Dobzhansky T (1967) The biology of ultimate concern. Yale University Press, New Haven

Dobzhansky T (1968) On some fundamental concepts of Darwinian biology. In: Dobzhansky T, Hecht MK, Steere WC (eds) Evolutionary biology, vol 2. Plenum Press, New York, pp 1–32

Dobzhansky T (1969) On Cartesian and Darwinian aspects of biology. In: Morgenbesser S, Suppes P, White M (eds) Philosophy, science, and method: essays in honour of Ernest Nagel. St. Martin's Press, New York, pp 165–178

Dobzhansky T (1970a) Genetics of the evolutionary process. Columbia University Press, New York/London

Dobzhansky T (1970b) Theodosius Dobzhansky papers. Mss.B.D65. American Philosophical Society, Philadelphia

Dobzhansky T (1974) Chance and creativity in evolution. In: Ayala F, Dobzhansky T (eds) Studies in the philosophy of biology. University of California Press, Berkeley, pp 307–338

Dobzhansky T (1975) Evolution and man's self-image. In: Goodall V (ed) The quest for man. Phaidon Press, New York, pp. 189–220

Dobzhansky T (1980) The birth of the genetic theory of evolution in the Soviet Union in the 1920s. In: Mayr E, Provine WB (eds) The evolutionary synthesis: perspectives on the unification of biology. Harvard University Press, Cambridge/London, pp 229–242

Dobzhansky T, Bridges CB (1928) The reproductive system of triploid intersexes in *Drosophila melanogaster*. Am Nat 62:425–434

Dobzhansky T, Hunter AS, Pavlovsky O, Spassky B, Wallace B (1963) Genetics of an isolated marginal population of *Drosophila pseudoobscura*. Genetics 48:91–103. [GNP XXXI: 728-39]

Dobzhansky T, Levene H (1948) Proof of operation of natural selection in wild populations of *Drosophila pseudoobscura*. Genetics 33:537–547. [GNP XVII: 515–525]

Dobzhansky T, Levene H (1955) Developmental homeostasis in natural populations of *Drosophila pseudoobscura*. Genetics 40: 797–808. [GNP XXIV: 620–631]

Dobzhansky T, Queal ML (1938) Genetics of natural populations. I. Chromosome variation in populations of *Drosophila pseudoobscura* inhabiting isolated mountain ranges. Genetics 23:239–251

Dobzhansky T, Spassky B (1953) Genetics of natural populations. XXI. Concealed variability in two sympatric species of *Drosophila*. Genetics 38:471–484

Dobzhansky T, Spassky B (1954a) Genetics of natural populations. XXII. A comparison of the concealed variability in *Drosophila prosaltans* with that in other species. Genetics 39:472–487

Dobzhansky T, Spassky N (1954b) Environmental modification of heterosis in *Drosophila pseudoobscura*. Proc Natl Acad Sci USA 40:407–415

Dobzhansky T, Spassky B (1968) Genetics of natural populations. XL. Heterotic and deleterious effect of recessive lethals in populations of *Drosophila pseudoobscura*. Genetics 59:411–425

Dobzhansky T, Wallace B (1953) The Genetics of Homeostasis in *Drosophila*. Proceedings of the National Academy of Sciences of the United States of America 39: 162–171

Dubinin NP, Romaschoff DD (1932) Die genetische Struktur der Art und ihre Evolution. Biologichesky Zhurnal (Moscow) 1:52-95

Dubinin NP et al (1934) Eksperimental'nyi analiz ekogenotipov *Drosophila melanogaster*. [The experimental analysis of ecogenotypes of *Drosophila melanogaster*], parts 1 and 2. Biologicheskii Zhurnal 3:166–216

Filipchenko IA (1912) Razvitie izotomy [The development of isotomes]. St. Petersburg

Filipchenko IA (1919) Khromozomy i nasledstvennost [Chromosomes and heredity]. Priroda 7–9:327–350

Filipchenko IA (1922) Zakon Mendelia i zakon Morgana [Mendel's law and Morgan's law]. Priroda 10–12:51–66

Gould SJ (1982) Introduction. In: Dobzhansky T (ed) Genetics and the origin of species. Columbia University Press, New York, pp xvii–xli

Gould SJ (1983) The hardening of the modern synthesis. In: Grene M (ed) Dimensions of Darwinism. Cambridge University Press, Cambridge, pp 71–93

Gould SJ (2002) The structure of evolutionary theory. Harvard University Press, Cambridge, MA

Greene JC, Ruse M (1996) On the nature of the evolutionary process: the correspondence between Theodosius Dobzhansky and John C. Greene. Biology and Philosophy 11:445–491

Jordan K (1905) Der Gegensatz zwischen geographischer und nichtgeographischer Variation. Zeitschrift fuer wissenschaftliche Zoologie 83:151-210

Kolchinskii EI, Manoilenko KV, Konashev MB (eds) (2002) U istokov akademicheskoi genetiki v Sankt-Peterburge [At the roots of academic genetics in St. Petersburg]. Nauka, St. Petersburg

Konashev MB (1994) From the archives: Dobzhansky in Kiev and Leningrad. In: Adams MB (ed) The evolution of Theodosius Dobzhansky: essays on his life and thought in Russia and America. Princeton University Press, Princeton, pp 63–83

Lewontin R (1974) The genetic basis of evolutionary change. Columbia University Press, New York

Lewontin RC, Moore JA, Provine WB, Wallace B (eds) (1981) Dobzhansky's genetics of natural populations I-XLII. Columbia University Press, New York. [abbreviated GNP]

Marsh FL (1944) Frank Lewis Marsh Papers: Box 1, folder 5, Frank Lewis Marsh Papers (collection 17), Center for Adventist Research, James White Library, Andrews University, Berrien Springs, Michigan

Mayr E (1980) Prologue: some thoughts on the history of the evolutionary synthesis. In: Mayr E, Provine WB (eds) The evolutionary synthesis: perspectives on the unification of biology. Harvard University Press, Cambridge, pp 1–48

Millstein RL (2002) Are random drift and natural selection conceptually distinct? Biol Philos 17:33–53

Millstein RL (2017) Genetic drift. In: Zalta EN (ed) The Stanford encyclopedia of philosophy (Fall 2017 edition). https://plato.stanford.edu/archives/fall2017/entries/genetic-drift/

Muller HJ (1956) Genetic principles in human populations. Am J Psychiatry 113:481–491. Reprinted in Science Monthly 83 (1956): 277–286

Muller Papers (1957) Lilly Library. Indiana University, Bloomington, IN

Philiptschenko J (1929) Gene und Entwicklung der Ährenform beim Weizen. Biologisches Zentralblatt 49:1–16

Powell JR, Dobzhansky T, Hook JE, Wistrand HE (1976) Further studies on rates of dispersal of *Drosophila pseudoobscura* and its relatives. Genetics 82:493–506. [GNP XLIII]
Provine WB (1971) Origins of theoretical population genetics. University of Chicago Press, Chicago
Provine WB (1986) Sewell Wright and evolutionary biology. University of Chicago Press, Chicago
Rensch B (1947) Neuere Probleme der Abstammungslehre. F. Enke, Stuttgart
Rumiantsev B (1928) Trudy Leningradskogo Obshchestva Estestvoispytatelei 58(2):45–64
Ruse M (1996) Monad to man: the concept of progress in evolutionary biology. Harvard University Press, Cambridge/London
Schmalhausen II (1949) Factors of evolution: the theory of stabilizing selection. The Blakiston Company, Philadelphia/Toronto
Simpson GG (1944) Tempo and mode in evolution. Columbia University Press, New York
Simpson GG (1949) The meaning of evolution: a study of the history of life and of its significance for man. Yale University Press, New Haven/London
Timoféeff-Ressovsky NW (1940) Zur Analyse des Polymorphismus bei *Adalia bipunctata* L. Biologisches Zentralblatt 60:130–137
Timoféeff-Ressovsky HA, Timoféeff-Ressovsky NW (1927) Genetische Analyse einer freilebenden *Drosophila melanogaster* population. Wilhelm Roux' Archiv für Entwicklungsmechanik der Organismen 109:70–109
van der Meer JM (2007) Theodosius Dobzhansky (1900-1975). In: Rupke N (ed) Eminent lives in twentieth-century science and religion. Universitätsverlag, Göttingen, pp 79–101
Vetukhiv MO (1953) Viability of hybrids between local populations of *Drosophila pseudoobscura*. Proc Natl Acad Sci U S A 39:30–34
Wallace B (1957) Some of the problems accompanying an increase of mutation rates in Mendelian populations. In: Effect of irradiation on human heredity. World Health Organization, Geneva
Wright S (1931) Evolution in Mendelian populations. Genetics 16:97–159
Wright S (1932) The roles of mutation, inbreeding, crossbreeding, and selection in evolution. Proceedings of the Sixth International Congress of Genetics, vol 1, pp 356–366

Part V
New Lights on Recent Developments

Chapter 12
What's Natural About Natural Selection?

Jeffrey H. Schwartz

Abstract As Darwin originally conceived it, natural selection not only chose features that, in retrospect, would allow those with them to be better adapted to their immediate circumstances—it was also the agent provocateur of the pool of variation from which it chose these better-adapted features. Gradually, over immense amounts of time, the accumulation of these "infinitesimally small" adaptive changes would yield a new species. After Morgan successfully substituted gene mutation for natural selection as the basis of morphological change, for many evolutionists (but not those following in the intellectual footsteps of Darwin's contemporary saltationists Mivart, Huxley, and Galton) natural selection remained the guiding "force" that, by choosing from among available variants, continually reshaped features and, thus ultimately, species. The late twentieth century-to-present language of evolution has retained the notion of natural selection having the power to provoke change, even at the genetic level. Here I review the history of these and alternative, non-Darwinian ideas—focusing on the inclusion as well as absence of development and developmental thinking especially in terms of conceptions of heredity and species formation—and point out how much of what is assumed to be true of evolution by means of natural selection remains assumption and suggest that it is time to re-evaluate the utility of invoking selection to explain evolutionary phenomena.

Keywords Natural selection · Gradualism · Saltation · Development · Population genetics · Origin of species · Survival of species

J. H. Schwartz (✉)
Department of Anthropology, University of Pittsburgh, Pittsburgh, PA, USA
e-mail: jhs@pitt.edu

R. G. Delisle (ed.), *Natural Selection*, Evolutionary Biology – New Perspectives on Its Development 3, https://doi.org/10.1007/978-3-030-65536-5_12

12.1 Introduction

As an undergraduate and beginning graduate student at Columbia University during the 1960s, I took courses on evolution from Ernst Mayr's only Ph. D. student Walter Bock, including one he co-taught with the Dobzhanskian population geneticist, Frederick Warburton. Prominent in the required reading were the keystone publications of the "modern evolutionary synthesis" by Theodosius Dobzhansky, Mayr, and George Gaylord Simpson (see discussion below). This, with the influence of my thesis advisor, Ralph Holloway, who had been a student of Sherwood Washburn, the "father" of the "new physical anthropology" (an attempt to bring the discipline into the twentieth century by aligning it with the "modern synthesis") (Washburn 1951), shaped my understanding of evolution, i.e., how it "worked" (gradually via the accumulation of small mutations that followed an adaptive path guided by natural selection) and how one engaged in systematics and phylogenetic reconstruction (overall similarity, between living species, extinct species, and proposed ancestors and descendants).

I spent my second graduate year at the Natural History Museum [formerly the BM (NH)], London, studying early fossil primates and extant prosimians. As a good mammalian systematist, I learned the details of cranial and especially dental morphology—often the only representation in the fossil record of small extinct mammals. But I also became interested in tooth development as a potential window on the systematics and phylogenetic relationships of prosimians and realized that an understanding of the development of the adult anatomical structures that were the basis of comparative paleontological and neontological inquiries was essentially lacking, especially for mammals. As luck had it, Geoff Osborn, who was applying Lewis Wolpert's (1969) spatial pattern formation model based on chick limb formation to teeth (Osborn 1971), was at Guy's Hospital, London. He agreed to meet and discuss this novel way of understanding development, especially of repeated or meristic structures, such as limb bones and teeth: The "instructions" for developing complex morphology came not from morphogens "telling" naïve cells what to create [as Percy Butler (1939, 1978) argued for teeth] but from within cells via a hierarchical expression of "information" in developmental time and cellular space. Clearly, I had to keep learning about the development of structure and have continued to do so.

Also while at the BM(NH), I was exposed to an alternative approach to phylogenetic reconstruction—cladistics—which had been articulated by the German entomologist Willi Hennig (1950, 1966) and adopted by most of the Museum's paleontologists: Rather than rely on degree of overall similarity between, or the stratigraphic chronology of specimens (species), a cladist sought to distinguish between features that many taxa shared (primitive retentions), those that were unique to (derived for) individual species, and others that were distinctive of and also derived for a hierarchy of nested sets of species (clades and their subclades). It seemed that understanding the development of the adult features that were the traditional foci of systematics and phylogenetic reconstruction would be complementary to these endeavors.

After returning to the USA in 1972, I met Niles Eldredge, who had just published his model of "punctuated equilibria" with Stephen Jay Gould (Eldredge and Gould 1972), which hypothesized that after the geologically speedy emergence of a new species, it ceased changing. The picture of evolution was therefore not "change in continual motion" from one species to another but periods of "stasis" interrupted or punctuated by rapid speciation events. It was a schizophrenic time for me, going from lunch with Niles near the American Museum of Natural History to meet with Walter Bock, who saw "punk eke," and cladistics, as anathema. After weighing all options, I embraced cladistics and "punk eke."

As I later realized while pursuing a history of evolutionary thought that intertwined paleontology, population and developmental genetics, and comparative and developmental anatomy (Schwartz 1999b, 2009, 2011), there wasn't much difference between "punk eke" and Simpson's (1944) "quantum evolution" (see discussion below), i.e., "quantum evolution" sometimes allowed that the transition from one species to another could be rapid. However, in keeping with the "synthesis," "quantum evolution" posited that these emergent species would continue to change (but now gradually) to become different species; in "punk eke," they did not. Yet, both models posited natural selection as the driving force behind species changing via a stream of intermediates. So, why the often hostile response to "punk eke" [e.g., see Charlesworth et al. (1982) and references in Gould and Eldredge (1977)]? Probably because "punk eke," with its insistence on periods of stasis, wasn't "synthesis" enough. Although I also came to realize that Eldredge and Gould's examples of continuous transformation from one species of trilobite or Pleistocene snail to another actually portrayed the replacement of one species by another, the point is that, as presented, "punk eke" followed some of the "rules"—geographic isolation followed selection-guided continuous change—but not all of them. Apparently, Simpson's "quantum evolution" was as radical as the "synthesis" could get.

When I was invited to contribute to this volume, I knew I wanted to use a historical review to highlight the coming and going of alternative ideas. I also thought it important to explore "why" and "how" suggestions that now seem worthy of attention were submerged and their alternatives raised to the level of "truth" and "received wisdom." As I began writing, I came to think that the intellectual changes during an individual's career, as well as the ways in which some theories survived and their alternatives suppressed and omitted from revisionist histories, would be best presented by using the authors' own words and, by doing so, hopefully avert criticism for misrepresenting anyone's favorite authorities.

I also see this contribution as complementing others not only in "fleshing out" individuals and ideas but through focusing on the place as well as the absence of developmental thinking in the history of evolutionary thought, especially in conceiving natural selection as well as heredity and its relation to natural selection. As I will review, the sidelining of development, and of an understanding of the morphology that systematists compare and upon which selectionists and adaptationists muse, began in the early twentieth century with Thomas Hunt Morgan and his founding of population genetics and was continued by the mathematical population geneticists and then the architects and scions of the "modern evolutionary synthesis." By focusing on genes and alleles, the "organism" was essentially disappeared. Heredity became a matter of tracking the transmission of the adult features whose existence and function were interpreted through the lens of natural selection and adaptation. If discussed, species formation came to be thought of in these terms. By reducing heredity to genes and alleles, species diversification became the result of interfering with their transmission sufficiently for natural selection to work its magic.

If this chapter was a historical review of developmental thinking alone, not only would large periods of time be blank, I am certain critics would find it lopsided if not biased in being inconsistent with the popularized history of evolutionary thought, from Darwin to the "synthesis." Yet, the "story" of evolutionary thought is multi-faceted—more like a patchwork quilt than a seamless web—with its "storytellers" typically emphasizing only some aspects of an organism's biology or of a larger picture or pattern of life. Since none of us is exempt from this criticism, I will try to stitch together some of this intellectual quilt from the perspective of the pieces that have been sewn into received wisdom and those that have not. In doing so, I hope it will become obvious which pieces do or do not fit together and which should be replaced entirely.

12.2 Charles Darwin

12.2.1 Natural Selection, Gradualism, and the Origin of Species

In Notebook B, which he wrote between 1837 and 1839, Charles Darwin began laying the foundation for his evolutionary ideas with thoughts on the transmutation of species via descent with modification. There, he presented his conception in a diagram with an ancestral species at one point and an increasing number of descendants emanating from it, which he depicted in an irregular and somewhat hierarchical branching pattern (p. 36). Darwin would later refine this diagram, which was the only illustration in the first (Darwin 1859) and all subsequent editions of *On the Origin of Species by Means of Natural Selection* [the sixth and last published in 1872 (Darwin 1872)].

In 1838, while still writing in Notebook B, Darwin composed Notebook D, in which he noted (p. 135) his reading Malthus's 1826 edition of *An Essay on the Principle of Population* (Malthus 1826), which informed the naturalist's conceptualization of transmutation via natural selection. It was also in Notebook D that Darwin argued strongly against saltation and for change via the gradual accumulation of infinitesimally small changes over long periods of time.

But, why were infinitesimally small changes necessary to Darwin's conception of evolution? Because of his fear of the antithesis: If differences between species arise suddenly ("per saltum"), as the fossil record even then recorded, Special Creation became a viable explanation of the morphological "gaps" between species. Given this faith-based alternative to evolutionary thinking, Darwin had to justify why the fossil record did not preserve evidence of smoothly continuous, gradual change. Further, because he envisioned a world in which differences between individuals, the sexes, and species formed a morphological continuum synchronically as well as diachronically (see detailed discussion in Delisle this volume), Darwin had to justify why reality did not conform to his conception of a seamless gradation between individuals at any point in time, as well as between individuals from the past to the present. The simple scenario, which he (Darwin 1859) elaborated at great length, was that these intermediates had "disappeared" through various geological and environmental processes, that is, while continuous variation had linked organisms in a seamless web, ecologic, taphonomic, and geological processes created and continue to create a false picture of morphological and taxic discreteness.

This argument notwithstanding, Darwin still faced the dilemma of identifying the source of the almost-imperceptible differences between individuals that his model required. His primary agent provocateur was Natural Selection, to which he added blending inheritance and use-disuse, the latter reflecting the notion that an organism's desires could empower it to change itself and that these changes would be passed on to its offspring. With a seemingly limitless source of variation from which natural selection favored individuals with the momentarily and situationally better-adapted features, the cloud of variation shifted slowly, trudgingly, and inexorably through time, gradually transforming one species into another. However, in response to Fleeming Jenkin's (1867) observation that the overwhelming majority of individuals with the less well-adapted variants would always outnumber and, through mating, swamp those with the better-adapted variants, Darwin conceived a model of inheritance—Pangenesis (Darwin 1868)—that emphasized the roles of use-disuse and blending inheritance in generating individual variation. He also produced a greatly revised fifth edition of the *Origin* in which he downplayed natural selection in favor of use-disuse as a major source of variation (Darwin 1869).

12.2.2 Natural Selection and Development

The place of organismal development in Darwin's model of evolutionary change is interesting. Although he (Darwin 1859) cited the work of the comparative developmental biologist Karl Ernst von Baer (1828), for which he is credited with a modern perspective (Bowler 1989), his conception of the roles of natural selection and use-disuse in engendering change is incompatible with it. Consider the following.

Although a reliable and repeatable system of classifying organisms was not forthcoming until the eighteenth century, with Linnaeus' *Systema Naturae* (Linnaeus 1735), the art of taxonomy in organizing Nature to reflect the sequence of

Creation (a taxonomic "Great Chain of Being") was in force centuries earlier, as is evidenced in classifications such as those proposed by Konrad Gesner (1565–1566) (who created the albeit ill-defined rank of "genus," which, with species, became the cornerstone of Linnaean binomial classification). Although fossils, which were recognized in the seventeenth century as the remains of extinct organisms and seen as reflecting a picture of life in the past no less diverse than in the present, provided taxonomists with "missing links" for their Great Chains of Being, efforts in the eighteenth century to reveal the finer details of the Creation story turned to embryology and gross organismal development (Schwartz 1999b). In addition to seeking evidence in malformed, teratological, and prematurely aborted fetuses, early comparative developmental anatomists interpreted the "stages" of an organism's ontogeny as, like Adam and Eve, the fully formed adult iterations of the sequence of Creation as recounted in *Genesis*. Indeed, in the battle between the preformationist spermists and ovists, the common belief was that, regardless of which sex cell contributed to the next generation, it housed all subsequent generations in the form of the adult (Schwartz 1999b). This belief was so entrenched that, as studies on chicks became commonplace, it was assumed that the adult form lay invisibly in the egg. The fact that the emerging chick lacked adult features and proportions did not dissuade preformationists of this conviction.

However, contrary to the widespread belief that the developmental stages of an organism represented a sequence of adults that reflected the history of Creation, from the lowest and most base to the highest of organisms, von Baer (1828) argued that the ontogenies organisms shared were only common embryonic stages. Thus, for example, features such as supposed gill slits in mammal embryos were not reiterations of adult fish. Only after an individual parted company with, or, as von Baer expressed it, "deviated" from the common vertebrate developmental Bauplan did it proceed along the ontogenetic trajectory through which the features of its larger and increasingly smaller taxonomic groups, and eventually those specific to its species, emerged and became incorporated into the total morphological make-up of the adult. Consequently, once an individual diverged from the common vertebrate ontogenetic pathway, and unless derailed from it, it was developmentally committed to become an adult with a set of hierarchically relevant features that reflected its position in an also hierarchically arranged classification, from the most inclusive to the most specific taxonomic rank.

How, then, would natural selection and use-disuse enact the kind of evolutionary change Darwin sought? Not via an embryo or larva, because it (and its features) would not be available to natural selection and use-disuse. But they would be for the post-embryonic and post-larval individuals that were committed to a certain developmental course and interactive with their environment, other species, and others of their own kind. Indeed, for natural selection to do its "work," individuals would have to embody the versions or variations of their species' features in sufficient functional form that it could "choose" between momentarily and situationally better and less well-adapted variants.

Similarly, for use-disuse to be an effective mechanism of change, an individual would have to be at a stage in its growth and development whereby it would be affected by factors external to it. Consequently, reproductive transmission of a history of postnatal or post-larval (post-metamorphic) acquired changes, in concert with blending inheritance, creates the pool of variation that an embryo or lava receives and expresses either immediately or sequentially during its lifetime, recapitulating its parents' histories of development (e.g., the expression of secondary sexual characteristics) and acquisition of characteristics. In the world of use-disuse, the adult creates and transmits novel morphology to the next generation, whereas, as von Baer's laws require, change affecting embryonic development would produce adults with novel morphology or variations on a morphological theme who would pass this to the next generation via their fetuses or larvae. In this regard, one might gain insight into why Ernst Haeckel (1866, 1868) quickly embraced Darwin and the *Origin* (see Levit and Hossfeld's first contribution in this volume). For in his Biogenetic Law ("ontogeny recapitulates phylogeny"), the recapitulation that unfolds during an individual's ontogeny is not as von Baer envisioned it but a sequence of adult ancestors. Further, use-disuse is the result of the *behavior* of the post-natal/post-larval individual provoking morphological change—which is embodied in the phrase sometimes used to characterize it, "an organism's desires can engender change" (Schwartz 1999b). To use one of Darwin's examples, as a gazelle would wish to outrun a cheetah and the cheetah the gazelle, so, too, would each become faster because its morphology (more likely physiology) would change accordingly. Yet, while some may be fleeter of foot, the gazelle remains a gazelle, and the cheetah a cheetah.

12.3 Saltation and Development: Evolution and Non-Darwinian Thought

12.3.1 Thomas Henry Huxley: Darwin's Slavish "Bulldog"?

The year after Darwin published the first edition of the *Origin*, Thomas Henry Huxley reviewed it in *The Westminster Review* (Huxley 1860). Although popular history portrays Huxley as defending Darwin's conception of evolutionary change in its entirety, this is a misrepresentation. As he wrote in *Evidence as to Man's Place in Nature* (Huxley 1863a): "[T]he last position in which I wish to find myself is that of an advocate of Mr. Darwin's, or any other views—if by an advocate is meant one whose business it is to smooth over real difficulties, and to persuade where he cannot convince" (p. 127).

Indeed, upon reading a draft of the *Origin*, Huxley admonished Darwin in a letter dated 23 November 1859: "You have loaded yourself with an unnecessary difficulty in adopting *Natura non facit saltum*" (nature does not make leaps). Further, while Huxley's reported comment "How stupid of me not to have thought of that" (which a descendant of Huxley's told me is part of family lore) has been taken as evidence of his wholehearted embrace of Darwin's conception of evolutionary change, this letter and his review of the *Origin* strongly suggest that what he, a fellow evolutionist, meant was, "How stupid of me not to have thought of the phrase 'descent with modification'"—which is applicable to both gradual and saltational models of change.

Huxley began his review of the *Origin* praising the influence of Darwin's opus on individuals of all persuasions, from the religious to the scientific, the philosophical to the naturalistic, for ". . .whatever their opinions as to the ultimate fate of the doctrines put forth, [they] acknowledge that the work in which they are embodied is a solid contribution to knowledge and inaugurates a new epoch in natural history" (p. 541). But, he admitted, "'The Origin of Species' is by no means an easy book to read if by reading is implied the full comprehension of an author's meaning" (ibid.). More critically:

> We do not speak jestingly in saying that it is Mr. Darwin's misfortune to know more about the question he has taken up than any man living. . .But this superabundance of matter must have been embarrassing to a writer who, for the present, can only put forward an abstract of his views; and thence it arises, perhaps, that notwithstanding the clearness of the style, those who attempt fairly to digest the book find it much of a sort of intellectual pemmican, a mass of facts crushed and pounded into shape, rather than held together by the medium of an obvious logical bond; due attention will, without doubt, discover this bond, but it is often hard to find.
>
> Again, from sheer want of room, much has to be taken for granted which might readily enough be proved; and hence, while the adept, who can supply the missing link in the evidence from his own knowledge, discovers fresh proof of the singular thoroughness with which all difficulties have been considered and all unjustifiable suppositions avoided, at every reperusal of Mr. Darwin's pregnant paragraphs, the novice in biology is apt to complain of the frequency of what he fancies is gratuitous assumption. (p. 542)

Received wisdom notwithstanding, Huxley was skeptical of Darwin's conception of evolution and critical of his colleague's rejection of "monstrosities" and "sports" or "sports of Nature" as being relevant to understanding the origination of morphological novelty and, thus, of new species. Indeed, it was common knowledge that breeders would create a new breed or variety of animal or plant by using as its progenitor a "monstrosity" or "sport"—a single offspring that differed, not even necessarily greatly, from the parental stock in agriculturally (or societally) relevant and useful ways. One of two examples Huxley (Huxley 1860) cited in his review of the *Origin* was the case of Ancon or Otter sheep, whose long body and short legs make it impossible for them to leap over fences. As he (Huxley 1860) recounted, in 1791 a male lamb with these attributes was the unexpected offspring of a "normal" ewe and ram owned by a farmer in Massachusetts. This lamb arose, Huxley explained (p. 547), per saltum, as did progenitors of other breeds and varieties of animal and plant (e.g., Niata cattle and budding sprouts, respectively). Further, the reason a single individual with novel features could serve as the progenitor of a new breed or variety of animal or plant was because it could be bred with "normal" individuals to produce more of its kind (which, in and of itself, is an important but long-overlooked fact relevant to objections to "saltational" change).

The implication of "monstrosities" and "sports" for understanding evolutionary change, which Mivart (1871) subsequently built upon (see below), is that morphological novelty results from the alteration of an individual's development prior to the emergence of features typical of its parents. And the implication of this is that neither natural selection nor use-disuse play roles in the creation of morphological novelty. Rather, the emergence of morphological novelty is abrupt and unexpected, with the novel feature(s) being in a form sufficiently functional that their bearer could both survive and reproduce, not only with individuals like it but with members of its parental stock. Accordingly, the emergence and persistence of morphological novelty are two different phenomena. Thus, if natural selection were to play any role in an evolutionary context, it is with the latter, not former, phenomenon. Nevertheless, and even though he (Darwin 1868) documented hundreds of examples of domesticated animal and plant variants and, when known, acknowledged their "saltational" beginning, Darwin rejected their relevance to illuminating evolutionary change on the grounds that while "monstrosities" and "sports" prospered under the supervision of the breeder, they would not survive in the wild, in large part because they were not the products of a history of preceding generations adapting to the environmental circumstances in which they found themselves.

12.3.2 Thomas Henry Huxley and Development

In *Man's Place in Nature*, Huxley (1863a) continued questioning natural selection as an evolutionary "force," especially in the formation of species and of the morphological uniquenesses on which their taxonomic identification relies. In addition to expressing these doubts in the chapter "On the relation of Man to the lower animals," he implicitly rejected Darwin's insistence on natural selection *über alles* through his application of von Baer's laws to the matter of "Man's place in Nature."

Although Linnaeus (1735) was the first taxonomist to place humans in a group with other animals—Anthropomorpha, which he changed to Primates (Linnaeus 1758) (thereby angering fellow Creationist taxonomists and naturalists who allocated humans to their own group)—Huxley was the first to back this up with detailed comparative gross and developmental anatomy (Schwartz 2005b).

In mounting a morphological argument for placing humans within the Order Primates and, within Primates, closer to the large-bodied apes, Huxley used the gorilla as the foil—since, he stated, most folks who were familiar with primates, and apes especially, knew of the gorilla (Huxley 1863b). Further, because the gorilla more closely approaches humans in overall body proportions, he would compare humans with gorillas and gorillas with other primates to assess similarities between the former two (Schwartz 2005b). From here, Huxley turned to hard- and soft-tissue morphology, demonstrating that humans were like the gorilla and, when gorilla comparisons were not available, also like one of the other large-bodied apes (orangutans or chimpanzees) in ways in which they, like the gorilla, differed from other primates. Although paleoanthropologists and others continue to misrepresent Huxley as demonstrating that humans and the African apes constitute a close evolutionary group, he actually concluded that the large-bodied apes and humans should be classified in separate taxonomic groups (Schwartz 2005b). However, since development is the focus, I will dwell on it more fully.

Huxley was clearly von Baerian:

> There is not much apparent resemblance between a barn-door Fowl and the Dog who protects the farm-yard. Nevertheless, the student of development finds, not only that the chick commences its existence as an egg, primarily identical, in all essential respects with that of the Dog, but that the yelk [yolk] of this egg undergoes division ... by precisely similar methods, into a young chick, which ... is so like the nascent Dog, that ordinary inspection would hardly distinguish the two.
>
> The history of the development of other vertebrate animals, Lizard, Snake, Frog, or Fish, tells the same story... [T]here is a period in which the young of all these animals resemble one another, not merely in outward form, but in all essentials of structure, so closely, that the differences between them are inconsiderable, while, in their subsequent course, they diverge more and more widely from one another. And it is a general law, that, the more closely any animals resemble one another in adult structure, the longer and the more intimately do their embryos resemble one another.
>
> Thus the study of development affords a clear test of closeness of structural affinity, and one turns with impatience to inquire what results are yielded by the study of the development of Man. Is he something apart? Does he originate in a totally different way... thus justifying those who assert him to have no place in nature and no real affinity with the lower world of animal life? Or does he originate in a similar germ, pass through the same ... progressive modifications ... ? The reply is not doubtful for a moment... Without question, the mode of origin and the early stages of the development of man are identical with those of the animals immediately below him in the scale;—without a doubt, in these respects, he is far nearer the Apes, than the Apes are to the Dog. (pp. 79–81)

In this correct application of von Baer's law of deviation, Huxley reaffirms the utility of comparative developmental studies in deciphering the evolutionary relationships of organisms, from the largest to the smallest inclusive taxonomic groups, e.g., vertebrate, mammal, primate, and human. He also reaffirms from an ontogenetic perspective the relevance of "monstrosities" and "sports" to understanding how evolution "works." As natural selection has nothing to do with the emergence of "monstrosities" and "sports," it also has nothing to do with the ontogenetic pathway that either maintains, from one generation to the next, the morphological integrity of individuals of the same species, or leads to the emergence of offspring that differ perceptively from their parents.

Although not stated outright, through his earlier example of Ancon sheep and appreciation of the relevance of "monstrosities" and "sports" to understanding morphological and thus evolutionary innovation as the "stuff" of species—which, given his acknowledgement of "like begets like," would at least hint at the converse, i.e., developmental constraints maintain morphological continuity from one generation to the next—Huxley makes clear that the origin of morphological novelty and thus of species is divorced from whatever effects natural selection might have on individuals, populations, and species. That is, if the hierarchy of development from establishing the basic Bauplan of one's larger evolutionary group to the morphological fine-tuning that defines one's species [which developmental biologists Davidson and Erwin (2006) arrange hierarchically from gene regulatory networks (GRNs) to developmental gene batteries (GRBs)] is consistent over time, the question becomes: How is this developmental continuum modified to permit the emergence of morphological novelty without extinguishing its bearer(s)? Although Huxley did not address this question, his fellow saltationist St. George Mivart (1871) did.

12.3.3 St. George Mivart, Saltation, and Development

Mivart (1864, 1868, 1873, 1882) was one of England's most respected and thoughtful comparative anatomists. Like Huxley, he (Mivart 1871) saw merit in the evolutionary implications of the sudden and unexpected appearance of "monstrosities" and "sports" (Schwartz 2005a). He also doubted that features that were crucial to an individual's survival, or that needed to be fully functional to be effective (such as a brain, sex organ, and even the "horns" of sexually dimorphic beetles), would have arisen gradually through a process in which natural selection played a role in both their emergence and subsequent persistence.

Whereas Darwin (1868) had compiled hundreds of examples of variation to support his argument for gradual change, in *On the Genesis of Species*, Mivart (1871) listed case after case in which features necessary to the survival and procreation of organisms would only have significance if they emerged per saltum and, if not undermining the survival of their bearers, were transmitted to offspring. For Mivart, the emergence of morphological novelty, and thus of different species, was a process different from the survival of species (Schwartz 2005a). That is, while Darwin envisioned the emergence of morphological novelty and of new species as the result of the daily grind of natural selection extended over immense amounts of time (i.e., adaptation writ large), for Mivart, morphological novelty resulted (to use more modern language) from the derailment of an otherwise constrained developmental pathway (Maresca and Schwartz 2006). In short, if a novelty didn't kill its bearer, it existed because its bearer did. Consequently, if natural selection was a biological reality, it served to fine-tune the novelty, from generation to generation. In this regard, one could more sensibly rename Darwin's opus *On the Origin of Variation by Means of Natural Selection* (Schwartz 1999b).

In support of the possibility that morphological novelty could arise per saltum, through the alteration of developmental pathways, Mivart turned to another saltationist, Darwin's cousin, Francis Galton.

Although Galton (1869) may be known best as the first to offer a statistical prediction of heredity ("ancestral heredity")—and, unfortunately, for introducing eugenic thinking—he was a polymath, who excelled as much in music as in science (Schwartz 1999b). His (Galton 1871) experiments in transfusing blood in flop-eared rabbits in attempting to test Darwin's theory of Pangenesis resulted in conflict between the two, which, I suggest, would have existed anyway because Darwin was a gradualist who embraced the notion of blending inheritance, while Galton was a saltationist who, after his experiments to demonstrate Pangenesis failed, proposed the first theory of particulate inheritance (the "stirp" theory), which also served to reject use-disuse (Galton 1875, 1876). Among Galton's laudable endeavors was his support of Mendel's first champion, William Bateson. Indeed, it was Galton who, on behalf of the Royal Society, asked Bateson to expand Mendel's principles from plants to animals, which he did with his colleague Edith Saunders (Bateson and Saunders 1902).

The concept Mivart (1871) took from Galton in mounting a developmental argument for the abrupt emergence of morphological novelty came from the latter scholar's observations of, and extrapolations from, the inorganic world. Specifically, as Mivart summarized, Galton realized that differences between crystals (e.g., quartz) were always in integers, complete numbers, such as 4 versus 5, and not, for example, 4, 4.22, 4.49, 4.71, etc. To explain this, Galton suggested that the transition from one crystalline configuration to another involved not a series of intermediate steps but a "flip" from one state of inorganic equilibrium to another. This Mivart used to model the emergence of organic novelty: not the gradual transformation of a sequence of "intermediates" but a "flip" from one state of developmental equilibrium to another, with the understanding that this "flip" may not result in a viable organism.

But while the cause of this "flip" may be unknown, it had to occur early enough to affect an organism's development, exempt from the whims of natural selection. Consequently, if there was a process involving natural selection, it was separate from, and secondary to, the process that brought about change, not the other way 'round (Schwartz 2005a). In other words, a feature had to exist, with its bearers presenting slightly different variants of it, before—if true—natural selection could favor an individual with one variant over an individual with a slightly different variant.

12.3.4 When Is an "Intermediate" Not an "Intermediate"?

The year after Mivart's (1871) saltationist manifesto, Darwin (1872) published the sixth and last edition of the *Origin*. A close read of it and in comparison with the fifth edition (Darwin 1869), which, to repeat, Darwin substantially revised in reaction to Fleeming Jenkin's criticism, reveals that his primary goal was to counter, dismiss, and even denigrate with words verging on the ad hominem Mivart's case for saltation.

The object of discussion was the feathered bird-like reptile *Archaeopteryx*, the first specimen of which was discovered in Germany sometime in 1860 or 1861. Although seemingly "intermediate" between reptiles and birds (e.g., *Archaeopteryx* had feathers and teeth), this did not mean, Mivart argued, that *Archaeopteryx* was evolutionarily intermediate between reptiles and birds and thus that paleontologists would eventually find a trail of intermediates leading to it from full-fledged reptiles and from it to full-fledged birds. Having appeared abruptly, *Archaeopteryx* was just "there." Further, Mivart suggested, just because *Archaeopteryx* may appear to be "intermediate" between reptiles and birds, it could simply have been an evolutionary "dead end," a taxon without "issue."

Darwin (1869), however, saw in *Archaeopteryx* the intermediate he needed to bolster his case for large-scale change resulting from the accumulation, from one generation to the next, of infinitesimally small changes over immense amounts of time under the guidance of natural selection. But his rejection of saltation rested primarily on the argument that Mivart would have us believe that a single individual, so different from its parental stock, would be the progenitor of an entirely new species. Unfortunately, the question "How could a new species arise from a single individual?" would be marshaled time and time again by those defending a Darwinian model of gradual change against anything approaching a saltationist counter-explanation [e.g., witness the ad hominem attacks by two of the founders of the "modern evolutionary synthesis" (Dobzhansky 1941; Mayr 1942) on Goldschmidt (1940) and his notion of "hopeful monsters"]. Indeed, Darwin reiterated, only individuals of the same species differing in only the slightest ways would be able to interbreed. Consequently, the picture of evolution is of large numbers of individuals changing gradually over time.

Yet Darwin's rebuttal of Mivart is circular: Since only individuals that are virtually identical can interbreed, the only way in which morphological change can occur is by morphologically similar individuals interbreeding over long periods of time, with natural selection steering this continuum of passive participants through the Scylla and Charybdis of environmental change, competition, struggle for existence, etc. Unfortunately, neither Huxley nor Mivart jumped back into the fray to remind that new breeds and varieties of animal and plant always began with the appearance of an individual who differed from its parents in ways a breeder thought would be advantageous for practical purposes and that this individual served as the progenitor of this new breed or variety by breeding it with "normal" individuals until there were enough individuals with its qualities to form a breeding colony. Indeed, while still popular in certain circles, Darwin's dismissal of saltation was based on filling the gaps in the fossil record and the discontinuities between extant species and rejecting the results of animal and plant husbandry on the grounds that this would not work in the wild—something Thomas Hunt Morgan (1916) would later echo in dismissing large-scale mutations observed in laboratory fruit flies as being viable in nature.

12.3.5 William Bateson

In the 1880s, the British developmental biologist William Bateson, who had done his dissertation research on the hemichordate *Balanoglossus* to try to understand how, in the absence of intermediate forms, chordates may have evolved from an invertebrate ancestor, decided he would bring to evolutionary biology what Darwin had not (Bateson 1894): an answer to the question "How does variation arise?" He went first to saline lakes in Central Asia, which, he thought, would provide insight into this matter because of their dramatic swings in levels of salinity. However, rather than observing the emergence of new variation, the organisms he studied always expressed the already-present variation but in different relative proportions depending on the level of salinity. A similar study in Egypt gave the same results.

Frustrated but spurred on by these "failures," Bateson immersed himself in a study of variation, which culminated in a nearly 600-page tome in which he catalogued and described hundreds of morphological variants in plant and animal, whether radially or bilaterally symmetrical (Bateson 1894). The differences between individuals of the same species were expressed as additions to, or subtractions from, anatomical regions that represented series of repeated parts in bilaterally and radially symmetrical organisms (e.g., respectively, hand and foot bones, vertebrae, mammary glands, leaves on a stem, and fern fronds, or starfish "arms," sea urchin "segments," and flower petals). Bateson defined these arrangements of repeated, meristic parts as "homeotic" series and their alteration as "homeosis" (which in the latter twentieth century was adopted in identifying "homeobox genes," whose differential expression in developmental time and cellular space establishes the basic body plans of plants and multicellular animals). Further, as he pointed out, the typical pattern of "more" or "fewer" elements in a homeotic series was expressed by the addition or absence not of fractions of but of entire structures (e.g., a mammary gland, one or more vertebrae, a toe with all bones, petals, starfish "arms," echinoderm segments). This, he argued, meant that variation was not, as Darwin claimed, continuous but discontinuous.

To explain this pattern of gain/loss within a homeotic series, Bateson hypothesized what would be recognized as "developmental thresholds" [e.g., Goldschmidt (1940) and Waddington (1940)], wherein the strength of the developmental "wave" along a meristic or homeotic series determined both the number of elements (always in integers) in it and their size/shape relationships (e.g., as in a vertebral column). (Bateson called this his "vibratory theory" and likened it, for example, to the effect of the intensity of waves on the spacing of ripples on the shore.) Like Mivart, Bateson concluded that natural selection has nothing to do with change—in this case, in a homeotic series—which also led him to conclude that the origin of morphological novelty and of species (the identification of which he discussed at length) was a process separate from the persistence of the novelty and the survival of individuals with it (see Ochoa first contribution in this volume).

When, in 1901, Bateson read Mendel's article (1866), he saw in the Austrian monk's demonstration through breeding experiments with sweat peas of features coming and going as "units," and of a one-to-one relationship between "features" and "factors of heredity," support for his observations on morphological variation. This, in turn, led to his establishing a laboratory on his own property dedicated to breeding experiments, especially of fowl, from which he and Edith Saunders extended Mendel's principles of heredity from plants to animals (in which, among other things, they created the system P = parental, F = filial, etc. and, from "allelomorph," the notion of alleles) and from which he identified "co-dominance" (Bateson and Saunders 1902). Bateson also saw in Mendel's demonstration in sweet peas of discrete hereditary factors support for discontinuous variation and the abrupt appearance of novel features, rather than for ongoing, smoothly continuous, natural selection-mediated variation as the basis for morphological change. (Bateson used "variation" to mean differences between individuals as well as between species.)

After Bateson, the conception of there being discrete "factors for" specific features led to the division of the early twentieth-century world of evolutionary inquiry into two camps: (1) the Mendelians, who saw the discreteness of "factors" of heredity as indicating that processes leading to morphological difference created discrete features, and thus also discrete and distinctly different species, without intermediate stages, and who distinguished between the origination of morphological novelty and its persistence, and (2) the Darwinians, who embraced the expectation of demonstrating an unbroken, gradual transition of one species into another through an ongoing process of natural selection favoring, from one generation to the next, the better-adaptive variants (Schwartz 1999b).

12.4 Thomas Hunt Morgan

12.4.1 Saltationist and Anti-Darwinian

Early in his career, Morgan (1903) was skeptical of the "power" of natural selection to produce evolutionarily significant change by first provoking the emergence and then directing the gradual accretion of a stream of infinitesimally small variations on a morphological theme. To begin with, he objected to the utilitarian notion of adaptation in which Darwin's theory was firmly planted, i.e., that features arise in order to allow their bearers to adapt to their environmental circumstances, and natural selection favors individuals whose features make them better adapted, or more fit, than others. As he put it: "The destruction of the unfit, because they can find no place where they can exist, does not explain the origin of the fit" (p. 462).

Morgan also rejected Darwin's analogizing natural selection with breeders (artificially) selecting individuals with "useful" attributes to create new breeds of animal, for "new species comparable in all respects to wild ones have not been formed, even in those cases in which the variation has been carried farthest" (p. 405). Further, while Darwin invoked sexual selection to account for features for which he could not conceive an adaptive advantage (such as a peacock's tail), Morgan questioned whether one could ever determine what is, or is not, adaptively advantageous for any given individual: "If. . .we assume that the origin of the responses has nothing to do with their value to the organism, we meet with no difficulty in those cases in which the response is of little or of no use to the organism" (p. 405).

Like Bateson, Morgan (1903) interpreted Mendel's work as indicating that morphological features were discrete entities, not fuzzy elements of a cloud of variation linking individuals, sexes, and species. By embracing the Dutch botanist Hugo de Vries' (1910) "mutation theory," which postulated that new features arise suddenly, Morgan believed that so, too, could new species emerge abruptly. Consequently, like Mivart, Bateson, and de Vries, as Morgan saw it, if natural selection played any role in "evolution," it was only after the emergence of the novel morphology.

Perhaps not surprisingly, Morgan (1903) took issue with the practice of aligning specimens in a sequence that was then taken as representing an evolutionary transformation series—which, of course, is relevant to Darwin's claim that *Archaeopteryx* demonstrated that evolutionary change occurrs via the transformation of intermediates over time, from the simpler to the more complex, under the guidance of natural selection: "It does not follow because we can arrange such series ... that the more complex conditions have been ... formed in exactly this way..." (p. 415) (see Ceccarelli this volume for discussion of Morgan versus the paleontologist Osborn). Even in 1925, when he contrived a way to incorporate discrete factors of heredity into a model of gradual change, Morgan (1925) was critical of how paleontologists typically interpreted the fossil record:

> When the biologist thinks of the evolution of animals and plants... [h]e thinks of series of animals that lived in the past . . .whose bones and shells have been preserved in the rocks . . . He thinks of these animals as having in the past given birth, through an unbroken succession of individuals, to the living inhabitants of the earth today. He thinks that some of the simpler types of the past have in part changed over into the more complex forms of the present time.
>
> He is thinking as the historian thinks, but he runs the risk of thinking that he is explaining evolution when he is only describing it. (pp. 1–3)

Morgan (1903) also understood that anticipating intermediate forms in the fossil record could be fraught with error—which he exemplified with the history of the Niata breed of cattle:

> In Paraguay, during the last century (1770), a bull was born without horns, although his ancestry was well provided with these appendages, and his progeny was also hornless, although at first he was mated with horned cows. If the horned and the hornless were met in a fossil state, we would certainly wonder at not finding specimens provided with semidegenerate horns, and representing the link between both, and if we were told that the hornless variety may have arisen suddenly, we should not believe it and we should be wrong. (p 315)

Further, because of its potential for conceiving human evolution in terms of pitting one "race" against another, Morgan's greatest objection to Darwin's theory of evolution was its reliance on competition and a "struggle for existence," which, through survivorship, served to distinguish between the "better" and "less well" adapted individuals.

As Darwin (1859) framed it:

> It is good thus to try in our imagination to give any form some advantage over another . . . All that we can do, is to keep steadily in mind that each organic being is striving to increase at a geometrical ratio; that each at some period of its life, during some season of the year, during each generation or at intervals, has to struggle for life, and to suffer great destruction. When we reflect on this struggle, we may console ourselves with the full belief, that the war of nature is not incessant, that no fear is felt, that death is generally prompt, and that the vigorous, the healthy, and the happy survive and multiply. (p. 79)

To which Morgan (1903) replied:

Th[is] kindliness of heart . . . may arouse our admiration for the humanity of the writer, but need not, therefore, dull our criticism of this theory. For whether no fear is felt, and whether death is prompt, or slow, has no bearing on the question at issue—except as it prepares the gentle reader to accept the dreadful calamity of nature, pictured in this battle for existence, and make more contented with their lot "the vigorous, the healthy, and the happy". (p. 116)

And 6 years later in his presidential address to the American Society of Naturalists (Morgan 1909):

Is the battle always to the brave—for the brave is sometimes stupid—or the race to the swift, rather than to the more cunning? . . . [A]n individual advantage in one particular need not count much in survival when the life of the individual depends on so many things—advantages in one direction may be accompanied by failures in others, chance cancels chance. (p. 209)

Morgan (1903) summarized his conception of evolution thusly:

If we suppose that new mutation and 'definitely' inherited variations suddenly appear, some of which will find an environment to which they are more or less fitted, we can see how evolution may have gone on without assuming new species have been formed through a process of competition. Nature's supreme test is survival. She makes new forms to bring them to this test through mutation, and does not remodel old forms through a process of individual selection. (p. 464)

12.4.2 Natural Selection

Not long afterward, Morgan moved from Bryn Mawr College to Columbia University and established the first laboratory to study population genetics through breeding experiments using the common fruit fly, which he chose because they have only 4–6, large, easily visualized chromosomes and reproduce every 10–12 days (Morgan 1916). Without going into the details of these experiments, their success rested on Morgan and his collaborators Muller, Sturtevant, and Bridges accepting the "chromosome theory," which posited that hereditary "factors" were situated in chromosomes rather than the alternative that Bateson and others embraced, in which these factors lay elsewhere in the cell. Because chromosomes were known to break during cell replication, with their fragments recombining in different orientations and combinations, Morgan and his colleagues figured that heredity factors that were transmitted together must lie close enough to one another on a chromosome that breakage would not dissociate them. Thus, the features these factors underlay would be expressed together in offspring. By selectively breeding individuals with specific features, and noting whether they were expressed in tandem in offspring, they produced the first gene maps (Schwartz 1999b).

In 1916, in *A Critique of the Theory of Evolution*, Morgan (1916) discussed implications of this research: (1) only mutation and not natural selection produced morphological change; (2) whatever the natural cause of mutation, it is the same in the wild and in the laboratory; (3) inheritance in wild and laboratory populations follows Mendel's laws (segregation and independent assortment); (4) small- (e.g., thoracic bristle number, eye color) and large-scale (e.g., absence of eyes or wings) morphological change occurs via mutation, not natural selection; (5) whether small- or large-scale, the mechanism underlying mutation is the same; and (6) large-scale morphological change may arise in a single step or through a slow process.

Here and in subsequent monographs on evolution, Morgan (1925, 1935) maintained his rejection of competition and "struggle for existence." Further, by replacing natural selection with mutation as the cause of morphological change, he redefined the former in various ways. For example, after rejecting natural selection as a creative process that either produces something out of nothing or "determines the direction in which variation occurs" (p. 192), Morgan (1916) responded to his own questions [Does selection play any role in evolution, and, if so, in which sense ... Does elimination of the unfit influence the course of evolution, except in the negative sense of leaving more room for the fit?" (p. 187)]: "[I]f you mean by a creative process that by picking out a certain kind of individual and multiplying its numbers a better chance is furnished that a certain end result will be obtained, such a process may be said to be creative" (p. 193). Morgan summarized his thoughts on natural selection thusly:

> Natural selection as here defined means both the increase in the number of individuals that results after a beneficial mutation has occurred (owing to the ability of living matter to propagate) and also that this preponderance of certain kinds of individuals in a population makes some further results more probable than others. More than this, natural selection can not mean, if factors are fixed and are not changed by selection. (p. 194)

These quotes are interesting in light of his seemingly contradictory comments a few pages earlier:

> If through a mutation a character appears that has a *beneficial* influence on the individual, the chance that the individual will survive is increased, not only for itself, but for all of its descendants that come to inherit this character. It is this increase in the number of individuals possessing a particular character, that might have an influence on the course of evolution. This gives a better chance for improvement by several successive steps; but not because the species is more likely to mutate again the same direction. An imaginary example will illustrate how this happens: When elephants had trunks less than one foot long, the chance of getting trunks more than one foot long was in proportion to the length of trunks already present and to the number of individuals; but increment in trunk length is no more likely to occur from an animal having a trunk more than one foot long than from an animal with a shorter trunk.
>
> The case is analogous to tossing pennies. At any stage in the game the chance of accumulating a hundred heads is in proportion to the number of heads already obtained, and to the number of throws still to be made. But the number of heads obtained has no influence on the number of heads that will appear in the next throw. (pp. 189–190)

That is, since mutation is random, even if the resulting feature is deemed "beneficial" [which, in 1903 (see above), he doubted could be ascertained], there is no biological reason for another to follow in its footsteps, leading to directionality. Nevertheless, Morgan proposed a way in which directionality can be achieved:

> Owing... to this property of the germ plasm to duplicate itself in a large number of samples not only is an opportunity furnished to an advantageous variation to become extensively

multiplied, but the presence of a large number of individuals of a given sort prejudices the probable future result. (p. 192)

In other words, the more individuals there are with a variant that is more "beneficial" than others, the more likely it is that this variant will become more prevalent in successive generations. But while this appears to make some sense, it is a statement about which variant in an existing field of variation becomes more prevalent, not about the emergence of morphological novelty, either abruptly and saltationally or by the gradual transformation of something into something slightly different.

There are other contradictions in Morgan's presentation. First, the example of trunk-length change rests on the *reconstruction* of soft tissue onto the skulls of different species, if not also genera, of extinct elephants, and on the final "product" being the long trunk of extant elephants. Even so, as Morgan acknowledges, and reminiscent of Mivart and of Morgan earlier self, just because one can arrange species in a series, this doesn't mean this represents a true evolutionary sequence. Second, even if this was a real evolutionary sequence, change in trunk length is stepwise. Third, whether real or imagined, differences in trunk length between extinct and extant elephants are those between different taxa, not individuals of the same species. Thus, Morgan is conflating taxically significant difference with individual variation, i.e., differences *between* kind versus differences *within* kind. And fourth, either the "fit" exist because the "unfit" don't (the position he takes in all monographs, including in 1903), or the "fitter" exist because they possess the more "beneficial" variants of a feature (a position he also takes). You can't, however, have it both ways. Further, there is a significant difference between the concepts of eliminating the less fit and of advantaging the more fit. In the former instance, the "fit" and their more "beneficial" features are not selected, whereas in the latter, they are.

Another contradiction in Morgan's (1916) presentation is his argument that change in trunk length is both random and not necessarily sequential and yet also incremental and for the improvement of the species. As quoted above: "Evolution of wild species appears to have taken place by modifying and improving bit by bit the structure and habits that the animal or plant already possessed" (pp. 86–87). Further at odds is that while his discussion of trunk length relies on the existence of elephant taxic diversity, he never addresses how taxic diversity might arise (i.e., speciation). Indeed, as he (Morgan 1916) reiterates on the page after the quote immediately above: "[T]he emphasis may be placed less on the competition between the individuals of a species (because the destruction of the less fit does not *in itself* lead to anything that is new) than on the appearance of new characters and modifications of old characters that become incorporated in the species, for on these depends the evolution of the race" (p. 88).

12.4.3 Morgan Goes Gradual

As mentioned above, Morgan and his colleagues observed in their fruit fly colonies that spontaneous and unexpected mutation could produce small- as well as large-scale morphological change in the space of a single generation. As he (Morgan 1916) wrote, "[S]ome of the changes [are] so slight that they would be over-looked except by an expert, others so great that in the character affected the flies depart far from the original species" (p. 117). Further, in addition to suggesting that the mechanism underlying mutation was the same whether the morphological result was profound or trivial, his collaboration with Muller, Sturtevant, and Bridges (Morgan et al. 1926) recorded that (1) mutations appear independently and several times; (2) most mutations arise in the recessive state [as Bateson (1913) had suggested, because dominant mutations often have negative effects]; (3) the unexpressed mutation spreads silently through the colony, increasing the number of heterozygotes with it, who then produce (many) homozygous offspring with the novel trait(s); (4) individuals with these novel traits breed true, producing more like them (like Niata cattle); (5) even though differing greatly from individuals with the normal or "wild" parental condition, individuals with novel traits can be crossbred with them; and (6) offspring of these crosses are not always sterile and express either the novel or the normal (wild) trait, not something intermediate between them.

These observations demonstrate that (1) rather than being gradual, change is abrupt; (2) more than one individual can express the novel, mutation-induced morphology; (3) these individuals can interbreed and produce more like themselves; and (4) although some offspring may be sterile, crosses between mutated and normal individuals do produce reproductively viable progeny. (Clearly, these phenomena were known, at least intuitively, to and used by nineteenth-century animal breeders.) Further, whether small- (e.g., bristle number, wing venation pattern, relative wing length) or large-scale (eyeless, wingless, bithorax), change from a wild to mutated state involves the entirety of the novel feature. Given these observations, one wonders why Morgan (1925) came to reject a saltational model and promote a gradual model of change. An answer lies in his colleague Muller's manipulation of average fruit fly wing length (Schwartz 2006).

Typically, fruit fly wing length is slightly longer than thorax length. Of course, some individuals develop wings that are shorter and some longer than the thorax. In his experiment, Muller began with individuals whose wings were slightly shorter than the thorax. By selecting from each generation individuals with the shortest wings, he gradually increased the *prevalence* of individuals with wings that were much shorter than the thorax (there were always individuals with relatively longer wings). Although Morgan did not articulate this, I suggest he extrapolated from Muller's experiment that, were the process to continue, winglessness would be achieved (Schwartz 2006). If true, Morgan would still be confronted with the observation that winglessness could occur in the space of a single generation. Nevertheless, notwithstanding this and similar observations of major morphological change from one generation to the next, and of offspring with these changes breeding successfully with others like themselves as well as with individuals presenting the normal or wild condition, Morgan (1925), like Darwin before him, decided what would and would not be viable in nature:

> While it is true that many of the mutant characters that are preserved by geneticists for a study of heredity are extreme departures from the original type, these are utilized rather than smaller differences because they are easily observed and their classification readily determined. Those characters that depart little from the parent type, or are difficult to distinguish from the fluctuating variations of that type, are more difficult to study and are neglected. Hence has arisen in some quarters the erroneous idea that all mutant types are defectives and incapable of competing with the original form.
>
> In this connection it is important to keep in mind the fact that animals and plants are highly adapted to the conditions of life in which they live. Any change, and especially any great change in them, is far more likely to throw them out of balance with their environment than to bring to them an advantage, but it is possible, nevertheless, that some of the changes, however slight, might be beneficial, especially those that add to or diminish slightly some important character or function already present. These changes might furnish materials for evolution. (pp. 37–38)

As Darwin had argued that only small-scale differences between individuals of domesticated breeds of animal and varieties of plant would be viable in nature, Morgan made a similar case for small mutations: Although individuals with small- and large-scale morphological changes, reflecting small- and large-scale mutations, emerged in his fruit fly colonies, only those underlain by small mutations would be viable in nature. The problem, however, that Morgan now had to deal with was Mendelian inheritance and the significance of factors of heredity being discrete entities that, in turn, produced discrete morphologies. That is, the concept of discrete factors of heredity and of their discrete morphological products supported a model of discontinuous variation, not one of Darwinian continuous variation. Morgan's (1925) solution: The mutations were so small, and the changes they induced so slight, the result was effectively continuous. Further, evolutionary change was gradual, "bit by bit." Thus, in one fell swoop, Morgan melded Mendelism with Darwinian selection-mediated gradualism and provided the foundation for subsequent Darwinian models, especially those of the architects of the "modern evolutionary synthesis" (Dobzhansky, Mayr, and Simpson) and the British school (led by J. Huxley).

Morgan's (1925) intellectual relationship with natural selection and adaptation, however, remained conflicted. On the one hand, he conceived change as improving the "race" "bit by bit" via the acquisition of "beneficial" changes—which, because he believed that the "fit" would prevail and the prevalence of a beneficial trait predispose mutation and thus morphological change to follow along the same trajectory, one can reasonably infer that he meant "adaptation to one's environmental circumstances." On the other hand, he (Morgan 1925) argued that change and one's environmental circumstances were decoupled:

> The duality of the evolution process from the point of view of natural selection has not always been sufficiently emphasized. A series of events that can be given a strictly causal interpretation leads to the occurrence of a new individual, which, through other properties inherent in living matter, can reproduce a group of individuals like itself. Another and

> entirely unconnected series of events in the outer world has produced another situation as when the land was lifted above the water. If the new type *happens* to come into relation with the new world it may perpetuate itself there. This is adaptation—the fortuitous coming together of the results of two processes that have developed independently of each other. The fitness of the animal or plant to an environment that it finds existing, gives the false impression that its relation to the environment, its adaptation, has come about through a response to the environment. The central idea of natural selection, as generally understood at the present time, is that the relation is purely fortuitous. The organism had been produced by one series of events, the environment by another; the relation of the two is secondary. (pp. 150–151)

Here and elsewhere, Morgan does not discuss the origin of taxic diversity, that is, the multiplication of taxa. Further, now, Morgan's "adaptation" and "natural selection" are divorced from any connection the organism has with its environment. Indeed, although having argued that species change "bit by bit" with a relation to natural selection that improves the "race," the advent of the initial novelty remains unpredictable, such that, while the coming together of organisms and environments is "fortuitous," this, somehow, is adaptation. Interestingly, it seems that Morgan is not too far from the theme first proposed by Mivart: The advent of morphological novelty has nothing to do with the environmental circumstances in which organisms find themselves. Curiously, while Mivart allowed the possibility of natural selection playing a role in the survival of species (via the persistence of the novelty), in the quote immediately above, Morgan's natural selection is a silent observer.

12.5 Toward an Evolutionary Synthesis?

As the early twentieth century witnessed a conflict between Mendelism, discontinuous variation, and the possibility of saltational change and Darwinism, continuous variation, and gradual change, statistics entered the domain of genetic studies. In 1902, George Udny Yule (1902) proposed the "multiple factor hypothesis," in which he argued mathematically that, although discrete, the differences between variants in a population were so slight that variation was essentially continuous. In doing so, he also reconciled Mendel's laws of inheritance, Darwin's model of gradual change via natural selection, and Galton's law of ancestral heredity (Schwartz 1999b). However, the question remained: How to statistically predict the effect of selection over time on gene/allele frequencies? In 1908, and independently, Godfrey Harold Hardy and Wilhelm Weinberg proposed mathematically how equilibrium could be reached in a single generation (Provine 1971). The problem was that the "Hardy-Weinberg equation" dealt only with maintaining the stability of allele frequencies after equilibrium had been achieved. Seven years later, H. J. Norton developed a theoretical mathematical model that predicted that even low-intensity selection pressure could shift gene frequencies significantly and in a matter of a few generations (Provine 1971). Genetics had gone statistical.

Central to the ensuing mathematic modelling of genetics and natural selection was the work of William Castle (1913), whose breeding experiments with hooded mice focused on the connection between Mendelian genetics and evolution via testing the effects of selection on Mendelian characters. Through experiments in which he interbred relatives, Castle became interested in the consequences of inbreeding—something animal breeders had exploited in their creation of new breeds:

> In the production of pure breeds of sheep, cattle, hogs, and horses inbreeding has frequently been practiced extensively, and where in such cases selection has been made of the more vigorous offspring as parents, it is doubtful whether any diminution in size, vigor, or fertility has resulted. Nevertheless it very frequently happens that when two pure breeds are *crossed*, the offspring surpass either pure race in size and vigor.
>
> Inbreeding, also, by its tendency to secure homozygous combinations tends to bring to the surface latent or hidden recessive characters. If there are in nature defects of weaknesses of the organism, such as albinism and feeble-mindedness in man, then inbreeding is distinctly bad . . . [C]ontinual crossing only tends to *hide* inherent defects, not to exterminate them; and inbreeding only tends to bring them to the surface, not to *create* them. We may not, therefore, lightly ascribe to inbreeding or intermarriage the *creation* of bad racial traits, but only their manifestation. Further, any racial stock which maintains a high standard of excellence under inbreeding is certainly one of great vigor, and free from inherent defects. (pp. 149–151)

12.5.1 Ronald A. Fisher

In 1930, R. A. Fisher (1930) published the first edition of *The Genetical Theory of Natural Selection*. The preface laid out the premise:

> Natural Selection is not Evolution. Yet, ever since the two words have been in common use, the theory of Natural Selection has been employed as a convenient abbreviation for the theory of Evolution by means of Natural Selection, put forward by Darwin and Wallace. This has had the unfortunate consequence that the theory of Natural Selection itself has scarcely ever, if ever received separate consideration.
>
> [A]dvocates of Natural Selection have not failed to point out, what was evidently the chief attraction of the theory to Darwin and Wallace, that it proposed to give an account of the means of modification in the organic world by reference only to "known," or independently demonstrable, causes. The alternative theories of modification rely, avowedly, on hypothetical properties of living matter inferred from the facts of evolution themselves... The present book, with all the limitations of a first attempt, is at least an attempt to consider the theory of Natural Selection on its own merits. (p. vii)

For Fisher, the problem of presentation lay not in combining Mendelism and mathematics but in differences in imaginative thinking between biologists and mathematicians: "I can imagine no more beneficial change in scientific education than that which would allow each to appreciate something of the imaginative grandeur of the realms of thought explored by the other" (p. ix). From this, as Darwin had done liberally throughout the *Origin*, Fisher embarked on a "thought experiment," using his imagination to convince the imaginations of his audience of his perception of reality: Evolutionary change is progressive and the result of the gradual accumulation of small, continuous variants (Schwartz 1999b). Since evidence was lacking on how complex morphology, such as the vertebrate eye, emerged, Fisher used *his* imagination to develop mathematical definitions and formulae in accordance with how he thought evolution worked.

As Fisher envisioned it, evolutionary progress was evidenced in both a steady stream of increasingly complex organisms and an organism's fitness to survive, which was reflected in its reproductive success (= the number of offspring an individual of a certain age provides to the next generation). Analogizing the birth of an individual with the loan to it of life, its offspring constituted payment of the debt. The ability of an individual to pay the "debt" was correlated with its genetic variance or genetic potential, which Fisher determined by averaging the positive and negative effects of that individual's genes and evaluating this against the average gene effects in the population. Thus, he (Fisher 1930) reasoned, "Any net advantage gained by an organism will be conserved in the form of an increase in population" (p. 47).

Fisher (1930) defined natural selection thusly: "The rate of increase in fitness of any organism at any time is equal to its genetic variance in fitness at that time" (p. 35). As he explicated in the second edition of *The Genetical Theory* (Fisher 1958), the (i.e., his) conception of natural selection "refers only to the variation among individuals (or co-operative communities), and to the progressive modification of structure or function only in so far as variations in these are of advantage to the individual, in respect to his chance of death or reproduction"; this way of thinking "affords a rational explanation of structures, reactions, and instincts which can be recognized as profitable to their individual possessors" (p. 49).

In responding to critics' claim that Darwin's conception of evolution left too much to chance, Fisher (1930) offered the following analogy:

> The income derived from a Casino by its proprietor may, in one sense, be said to depend upon a succession of favourable changes, although the phrase contains a suggestion of improbability more appropriate to the hopes of the patrons of his establishment. It is easy without any profound logical analysis to perceive the difference between a succession of favourable deviations from the laws of chance, and on the other hand, the continuous and cumulative action of these laws. It is on the latter that the principle of Natural Selection relies. (p. 37)

Regarding adaptations, Fisher (1930) considered them "the consequences to the organic world of the progressive increase of fitness of each species of organism" (p. 38). However, consistent with his embrace of change being gradual and incremental, and like Darwin and Morgan in rejecting large-scale change as being viable in nature, Fisher claimed that too large a change would negatively affect the progress and improvement of a species' members. Concerned that evolutionary biologists were defining adaptation post hoc by giving examples of it (e.g., a giraffe's long legs and neck constitute an adaptation for feeding high up in the trees), Fisher (1930) provided his own, albeit circular, definition:

> An organism is regarded as adapted to a particular situation, or to the totality of situations which constitute its environment, only in so far as we can imagine an assemblage of slightly different situations, or environments, to which the animal would on the whole be less well

adapted; and equally only in so far as we can imagine an assemblage of slightly different organic forms, which would be less well adapted to that environment. (p. 38)

In considering genetics, adaptation, and natural selection, Fisher (1930) admitted that, while random mating and independent assortment may redistribute alleles, resulting in an essentially limitless source of variation for natural selection to exploit, only mutation provided the material for new variants. The difficulty Fisher had to confront was the increasingly inescapable fact that most mutations arise in the recessive state, and, unless they become dominant, are expressed only when transmitted across enough generations to produce the heterozygotes that would then produce homozygous offspring. For, as he conceived it: (1) only when a mutation is in the dominant state will it have evolutionary import, and (2) heterozygosity always has a selective advantage over homozygosity because it provides greater genetic variance to the species. In acknowledging that most non-deleterious mutations arise in the recessive state, and in attempting to reconcile this with his insistence on heterozygosity being the preferred allelic state, Fisher (1930) conceived the "dominance theory," in which he postulated that, although most mutations arise in the recessive state and then, through heterozygosis, spread silently from one generation to the next, this process was short-lived because a "modifier" would soon convert the recessive to the dominant state, thereby maintaining heterozygosity. However, as Wright (1930) pointed out in his critical review of *The Genetical Theory*, there is no reason why such a "modifier" would only affect Fisher's preferred recessive mutations and not most, if not all, alleles, whether recessive or dominant.

Fisher's embrace of Darwinian gradualism and the selective advantage of heterozygosity aside, he (Fisher 1930) did tackle the matter of species diversity, that is, how to get two species from one. First, he discussed the nature of sexually reproductive species: (1) members of a species are specifically adapted to a specific niche; (2) although members of a species present slightly different genotypes, they share most of their genetic loci because of common descent; and (3) in their genetic commonality, individuals have a potential if not real reproductive and, thus, genetic relation. In considering how to get two from one species—speciation or, as Fisher called it, the "fissioning" of species—the question was how to interfere with, or interrupt, the exchange of genes, i.e., gene flow. Fisher's answer is a geographic barrier that would prevent some individuals from mating with others of its species (subsequently referred to as allopatric speciation):

> In many cases without doubt the establishment of complete or almost complete geographical isolation has at once settled the line of fission, the two separated moieties thereafter evolving as separate species, in almost complete independence, in somewhat different habitats, until such time as the morphological differences between them entitle them to "specific rank". (p. 125)

In addition to geographic isolation, Fisher envisioned environmental instability [which he called "environmental heterogeneity" (p. 126)], causing genetic instability, as a potential source of species fissioning. Environmental instability could lead to adaptations that might cause stress within the species that could interfere with mating and reproduction, e.g., change in timing of seasonal reproduction. (Although not identified as such, this is consistent with what would later be called sympatric speciation.) Environmental instability and stress could also be due to the adaptations of members of a widespread species to local conditions that would not be transferrable throughout the species via gene flow. However, whatever the mode of fissioning, Fisher's conception of speciation was that of a geographically widespread species cleaved into relatively large populations, with natural selection acting on their respective, independently acquired novelties. This and Fisher's conception of species as conglomerates of individuals that can reproduce with one another lay the groundwork for the "modern evolutionary synthesis" (Schwartz 1999b). Interestingly, Fisher made no mention of J.B.S. Haldane's articles of the 1920s and only cited Wright's (Wright 1925) 1925 study of Guinea pig genetics, which the latter geneticist undertook while pursuing his Ph.D. with Castle. Fisher did cite Castle's (1913) major opus, but only insofar as he thought Castle's experiments on Mendelian inheritance were compatible with the genetic changes that he, Fisher, considered viable (p. 165). Given Castle's experimental studies and Haldane and Wright's mathematical modeling, in which homozygosity was a consequence of inbreeding and a major actor in their conceptions of evolutionary change, it is not surprising that Fisher ignored their contributions.

12.5.2 J.B.S. Haldane

While Fisher's insistence on the selective advantage of heterozygosity led him to propose his "dominance theory," J.B.S. Haldane (1932a) pursued a different approach to understanding how evolution "works."

Although he began where Fisher had begun—most mutations arise in the recessive state (Haldane 1927)—Haldane saw no impediment to letting the process play out, first producing heterozygotes and then heterozygotes producing homozygotes expressing the mutation. At some point, the mutation may be converted to the dominant state. Haldane (1924, 1926) also differed from Fisher in suggesting that a certain amount of inbreeding would speed up this process. Furthermore, it would be expected that, in the wild, individuals in proximity to one another were more likely be potential mates than those geographically farther apart.

Haldane also differed from Fisher in the size of the populations necessary for the emergence of new species. Rather than a widespread species separating into two relatively large groups, he (Haldane 1932b) proposed that the process of speciation began with small isolated groups:

> It is suggested that in many cases related species represent stable types such as I have described, and that the process of species formation may be a rupture of the metastable equilibrium. Cleary such a rupture will be specially likely where small communities are isolated. (pp. 141–142)

The implications are significant: (1) such a group of individuals would likely differ from most others of its species in expressing variations toward the extremes of their normal distribution; (2) the chance of inbreeding would be greater in a small group of individuals; and (3) species need not always be changing, but may remain in genetic (and thus morphological) stasis. Regarding (3), and as far as I can tell, Haldane's "metastable species" model is not cited elsewhere, even by Eldredge and Gould (1972), whose model of "punctuated equilibria," albeit derived from their studies of trilobites and Pleistocene Bermudan snails, is similar in presenting the larger picture of evolution.

Being open to all potential evolutionary scenarios, Haldane thought that single mutations could lead to the abrupt appearance of novel morphologies, but, like Mivart, he rejected the notion that complex structures, such as the vertebrate eye, evolved gradually. Further, even though he was open to the possibility of evolutionary change being gradual, he (Haldane 1932a) cautioned against conflating gradual change with continuous variation:

> [M]atters would have been easier if heritable variations really formed a continuum, as Darwin apparently thought, i.e. if there were no limit to the possible smallness of a variation. But this is clearly not the case when we are considering meristic characters. Mammals have a definite number of neck vertebrae and chromosomes, most flowers a definite number of petals, exceptional organisms being unhealthy. And the atomic nature of Mendelian inheritance suggests very strongly that even where variation is apparently continuous this appearance is deceptive. On any chemical theory of the nature of genes this must be so. (p. 57)

Haldane also embraced Lotsy's (1916) studies on hybridization and his conclusion that hybridization between members of different species could result in the abrupt appearance of new species. However, while echoing Lotsy in emphasizing the role of recessives in bringing about evolutionary significant novelty (two quotes down), Haldane made no mention of the former geneticist's conception of natural selection. Reminiscent of the early Morgan (1903), Lotsy (1916) embraced the notion of eliminating the "unfit," not the selection of the "fit":

> Selection, if not interfered with by crossing, inevitably ends with extinction because selection can act through extermination only, and by thus diminishing the number of differently constituted types, decreases the possibility of crossing and thus the formation of new types, which formation not only means the origin of new species, but also the possibility of adaptation, as the only way in which a Linneon [a group of related as well as transitively-related organisms] can adapt itself to new circumstances, is a putting into the world of new types. (p. 135)

In discussing natural selection, Haldane (1932a) acknowledged that one can never observe its long-term effects because it acts so slowly. But one could gain insight by studying short-term effects [an argument that Dobzhansky (1935a) would later make and upon which the "modern evolutionary synthesis" would be based (Dobzhansky 1937, 1941)]. One example (pp. 51–52) of short-term effects that Haldane (1932a) used involved the moth *Oporbia autumnata*, whose original forest habitat of pine, birch, and alder had, about 1800, been divided in half with the planting of a swath of heather. After a storm in 1885, birch replaced pine in the southern part, while pine became dominant in the northern portion. By 1907, the dark variety of moth was most prevalent in the northern sector (96%), while the light variety was dominant in the south (85%). The suggested probable causes were predation and different timing of egg laying. Nevertheless, Haldane confessed, "in these observations no new characters appear in the species as a result of selection ... Novelty is only brought about by selection as the result of the combination of previously rare characters" (p. 52). Further:

> The effect of selection of a given intensity depends entirely on the type of inheritance of the character selected and the system of mating... [S]election is rapid when populations contain a reasonable proportion of recessives, but excessively slow, in either direction, when recessives are very rare ... When the character is due to several rare genes the effect of selection is also very slow even if the genes are dominant. But however small may be the selective advantage the new character will spread, provided it is present in enough individuals of a population to prevent its disappearance by mere random extinction. (p. 54)

Among his list of "causes" of evolution, Haldane (1932a) discussed mutation pressure specifically and selection pressure indirectly. Regarding mutation pressure, which is a type of natural selection he considered less important than did Fisher, Haldane acknowledged Fisher for pointing out that "mutation alters the environment in which other genes are placed, and thus the incidence of selection" (p. 193). Haldane's thought experiment on how this would play out is as follows: Individuals possess allele A_1 or A_2 and allele B or B$'$; B$'$ is less harmful in the presence of A_2 than A_1; consequently, A_2 has a selective advantage over and will tend to replace A_1 in the population. Haldane concluded: "Mutation pressure must be a slow cause of evolution, but it certainly cannot be neglected when organisms are in a fairly constant environment of long periods of time" (p. 194).

Perhaps more interesting, however, is the notion of "selection pressure," a term that Morgan and then Sewall Wright used and that gained traction in evolutionary discourse (discussed later). With his example of change being possible when selection pressure weakens, Haldane suggests that natural selection is an active force in producing organismal change:

> [W]here natural selection slackens, new forms may arise which would not survive under more rigid competition, and many ultimately hardy combinations will thus have a chance of arising. Ford (1930) describes a case which may be interpreted in this way in the butterfly *Melitaea aurinia*. This seems to have happened on several occasions when a successful evolutionary step rendered a new type of organism possible, and the pressure of natural selection was temporarily slackened. Thus the distinction between the principal mammalian orders seems to have arisen during an orgy of variation in the early Eocene ... Since that date mammalian evolution has been a slower affair, largely a progressive improvement of the types originally laid down in the Eocene. (pp. 57–58)

Since, in earlier publications, Haldane discussed "selection intensity," it may not be inappropriate to suggest that the subsequently used term "pressure" is a suitable synonym. As he (Haldane 1931) wrote:

> The assumption is often made that when competition is extremely intense at any stage in a life cycle, natural selection is bound to be intense also. This assumption will be examined

> quantitatively and it will be shown that the intensity of selection may diminish and become negative at high rates of elimination, while at its best its increase is extremely slow. (p. 131)

Further, here and elsewhere, Haldane (1924, 1931) links selection intensity with competition, with the former being the ratio of individuals that are eliminated relative to those that survive. In his first article (Part I) on a mathematical theory of natural and artificial selection, in Sect. 2 ("Specification on the Intensity of Selection"), Haldane (1924) offers this:

> If [the total number of a population] is limited by the environment, natural selection may cause it to increase or diminish. It will for example tend to increase if selection renders the organism small or fitter to cope with its environment in general. If on the other hand selection increases its size, or merely arms it in the struggle with other members of its species for food or mates, the population will tend to diminish or even to disappear. (p. 20)

Haldane (1932a) summarizes:

> We have seen that natural selection is a reality. That the facts of variation, though different from what Darwin believed them to be, are yet such as to yield a raw material on which natural selection can work. We have also seen that variation directly induced by the environment is not in itself competent to explain the known facts of evolution. But we know very little about what is actually selected, and any attempt to give a concrete account of natural selection at work must be decidedly speculative. (p. 111).

> To sum up, it would seem that natural selection is the main cause of evolutionary change in species as a whole. But the actual steps by which individuals come to differ from their parents are due to causes other than selection, and in consequence evolution can only follow certain paths. These paths are determined by factors which we can only very dimly conjecture. Only a thorough-going study of variation will lighten our darkness. (pp. 142–143)

In one way, Haldane was quite eclectic in allowing for more than one tempo and even mode of evolutionary change: gradual, rapid, even stasis followed by change. His proposing that new species would more probably arise from small, isolated populations—rather than the cleaving of one large population into large sub-groups—and recognizing that most non-lethal/deleterious mutations arise in the recessive state and must undergo generations of heterozygosis before being expressed in homozygotes, should have had an impact on future evolutionary speculation. But it didn't. What legacy Haldane did leave were the notions of "mutation pressure" and "selection intensity." Further, notwithstanding his own admonitions against imbuing selection with the power to enact evolutionary change and his discussion of the elimination of the unfit as the reason for the persistence of the fit, Haldane ends up promoting natural selection as the creator of evolutionary change.

12.5.3 Sewall Wright

Like Haldane, Sewall Wright (1929, 1932, 1934) relied on the facts that (1) most mutations occur in the recessive state, and thus must reach homozygosity in the population to be expressed, and (2) as known from breeding experiments, a certain amount of inbreeding will speed up this spread and proliferation of individuals with a newly expressed novelty (or novelties); [Wright (1923) used this knowledge to explain the rapidity with which the Duchess Shorthorned breed of cattle was established]. In part echoing Haldane, Wright thought that a smaller number of individuals situated peripherally in its species' range, and thus possessed of a skewed representation of variants, would better provide fodder for evolutionary novelty than large subgroups of a species.

Through breeding experiments on multi-gene effects, Wright demonstrated that, rather than there being a one-to-one relationship between a gene and a specific physical characteristic, different gene combinations led to different gene interactions that manifested themselves differently. Further, he proposed, gene substitution and allelic mutation contributed to the sphere of potential gene interaction. Together, these factors would make possible a limitless number of combinations of interactive genetic systems. Wright (1932) incorporated all of this into his "shifting balance theory" (later called "selective topography" or "genetic drift"), which he represented as a topographic map with high and low peaks. The high peaks corresponded to interactive allelic systems with higher selective values and the latter to allelic systems of lower selective value; he identified the valleys in between as "saddles." Wright (1932) explains:

> In a rugged field of this character, selection will easily carry the species to the nearest peak, but there may be innumerable other peaks which are higher but which are separated by "valleys". The problem of evolution as I see it is that of a mechanism by which the species may continually find its way from lower to higher peaks in such a field. In order that this may occur, there must be some trial and error mechanism on a grand scale by which the species may explore the region surrounding the small portion of the field which it occupies. To evolve, the species must not be under strict control of natural selection. Is there such a trial and error mechanism? (pp. 356–357)

After discussing various scenarios in which he changed the parameters of mutation, group size, and selection pressure, Wright (1932) answered his question:

> Finally ... let us consider the case of a large species which is subdivided into many small local races, each breeding largely within itself but occasionally crossbreeding. The field of gene combinations occupied by each of these local races shifts continually in a nonadaptive fashion (except in so far as there are local differences in the conditions of selection) ... With many local races, each spreading over a considerable field and moving relatively rapidly in the more general field about the controlling peak, the chances are good that one at least will come under the influence of another peak. If a higher peak, this race will expand in numbers and by crossbreeding with the others will pull the whole species toward the new position. The average adaptiveness of the species thus advances under intergroup selection, an enormously more effective process than intragroup selection. The conclusion is that subdivision of a species into local races provides the most effective mechanism for trial and error in the field of gene combinations. (p. 361)

Here, Wright favors small subsets of a species as the source of change of an entire, and geographically widespread, species. While these are not isolated groups, their differing circumstances vis-à-vis different gene combinations relative to different adaptive peaks could result in the group with the "best" gene combinations infusing the rest of its species with them. Like Lotsy, and citing other literatures supporting this notion (e.g., studies on Hawaiian and Tahitian land snails, fish, garter snakes, bird lice, deer mice, gall wasps), Wright views differences between subgroups as non-adaptive: "Many of these authors insist on the nonadaptive characters of most of the differences among local races... Others attribute all differences to the environment, but this seems to be more an expression of faith than a view based on tangible evidence" (p. 362). Although Wright's discussion of "intergroup" and "intragroup" selection is between subgroups versus individuals of the same species, his emphasis on the former ran contrary to the Darwinian model of selection acting on individuals. Indeed, this hardline Darwinian doctrine, subsequently championed, for example, by G.C. Williams (1966, 1992), formed an impenetrable intellectual barrier to reasonable arguments in the 1970s for considering "group" or "species selection" as playing a role in evolutionary change (Gould and Eldredge 1977; Stanley 1975).

While the model cited immediately above only relates to the transformation of a species, like Haldane, Wright (1932) did consider multiplication of species via isolated groups, albeit with the caveat that the change need not be adaptive and with "hybrid sterility" an element of the process:

> It need scarcely be pointed out that with such a mechanism complete isolation of a portion of a species should result relatively rapidly in specific differentiation, and one that is not necessarily adaptive ... Such isolation is doubtless usually geographic in character at the outset but may be clinched by the development of hybrid sterility. The usual difference of the chromosome complements of related species puts the importance of chromosome aberration as an evolutionary process beyond question, but, as I see it, this importance is not in the character differences which they bring... but rather in leading to the sterility of hybrids and thus making permanent the isolation of the two groups. (p. 361)

"Sterility" had been a criterion Great Chain of Being taxonomists used in defining species. Since each had been specially created by a Divine Being, whose actions were unquestionable, and it would be anathema for individuals of specially created species to mate, the test of the validity of correct species identification was sterility of unholy hybrid offspring (Schwartz 1999a, b).

Wright (1932) summarized his evolutionary perspective:

> The course of evolution through the general [topographic] field is not controlled by the direction of mutation and not directly by selection, except as conditions change, but by a trial and error mechanism consisting of a largely nonadaptive differentiation of local races (due to inbreeding balanced by occasional crossbreeding) and a determination of the long time trend by intergroup selection. The splitting of species depends on the effects of more complete isolation, often made permanent by the accumulation of chromosome aberrations, usually of the balanced type. Studies of natural species indicate that the conditions for such an evolutionary process are often present. (pp. 366–367)

Although Wright was prolific for decades thereafter (Provine 1986), these themes, articulated toward the beginning of his career, dominated his publications [as in his last publication (Wright 1978)]. While he does refer specifically in an early article to "mutation pressure" and "selection pressure" (Wright 1931), one is left wondering what, exactly, he meant. Regarding mutation pressure, he merely stated that "[t]he effects of different simple types of evolutionary pressure on gene frequencies are easily determined" (p. 100), leaving one to divine what "evolutionary pressure" is. His conception of selection, and selection pressure, is equally slippery (Wright 1931):

> Selection, whether in mortality, mating or fecundity, applies to the organism as a whole and thus to the effects of the entire gene system rather than to single genes. A gene which is more favorable than its allelomorph in one combination may be less favorable in another. Even in the case of cumulative effects, there is generally an optimum grade of development of the character and a given plus gene will be favorably selected in combinations below the optimum but selected against in combinations above the optimum … [T]he greater the number of unfixed genes in a population, the smaller must be the average effectiveness of selection for each of them. The more intense the selection in one respect, the less effective it can be in others. (p. 101)

Although not ascribing to selection "creativity" in evolutionary processes, Wright's use of the word does present it as an active participant, which would seem to contradict (or at least be incompatible with) the argument throughout his publications that the differences that emerge between semi- and fully isolated groups are non-adaptive, being the fortuitous representation of gene combinations whose existence is not correlated with the environmental circumstances in which these small groups find themselves.

12.6 Was the "Modern Evolutionary Synthesis" a Synthesis?

In 1909, Thomas Hunt Morgan (1910) delivered the presidential address at the annual meeting of the American Society of Naturalists in which he argued that no discipline purporting to study evolution and adaptation—whether natural history, comparative anatomy, paleontology, or embryology—had shed any useful light on the topic. He also criticized what he defined as the two schools of thought attempting to understand organismal evolution: selectionists and evolutionists. The former, he stated, believed that variation emerged by chance, and then natural selection acted on it, "deciding" if it was adaptive. Evolutionists, on the other hand, envisioned a one-to-one relationship between the origin of variation and its utility or adaptiveness. In his address, Morgan argued that the only way to achieve an understanding of evolution and adaption was through experimentation—which he proceeded to do through his studies of fruit fly genetics, coming to side with the "selectionists" insofar as mutation provided the variants on which natural selection might act. The apparent contradiction, as reviewed earlier, was his assuming that a mutation in one direction would predispose subsequent mutations to follow the same course, i.e., whatever selection considers "beneficial" foreshadows (and predicts) an adaptive trend over time.

Morgan's sidelining comparative anatomy and paleontology would be reflected in the "modern evolutionary synthesis," in which Dobzhansky (1937, 1941) and

Mayr (1942) felt they had license, through studies in population genetics, to discuss "the origin of species" (the phrase being in their monographs' titles), while mammalian paleontologist Simpson (1944) had to be content commenting on the gross, and incomplete, pattern of change and diversification represented in the fossil record (reflected in the title of his monograph, *Tempo and Mode of Evolution*). That Morgan also denied to embryology and, more broadly, the study of organismal development a central role in understanding evolutionary change, is interesting considering this was his area of research until he became a fruit fly population geneticist. Nevertheless, an appreciation of the importance of development to understanding evolution never disappeared.

12.6.1 Gavin de Beer: Development Versus Genetics

Although Haldane did not incorporate development into his evolutionary considerations, he (Haldane 1932a) did refer to the work of Britain's pre-eminent developmental biologist, Gavin Rylands de Beer (1930), and specifically to de Beer's categories of "clandestine evolution," wherein embryos diverge from adult form, and "neoteny," which is one expression of pedomorphosis (the retention in adults of juvenile features) and the mechanism Louis Bolk (1926) had invoked to explain why humans differ from apes in presenting, for example, a large neurocranium, small face, flexed basicranium, and "hairlessness."

In *Embryology and Evolution*, de Beer (1930) argued that, in order to understand organismal evolution, one must first understand the relationship between ontogeny and morphological change, which he referred to as "heterochrony" and conceived as an interplay between the differential rates of attaining sexual versus physical maturity. From his perspective, the totality of an organism's somatic being was the result of an ontogenetic hierarchy, with phylogenetically significant changes or differences occurring early, and the slight differences that constitute variation emerging late in an organism's development: "The substitution of one character for another in the adult does not usually involve heterochrony, and produces only small phylogenetic effects" (pp. 74–75).

Although he was well-informed on the current state of genetics, de Beer did not see how population genetics had much, if anything, to do with understanding the mechanism(s) of evolutionary change (Schwartz 1999b). Rather, thinking in terms of the entire organism and morphological constancy from one generation to the next (T. H. Huxley's "like begets like"), he distinguished between an organism's "internal" and "external" factors, with the former "ensur[ing] that if the external factors are normal and do evoke any response in development and produce an animal at all, that animal will develop along the same lines as its parent"; in other words, ontogeny "require[s]. . . the transmission by internal factors of the capacity to react . . . to new stimuli which evoke one response after another" (p. 17). Without identifying it as such, de Beer was proposing an epigenetic model of change—one without selection playing a role in the outcome of the individual or its species. In brief, the origination of evolutionary novelty—evolution—is the result of an alteration of internal factors and the consequent modification of developmental rates.

As to "how" internal factors produce change if one assumes that external factors tend to be constant, de Beer was critical of current genetic explanations:

> These internal factors have recently been subjected to very intense study, but only as regards the mechanism of their transmission from parent to offspring. They are now called Mendelian factors or genes, and are regarded as discrete units situated in or on those universal constituents of the nuclei of cells—the chromosomes. A change induced in one of these genes is called a mutation, and the gene is then fixed in this changed condition until it mutates again. It is now usual to regard these mutations as responsible for the appearance of novelties in evolution.
>
> The way in which the genes are sorted out and distributed between parent and offspring is well known as a result of Morgan's continuation of Mendel's lead, and forms the subject-matter of the science of genetics. In the process of their transmission the genes are carried in the germ-cells, the egg, and the sperm. Germ-cells, like all cells, only arise by the division of pre-existing cells, and the production of ripe germ-cells occupies two cell-generations. It is during these two cell-generations (which culminate in fertilization) that the distribution of genes takes place, and so the geneticist who studies this distribution in hereditary transmission is really following the genes through two cell-generations only. But it takes fifty-six generations of cells to reproduce a body like that of a man out of a fertilized egg (itself a single cell), and during these fifty-six generations, the genes are playing their part in company with the external factors in moulding the animal through the successive stages of ontogeny. We now want to know how this part is played. (pp. 19–21).

It is noteworthy that, throughout *Embryology and Evolution*, de Beer never invoked or even used the terms "natural selection" and "adaptation" in his modeling of evolutionary change. Rather, the effect of the interaction between "external" on "internal" factors was to potentially alter the timing of development, which has nothing to do with the result being adaptive (or non-adaptive) or susceptible to the whims of selection.

Further, and reminiscent of von Baer's law of deviation, Huxley's (1863b) developmental argument for "Man's relation to the lower animals," and Morgan's (1903) comment on the artificiality of lining fossils up as if this would reflect a true evolutionary sequence, de Beer (1930) explains why this practice, commonplace among paleontologists and especially human paleontologists, is invalid:

> Phylogeny is the rehabilitated scale of beings, a row (or rather a number of rows) of adult forms which are related to one another, not from adult to adult, but from the fertilized egg which gives rise to one adult to the fertilized egg which produces the next. The adult forms in phylogeny are therefore disconnected. (p. 28)

> Phylogeny is a series of adult forms which are disconnected and causally unrelated to one another; each adult form being the result of an ontogeny which differs from the previous one ... Successive ontogenies are related to one another by the transmission of internal factors from fertilized egg to fertilized egg ... Modifications in ontogeny (in a constant environment) are due to changes in the internal factors... Phylogeny is therefore due to modified ontogeny. (p. 107)

In making his case for the role of internal factors and the importance of developmental timing in producing organismal change, de Beer referred to the developmental geneticist Richard Goldschmidt's (1923, 1927) early experimental studies on intersexuality in gypsy and nun moths (genus *Lymantria*). In brief, Goldschmidt demonstrated that the determination of sex depends on the relative speed with which "male-producing" and "female-producing" genes—each with specific qualitative and quantitative values and reaction rates—are expressed during development. For example, by experimentally fertilizing an egg from a "race" of gypsy moth that had a weak female-producing gene with sperm from a "race" that had a strong male-producing gene, a potential female can be converted, depending on the strength of influence of the gene, either partially or completely, into a male. This happens because, in this example, while the individual had been developing toward being female, the influence of the male-producing gene overtook that of the female-producing gene. The stronger the effect of the male-producing gene, the earlier in development the conversion from female to male takes place and vice versa. [As evidenced in a study on the red-eared slider turtle (Weber et al. 2020), sex determination is temperature dependent, with higher temperatures blocking the male-producing epigenetic molecular pathway.] From Goldschmidt's work and other experimental studies [e.g., Ford and Huxley (1927) on altering the development of brackish-water shrimp eye color by changing the temperature in which individuals developed], de Beer (1930) concluded that "various organs develop in a certain definite order in time" and that "by acting at different rates the genes can alter the time at which certain structures appear" (p. 23). Natural or any other brand of selection had nothing to do with the final outcome.

12.6.2 Theodosius Dobzhansky: The Eclectic Evolutionist

Although many students of the history of evolutionary thought would probably think of Dobzhansky in the context of combining population genetics with natural selection as the basis of the "modern evolutionary synthesis" (Dobzhansky 1941), the geneticist's early work focused on the behavior of chromosomes in relation to the sterility versus fecundity of fruit fly hybrids and, like Ford and Huxley, the effects of temperature on development, specifically the determination of sex in fruit flies (Schwartz 1999b). In 1935, however, Dobzhansky (1935a, b) turned to the question of species determination and identification.

Differing from Morgan, but like Bateson, whom he cited, Dobzhansky thought that the separateness of chromosomes indicated that discontinuous rather than continuous variation was the rule in nature. From this perspective, he (Dobzhansky 1935a) was critical of evolutionists who strove to find specimens that would fill the gaps between discontinuous forms rather than try to understand the nature of the gaps: "Discontinuous variability constitutes a foundation of the biological classification" (p. 347). However, based on the assumption that "discontinuity of the living world is constantly emerging from a continuity" (p. 353), Dobzhansky conceived species as originating from a widespread, freely interbreeding, continuously variable parent species that "becomes split into two or more populations that interbreed with each other no longer" (p. 348). Recognizing the difference between the "potential" and the "act" of interbreeding and that, in the wild, individuals in proximity may not interbreed, Dobzhansky proposed that, in addition to geographic separation, physiological mechanisms may also lead to the isolation of subgroups, i.e., because either (1) individuals fail to produce hybrid offspring or (2) hybrids are sterile.

In the first edition of *Genetics and the Origin of Species*, Dobzhansky (1937) revealed his approach to understanding evolution as a geneticist, uniformitarian, and gradualist:

> Since evolution is a change in the genetic composition of populations, the mechanisms of evolution constitute problems of population genetics... Experience seems to show ... that there is no way toward an understanding of the mechanisms of macro-evolutionary changes, which require time on a geological scale, other than through a full comprehension of the micro-evolutionary processes observable within the span of a human lifetime. (pp. 11–12)

> The theory of evolution asserts that the beings now living have descended from different beings which have lived in the past; that the discontinuous variation observed at our time-level, the gaps now existing between clusters of forms, have arisen gradually, so that if we could assemble all the individuals which have ever inhabited the earth, a fairly continuous array of forms would emerge: that all these changes have taken place due to causes which now continue to be in operation and which therefore can be studied experimentally. (p. 7)

While his references to Bateson in the early chapters of *Genetics* would suggest that he conceived discontinuity similarly, about midway in, Dobzhansky states that while there are special cases in plants where multiplication of chromosome number produces large-scale change, in general, and echoing Morgan, the mutations that accumulate to produce change are minute. Regarding selection, Dobzhansky allowed that it might not always play a role in change, including the differentiation of races or varieties: "The action of selection interwoven with and following after that of isolation becomes... more effective than selection alone is likely to be" (p. 148). Further, regarding natural selection:

> In its essence, the theory of natural selection is primarily an attempt to give an account of the probable mechanism of the origin of the adaptations of the organisms to their environment, and only secondarily an attempt to explain evolution at large. (p. 150)

> Whether the theory of natural selection explains not only adaptation but evolution as well is quite another matter. The answer here would depend in part on the conclusion we may arrive at on the problem of the relation between the two phenomena. (pp. 150–151)

> Natural selection is probably most important when the environment undergoes changes, for it is the sole known mechanism capable of producing a reconstruction of the genetic make-up of the species population from the existing elements. Such a reconstruction may be necessary in order that the species remain attuned to the demands of the environment and escape extinction. (p. 186)

Although in these quotes and throughout *Genetics* Dobzhansky (1937) promotes natural selection as having creative agency, in apparent contradiction, he also provides a version of selection that is more akin to earlier conceptions [e.g., Morgan (1903)] in which the "fit" exist by virtue of elimination of the "unfit": "Evolutionary plasticity can be purchased only at the ruthlessly dear price of continuously sacrificing some individuals to death from unfavorable mutations. Bemoaning this imperfection of nature has, however, no place in a scientific treatment of the subject" (p. 127).

Following the quote from page 186 (above), Dobzhansky devoted the next and last five pages of this chapter to Wright's ["very interesting" (p. 186)] "shifting balance theory." In addition to summarizing its mains points, Dobzhansky reproduced Wright's graphs depicting the influence of population size on the distribution of gene frequencies and the diagrams with which Wright illustrated six different fates of populations of differing sizes and gene combinations relative to the peaks and valleys portrayed in the general topographic map presenting his shifting balance theory.

Dobzhansky's focus on Wright's shifting balance theory also reflected the importance he placed on physiological isolating mechanisms rather than geographic separation alone as instigating the separation of subgroups of species that could lead to the emergence of new species (he did not in this edition of *Genetics* use the term "speciation"). Dobzhansky also extensively cited Goldschmidt's studies on intersexuality in *Lymantria* and the roles of multi-gene interaction and genetic/developmental thresholds, as well as his own work on chromosomal breakage and rearrangement (Dobzhansky 1935a, b), to argue that simple, single-gene mutation was insufficient to engender organismal change. Dobzhansky also referred to Goldschmidt's studies on temperature-dependent sex determination in gypsy moths as a major example of adaptation and comes across as being in intellectual synchrony with Goldschmidt (1933) in rejecting subspecies as the de facto antecedent to a new species. To wit (Dobzhansky 1937):

> The discontinuous or particular nature of the germ plasm is of course beyond doubt ... A rapidly growing amount of evidence indicates, however, that the genes are not quite so impregnable and impervious to the influence of their neighbors as has been thought. The effect of a gene on development is a function of its own structure as well as of its position, in the chromosome. A change in the linear order of the genes in a chromosome may then leave the quantity of the genes unaffected, and yet the functioning of the genes may be changed. (p. 115)

> The present writer has therefore proposed (Dobzhansky 1935a) to define species as that stage of evolutionary process, "at which the once actually or potentially interbreeding array of forms becomes segregated in two or more separate arrays which are physiologically incapable of interbreeding". (p. 312)

Dobzhansky's (1937) only reference to Goldschmidt that might be interpreted as having a negative spin is in his discussion of physiological isolating mechanisms. However, it is not an attack on Goldschmidt but a begrudging acknowledgement of the current understanding of the species problem:

> It is a fair presumption that the pessimistic attitude of some biologists (e.g., Goldschmidt 1933), who believe that genetics has learned a good deal about the origin of variations with a species, but next to nothing about that of the species themselves, is due to the dearth of information on the genetics of isolating mechanisms. The maintenance of the separation between species is due to the presence of physiological isolating mechanisms that hinder their free interbreeding; races of a species are as a rule not so isolated, or show only rudiments of isolation. So long as the genetics of the isolating mechanisms remains almost

a terra incognita, an adequate understanding, not to say possible control, of the process of species formation is unattainable. (p. 232)

But Dobzhansky's somewhat expansive perspective on evolutionary change, in which he denied significance to simple mutation and subspecies as the progenitors of new species, embraced developmental genetics and did not rule out rapid major change, soon turned to the more limited view he espoused in the revised, 1941 edition of *Genetics*, i.e., new species emerge through the gradual accumulation of slight mutations inducing selection-mediated small morphological and physiological change in subgroups of species that were initially geographically isolated and, because of this process of accretional change, eventually became physiologically and thus reproductively isolated from one another (Dobzhansky 1941) (see van der Meer this volume for further discussion of Dobzhansky).

12.6.3 Richard Goldschmidt, Dobzhansky's Chromosomal Rearrangement, and the Narrowing of Evolutionary Possibility

The reason Dobzhansky revised *Genetics and the Origin of Species* and constrained the realm of evolutionary possibility for the origin of species—including turning his back on his own work on chromosomal rearrangement and its potential significance in species formation—was, I suggest (Schwartz 1999b, 2011), in large part because of his anger toward Goldschmidt for grounding his theory of the origin of species through systematic mutation in his, Dobzhansky's, work on chromosomal rearrangement (Goldschmidt 1940). Indeed, while Dobzhansky (1941) did increase the citations of studies on population genetics, especially regarding subspecies and clines, in the second edition of *Genetics*, his references to Goldschmidt were primarily an attack on him for his "iconoclastic" theory of evolution, in which he (like many before him) distinguished between the origin of species and the persistence of species. The former, which Goldschmidt (1940) referred to as "micromutation" and "microevolution" (see Adams this volume for a historical account of these terms), was demonstrated by population genetic studies on variation within species and experiments in which variation was manipulated via artificial selection, which served as a surrogate for "natural selection." Further, the notion held by neo-Darwinians of subspecies evolving into species through the gradual accumulation of small mutations under the guidance of selection demanded the existence of missing links to fill the gaps between species, both living and extinct, which Goldschmidt's theory of the origin of species via systematic mutation deemed unnecessary. [One example Goldschmidt (1933, 1940) used was flatfish, e.g., flounder, and the absence of intermediates between them, with the developmental migration of an eye to the other side of its otherwise bilaterally symmetrical body and typical bilaterally symmetrical fish.] Goldschmidt (1940) was also critical of the mathematical geneticists, especially Wright and his appeal with neo-Darwinians, which would include Dobzhansky, in modeling evolutionary change via the accumulation of mutations, with or without the imposition of selection (the latter case constituting "genetic drift"). In this regard, Goldschmidt (p. 137) reflected the sentiment of Wilhelm Johannsen [who coined the term "gene" (Johannsen 1909)], namely, biology must be studied *with* mathematics, but not *as* mathematics.

But, the origin of species was another matter altogether and the gaps between species, living and extinct, "unbridgeable" by current genetic studies [as even Dobzhansky (1937) acknowledged in quoting Goldschmidt (above)]. Skeptical of genes being, as Morgan (1916, 1925) suggested, like "beads on a necklace," and given his research on gypsy moth intersexuality, multi-gene interaction, and developmental velocities and thresholds, Goldschmidt (1940) sought a more comprehensive model of evolutionarily significant change and the origin of species in the reorganization of already-present genetic material into new and different multigene interactions that, in turn, led to different morphological and physiological configurations ("macromutation" and "macroevolution"):

> I have come to the conclusion that all the recent developments of genetics tend to show that the classical theory of the gene as an actually existing unit, lying in the chromosome like a bead in a string of beads, is no longer tenable; that the linear order of the loci in a chromosome is an internal pattern of integrating elements which does not necessarily involve the existence of separate units of a molecular order; that the mutational change at a definite locus, which alone informs us of the existence of this locus, does not prove that a particle is located at this point, a break [of a chromosome—added comment], for example, being also able to account for the changed action. In spite of all opposition which such a view point is bound to encounter, I am fully convinced that geneticists will have to accept it eventually ... We have already tried to visualize the effects of repatterning (a fact that has to be understood in terms of its effect) by models which help in forming a mental image. We may now use one of these models as a simile for the systemic pattern mutation. Let us compare the chromosome with its serial order to a long printed sentence made up of hundreds of letters of which only twenty-five different ones exist ... [A] misprint of one letter here and there will not change the sense of the sentence; even the misprint of a whole word (*rose* for *sore*) will hardly impress the reader. But the compositor might arrange the same set of type into a completely different sentence with a completely new meaning, and this in a great many different ways, depending upon the number of permutating letters and the complexity of the language (the latter acting as "selection"). To elevate such a model to the level of a biological theory we have ... to restate it in chemical terms.
>
> Let us compare the chromosome to a very long chain molecule of a protein. The linear pattern of the chromosome is then the typical pattern of the different amino-acid residues. Let us assume that this chain molecule acts as an autocatalytic proteinase ... As it is known that each protein ... is characterized by the length of the chain, the type of amino acid residues, and the specific order or pattern or rhythm of the repetition of these residues along the chain, innumerable types of protein may be obtained by permutation of these three variables, without any change within the individual residues, the loci of the chain ... The mechanism of the possible changes from one type of protein to another by a pattern change involving the three variables may be described in terms equivalent to the words breakage, inversion, translocation, deletion, rearrangement ... As soon as this transformation is completed, a new protein, proteinase, chemical system has been achieved ... [L]arger and complete repatterning effects, producing a new chemical system though using nothing but the same residues, would be the model for those complete pattern changes within the chromosome, or systemic mutations, which account, as we believe, for the two major steps of macroevolution ... [*A*]*n unbiased analysis of ... pertinent facts shows that*

> *macroevolution is linked to chromosomal repatterning and that the latter is a method of producing new organic reaction systems,, a method which overcomes the great difficulties which the actual facts raise for the neo-Darwinian conception as applied to macroevolution.* (pp. 247–249)

Reminiscent of Mivart's reference to Galton, Goldschmidt (1940) described evolution in these terms: "The germ plasm controls the type of the species by controlling the developmental processes of the individual ... Evolution, therefore, means the production of a changed process of development, controlled by the changed germ plasm, as well as the production of a new pattern of germ plasm" (p. 251). In support of his genetic argument, Goldschmidt cited D'Arcy Thompson (1917) on growth and form and the potential for achieving large-scale morphological change via differential growth and "velocities" of growth patterns and the refinements to Thompson's theory by de Beer (1930) and Julian Huxley (1932).

In 1941, Wright (1941) reviewed *The Material Basis of Evolution*. Although he had objections, he was evenhanded in his presentation of Goldschmidt's ideas, e.g., micro- versus macro-evolution and mutation. However, Wright did reject Goldschmidt's suggestion that developmental alterations affecting intersexuality could begin the process of speciation in favor of his own notion that isolation of subspecies followed by genetic drift and accumulated change would lead to interspecies sterility or sterile hybrids. Besides, he commented, neo-Darwinians knew that the action of a gene was not connected to its chromosomal position and, in any case, the effect of chromosomal rearrangement in engendering species change was at odds with neo-Darwinian tenets. Nevertheless, Wright did acknowledge Goldschmidt's significant contributions in the fields of developmental genetics and evolutionary theory and concluded his review on a positive note:

> While the reviewer radically disagrees with the author's central thesis, he wishes to testify to the importance of the book. A great store of well-selected data have been assembled from diverse sources, fairly presented and discussed from view-points which must be carefully considered by any one interested in the problem of evolution. (p. 168)

Aside from Schindewolf's (1993) embrace of Goldschmidt's distinction between micromutation/microevolution and macromutation/macroevolution as being consistent with his theory of stepwise evolutionary change, Wright's review of *The Material Basis of Evolution*, constituted the last neutral commentary Goldschmidt would receive.

Although Dobzhansky (1941) and then Mayr (1942) attacked him via his monograph, *The Material Basis of Evolution* (Goldschmidt 1940), Goldschmidt (1933) had earlier presented an outline of micromutation/microevolution versus macromutation/macroevolution at the 1933 meeting of the American Association for the Advancement of Science. For example, he commented, there is "no reliable fact known which would force us to assume that geographic variation or formation of subspecies has anything to do with speciation" and "[t]he transformation of one species into another is possible only if permanent changes in the genetic make-up occur, and if the changed forms stand the test of selection" (p. 543). Goldschmidt also discussed the emergence of morphological novelty in the context of "preadaptation," meaning "adaptations are not originated in the surroundings in which they are found and also not caused by whatever action of these surroundings; moreover, adaptive characters appear as chance mutations, without any relation to their future adaptational value, as preadaptations. But these changes allow the organism to migrate into new surroundings, into which it will fit on the basis of its preadaptation" (p. 541). [Decades later, without citing Goldschmidt, Gould and Vrba (1982) would promote the same idea for which they coined the term "exaptation."]

Consistent with de Beer (1930) and J. Huxley's (1932) conception of evolutionarily significant organismal change [and the notion "if a new feature doesn't kill you, you have it"], Goldschmidt (1933) presented the case succinctly:

> Development . . .within a species is . . . considerately one-tracked. The individual developmental processes are so carefully interwoven and arranged so orderly in time and space that the typical result is only possible if the whole process of development is in any single case set in motion and carried out upon the same material . . . [C]hanges in this developmental system leading to new stable forms are only possible as far as they do not destroy or interfere with the orderly progress of developmental processes. Of course, everybody knows that this is the reason why most mutations are lethal. (p. 543)

> [T]he most probable mutational change with a chance to lead to a normal organism is a change in the typical rate of certain developmental processes. (p. 544)

It is clear from these quotes alone that what Goldschmidt was suggesting would result in what nineteenth-century animal and plant breeders, Darwin, Huxley, and Mivart referred to as "monsters" or "sports." And, indeed, in this historical context, Goldschmidt (1933) used the term correctly:

> Of course, in most cases such a shift of a partial process would lead to the production of monstrosities . . . But we must not forget that what appears to-day as a monster will be to-morrow the origin of a line of special adaptations. The dachshund and the bulldog are monsters. But the first reptiles with rudimentary legs or fish species with bulldog-heads were also monsters . . . [O]ccasionally, though extremely rarely, such a mutation may act on one of the few open avenues of differentiation and actually start a new evolutionary line . . . Here, then, an avenue would be open to considerable evolutionary change with a single basic step, provided that the new form could stand the test of selection, and that a proper environmental niche could be found to which the newly formed monstrosity would be preadapted and where, once occupied, other mutations might improve the new type. (pp. 544–555)

> [R]are but extremely consequential mutations affecting rates of decisive embryonic processes . . . might give rise to what one might term hopeful monsters, monsters which would start a new evolutionary line if fitting into some empty environmental niche. (p. 547)

One would think, if Dobzhansky (1937) had had a problem with Goldschmidt's use in 1933 of the concept of "hopeful monsters," he would have stated as much in the first edition of *Genetics and the Origin of Species*. But he did not, which would seem to be consistent with his then more open-minded approach toward evolutionary possibility. However—and this I believe supports my (Schwartz 2011) suggestion that Dobzhansky was outraged by Goldschmidt's incorporation of his work on chromosomal rearrangement into the theory of systemic mutation—in many instances in the second edition of *Genetics*, from nearly the beginning of the book, Dobzhansky (1941) set upon Goldschmidt, using the argument that (1) systematic mutations have not been observed (which, however, doesn't disprove anything but is in keeping with Dobzhansky's uniformitarian position that, since the processes we observe today are the same as in the past, and one can never study the origin of species during one's lifetime, one is justified in extrapolating from the study of living organisms), (2) following in Morgan's footsteps, only small, gradually accumulating, minuscule mutations will allow multiple individuals to reproduce and yield sufficient numbers of offspring per generation to maintain this trajectory; (3) subspecies are often the progenitors of new species; and (4) the viable scenario for speciation is geographic followed by physiological and then reproductive isolation. Further, the implication of Dobzhansky's dismissal of all aspects of Goldschmidt's theory and the extreme language he used to do so (e.g., "cataclysmic") would appear to be "if one is not Darwinian, one is not a serious evolutionary biologist." For example:

> According to Goldschmidt, all that evolution by the usual mutations—dubbed "micromutations"—can accomplish is to bring about "diversification strictly within a species, usually, if not exclusively for the sake of adaptation of the species to specific conditions within the area which it is able to occupy." New species, genera, and higher groups arise at once, by cataclysmic saltations—termed macromutations or systemic mutations—which bring about in one step a basic reconstruction of the whole organism. The role of natural selection in this process becomes "reduced to the simple alternative: immediate acceptance or rejection". A new form of life having been thus catapulted into being, the details of its structures and functions are subsequently adjusted by micromutation and selection. It is unnecessary to stress here that this theory virtually rejects evolution as this term is usually understood (to evolve means to unfold or to develop gradually), and that the systemic mutations it postulates have never been observed. It is possible to imagine a mutation so drastic that its product becomes a monster hurling itself beyond the confines of species, genus, family, or class. But in what Goldschmidt has called the "hopeful monster" the harmonious system, which any organism must necessarily possess, must be transformed at once into a radically different, but still sufficiently coherent, system to enable the monster to survive. The assumption that such a prodigy may, however rarely, walk the earth overtaxes one's credulity, even though it may be right that the existence of life in the cosmos is in itself an extremely improbable event. (pp. 32–33)

> We need not consider here the fact that systemic mutations have never been observed, and that it is extremely improbable that species are formed in so abrupt a manner. Sufficient evidence for a rebuttal of Goldschmidt's and kindred views is available. (p. 80)

> Goldschmidt's views sound iconoclastic indeed. In the opinion of the present writer, their cutting edge is, however, very much blunted if they are stripped of their revolutionary phraseology, and if the belief, which is certainly unwarranted by facts, that all mutations are accompanied or caused by structural changes in the chromosome are rejected. (p. 110)

> [T]he reproductive biology of any one species is so organized as to insure procreation of a number of offspring sufficient to maintain it on a survival level, and yet the same reproductive biology militates against the production of offspring from crosses with other species. It is not easy to visualize how such a state of affairs develops ... Genetic changes engendering isolation must not only prevent the crossbreeding between the mutant and the original type, but must simultaneously insure a normal reproduction of the mutants ... It follows from the above consideration that a great difficulty is encountered in the establishment of any isolating mechanism in a single step ... As far as the present writer is able to see, this consideration is fatal to Goldschmidt's (1940) theory of evolution by "systemic" mutations,

since these mutations are supposed to induce at once a complete isolation of the newly emerged species from its ancestor. (pp. 280–281).

Darwin's explanation continues to be accepted by a tremendous majority of biologists ... Evolutionary theory has been, however, recently challenged by Goldschmidt (194). According to his view, species do not evolve; they arise at once through catastrophic systemic mutations. The necessary corollary of this view evidently is that, contrary to Darwin, species do not come from races. The incipient species, the races that stand on the threshold of species, are a delusion ... The proposition that races may be ... incipient species has become so thoroughly a part of the biologist's convictions that it is taken almost for granted ... According to Goldschmidt ... [s]pecies are separated by "bridgeless gaps" ... To prove Goldschmidt's thesis it is absolutely essential to show that the *entire* living world is divisible into groups separated by "bridgeless gaps" ... Since Goldschmidt himself not only did not attempt to demonstrate the feasibility of such a paragon of systematics but has eschewed giving any indication of a method whereby the "bridgeless gaps" can be discovered, the main premise of this theory is wholly unacceptable. (pp. 369–371)

One scientist who Goldschmidt (1940) cited as bringing support to his theory of systemic mutation and the reality of "bridgeless gaps" between species was the vertebrate and invertebrate paleontologist, Otto Schindewolf (1936), who had embraced his work on intersexuality as being consistent with the ever-present gaps in the fossil record between species and between major clades. This Schindewolf took as evidence of the abrupt appearance of evolutionary/morphological novelty and the basis for his saltational, step-wise theory of species origination. Curiously, while Schindewolf [the 1950 German edition, translated in 1993 (Schindewolf 1993)] supported Goldschmidt's theory, Dobzhansky (1941) managed to embrace the paleontologist's ideas while rejecting Goldschmidt's. He did so (pp. 336–344) by invoking the work of Sewall Wright, after making this comment:

As pointed out in Chapter I, genetic studies are confined to microevolutionary changes, that is, to alterations in the composition of populations observable within a human lifetime. Inferences regarding macroevolution which may be drawn from genetic data are of necessity extrapolations. Nevertheless, a geneticist can hardly eschew trying to sketch some sort of a general picture of evolution, the more so since Goldschmidt (1940) published his challenging theory of systemic mutations. (pp. 336–337)

A circumstance that is most irksome to a student of phylogenetic histories is that major evolutionary advances, the first appearances of new families, orders, and classes, seem to occur suddenly, with few or no intermediates between the new groups and their putative ancestors being preserved as fossils. The "missing links" are, indeed, seldom found. Schindewolf (1936), in his brilliant and provocative discussion of the bearing of palaeontological findings on theories of evolution, stresses this sudden emergence of the radically new organic forms and infers that evolution takes place in part by what seems to correspond to Goldschmidt's systemic mutations. The fact at hand can, however, be accounted for without recourse to Goldschmidt's assumptions. (p. 343)

Like Darwin before him, and others subsequently, Dobzhansky filled in the "gaps" in the fossil record with the scenario that there had been an unbroken chain of intermediates, but they did not make it into the fossil record. Why Dobzhansky didn't include Schindewolf in this "refutation" is a mystery, ever more so because, in his theory of stepwise evolutionary change, Schindewolf was clearly non- (if not anti-) Darwinian, even to the extent of suggesting that the first bird would have hatched from the egg of a reptile.

Dobzhansky then seeks rebuttal of Goldschmidt in Wright's model of adaptive peaks:

> [T]he roads leading from one adaptive peak to others usually cross adaptive valleys. The "missing links" are, therefore, expected to be creatures able to subsist only in certain special environments, and hence rare in nature. Rare species are evidently seldom preserved and discovered as fossils. (pp. 343–344)

Since Wright (1945) held nothing back when criticizing Simpson (1944) for misrepresenting his "shifting balance theory," I would not be surprised if he had in some way responded similarly to Dobzhansky. I have not found printed evidence of this, but since Wright later attempted to incorporate Goldschmidt's conception of "macromutation" into his "shifting balance theory" (Dietrich 2000), I wouldn't be surprised if he had expressed displeasure in some manner.

As Goldschmidt emphasized, and was clearly understood by Thompson, de Beer, J. Huxley, and Waddington (1940) (whose monograph on genetic competence, multi-gene effects and interaction, and relative growth rates was published the same year as Goldschmidt's), the gradual accumulation of simple, single-allele/gene mutations would not produce complex morphology. Even Haldane (see above) was unconvinced that complex morphology arises in this manner. Although Goldschmidt was wrong about chromosomal rearrangement and a consequent alteration of gene spatial relationships being the source of novel multi-gene interactions that produced species differences [although it does remind of the effects of alternative intron splicing (Ast 2005)], his distinction between population genetics underlying individual variation, and something more profound as underlying taxically significant difference, provided a different perspective on mutation. Yes, mutation could simply be a change in an allele or gene, but it could also simply be (but with major consequences) an alteration, or more simply a different exploitation of the existing genome to produce a novel field of multi-gene interaction and developmental threshold effects, i.e., differential gene expression in developmental time and cellular space (Maresca and Schwartz 2006).

Although Dobzhansky insisted that the term "evolution" only means "to unfold or develop gradually," and thus evolutionary change must also proceed gradually, this does make it true. In fact, because the term "evolution" had been used by Great Chain of Being embryologists to refer to the "unfolding" of an organism's development, Darwin (1859) used "evolution" only once in the *Origin*. Nevertheless, Dobzhansky's definition served and continues to deny relevance to understanding a historical phenomenon we will never actually know from anything but a Darwinian explanation of evolution as selection-mediated adaption over long periods of time.

12.6.4 A "Modern Evolutionary Synthesis"?

Among the topics conspicuously absent from both editions of *Genetics and the Origin of Species* is the one thing on which Fisher, Haldane, and Wright could all agree: Most mutations arise in the recessive state and, through the process of heterozygosis, spread silently through the population until, if at all, they convert to the dominant state; meanwhile, heterozygote mating leads, among other possible results, to the emergence of multiple individuals who, being homozygous, will express the morphological or physiological novelty; homozygotes mating will produce more like them, as will the mating of heterozygotes. Although Dobzhansky did discuss heterozygosity and, like Haldane especially, believed that a certain amount of homozygosity in addition to heterozygosity is necessary to maintain a pool of variation that selection would have available to allow a species to adapt to its ever- and unpredictably changing environmental circumstances, the notion of a gradual accumulation of minute mutations ultimately rests on these mutations being immediately available to selection and thus, although not stated outright, most often emerging in the dominant, i.e., expressed, state—even though informed geneticists knew that most mutations that arise as dominants are deleterious if not lethal.

Further absent from the revised edition of *Genetics* is meaningful and substantive reference to Julian Huxley's (1940b) edited volume published a year earlier, *The New Systematics*. Indeed, Dobzhansky (1941) refers only to an earlier publication of Huxley's and to his introductory chapter in *The New Systematics* in one sentence: "Such regular geographical gradients or 'clines' (Huxley 1938a, 1940) in the frequencies or in the expression of variable characters are a common situation in many animal and plant species" (p. 67).

The reason this omission is historically significant is because, in contrast to Dobzhansky's (1941), Mayr's (1942), and Simpson's (1944) narrow view of evolution, grounded in population genetics [with Simpson trying to make paleontology compatible by incorrect use of Wright's "shifting balance theory"], *The New Systematics* included chapters by biologists working on a diversity of organisms (fungi, plants, animals) in a diversity of disciplines (genetics, embryology ecology, paleontology, experimentation), all trying to tackle the problems of systematics and the origin and identification of species, but recognizing that a complete "synthesis" of all life forms was not possible. In their chapters, Huxley (1940a) and, of course, de Beer (1940) emphasized the place of development and the myriad factors (internal and external to the organism) that influence development, as essential to understanding evolutionary processes and to doing systematics properly.

One reason for Dobzhansky ignoring Huxley could be that, rather than isolation, the latter evolutionary biologist emphasized natural selection as the primary agent provocateur of species formation, which he (Huxley 1942) elaborated upon a year later in his monograph, *Evolution: The Modern Synthesis* (Schwartz 2011). Another may be that, although the scientists Huxley invited to the conference he organized in 1939, which led to *The New Systematics*, were primarily British, the two animal geneticists were not: a Russian based in Berlin, Timofeeff-Ressovsky, and an American, Wright, who was then at the University of Chicago. For, while these population geneticists had published extensively, but in articles, Dobzhansky had produced a "synthesis" of population genetics with the first edition of *Genetics and Origin of Species* (Schwartz 2011). However, perhaps the most likely reason Dobzhansky ignored Huxley and fellow contributors to *The New Systematics* comes from Huxley's (Huxley 1940a) introduction to that volume:

> Dobzhansky (1937) has drawn attention to the fact that they [species] are the most stable units in taxonomic practice, as compared either with infra-specific categories such as variety or subspecies, or supra-specific ones such as genus, subfamily, or family. When, however, he tries to define the species-level as that stage in taxonomic differentiation after which fertile interbreeding is impossible, he goes far beyond the facts. It is certainly right to attempt a dynamic, in place of a static, definition by thinking of subspecies and species as stages in a process of evolutionary diversification: but it is impossible to insist on infertility as the sole criterion of this stage. Many groups, especially among plants, universally recognized as species by taxonomists, are capable of fertile intercrossing, and in many others we find sterility between mere strains of obvious species. Thus either Dobzhansky's definition is untrue, or, if true, taxonomic practice must be so re-cast as to rob the term species of its previous meaning. (pp. 16–17)

Huxley's criticism may also underlie why Dobzhansky (1941) only briefly mentions systematics, lumps it together with the study of morphology, and then denigrates both as being "predominantly descriptive and observational disciplines [that] took precedence among biological sciences during the eighteenth and nineteenth centuries" (p. 6). In contrast, Dobzhansky declares, "[g]enetics [as] a branch of physiology concerned in part with the problem of organic diversity, is a nomothetic [law-creating] science" (p. 7). As I (Schwartz 2011) commented on this situation:

> What an historical irony it is then that those who acknowledge J. Huxley as the author of the phrase, "the modern synthesis", do so as if he had coined it specifically to characterize Dobzhansky's brainchild. In reality, however, Huxley's endeavors represent a real attempt at synthesizing a diversity of biological realms, while Dobzhansky's unidimensional and unilateral melding of Darwinian notions of gradualism and selection with interpretation of experiments in population genetics won out and consequently became *the* constraint under which a diversity of biological disciplines were intellectually incarcerated and forced to operate. Thus, in spite of Dobzhansky's (1941) seemingly synthetic declaration that "the present aim of the book is to review the genetic information bearing on the problem of organic diversity, and as far as possible to correlate it with the pertinent data furnished by taxonomy, ecology, physiology, and other related disciplines" (p. 7) [but not, however, morphology or the "purely morphological aspect of the problem" of organic diversity—comment added], it is difficult to gloss over what he really believed: namely, that the genetic tail not only wags the biological dog, [it] also determines what is correct and what is not in evolutionary thinking … As he informs the reader in the first of many iterations of this theme: "[I]t remains true that genetics has so profound a bearing on the problem of the mechanisms of evolution that any evolutionary theory which disregards the established genetic principles is faulty at its source." (p. 9) (pp. 8–9)

It is because of this history that I (Schwartz 2011) would situate the beginning of the "modern evolutionary synthesis" with the second edition of *Genetics and the Origin of Species*. Indeed, harkening back to de Beer's (1930) comment about the narrow, two-cell generation focus of population geneticists, the focus of the "synthesis" and its followers became the mechanism of transmission from parent to offspring of the genes that were believed to underlie traits and their variations, i.e., "population thinking."

With the publication of the second volume of the triumvirate from which the "synthesis" emerged—Ernst Mayr's (1942) *Systematics and the Origin of Species*—"population thinking" became inextricably entrenched and alternatives (i.e., Goldschmidt's) rejected. As Mayr (1949) subsequently wrote:

> Not individuals but populations are the units of evolution and such populations can drift apart decisively through an accumulation of small, often minute, genetic differences. Such *gradual* speciation is the norm while instantaneous speciation is the exception. (p. 285)

That "population thinking" became an accepted element of subsequent evolutionary discourse is interesting given that, even in Mayr's (1942) world view, natural selection is supposed to act on individuals and not on their genes.

Further, in contradistinction to the title of his monograph, with the emphasis being on conceiving the origin of species through the lens of populational variation, Mayr's *Systematics and the Origin of Species* was anything but a consideration of systematics. Indeed, even though he did discuss species as biological entities, but not as a systematist on how one might go about identifying species, Mayr's preoccupation was their transformation through a seamless, smoothly continuous process:

> First, there is available in nature an almost unlimited supply of various kinds of mutation. Second, the variability within the smallest taxonomic units has the same genetic basis as the differences between the subspecies, species, and higher categories. And third, selection, random gene loss, and similar factors, together with isolation, make it possible to explain species formation on the basis of mutation, without any recourse to Lamarckian forces. (p. 70)

> That speciation is not an abrupt, but a gradual and continuous process is proven by the fact that we find in nature every imaginable level of speciation, ranging from an almost uniform species at one extreme to one in which isolated populations have diverged to such a degree that they can be considered equally well as separate, good species at the other extreme. I have tried in a recent paper . . . to analyze this continuous process. . . A widespread species is more likely to represent the first stage of speciation than one with a narrowly restricted range. (p. 159)

Clearly, however, if evolutionary change is a continuum of transformation, species are unidentifiable.

Although he, like Dobzhansky, favored subspecies as the source of new species—indeed, subspecies are "incipient" species (p. 155)—Mayr placed more emphasis on geographic isolation of subspecies, followed by selection acting on their different genetic solutions to new environmental circumstances, i.e., "[g]eographical isolation is thinkable only, if subspecies are incipient species" (p. 159). Yet, isn't the concept of "incipient" an imaginary construct, as in believing that an alluvial plane is an incipient mountain (Schwartz 2005b)?

Nevertheless, as is no doubt reflected in Mayr's history of self-citing his conception of speciation as evidence of his actually having observed speciation in action (Schwartz 2011), his conviction of correctness, and contempt for alternative thinking, was played out in his attacks on Goldschmidt: "The fact that an eminent contemporary geneticist (Goldschmidt) can come to conclusions which are diametrically opposed to those of most other geneticists is striking evidence of the extent of our ignorance" (p. 65). He also accused Goldschmidt of purposefully manipulating differences between subspecies of gypsy moths in order "to deceive a specialist of moths" (p. 137) (although Goldschmidt was a specialist on gypsy moths). Further, in misrepresenting both the current state of understanding species genetically and Goldschmidt's theory of systemic mutation as change in a single gene, Mayr stated emphatically, "we now know that species differ by so many genes that a simple mutation would, except for some cases in plants, never lead to the establishment of a new species" (p. 137). Further on Goldschmidt:

> To him a species is like a Roman mosaic, consisting of thousands of bits of marble. A systemic mutation would be like the simultaneous throwing out of all the many thousands of pieces of marble on a flat surface so that they would form a completely new and intelligible picture. To believe that this could actually happen is, as Dobzhansky has said in a review of Goldschmidt's work, equivalent to "a belief in miracles." It seems to me not only that Goldschmidt did not prove his novel ideas, but also that the existing facts fit orthodox ideas on species formation so adequately that no reason exists for giving them up. (p. 155)

Curiously, and like Dobzhansky, Mayr was sympathetic toward Schindewolf (1936), even though the paleontologist's non-/anti-Darwinian theory of stepwise evolution rested on the fact that the fossil record is riddled with taxic discontinuity. Further confusing is Mayr (1942) citing Schindewolf for providing evidence in support of the gradual emergence of species: "Paleontologists have too many examples of perfect evolutionary series, leading without obvious breaks from species to species in subsequent horizons, to believe in the instantaneous creation of species" (p. 297). Nevertheless, in spite of his asserting as fact his speculation on how evolution works in the origination of species by means of natural selection, Mayr (1942) concludes:

> [W]e may say that all the available evidence indicates that the origin of the higher categories is a process which is nothing but an extrapolation of speciation. All the processes and phenomena of macroevolution and the origin of the higher categories can be traced back to intraspecific variation, even though the first steps of such processes are usually very minute. (p. 298)

Although the mammalian paleontologist Simpson started writing the third volume on which the "modern evolutionary synthesis" was based, *Tempo and Mode in Evolution* (Simpson 1944), in 1938, he had completed but not submitted it for publish when he went to serve in the US military in 1942; his wife and the paleontologist E. H. Colbert shepherded it through publication (Schwartz 2011).

Simpson began by admitting that, as a paleontologist, he could only discuss "evolutionary rates under natural conditions" (tempo) and "the way, manner, or pattern of evolution." That is, he was constrained to discuss "how populations became genetically and morphologically differentiated" (mode) (pp. *xvii–xviii*). Further, because his interest "[wa]s to discuss the 'how' ... and the 'why' of evolution, [and] not the 'what'," he would not discuss phylogeny or morphogenesis (ibid.). Nevertheless, he contended, paleontology does provide the larger picture, which laboratory studies cannot.

From study of the mammalian fossil record, as exemplified by horses, Simpson argued that evolutionary change was not always gradual, but sometimes rapid, for which he proposed the "quantum theory" of evolutionary change. In attempting to integrate this observation with population genetics, he adopted Wright's topographic representation of the "shifting balance theory," with rapid change occurring as a species (or subsets of it) traversed from one adaptive peak to another. After a spurt of "quantum" evolution and the rapid appearance of a new species, "normal" evolutionary rates, with "normal" effects of selective, would return, and the species would proceed to change gradually. Although the equid fossil record, as was typical of the fossil record in general, was discontinuous, presenting the proverbial "gaps" between potentially related taxa, Simpson, like Darwin and Dobzhansky, argued that intermediates in the continuum of transformation from one species to the next had existed but were taphonomically disappeared. His rationale is as follows: The number of individuals in this transformation series was so small, the chance of their becoming fossilized and then surviving in the fossil record was next to nil. As he (Simpson 1944) summarized it:

> The typical pattern involved is probably this: A large population is fragmented into numerous small isolated lines of descent. Within these, inadaptive differentiation and random fixation of mutations occur. Among many such inadaptive lines one or a few are preadaptive, i.e., some of their characters tend to fit them for available ecological situations quite different from those occupied by their immediate ancestors. Such groups are subjected to strong selection pressure and evolve rapidly in the further direction of adaptation to the new status. The very few lines that successfully achieve this perfected adaptation then become abundant and expand widely, at the same time becoming differentiated and specialized on lower levels within the broad new ecological zone. (p. 123)

Although Simpson tried to reconcile population genetics with his own field of study, in the end, and as Morgan had prophesized, paleontology took a back seat to the population genetics espoused by Dobzhansky and parroted by Mayr. As he (Simpson 1945) wrote the following year in recognition of this but in defense of paleontology, and in contrast to his earlier-stated disinterest in phylogeny and morphogenesis:

> Phylogeny cannot be observed. It is necessarily an inference from observations that bear on it, sometimes rather distantly, and that can usually be interpreted in more than one way... The most direct, but unfortunately not the most useful, approach to phylogeny of recent animals is through their genetics. The stream of heredity makes phylogeny; in a sense, it is phylogeny. Complete genetic analysis would provide the most priceless data for mapping of this stream, although it would only exceptionally provide unequivocal and conclusive criteria ... Aside from such rare cases, genotypical similarities and dissimilarities have to be interpreted phylogenetically in much the same way as phenotypic likenesses and differences.
>
> Genetical, physiological, embryological, and geographical data have been mentioned first [earlier in this publication], mainly because the orthodox classifier is likely to think of them last, but morphological data and paleontological data ... always have been and ... always will be the principal basis for the study of phylogeny. (p. 7)

Although Simpson (1944, 1945) tries to both embrace population genetics and maintain the relevance of paleontology to the study of evolution, he follows in Dobzhansky and Mayr's footsteps in dismissing if not denigrating the ideas of others to support his own perspective.

In *Tempo and Mode in Evolution*, Simpson (1944) cites *The New Systematics* in the bibliography but only refers in the text to Huxley's (1932) discussion in *Problems of Relative Growth* of the development of an antlers-to-body-size relation in deer. At the beginning of *The Principles of Classification and a Classification of Mammals*, Simpson (1945) mentions *The New Systematics* in support of his (Simpson's) embrace of "[g]enetical, physiological, embryological, and geographical data" but then dismisses it as insignificant and failing to contribute to the study of evolution: "[F]rom that book alone it is hardly possible to determine exactly what was new about its systematics or to draw up a formal statement of its principles, but it did make clear that there was a ferment working in the field" (p. 64). As far as Simpson was concerned, the way in which he went about doing systematics constituted "the *new* new systematics" (ibid.). However, while Dobzhansky and Mayr embraced at least some of Schindewolf's ideas, and did not treat him as they did Goldschmidt for those ideas with which they may have disagreed, Simpson (1949) was less kind. How he rejected Schindewolf's work via analogy with what had become the "synthesis's" conception of speciation is of historical interest.

In his review of the work of the German evolutionists Rensch, Zimmerman, and Schindewolf, Simpson (1949) writes:

> For some ten years Central Europe and the West have been separated by a highly effective intellectual isolating mechanism, only now beginning to break down. This barrier to inter-thinking virtually stopped the interchange of knowledge and ideas between the respective populations, within each of which the development of evolutionary theory continued independently of that within the other. The phenomenon is in some ways analogous to the result of cessation of inter-breeding, stopping the interchange of genes between populations and resulting in their independent biological evolution. Explicitly, aspects of parallelism, convergence, and divergence occur among both the intellectual and the biological results of isolation.
>
> Isolation has, of course, worked both ways; western students have been isolated from German scholars as much as the reverse. It is, however, assumed that readers of [the journal] EVOLUTION are by now quite familiar with the main trends in western evolutionary studies during this period. These trends have been surprisingly uniform. Although using, at times, radically different data and working widely distinct fields, western studies have in almost all cases tended toward a synthesis the core of which is the action of natural selection on genomes in populations.
>
> The third work under review here, that of Schindewolf (1945), is in radical disagreement and exemplifies the phenomenon of divergence in intellectual evolution. Schindewolf departed in the main, from somewhat the basis that has produced the synthesis and he was, indeed, a pioneer in one aspect of it (Schindewolf 1936, an attempted synthesis of genetical and paleontological data). Divergence began, however, even in this earlier work and Schindewolf has now developed a general theory of evolution profoundly different from that of Rensch, Zimmermann, most English and American students, and many in other countries. (p. 178)

It is beyond the scope of this chapter to go into the details Simpson brought to his rejection of Schindewolf's work. Suffice it to say, he did acknowledge that his review was "an opinion on the theory rather than as an unbiased account of it" (p. 182). However, the point of citing Simpson's (1949) review here is to illustrate how steadfastly he, like Dobzhansky and Mayr, held on to their views and how they went about proselytizing them at the expense of alternative theories and to the detriment of their authors. Further, as is obvious in his review of Schindewolf, only 8 short years after the second edition of *Genetics and the Origin of Species*, and 7 after *Systematics and the Origin of Species*, Simpson was promoting the idea that only those in the "west" (i.e., he, Dobzhansky, and Mayr) achieved a true evolutionary synthesis and that the only mechanism underlying evolutionary change is "the action of natural selection on genomes in populations." In truth, though, considering that developmental genetics, embryology, comparative and developmental biology, and, in many ways, paleontology were in essence and often in fact subsidiary to population genetics, one would otherwise be hard-pressed to state, precisely, what "the synthesis" was. Indeed, regardless of how "correct" they were, Goldschmidt, Schindewolf, and J. Huxley strove to be synthetic through the inclusion into their evolutionary considerations of many if not all the disciplines that, for the most part, Morgan, Dobzhansky, Mayr, and even Simpson tossed aside. At least one can understand what the former three evolutionists were trying to articulate. In the case of "the synthesis," because its architects went to such lengths to reject the ideas of others, one better knows what it wasn't than what it was.

12.7 Discussion and Conclusion

As this historical review hopefully illustrates, conceptions of "natural selection" have been anything but consistent [see also Esposito this volume and da Silva (2017)], although the dominant theme has been "selection for" something that is beneficial to the organism. Further, because these conceptions have been vague, definitions of natural selection have been limited only by one's imagination.

One important point in this history is Morgan supplanting the notion that selection is behind every aspect of evolutionary change, with the concept that mutation introduces the variation upon which selection acts. Nevertheless, while Morgan also argued that selection was not a creative force, the property of "creativity" has, to varying degrees, been part of most discussions of it, which, I suggest, is reflected in the term "selection pressure"; (if you think about it, "selection pressure" is another way of suggesting that "an organism's desires can engender change"). Further, as Darwin invoked sexual selection to explain observations he could not accommodate in his portrayal of natural selection, the notion that selection is an ever-present, active "force" has led to notions of "positive" and "negative" selection—the latter to explain observations that aren't easily accommodated by the former and vice versa. Indeed, one may well wonder if Morgan's (1925) comment on how a paleontologist conducts business is not more broadly relevant to evolutionary biologists in general: "[H]e runs the risk of thinking that he is explaining evolution when he is only describing it" (pp. 1–3). For, isn't the notion of selection an ad hoc, after-the-fact attempt to comprehend a picture and pattern of life whose formation was and remains unobservable? In this regard, Darwin's analogizing from breeders "selecting for" individuals with traits deemed beneficial (for the breeder or society for essentially economic reasons) to selection in nature would appear to be as inappropriate as his rejecting "monsters" and "sports" as providing a window on evolutionarily relevant change. In the former instance, artificial is not a surrogate for natural selection, which also begs the question: Do experimental selection studies faithfully reflect how things occur in nature? Further, are all experiments equal?

Take, for instance, studies that induce genomic shifts in bacteria by exposing them to different circumstances, e.g., nutrition, temperature, salinity, light, chemical, etc. [e.g., Barrick et al. (2009), Lenski and Travisano (1994)]. The focus of such studies is the rate of mutation and its effects on the population. Being bacteria, the numbers involved in such studies are enormous, which means that, by chance alone, multiple individuals will exhibit the same mutation and one that will allow them to survive and faithfully produce more like themselves (because a bacterial genome, which is a single circular chromosome, can replicate completely). Eventually the genomic signature of the affected population will change, and those bacteria that do not change do not survive or replicate. As authors of these studies correctly conclude, these bacteria are adapting, but they are doing so through the lens of the selection parameters of the experiment. They are not, however, as is also claimed, evolving. The reason this is true is that about 97–98% of a bacterial genome is coding (producing metabolically active proteins and enzymes, which are relevant to survival in and interaction with environmental circumstances), while the rest is non-coding or regulatory (Eisen 2000) (further, a change in a bacterium's regulatory region is reversible). Consequently, a change in a bacterial DNA sequence is likely to affect a nucleotide in the coding region. If this occurs at the third base position of a codon, the amino acid is not altered (Kimura 1968, 1985). If it does change the codon and its encoded amino acid, the resulting polypeptide (protein) will be different and so will the bacterium's relationship to its circumstances.

In contrast to bacteria, only 2–3% of a metazoan genome is coding (and relevant to adapting, as in, e.g., changes in oxygen-binding hemoglobin), while the rest is non-coding, consisting of developmentally regulated genes, large promoter regions, introns, etc., and is involved in the formation of complex, three-dimensional structure (i.e., topology) through RNA polymerase-directed alternative intron splicing, transcriptional regulation of developmental genes, etc. (Ast 2005; Eisen 2000; Gerhart and Kirschner 1997; Wray 2003). Being multicellular, mutation will more likely affect a metazoan's somatic cells, with no consequence to offspring. (This point is important since the language used in the literature casts mutation as always being transmitted to offspring, expressed, and available to selection.) If, however, mutation were to affect a potential sex cell prior to or during gametogenesis, it would more likely impact the non-coding than coding region—which could affect and thus derail developmentally regulated signaling pathways, leading not to "like begetting like" but to teratological conditions if not death as some point in the life cycle (Maresca and Schwartz 2006; Schwartz and Maresca 2006).

However, metazoan cells have myriad safeguards through the action of heat shock proteins to prevent as well as to eliminate the intrusion of mutational change into the genome. Therefore, mutation rates remain low (10^{-8-9}) and primarily the consequence of UV radiation (Maresca and Schwartz 2006). Thus, the question is "What mechanisms might overload the heat shock protein response, increase the *effective* mutation rate, and allow the possibility of organismal change?" (Maresca and Schwartz 2006). Since this topic has been discussed at length elsewhere (Maresca and Schwartz 2006), I will not do so here. Suffice it to say that bacteria are not evolutionary surrogates for metazoans (or plants). Bacteria adapt. Metazoans adapt as well (via the coding region), but they also evolve. Thus, while it may be commonplace to refer to experimental studies on bacteria as demonstrating "adaptive evolution" (Lenski and Travisano 1994), this does not reflect fact. The problem is linguistic: Because evolution has been equated with adaptation, but over geological time, and both adaptation and evolution can be described as "change," adaptation and evolution have long been conflated as one and the same, leading to the wholesale and interchangeable use of these terms, regardless of the biological entity or circumstance.

It may be more psychologically satisfying to think that an organism has certain characteristics because they are or were beneficial to their bearer and/or their bearer's ancestors, i.e., they were "selected for," making their bearer more "fit." However, as Morgan and others have argued, it may be no less likely that the picture of life we are presented with exists because features and their bearers didn't survive or contribute significantly to future generations. In further consideration of the notion of "selection for" "beneficial" traits, one might reflect on Fleeming Jenkin's (1867) criticism of Darwin's presentation, i.e., Why wouldn't the more numerous, less well-adapted individuals outbreed and swamp the less numerous, more and especially most well-adapted, individuals? Isn't this what a normal distribution represents, with the "most" and the "least" being fewer in number than "everyone else" and at the extremities of the distribution? Perhaps, as I suggested earlier, the way to think of the situation is, "If a feature, no matter how it came into being, doesn't kill you, you have it."

In this regard, let us return to the matter of how complex structure may emerge: Is it gradually, via natural selection choosing more "beneficial" versions of it, or abruptly, through developmental reorganization? Consider the vertebrate eye. From a Darwinian perspective, it emerged through a long process that began with the development of light sensitive cells (and in the correct locations in the head region), and selection over innumerable generations of individuals with greater numbers of these cells, invagination of the "correct" regions of the head, and the gradual differentiation of an eye's multifaceted components (Dawkins 1994). Clearly, not only must one embrace Morgan's contention that a mutation in one direction will be continued by subsequent mutations; one must also imagine a process in which the different components of an eye emerged independently and yet culminated in a functioning organ, replete with innervation, vascularization, a lens, muscles, etc.

But while Dawkins' scenario may have its appeal, it fails to acknowledge experimental demonstration decades earlier that, if the lens is extirpated early in development, an eye will not form—which is an example Goldschmidt (1940) used in making the case that complex structures must emerge in a state that is functional. That this phenomenon is true was recently demonstrated in greater detail: Although there are many developmentally regulated genes and transcription factors involved in the formation of the vertebrate eye, if one of them—the *Rx* gene—is inactivated, neither an eye with all its components nor its cranial socket develop (Mathers et al. 1997). In a different example of the integration of parts of a complex structure, when the degree of expression of one of the transcription factors—a bone-modifying protein—involved in the molecular signaling pathway underlying bird beak development is experimentally manipulated, not only is the morphology of the beak altered, so, too, is its attendant musculature and cranial skeleton, creating an integrated, functional whole (Abzhanov et al. 2004; Wu et al. 2004) (see Granovitch's discussion this volume on morphoprocess).

As discussed above, Dobzhansky's rejection of Goldschmidt's theory of systemic mutation on the grounds that a saltational appearance of a new species has never been observed applies equally to his claim that it is appropriate to extrapolate from experimental studies on selection to the formation of species. The difference between Goldschmidt's and Dobzhansky's, or between any non-gradualistic and gradualistic model of change leading to new species, is that, in one, adaptation via natural selection is dissociated from the origin of species, while, in the other, the origin of species is the result of selection-mediated adaptation over long periods of time. Dobzhansky's assertion notwithstanding, in neither case has species-relevant novel morphology and species formation been observed or experimentally generated. [Although extreme morphological disparity between parent and offspring can be induced experimentally, yielding phenocopies, this change is not permanent across generations (Waddington 1957).]

However, as, for example, Huxley, Mivart, de Vries, and the early Morgan appreciated, one might profitably extrapolate from the appearance of "monsters" and "sports," and their use by breeders to establish morphologically different breeds or varieties via cross-breeding with the parental ("normal," "wild") stock, to models of morphological change. In developmental terms, this means that the advent of morphological novelty is due, as de Beer phrased it, to internal factors, perhaps in coordination with external factors—which is generally consistent with, for example, Mivart, Bateson, Goldschmidt, Schindewolf, and Waddington's non-Darwinian theories of morphological innovation. This, however, was and is antithetical to Darwin's conception of natural selection and his reliance on "use-disuse," as well as to the model of speciation that underpinned the "synthesis," wherein organisms are naïve participants in a program dictated by external circumstances represented by the environment and natural selection (Schwartz 1999b).

One possible avenue for the introduction of morphological novelty was known to Fisher, Haldane, and Wright, as suggested by Bateson: Most mutations arise in the recessive state and spread silently through the population via heterozygosis until homozygotes express the physical manifestation of the mutations. This means that the provocation of the "mutation" which may indeed be environmental—causing, for example, the inability of heat shock proteins to eliminate DNA sequence errors introduced during replication, incompatibility between transcription factors and promoter regions, altered intron splicing patterns and molecular signaling pathways, increased or decreased transcriptional regulation, etc.—and its expression are separated by some number of generations. In turn, this means that the expressed change and the circumstances in which its bearers find themselves are, as Morgan even admitted, unconnected (Maresca and Schwartz 2006; Schwartz 1999b). (It further suggests that the notion of "exaptation"—wherein a "feature" is co-opted into a new or different functional role—is too narrow. That is, what one observes an organism doing, from which one confers function onto morphology, is likely one of a panoply of activities that morphology will allow its bearer to perform.)

That such a mechanism of inheritance and expressed morphological consequence exists was demonstrated by Johnson et al. (1998), who witnessed the sudden appearance of a litter of polydactylous mice born of a mother that was part of a colony of genetically determined mice used in medical experimentation. By working backward, they reconstructed that, at some point, there had been a mutation affecting the *Hoxd13* gene (one of many genes involved in limb formation) that, because it was recessive, had spread through the colony until heterozygote saturation led to the production of homozygote offspring expressing the mutation.

The simplicity of this mechanism notwithstanding, the question is, why did it not survive beyond Fisher, Haldane, and Wright into subsequent models of evolutionary change? Clearly, it would have been a response to the criticism of how more than one individual would be the bearer of abrupt morphological change. Further, since Dobzhansky, at least, was well-informed of the publications of these geneticists, he should also have been aware of this phenomenon. Yet, consideration of this mechanism is absent from evolutionary discourse from Dobzhansky on, without comment or attempted refutation—perhaps because it was inconsistent with the notion of the gradual accumulation of minute mutations.

Although the shift from thinking of "genes for" structure to hierarchical models of pattern formation began some time ago [e.g., Wolpert (1969)], the influence of "selection thinking" remains, even as the understanding of morphogenesis via tightly orchestrated molecular signaling pathways coupled with epigenetic influences has matured [e.g., see chapters in Pigliucci and Müller (2010)]. Indeed, one now sees in the literature the phrase "selection-induced phenotypic change" (Chaparro-Pedraza and de Roos 2020), which, of course, can only occur if there is "selection-induced" genetic and developmental change. Further, while there have been efforts to integrate developmental genetics, comparative and developmental morphology, and paleontology into a post-synthesis approach to understanding the emergence of morphological novelty and the origin of species [e.g., Erwin (2015), Schwartz (1999b)], endeavors such as those that resulted in "the extended synthesis" (Pigliucci and Müller 2010) have, like the architects of the "synthesis," focused on the "genetics" of evolution—albeit with a current understanding of developmental genetics and non-genetic factors influencing development. However, unlike Dobzhansky and Mayr, who, even with their circumscribed view of evolution, offered scenarios attempting to depict speciation, students of what's been called "evo-devo" (evolutionary developmental biology) have not yet achieved a synthesis of the diversity of elements that constitute evolutionary biology.

An aspect of the preceding historical review that I also hope has become clear is how, by squelching alternative ideas, the architects of the "synthesis" also squelched potential advances in evolutionary thinking. That is, even though the suggested underlying mechanisms of evolutionarily significant change may not have stood the test of time, either in part or full (e.g., Goldschmidt's theory of systemic mutation via chromosomal rearrangement), the alternatives to Darwinism, neo-Darwinism, and the "synthesis" provide a history of what might be called "evo-devo thinking." Yet, because the architects of the "synthesis" [see Simpson (1953), and especially Mayr (1963a, b, 1980, 1982, 1996), Mayr and Provine (1981)], also wrote histories of evolutionary thought that promoted their ways of thinking, alternative ideas remained submerged for decades. How unfortunate it is then that the theories of, for example, Bateson, de Beer, Goldschmidt, Schindewolf, Waddington, Thompson (1917) (the effects of gravity on development), and Løvtrup (1974) (the effects of cell-cell symmetry versus asymmetry and timing of gastrulation) were not elements of evolutionary discourse during the latter part of the twentieth century and into the twenty-first. Stephen Jay Gould's (2002) mammoth history of evolutionary thought, which did emphasize development, had little effect on most evolutionary biologists, probably because of antagonism toward the model of punctuated equilibria he co-authored with Niles Eldredge (Charlesworth et al. 1982).

Without doubt, it is difficult to shed received wisdom, even if parts of it are found to be inappropriate or even invalid. Thus, one continues to see in the developmental literature emphasis on natural selection inducing changes in developmental pathways to produce structure to fit an organism's needs (Kavanagh et al. 2007). Or, in the paleontological literature, for example, that the discovery of fossil flatfish with the migrating eye not quite on the side with the other eye demonstrates, contra Goldschmidt, the gradual evolution of flatfish under the guidance of natural selection (Friedman 2008).

In many ways, "natural selection" has merely supplanted the term "use-disuse," e.g., selectionist scenarios often contain the phrase "in order to" or convey intent. Yet, by referring to "natural selection" or just "selection," one is supposedly being scientific, not Lamarckian. Witness the case of an article on the origin of human bipedalism in which the argument was made that non-habitually bipedal ape ancestors standing upright on horizontal tree limbs with flexed knees would cause developmental changes that would yield the musculoskeletal morphology of fully bipedal humans (Thorpe et al. 2007) (now, this is "hopeful"). In response to criticism that this sounded like a "use-disuse" argument (Schwartz 2007), the authors pointed out they had "refer[red] to selection eight times" (p. 1065).

Perhaps the time has come to re-evaluate the utility and biological appropriateness of invoking selection as the de facto explanation for much of what is deemed evolutionary.

Acknowledgments I thank Richard Delisle for inviting me to contribute to this volume and thus provoking me to better explore the topic of, and question my own ideas about, natural selection.

References

Abzhanov A, Protas M, Grant BR, Grant PR, Tabin CJ (2004) *Bmp4* and morphological variation in beaks in Darwin's finches. Science 305:1462–1465
Ast G (2005) The alternative genome. Sci Am 292:58–65
Barrick JE, Yu DS, Yoon SH, Jeong H, Oh TK, Schneider D, Lenski RE, Kim JF (2009) Genome evolution and adaptation in a long-term experiment with *Escherichia coli*. Nature 461:1243
Bateson W (1894) Materials for the study of variation, treated with especial regard to discontinuity in the origin of species. Macmillan, New York
Bateson W (1913) Problems of genetics. Yale University Press, New Haven
Bateson W, Saunders ER (1902) Reports to the evolution committee of the royal society. Report 1. Experiments undertaken by W. Bateson, F. R. S., and Miss E. R. Saunders. Harrison & Sons, London
Bolk LL (1926) Das Probleme der Menschwerdung. Gustav Fischer, Jena
Bowler P (1989) Evolution: the history of an idea. University of California Press, Berkeley
Butler PM (1939) Studies of the mammalian dentition. Differentiation of the postcanine dentition. Proc Zool Soc London 109:1–36
Butler PM (1978) The ontogeny of mammalian heterodonty. Journale Biologique Buccale 6:217–227
Castle W (1913) Heredity in relation to evolution and animal breeding. D. Appleton, New York
Chaparro-Pedraza PC, de Roos AM (2020) Ecological changes with minor effect initiate evolution to delayed regime shifts. Nat Ecol Evol 4:412–418
Charlesworth B, Lande R, Slatkin M (1982) A neo-Darwinian commentary on macroevolution. Evolution 36:474–498. https://doi.org/10.1111/j.1558-5646.1982.tb05068.x
da Silva GO (2017) Is natural selection a chimera? Reflections on the 'survival' of a principle. Vibrant Virtual Braz Anthropol 14:1–21
Darwin C (1859) On the origin of species by means of natural selection, or the preservation of favored races in the struggle for life. John Murray, London
Darwin C (1868) The variation of animals and plants under domestication. John Murray, London
Darwin C (1869) On the origin of species by means of natural selection, or the preservation of favoured races in the struggle for life, 5th edn. John Murray, London
Darwin C (1872) On the origin of species by means of natural selection, or the preservation of favored races in the struggle for life, 6th edn. John Murray, London
Davidson EH, Erwin DH (2006) Gene regulatory networks and the evolution of animal body plans. Science 311:796–800
Dawkins R (1994) The eye in a twinkling. Nature 368:690–691
de Beer GR (1930) Embryology and evolution. Clarendon Press, Oxford
de Beer GR (1940) Embryology and taxonomy. In: Huxley J (ed) The new systematics. Clarendon Press, Oxford, pp 365–394
de Vries H (1910) The mutation theory (trans: Farmer JB, Dardbishire AD). Open Court, Chicago
Dietrich MR (2000) From hopeful monsters to homeotic effects: Richard Goldschmidt's integration of development, evolution, and genetics. Am Zool 40:28–37
Dobzhansky T (1935a) A critique of the species concept in biology. Philos Sci 2:344–355
Dobzhansky T (1935b) *Drosophila miranda*, a new species. Genetics 20:377–391
Dobzhansky T (1937) Genetics and the origin of species. Columbia University Press, New York
Dobzhansky TH (1941) Genetics and the origin of species, 2nd edn. Columbia University Press, New York

Eisen JA (2000) Assessing evolutionary relationships among microbes from whole-genome analysis. Curr Opin Microbiol 3:475–480
Eldredge N, Gould SJ (1972) Punctuated equilibria: an alternative to phyletic gradualism. In: Schopf TJM (ed) Models in paleobiology. Freeman, Cooper & Co., San Francisco
Erwin DH (2015) Novelty and innovation in the history of lfe. Curr Biol 25:930–940
Fisher RA (1930) The genetical theory of natural selection. Oxford University Press, Oxford
Fisher RA (1958) The genetical theory of natural selection, 2nd edn. Dover Publications, New York
Ford EB, Huxley JS (1927) Mendelian genes and rates of development in *Gammarus chevreuxi*. Br J Exp Biol 5:112–134
Friedman M (2008) The evolutionary origin of flatfish asymmetry. Nature 454:209–212
Galton F (1869) Hereditary genius: an inquiry into its laws and consequences. Macmillan, London
Galton F (1871) Experiments in pangenesis, by breeding from rabbits of a pure variety, into whose circulation blood taken from other varieties had previously been largely transfused. Proc R Soc (Biology) 19:393–404
Galton F (1875) A theory of heredity. Contemp Theatr Rev 27:80–95
Galton F (1876) A theory of heredity. J Anthropol Inst G B Irel 5:329–348
Gerhart J, Kirschner M (1997) Cells, embryos, and evolution: toward a cellular and developmental understanding of phenotypic variation and evolutionary adaptability. Blackwell, Malden, MA
Gesner K (1565–1566) De rerum fossilium, lapidium et gemmarum Maxime. Tiguri, Zürich
Goldschmidt R (1923) Einige Materialien zur Theorie der abgestimmten Reaktionsgeschwindigkeiten. Archiv für Mikroskopische Anatomie und Entwicklungsmechanik 98:292–313
Goldschmidt R (1927) Physiologische Theorie der Vererbung. Springer, Berlin
Goldschmidt R (1933) Some aspects of evolution. Science 78:539–547
Goldschmidt RB (1940) The material basis of evolution. Yale University Press, New Haven
Gould SJ (2002) The structure of evolutionary theory. The Belknap Press of Harvard University Press, Cambridge, MA
Gould SJ, Eldredge N (1977) Punctuated equilibria: the tempo and mode of evolution reconsidered. Paleobiology 3:115–151
Gould SJ, Vrba ES (1982) Exaptation – a missing term in science of form. Paleobiology 8:4–15
Haeckel E (1866) Generelle Morphologie der Organismen. Georg Reimer, Berlin
Haeckel E (1868) The history of creation: or the development of the earth and its inhabitants by the action of natural causes. A popular exposition of the doctrine of evolution in general, and that of Darwin, Goethe, and Lamarck in particular (trans: Lankester ER). Appleton, New York
Haldane JBS (1924) A mathematical theory of natural and artificial selection. Part I. Trans Cambridge Philos Soc 23:19–41
Haldane JBS (1926) A mathematical theory of natural and artificial selection. Part III. Proc Cambridge Philos Soc 23:363–372
Haldane JBS (1927) A mathematical theory of natural and artificial selection. Part IV. Proc Cambridge Philos Soc 23:607–615
Haldane JBS (1931) A mathematical model of natural and artificial selection Part VII. Proc Cambridge Philos Soc 27:131–136
Haldane JBS (1932a) The causes of evolution. Harper & Brothers, New York
Haldane JBS (1932b) A mathematical theory of natural and artificial selection. Part VIII: metastable populations. Proc Camb Philos Soc 27:137–142
Hennig W (1950) Grundzüge einer Theorie der Phylogenetischen Systematik. Deutscher Zentralverlag, Berlin
Hennig W (1966) Phylogenetic systematics. University of Chicago Press, Chicago
Huxley TH (1863a) Evidence as to man's place in nature. D. Appleton, New York
Huxley TH (1863b) On the relation of man to the lower animals. In: Huxley TH (ed) Man's place in nature. D. Appleton, New York
Huxley J (1932) Problems of relative growth. Methuen, London
Huxley J (1940a) Introductory: towards the new systematics. In: Huxley J (ed) The new systematics. Clarendon Press, Oxford, pp 1–46

Huxley J (ed) (1940b) The new systematics. Clarendon Press, Oxford
Huxley J (1942) Evolution: the modern synthesis. Harper & Brothers, New York
Jenkin F (1867) The origin of species. North Br Rev 46:149–171
Johannsen W (1909) Elemente der Exakten Erblichkeitslehre. Gustav Fisher, Jena
Johnson KR, Hope OS, Donahue LR, Ward-Bailey P, Bronson RT, Davisson MT (1998) A new spontaneous mouse mutation of *Hoxd13* with a polyalanine expansion and phenotype similar to human synpolydactyly. Hum Mol Genet 7:1033–1038
Kavanagh KD, Evans AR, Jernvall J (2007) Predicting evolutionary patterns of mammalian teeth from development. Nature 449:427–432
Kimura M (1968) Evolutionary rate at the molecular level. Nature 217:624–626
Kimura M (1985) Neutral theory of molecular evolution. Cambridge University Press, Cambridge
Lenski RE, Travisano M (1994) Dynamics of adaptation and diversification: a 10,000-generation of experiment with bacterial populations. Proc Natl Acad Sci U S A 91:6808–6814
Linnaeus C (1735) Systema naturae per regna tria naturae, secundum classes, ordines, genera, species cum characteribus, differentiis, synonymis, locis. Laurentii Salvii, Stockholm
Linnaeus C (1758) Systema naturae per regna tria naturae, secundum classes, ordines, genera, species cum characteribus, differentiis, synonymis, locis. Editio decima, reformata edn. Laurentii Salvii, Stockholm
Lotsy JP (1916) Evolution by means of hybridization. Martinus Mijhoff, The Hague
Løvtrup S (1974) Epigenetics: a treatise on theoretical biology. Wiley, New York
Malthus T (1826) An essay on the principle of populations; or, a view of its past and present effects on human happiness; with an inquiry into our prospects respecting the future removal or mitigation of the evils which it occasions. J. Johnson, London
Maresca B, Schwartz JH (2006) Sudden origins: a general mechanism of evolution based on stress protein concentration and rapid environmental change. Anat Rec B New Anat 289:38–46
Mathers P, Grinberg A, Mahon K, Jamrich M (1997) The Rx homeobox gene is essential for vertebrate eye development. Nature 387:604–607
Mayr E (1942) Systematics and the origin of species. Columbia University Press, New York
Mayr E (1949) Speciation and systematics. In: Jepsen GL, Simpson GG, Mayr E (eds) Genetics, paleontology, and evolution. Atheneum, New York, pp 281–298
Mayr E (1963a) Animal species and evolution. The Belknap Press of Harvard University Press, Cambridge, MA
Mayr E (1963b) Populations, species, and evolution. The Belknap Press of Harvard University Press, Cambridge, MA
Mayr E (1980) How I became a Darwinian. In: Mayr E, Provine WB (eds) The evolutionary synthesis. Harvard University Press, Cambridge, MA, pp 413–423
Mayr E (1982) The growth of biological thought: diversity, evolution, and inheritance. Belknap Press of Harvard University Press, Cambridge, MA
Mayr E (1996) The modern evolutionary theory. J Mammal 77:1–7
Mayr E, Provine WB (1981) The evolutionary synthesis. Bull Am Acad Arts Sci 34:17–32
Mendel G (1866) Experiments in plant hybridisation (trans: Bateson W). Harvard University Press, Cambridge, MA
Mivart SG (1864) Notes on the crania and dentition of the Lemuridae. Proc Zool Soc London 1864:611–648
Mivart SG (1868) Notes on the osteology of the insectivora. J Anat Physiol 2:117–154
Mivart SG (1871) On the genesis of species. John Murray, London
Mivart SG (1873) On *Lepilemur* and *Cheirogaleus* and on the zoological rank of the Lemuroidea. Proc Zool Soc London 1873:484–510
Mivart SG (1882) On the classification and distribution of the Aeluroida. Proc Zool Soc Lond 1882:135–208
Morgan TH (1903) Evolution and adaptation. Macmillan, New York
Morgan TH (1909) What are 'factors' in Mendelian explanations? Proc Am Breeder's Assoc 5:365–368

Morgan TH (1910) Chance or purpose in the origin and evolution of adaptation. Science 31:201–210
Morgan TH (1916) A critique of the theory of evolution. Princeton University Press, Princeton
Morgan TH (1925) Evolution and genetics. Princeton University Press, Princeton
Morgan TH (1935) The scientific basis of evolution. W. W. Norton, New York
Morgan TH, Sturtevant AH, Muller HJ, Bridges CB (1926) The mechanism of mendelian heredity. Henry Holt, New York
Osborn JW (1971) The ontogeny of tooth succession in *Lacerta vivipara* Jacquin (1787). Proc R Soc Lond B179:261–289
Pigliucci M, Müller GB (eds) (2010) Evolution: the extended synthesis. The M.I.T. Press, Cambridge, MA
Provine WB (1971) The origins of theoretical population genetics. The University of Chicago Press, Chicago
Provine WB (1986) Evolution: selected papers of Sewall Wright. The University of Chicago Press, Chicago
Schindewolf O (1936) Palaeontologie, Entwicklungslehre und Genetik. Bontrüager, Berlin
Schindewolf O (1993) Basic questions in paleontology: geologic time, organic evolution, and biological systematics (trans: Schaefer J). University of Chicago Press, Chicago
Schwartz JH (1999a) Can we really identify species, fossil or living? Anthropologie XXXVII(/ 3):221–230
Schwartz JH (1999b) Sudden origins: fossils, genes, and the emergence of species. Wiley, New York
Schwartz JH (2005a) Darwinism versus Evo-Devo: a late 19th c. debate. In: Mueller-Wille S, Reinberger H-J (eds) A cultural history of heredity III: 19th and early 20th centuries, vol Preprint 294. Max Planck Institute for the History of Science, Berlin, pp 67–84
Schwartz JH (2005b) The red ape: orangutans and human origins. Westview Press, Boulder, CO
Schwartz JH (2006) Decisions, decisions: why Thomas Hunt Morgan was not the father of "evo-devo". Philos Sci 73:918–929
Schwartz JH (2007) The origins of human bipedalism: comment. Science 318:1065
Schwartz JH (2009) Reflections on systematics and phylogenetic reconstruction. Acta Biotheor 57:295–305
Schwartz JH (2011) Systematics and evolutionary biology: uneasy bedfellows? Revista Mexicana de Biodiversidad 82:1–18
Schwartz JH, Maresca B (2006) Do molecular clocks run at all? A critique of molecular systematics. Biol Theory 1(4):1–15
Simpson GG (1944) Tempo and mode in evolution. Columbia University Press, New York
Simpson GG (1945) The principles of classification and a classification of the mammals. Bull Am Mus Nat Hist 85:1–350
Simpson GG (1949) Essay-review of recent works on evolutionary theory by Rensch, Zimmermann, and Schindewolf. Evolution 3:178–184
Simpson GG (1953) The major features of evolution. Simon and Schuster, New York
Stanley SM (1975) A theory of evolution above the species level. Proc Natl Acad Sci U S A 72:646–650
Thompson DAW (1917) On growth and form. Cambridge University Press, Cambridge
Thorpe SKS, Holder RL, Crompton RH (2007) Origin of human bipedalism as an adaptation for locomtion on flexible branches. Science 316:1328–1331
von Baer KE (1828) Über Entwicklungsgeschicte der Thiere: Beobachtung und Reflexion. Bornträger, Königsberg
Waddington CH (1940) Organisers and genes. Cambridge University Press, Cambridge
Waddington CH (1957) The genetic basis of the assimilated bithorax stock. J Genet 55:240–245
Washburn SL (1951) The new physical anthropology. Trans N Y Acad Sci 13:298–304
Weber C, Zhou Y, Lee JG, Looger LL, Qian G, Ge C, Capel B (2020) Temperature-dependent sex determination is mediated by pSTST3 repression of Kdm6b. Science 368:303–306

Williams GC (1966) Adaptation and natural selection: a critique of some current evolutionary thought. Princeton University Press, Princeton
Williams GC (1992) Natural selection: domains, levels, and challenges. Oxford University Press, New York
Wolpert L (1969) Positional information and the spatial pattern of cellular differentiation. J Theor Biol 25:1–45
Wray GA (2003) Transcriptional regulation and the evolution of development. Int J Dev Biol 47:675–684
Wright S (1923) Mendelian analysis of the pure breeds of livestock II: the Duchess family of Shorthorns as bred by Thomas Bates. J Hered 14:405–422
Wright S (1925) The factors of the albino series of Guinea pigs and their effects on black and yellow pigmentation. Genetics 10:223–260
Wright S (1929) Evolution in a Mendelian population. Anat Rec 44:287
Wright S (1930) Review of the genetical theory of natural selection by R.A. Fisher. J Hered 21:349–356
Wright S (1931) Evolution in mendelian populations. Genetics 16:97–159
Wright S (1932) The roles of mutation, inbreeding, crossbreeding and selection in evolution. Proceedings of the Sixth International Congress of Genetics 1:356–366
Wright S (1934) Physiological and evolutionary theories of dominance. Am Nat 68:25–53
Wright S (1941) The material basis of evolution (A review of the material basis of evolution, by Richard B. Goldschmidt). Sci Mon 53:165–170
Wright S (1945) Tempo and mode in evolution: a critical review (A review of tempo and mode in evolution, by George Gaylord Simpson). Ecology 26:415–419
Wright S (1978) The relation of livestock breeding to theories of evolution. J Anim Sci 46:1192–1200
Wu P, Jiang T-X, Suksaweang S, Widelitz RB, Chuong CM (2004) Molecular shaping of the beak. Science 305:1465–1466
Yule GU (1902) Mendel's laws and their probable relations to inter-racial heredity. New Phytol 1:194–238

Chapter 13
Natural Selection, Morphoprocess and a Logical Field of Evolutionary Concepts

A. I. Granovitch

Abstract Since the publication of *The Origin of Species* by Charles Darwin (1859), the mechanism of natural selection has invariably been considered as the main force of transformism. This tradition culminated in the paradigm of the Modern Synthesis. This study invites the reader to critically examine the precept that selectogenesis is the only possible framework for an evolutionary synthesis. Abundant data on the organisation of living systems at all levels, from molecules to ecosystems, indicate that various evolutionary mechanisms are possible, including those in which natural selection is not the main driving force. Time has come to take a broader look and develop a "logical field" of approaches allowing the explanation of evolution in all its diversity.

We will begin with the consideration of the mechanism of natural selection with its two main prerequisites, tychogenesis and selectogenesis. The first prerequisite, tychogenesis, seems a phantom in the light of the modern data on the canalised, limited, species-specific and dynamic range of the hereditary variation. The second prerequisite, selectogenesis, is extremely difficult to test. Further, the model of natural selection has several grave flaws. It underestimates the conditional nature of the importance of hereditary characters in the context of the organism's interaction with the environment; makes an unfounded extrapolation that the vector of selectogenetic impact is the same in the series of generations; misjudges the possibilities of the organism as a regulatory, self-organising system; and dismisses the possibility of transgenerational transfer of structural, dynamic and epigenetic information. Last but not least, it lacks the mechanism responsible for an increasing complexity of the morphofunctional features of organisms. Thus, overwhelming evidence indicates that the model of natural selection cannot be realistically considered as the major mechanism of transformism. Numerous diverse approaches to the explanation of evolutionary mechanisms have been suggested during the last two centuries. It is their combination rather than the elaboration of the selectogenetic narrative alone that should underlie the new evolutionary thinking. "Evolutionary

---

A. I. Granovitch (✉)
Department of Invertebrate Zoology, St. Petersburg State University, St. Petersburg, Russia
e-mail: a.granovich@spbu.ru

R. G. Delisle (ed.), *Natural Selection*, Evolutionary Biology – New Perspectives on Its Development 3, https://doi.org/10.1007/978-3-030-65536-5_13

syntheses" based on selectogenesis are no longer productive. What is needed is a total reset of evolutionary thought or, one may say, the development of the platform Evolution 2.0.

Keywords Evolutionary concepts · Morphoprocess and dissipative structures · Natural selection · Adaptogenesis and self-organisation · Models of microevolution

13.1 Introduction

Transformism—the idea of changeability of living creatures in the context of the historical development of the biota—is nothing new. The thought that forms change and creatures originate from other creatures was salient in certain religious and cosmological worldviews of Greek philosophers such as Anaximander, Empedocles, Xenophanes, and Titus Lucretius Carus (Sarton 1953; Berg 1977; Vorontsov 2004). A more concrete statement of the question about the mechanisms ensuring transformism took two thousand years, requiring a deeper insight into the diversity of the living organisms and the structure of the living systems.

Discussions about the driving forces of transformism welled up in the nineteenth century, when several algorithms/mechanisms explaining how organisms might change in the series of generations were suggested almost simultaneously. Some of them focussed on the abilities of individuals to respond to the impact of the environment (Lamarck 1809; Geoffrey 1818; Spencer 1893), while others centred around the regular character of the organisms' variability (Nägeli 1884; Agassiz 1874; Eimer 1890). Finally, a mechanism based on the selection of the fittest individuals in the population was proposed (Darwin 1872; Wallace 1889). This latter view implies a populational approach to the solution of the problem of the major mechanism of transformism.

It should be noted that all the mechanisms assumed that the living matter has a species structure. Correspondingly, the ideas about transformism in general narrowed down to the transformism of species or, to be more precise, to the changes of the morphofunctional modality of individuals composing this or that species.

Approaches to the comprehension of transformism are distilled in the following passage from Darwin's *The Origin of Species*, which sets forth three major potential mechanisms ensuring transformation of species. "The passage from one stage of difference to another may, in many cases, be the simple result of the nature of the organism and of the different physical conditions to which it has long been exposed; but with respect to the more important and adaptive characters, the passage from one stage of difference to another, may be safely attributed to the cumulative action of natural selection, hereafter to be explained, and to the effects of the increased use or disuse of parts" (p. 42, Darwin 1872).

Let us note that the emphasis in the formation of adaptive characters is on natural selection (NS) and preferential use of organs. Constructive features ("the nature of the organism...") are considered as an important but, as it were, a background factor.

This is so because it seemed that only NS and a direct adaptation of the organism ("...use or disuse of parts") could be associated with the formation of adaptive characters.

The calibre of a researcher is evidenced by the ability to perceive and register various aspects of a natural phenomenon, including seemingly contradictory ones. In this connection, one can only marvel at Darwin's attempt to tackle the mind-boggling problem of the mechanisms of transformism simultaneously, from several angles at once—and that despite the fact that the main message of *The Origin of Species* is, of course, the rationale for the hypothesis of NS, which was presented in this treatise in a definite form for the first time.

Regretfully, the subsequent fate of the evolutionary concept was governed by an entirely different logic: the choice of the leading notion accompanied by a violent criticism of the others and reminding the path of formation of religious doctrines. This approach culminated in the Modern Synthesis of the twentieth century (for review see Gould 2002), which, combining the logical scheme of NS with the advances of genetics in the matter of heredity, showed zero tolerance to the potential implementation of other mechanisms. It would seem that the only addition to the scheme of the functioning of NS made by the Modern Synthesis was the recognition of the importance of stochastic genetic processes in populations (Wright 1931).

The acceptance of the logical scheme of NS as the major mechanism of transformism entails the acceptance of a group of individuals, the population, as the main object of evolution. Further, we have to accept that changes of a purely group/populational parameter—the frequency of a certain allele, a certain character—are an elementary evolutionary phenomenon. The result is a peculiar logical collision. Evolution is thought of as a process transforming extremely complex highly integrated systems—organisms. At the same time, a system with an incomparably lower integrity—a group of individuals—is accepted as the main evolving object.

The evolutionary paradigm of the Modern Synthesis (MS), based on the ideas of population genetics and postulating NS as the main evolutionary mechanism, has faced numerous challenges in the last decades. A mere list of these challenges conveys the significance of the arising problems. (1) There are reasons to doubt that transformism may be explained with the help of mechanisms of population genetics (Lewontin 1974); (2) the punctuated equilibrium theory (Gould and Eldredge 1977; Gould 2002) implies that the laws of organisation of the living systems are important for the evolutionary process; (3) the new mechanistic philosophy questions the role of NS as a potential evolutionary mechanism (Barros 2008; Havstad 2011; Millstein 2013; Sharov and Igamberdiev 2014); (4) the phenomena of phenotypic plasticity and transgenerational inheritance are receiving increasingly more attention (Danchin et al. 2011; Kull 2014; Laland et al. 2014; Pigliucci 2010; West-Eberhard 2005; Jablonka and Lamb 2010); (5) our understanding of phenotypic manifestation of characters and their conditional nature is changing dramatically (Beklemishev 1994); (6) the importance of mechanisms of self-organisation of molecular complexes, cell organelles, cells, multicellular organisms, populations and communities is transpiring (Granovitch 2018; Johnson and Lam 2010; Kurakin

2007; Loose et al. 2008; Misteli 2001, 2007); and (7) the ubiquity and diversity of the mechanisms of genetic material transfer is becoming evident (Gladyshev et al. 2008; Koonin 2011; Skippington and Ragan 2011). In this light, those eager to explain the mechanisms of transformism find themselves on an entirely novel terrain.

Can our understanding of these mechanisms, amended and extended, remain within the framework of MS or shall we have to reconsider the entire evolutionary paradigm? Is it time for a revolution (a drastic reconsideration on a new basis) in the conceptual comprehension of the evolutionary phenomena (Kuhn 1962)? Let us not forget that evolutionary ideas are the core of the paradigm of general biology. Fundamental changes of views on evolutionary mechanisms, if they are to happen, may entail the reconstruction of the logic of the entire biological knowledge.

This study was carried out in response to the unfolding of a vast field of diverse potential evolutionary mechanisms, which is happening under our very eyes. It is an attempt to locate the traditional mechanisms of transformism in this field and to reveal their correlations with those that are totally at odds with the MS paradigm. In the beginning, we will consider the logical structure of the traditional conceptual model of NS. Then, based on the available data, we will define the type of the living systems to be considered as an evolving object. Finally, we will examine the ways and the mechanisms of transformation of these objects. This examination, stripped of the stifling restraints of a single dominant concept, will provide a more objective view on the potential diversity of evolutionary mechanisms.

13.2 Natural Selection

13.2.1 Natural Selection: Tychogenesis Plus Selectogenesis

(Tychogenesis [term: Osborn 1929, p. 813], from Greek *týche*—luck, fortune. Selectogenesis [term: Berg 1977; Lyubishchev 1982], from Latin *selectio*—choice, selection)

The logical scheme of the mechanism of NS is based on two prerequisites. The first is that the number of small changes of organisms transmitted to the next generation is unlimited. These undirected changes should potentially affect any characters of an organism; their emergence should not be associated with the impact of the environment and, thus, should not represent an adequate response to it. If so, variation is seen as a "homogenous" prerequisite allowing NS to "choose" from the available hereditary deviations. This view on variation was denoted by Osborn as "tychogenesis" (T) (Osborn 1929).

The second prerequisite of NS is an invariably unequal contribution of the carriers of different hereditary deviations into the next generation. This condition, in turn, has a prerequisite: a potentially greater number of offspring as compared with the number of parents, which may be achieved by means of differential fertility, or by means of differential viability, or both. In general, the sequence of events resulting in a differential contribution of different individuals into the genetic pool of the progeny is often referred to, following Darwin, as the "struggle for existence". However, within the framework of the model, it is not the interactions between individuals that make sense but only fitness, a differential contribution into the next generation. Thus, an emotionally more neutral term "selectogenesis" (S) corresponds better to the meaning of this logical model (Berg 1977; Lyubishchev 1982).

On the whole, the mechanism of NS is based on two key functional modules: tychogenesis and selectogenesis (T-S). Repeated iterations of the T-S scheme in the following generations under environmental conditions changing in a directed manner should, according to the model, result in corresponding directed changes of the morphofunctional features of the corresponding organisms.

Notes. (1) If the range of hereditary deviations is not random but is characterised by some degree of asymmetry ("non-homogeneity"), it is the constructive features of an organism that would determine the direction of the evolutionary process; selectogenesis in this case should be considered as a secondary, subordinate force (see for details Granovitch 2018). (2) The model may work if, and only if, there is a univocal correlation between the manifestation of characters in an organism with certain hereditary determinants; this correlation endures in the sequence of generations and is accompanied by tangible differences in the contribution of the carriers of characters into each of the following generations.

13.2.2 Potential Significance of Tychogenesis-Selectogenesis

In principle, the mechanism of T-S may be considered as a potential cause of transformism. Yet several questions arise: (1) Is it possible to verify the functioning of separate modules of T-S? (2) How realistic is the entire T-S scheme for the explanation of long-term changes of adaptive characters of species? (3) Does this logical scheme harbour a potential for a long-term increase of the organisation level of the organisms?

The following can be said about the testability of the separate modules of the T-S scheme. Abundant evidence of the organisms' variation shows that the first prerequisite of the action of NS, tychogenesis, is not fulfilled in nature. Hereditary variation is not random, is not homogenous and is not undirected. It is subject to stringent laws governing a preferential manifestation of characters; in other words, it is canalised. The analysis of the ways in which variation manifests itself at various levels indicates that it is regular, nomothetic (see for details Granovitch 2018). A reflection of the nomothetic nature of variation at the level of large taxonomic groups is convincingly supported by palaeontological data on numerous parallelisms in their evolutionary formation (Rozanov 1973; Ponomarenko 2005, 2008; Tatarinov 1976; Rozhnov 2005; Markov 2010). Moreover, it becomes clear after a close look on variation within large taxa that mathematically calculated morphospaces are very incompletely used by the real living forms (Thompson 1961; Thomas and Reif 1993; Raup 1966, 1967; McGhee, 1980, 2007; Ellison and Niklas 1988; Ubukata 2005; Brakefield 2010; Raup et al. 1978; Meyen 2007).

Further, both the range of variation and separate characters are manifested in a fairly conservative manner in representatives of species from the same genus and even family (Vavilov 1987; Meyen 2007). A non-homogeneous nature of the character space is indicated by a broad occurrence of agamic, autogamic and parthenogenetic species, whose representatives retain a stable species morphotype in independent genetic lineages (Grebelnyi 2008; Hausmann et al. 2003). Another indication is the ubiquity of cryptic species (Bickford et al. 2006), whose morphological conservativeness is at odds with genetic divergence.

An even more compelling evidence of the nomothetic character and the lack of stochasticity is provided by the analysis of the molecular level of formation and manifestation of the character space. The ideas about systemic organisation of genomes (Khesin 1984; Golubovskii 2000; Shapiro and Sternberg 2005) leave increasingly less room for the interpretations of molecular mechanisms underlying hereditary variation as stochastic processes. Reparation processes of the matrix synthesis (Radman 1974; Echols 1981; Karen and Hawley 2007), regulated changes of the level of mutations (Ponder et al. 2005; Galhardo et al. 2007) and the mechanism of horizontal transfer (Beaber et al. 2004; Hastings et al. 2004), various controlled and directed transpositions of mobile genetic elements (McClintock 1984; Gierl 1990; Fedoroff et al. 1995; Asis et al. 2008; Koonin and Wolf 2009; Koonin 2010) and the importance of epigenetic mechanisms for consideration of the range of hereditary variants (Jablonka and Lamb 2010; Allis et al. 2007) make up an impressive body of evidence of the directed character of the formation of the variation range. Moreover, the very definition of the range of hereditary variation reveals at least some connection with the specificity of the environmental impact upon an organism (Hall 1990, 1991; Beverley et al. 1984; Koonin and Wolf 2009; Paez-Espino et al. 2013; Koonin 2011; Asis et al. 2008; Kim et al. 2009; Hollick 2010; Landman 1991; Golubovskii 2000; Granovitch 2018).

We may conclude that in the light of the modern data tychogenesis is a phantom. What *is* productive is the study of the laws responsible for the canalised, limited, species-specific and dynamic range of the hereditary variation. Sadly, this means that the implementation of the logical model of T-S in the nature is unrealistic. It is the laws governing the formation of the range of characters that will have a decisive impact on the origin of forms, not the subsequent functioning of a hypothetic mechanism of selectogenesis. Even should its effectiveness be proven, selectogenesis will remain a subordinate factor, merely limiting the origin of forms.

The testability of selectogenesis, the second functional module of the T-S mechanism, is extremely limited. In effect, only its precondition—that progeny is more numerous than progenitors—can be reliably verified. Indeed, any populational study of any species group under near-optimal conditions shows that this "Malthusian" condition is fulfilled. The numbers of progeny are indeed greater (for many species, considerably or even by orders of magnitude greater) than the numbers of parent individuals. This is the essence of the immanent property of extended reproduction of living organisms used by the logical scheme of selectogenesis.

It seems that indications of selectogenesis at work can be observed with certainty only when the population contains individuals with lethal characters, non-viable under given conditions. Owing to their death, the potential variation range of a species will be realised incompletely, being limited by the degree of intensity of selectogenesis. The contribution of the carriers of lethal characters into the following generations will be equal to zero. At the same time, if the conditions change, the range of realised variation may be corrected. With the help of artificial selection/choice of breeders, one can sustain an extravagant part of the species variation range, as evidenced by varieties of cultivated plants and breeds of domestic animals.

The function of selectogenesis seems to be a rough adjustment of the potential variation range of a species. Its role is that of a negative, conservative factor. An intensive manifestation of selectogenesis (or, to be precise, a differential mortality/inability to leave progeny) leads to the narrowing of the realised variation range and the retardation of evolutionary processes. On the contrary, a diverse environment favourable for survival/reproduction promotes the manifestation of a considerable part of the variation range of a species. Potentially this situation contributes to an acceleration of evolutionary processes. In brief, from the viewpoint of the origin of forms, selectogenesis is a conservative, limiting factor; a decrease in the intensity of its impact reveals the potential diversity of intraspecies forms.

As for long-term processes of transformism of a species, the role of selectogenesis in them is apparently insignificant. This is due to several reasons. In the first place, the importance of separate characters, including hereditary ones, is conditional (Beklemishev 1994). This means that the manifestation of a character depends on the conditions of the environment and so does its potential significance for a given organism. Thus, it is in principle impossible to assess the significance of a hereditary character in terms of reproductive success. Moreover, each character influences the future interactions of the organism with the environmental conditions. This aspect—modification and an active choice of local conditions by the organisms—has often been mentioned in recent times in the context of the niche construction concept (Laland et al. 2016, 2017) as a phenomenon of potential evolutionary significance. Moreover, the conditional nature of characters may also be considered in a series of generations—that is, transgenerationally. The dynamics of environmental conditions (including that arising from the effects of niche construction), the changes in the manifestation of the characters themselves in subsequent generations and the functioning of populational compensatory mechanisms—all these—render unrealistic the stability of the selection vector in the series of generations (see, e.g. Grant and Grant 2002). The value of a character, assessed with the help of reproductive success, is dynamic, is being regulated by numerous feedbacks and cannot be strictly formalised. The conditional nature of the characters prevents a directed manifestation of selectogenesis.

A further complication of the selectogenetic interpretations of the evolutionary significance of characters arises from the ideas about the correlation of the organism's characters (Schmalhausen 1982) (the holisticity of the system). From this viewpoint, the organism is the result and the process of the operation of feedback mechanisms and compensatory mechanisms at all levels of its organisation. This prompts a fresh look on the problem of formation of adaptability (the correspondence of organisms to the environment) (see for details Granovitch 2018).

A demonstrative example of selectogenetic ideas is the adaptive landscape model (Wright 1932). The selection mechanism is seen as an instrument of the population's ascent to adaptive peaks in the series of generations. The multidimensionality of the real adaptive landscape implies the presence of vast neutral "areas" and "extradimensional bypasses" (Gavrilets and Gravner 1997; Gavrilets 2004). However, when we take into account the conditional nature of the characters, niche construction and self-organisation, the adaptive landscape appears as a dynamic relief or, rather, the water surface with dynamic wave peaks. At the same time, the wave crests are born from interactions of environmental conditions and the organism's response. It is evident that in such a situation a vectorised operation of selectogenesis is hardly possible.

Thus, the model of selectogenesis underestimates the conditional nature of the importance of hereditary characters for the organism in the context of its interaction with its environment; makes an unfounded extrapolation that the vector of selectogenetic impact is the same in the series of generations; misjudges the possibilities of the organism as a regulatory, self-organising system; and dismisses the possibility of transgenerational transfer of structural, dynamic and epigenetic information. Finally, the model of selectogenesis lacks the mechanism which drives the morphofunctional features of organisms towards increasing complexity. This is only possible under certain additional conditions, which are left out of the NS model.

The model is extremely difficult to verify in practice. The information available on the module that *can* be tested experimentally (the possibility of tychogenesis) disproves its postulates. Evolutionary possibilities of selectogenesis cannot be checked in a stringent experiment. So, unfortunately, the answers to the three questions put in the beginning of this section are negative and indicate that the model of NS cannot be realistically considered as the major mechanism of transformism.

The difficulty of verifying the NS model in general has long been known. It was in connection with the central place of NS in Darwinism that Karl Popper called it "...not a testable scientific theory, but a metaphysical research programme..." (Popper 2009, p. 167). Yet NS never came to play this role either. A research programme opens a direction of studies, suggesting a blueprint of experimental work, developing series of tests and marshalling evidence into a consistent body. On the contrary, the concept of NS closed many avenues of experimental research of evolutionary mechanisms. As noted above, it is as good as impossible to verify the functioning of NS, and so most experimental studies merely declare their allegiance to the concept and ignore all other mechanisms of transformism. Thus, the idea of evolution by NS became not a research programme but something entirely different—an axiom, a dogma or, shall we say, an article of faith.

13.3 The Search for an "Evolving" System

13.3.1 Hierarchy of Living Systems and the Object of Evolution

Transformism implies changes of the species characteristics by means of operation of hypothetic evolutionary mechanisms. However, the species characteristics themselves are morpho-anatomical, functional and ecological features of individuals, unicellular or multicellular organisms. So, what is it that changes in the process of operation of the mechanisms of transformism? In what way is one species different from the other? What makes the daughter species unlike the mother species? The difference resides in the modality of the distribution of characters of individuals. Similarly, species characteristics of a population are determined by qualitative differences of the constituent individuals.

The systems of supraorganismic level—populations and communities—are characterised by a considerably lesser integration, systemic integrity, than their constituent elements, individuals. To emphasise this feature, such systems are referred to as distributive ones (Serzhantov 1972). Both biocenoses and populations are characterised by a low degree of physiological and morphogenetic closeness (Beklemishev 1970). Therefore, on the whole, too, the transformism of the supraorganismic systems is mediated by the mechanisms of transformation of organisms; it is a reflection of fundamental processes of transformism of individuals at the level of the population and at the level of the species.

This line of reasoning also holds true during the search for the key evolving systems among the systems of the suborganismic level. All systems of this kind, from molecular complexes to the systems of organs of multicellular organisms, are elements of the morphofunctional organisation of an organism. Their autonomy is limited inasmuch as they are inbuilt into the maintenance of the organism's integrity. Their regulation is carried out in accordance with the demands of a more strongly integrated system of a higher rank, the organism. Their individualisation (Beklemishev 1970) is thus always lower than that of a system of an organismic rank. Formation and existence of the systems that are elements of a multicellular organism (organelles and their systems, cells, organs and systems of organs) are chiefly regulated by the organism itself (a highly integrated whole regulating its parts).

A thorough criticism of the attempt to reduce evolutionary mechanisms to primary replicators (genes) (see Williams 1966; Dawkins 1976, 1982) has been presented by Gould (Gould 2002, pp. 614–644) from the viewpoint of the integrity of an organism as a unit interacting with the environment. Correspondingly, it is the organism, not its part, that is the unit of selection (Gould 2002).

Thus, the search for suitable candidates for the role of evolving objects in the hierarchical series of living systems invariably leads us to the most highly integrated and the most individualised systems characterised by the greatest number of feedbacks, homeostatic mechanisms. These are, undoubtedly, the systems of the organismic level: unitary organisms with a high physiological autonomy, a complex behaviour and a high degree of morphogenetic closedness (Beklemishev 1970; McShea 2001; Levit and Scholz 2002; Granovitch et al. 2010).

All these considerations highlight the systems of the organismic rank—multicellular organisms and cells-organisms—during the search for the key evolving systems. And here we stumble upon another difficulty, which is: the study of evolutionary phenomena requires long-term observation of the evolving systems. At the same time, the organisms are finite. Their existence is confined to the short space of time between the formation of the zygote and death (for multicellular organisms) or else to the gap between the division of the mother cell and that of a daughter cell (for unicellular organisms). Surely such intervals are insufficient for consideration of evolutionary events. To get out of this predicament, let us slightly alter our ideas about organisms by assigning them a transgenerational meaning.

13.3.2 Morphoprocess as a Dynamic Evolving System

Living systems of the organismic rank are dynamic. The form of an organism is realised on the basis of a constant change of the composition of its constituent parts. This aspect is emphasised in the concept of the "organic morphoprocess" (Beklemishev 1994). Developing the ideas first suggested by Georges Cuvier (1817), V.N. Beklemishev suggested to consider the organism as being simultaneously an object and a process (Levit and Scholz 2002; Granovitch et al. 2010), a form enduring in a metabolic flow. This is a special system, maintaining a relative stability of its specific properties owing to the changes of the components. Thus, the organism is considered as a dynamic system—the morphoprocess. At the same time, the form of an organism (as an epiphenomenon of processes ensuring its existence), while seemingly constant during short time intervals, does undergo regular and species-specific growth, propagation and necrotic changes in the process of its ontogenesis.

A considerable progress in the understanding of these phenomena was achieved by analysing the properties of open systems, which absorb and give up matter and energy into the environment. Physical-chemical systems of this kind may exist far from the point of energy equilibrium (i.e. the maximum entropy point) for a relatively long time (Prigogine 1978; Prigogine and Stengers 2018; Cramer and Prigogine 1993). The flow of matter and energy through such a system results in the emergence of an epiphenomenon of a relatively stable structure. These enduring well-ordered complexes, whose morphofunctional integrity and stability is dynamic and ensured by the passage of the energy flow through the system, are referred to as dissipative structures (Prigogine 1978; Prigogine and Stengers 2018; Cramer and Prigogine 1993). This concept, first suggested to describe certain chemical reactions and physical processes, may be successfully applied to biological objects and even social phenomena (Prigogine 1978; Prigogine and Stengers 2018; Cramer and Prigogine 1993; Karsenti 2008; Artigiani 1987). Very possibly, the behaviour of nonlinear open systems under non-equilibrium conditions is a universal phenomenon.

It is significant that this concept makes the pioneering assumption—based on observations of the behaviour of thermodynamically non-equilibrium open systems including organisms—that the formation of dissipative structures is governed by physical and chemical laws. In turn, the law-governed dynamics of formation and stability of these structures accentuates the importance of self-assembly and self-organisation for the living systems (Sazer and Schiessel 2018; Bizzarri et al. 2018; de Paoli et al. 2017; Werner et al. 2017; Simunovic and Brivanlou 2017; Johnson and Lam 2010).

The dynamic stability of biological systems is described with the help of the term "a metastable pattern". It refers to the molecular dynamics of DNA oligomers (Horenko et al. 2008), the organisation of chromosomes (Sazer and Schiessel 2018), the organisation of cellular systems into tissues (Jakab et al. 2004) and molecular dynamics of cellular processes (metastable network processing) (Kurakin 2007). From the viewpoint of the dynamics of the system elements (its constituent parts) and energy flows through the system, it is quite possible to refer to dissipative structure as processes (metastable processes, Johnson and Lam 2010). This term emphasises, on the one hand, their nature as a dynamic process, simultaneously absorbing external elements and expulsing its own elements into the environment (process). On the other hand, it highlights the ability of dissipative structures to maintain their relatively constant form under certain conditions and to shift to a qualitatively different form of the process under others (metastable).

To note, this view on dissipative structures (a form enduring in a flow of matter and energy exchange with the environment) and the term "metastable process" itself are almost identical to the definition of the morphoprocess *sensu* Beklemishev (Beklemishev 1994) as applied to the systems of the organismic level of organization.

So, we have emphasised the dual nature of the organism as an object process and its mode of existence as a complex dissipative structure in the flow of matter and energy exchange with the environment. However, this was merely a prerequisite to the consideration of the morphoprocess at an evolutionarily significant time interval. Next we have to examine the three aspects of the morphoprocess, successively expanding our understanding of the temporal framework of the existence of systems of this kind. These aspects are the instant, the partial and the general morphoprocess.

An organism in a given time moment may be referred to as the *instant morphoprocess*. This notion emphasises the mode of existence of a dynamic form: a living organism exists only in the present moment; an epiphenomenon of its existence in a vanishing lapse of time is an "instant form". This notion—a snapshot, a captured momentary lapse of time—illustrates the uniqueness of the form reflecting the interaction of structural elements, energy flows and information flows (Granovitch et al. 2010).

Partial morphoprocess is the "first derivative" of the instant morphoprocess. It is the existence of an organism from its origin (the merging of gametes or cell division) to its death or the next division. The partial morphoprocess is the totality of instant morphoprocesses of a given organism. Its morphofunctional characteristics are not constant. They are dynamically stable at certain stages (propagation of the form) and change in a regular manner at other stages (Granovitch et al. 2010). So, in the partial morphoprocess we can identify changes associated with the dynamics of embryonic and postembryonic development. They are followed by changes associated with the formation of propagation elements (the structures ensuring reproduction). After that, "necrotic" changes, associated with the organism's senescence, set in. Two or several "stable" morphofunctional states connected by fast transitions can often be identified in the partial morphoprocess, indicating its metastability. A vivid example is species, in which the morphofunctional state of the larva is followed, after a fast transition by metamorphosis, by that of the imago (see for details Granovitch et al. 2010).

General morphoprocess is the "second derivative" of the instant morphoprocess. It is a rhythmic repetition of the partial morphoprocess at a sufficiently large time scale. Similar to the partial morphoprocess, the general morphoprocess is characterised by a dynamic stability of form, but, in contrast to the partial morphoprocess, it is also characterised by repeatability (reproducibility, cyclic nature) in a series of ancestor-progeny generations. A cycle of the general morphoprocess consists of one partial morphoprocess (Granovitch et al. 2010), and the features of the latter set the rhythm for the former.

In most cases the typology of morphoprocesses allows one to consider these systems within similar groups called species (to be more exact, species morphoprocesses). Intraspecific variability of the morphoprocesses is set by the plasticity of the morphoprocess and by the multiple genetic variants representing species morphoprocess (genetic heterogeneity). Despite its intraspecific variability, the group of species partial morphoprocesses is characterised by a stable manifestation of a complex of species characters. In this way, it reveals its modality, that is, the most steadily reproducing complex of characters of the partial morphoprocess of a species. Thus, general morphoprocesses of species, as series of constituent partial morphoprocesses, rhythmically reproduce the modal properties of a species.

In the light of the recent data, one more refinement of the notion of the morphoprocess is in order. I mean the data indicating the ubiquity of symbiotic associations (see for review McFall-Ngai 2008; Gilbert et al. 2012, 2015; Gilbert and Tauber 2016). An entire complex of organisms (prokaryotes, unicellular eukaryotes and, in some cases, even multicellular organisms), characterised by various symbiotic relationships with the host, are important for the implementation of the morphoprocess. These ideas are summarised in the concept of the holobiont (Zilber-Rosenberg and Rosenberg 2008; Bordenstein and Theis 2015; Rosenberg and Zilber-Rosenberg 2016; Sanchez-Canizares et al. 2017). The morphoprocess of a multicellular organism is that of a holobiont, the interaction of the organism proper with its symbiotic microbiota. Symbionts of a holobiont may determine the key features of the entire symbiotic system. Here are a few demonstrative examples: the feeding of anthozoans (Cnidaria: Anthozoa) is intimately linked with symbiotic dinophytes (*Symbiodinium* spp.) (van de Water et al. 2018; Sorek et al. 2018); termites and some cockroaches cannot feed on wood without unicellular symbionts

Trichomonadida, which, in turn, harbour bacterial microflora that breaks up cellulose (Tai et al. 2016; Waidele et al. 2016; Duarte et al. 2017); and parasitic filarial nematodes (Nematoda: Filariidae) cannot develop and reproduce without intracellular symbionts *Wolbachia* (Taylor et al. 2005). Data are accumulating on the dependence of normal morphogenetic processes on the composition and activity of the microflora (sponges, bryozoans, choanozoans) (Dayel et al. 2011); determination of the crucial characters of the holobiont, including sex determination, by microorganisms (Dunbar et al. 2007; Moran and Yun 2015; Tsuchida et al. 2010; Oliver et al. 2009; Cordaux et al. 2004); and the considerable influence of symbionts on the assortativity of mating and the formation of reproductive isolation (Sharon et al. 2010; Brucker and Bordenstein 2013).

It is fairly evident that the ideas about the morphoprocess of the holobiont are of paramount importance not only for understanding the functioning of the entire system within the framework of the partial morphoprocess but also for the refinement of the notion of the general morphoprocess in the context of its potential evolutionary transformations. The species properties of a holobiont are born in the crucible of interactions between its micro- and macroorganisms. A change of a component of the system may trigger the emergence of new species properties (for a detailed discussion of the evolutionary potential of interactions in a symbiotic system, see Gilbert et al. 2015).

This conceptual scheme allows a more meaningful discussion of evolutionary phenomena as it points to the category of living systems most promising for the evolutionary analysis. *The object of evolution is a concrete general morphoprocess of a holobiont*, that is, the ancestor-progeny line considered as an object-process. From this viewpoint, *an evolutionary phenomenon is a stable modification of the general morphoprocess in the process of its cyclic implementation.*

13.4 A Logical Field of Evolutionary Concepts

Having identified the symbiotic system of the organismic rank, the holobiont, as the object of evolution with which the key evolutionary events are associated, I would like to repeat that the changes of this system are certainly reflected both in the composition and the interactions of its components as well as in the populations and ecosystems of which it is part. In other words, we can trace evolutionary changes both at the suborganismic and at the supraorganismic level. However, the primary, the key, driving mechanisms of evolution should be looked for in the interaction of the organism/holobiont with its environment.

The *logical field of evolutionary concepts* featured in the title of this section and that of the entire work invites the reader to imagine all the possible variants/means of transformation of the object of evolution in all potentially possible ways. Moreover, we will also bear in mind the broadest possible range of potential ways of transformation of this evolutionarily significant object.

13.4.1 Three Principal Logical Schemes of Transformation of the General Morphoprocess

There are three principal logical schemes of transformation of the general morphoprocess. Historically, they correspond to the three conceptual approaches to the analysis of evolutionary mechanisms. All the three were considered by Darwin (see Introduction, Darwin 1872). From the point of view of the general morphoprocess, they may be defined as follows.

A. Constructional transformism (CT) (Granovitch 2018), syn. orthogenesis (term—Haacke 1893, cited in Popov 2018; Kellogg 1907; Gould 2002; Popov 2018), from the Greek ὀρθός (straight), syn. formalism/structuralism (Gould 2002)

The key point of CT is that the transformation of the general morphoprocess (evolutionary changes) is determined by its internal structure and laws of functioning. The environment merely triggers the manifestation of new structural-functional properties of the general morphoprocess. Evolution is predetermined by the laws of functioning of the general morphoprocess. This view, which highlights the structural-functional limitations of the morphoprocess and turns to general laws of physics and chemistry, is characteristic of the orthogenetic evolutionary concepts (Hyatt 1897; Cope 1887; Bateson 1894; Berg 1977; Schindewolf 1993; Popov 2018).

This view finds a solid support in numerous studies of the last 20 years devoted to self-organisation at various levels. The distinction between self-assembly (the minimisation of energy in a system approaching the state of thermodynamic equilibrium) and self-organisation (the formation of elaborate complexes in a system accompanied by energy expenditure) is based on the energy dependence (Werner et al. 2017; Halley and Winkler 2008; Gerhart and Kirschner 1997). In the context of this study, we are interested in the universal nature of the principle of autonomous origin of ordered structures and so may safely ignore the issue of energy provision of their elements.

Evidence of self-assembly and self-organisation of molecular complexes and organoids of the cells is overwhelming. Not only the genetic apparatus but all the other major structures of the cell are involved into these two processes. Self-assembly is characteristic of membranes, systems ensuring motility of cellular elements, organoids, cell division apparatus, the formation of vacuoles and, last but not least, the spatial organisation of chromosome material (Attardi and Schatz 1988; Palade 1983; Mitchison 1992; Warren and Wickner 1996; Inagaki and Katsuno 2017; Carazo-Salas and Nurse 2006; Cortes et al. 2006; Denton et al. 2003; Nedelec et al. 2003; Papanikou and Glick 2009; Svetina 2009; Sazer and Schiessel 2018; Misteli 2001). The entire morphofunctional organisation of the cell is a tight knot of coordinated self-organising processes (Misteli 2007; Kholodenko 2006; Johnson and Lam 2010; Karsenti 2008; Kurakin 2005). The cell itself may be considered as a morphoprocess of a unicellular organism implemented in a regular manner under given physical conditions.

The morphoprocess of multicellular animals is also a chain of mechanisms of self-organisation. Some examples of this are self-organisation of cellular aggregates into regular multicellular structures, including canalised development of embryos at the early stages of embryogenesis (Simunovic and Brivanlou 2017; Bizzarri et al. 2018; van den Brink et al. 2014; Turner et al. 2016; Baillie-Johnson et al. 2015; Bedzhov et al. 2015; Ozone et al. 2016), and autonomous patterning and self-organisation of different types of cells during the formation of tissues and complex systems of organs of a multicellular organism (Walton et al. 2016; Sheth et al. 2012; Onimaru et al. 2016; Raspopovic et al. 2014). Further evidence is the formation of specific architectonics of organs by means of interaction of differentiating cells, including the formation of complex organ-like aggregates from pluripotent iPS cells (Koehler et al. 2017; Dye et al. 2015; McCauley and Wells 2017; Sakaguchi et al. 2015; Taguchi et al. 2014; Takasato et al. 2015; Guye et al. 2016). The patterns of formation of repeated structures are autonomous and based on the physical-chemical reaction-diffusion model (Turing model, Turing 1952) (Cooke and Zeeman 1976; Sheth et al. 2012; Tsiairis and Aulehla 2016; Liao and Oates 2017). The work of the systems switching the development to new stable trajectories based on dynamic modules (Newman 2010) also represents a range of variants of self-organisation of the morphoprocess.

Finally, the formation of the symbiotic system/holobiont is also based on multiple feedbacks: mutual reactions of symbionts and the host organism. The result is a stable morphoprocess of the holobiont.

However, there is a flaw in this orthogenetic view on the morphoprocess as the result of self-assembly. It is the lack of a logical explanation of the formation of the adaptive nature of the morphoprocesses, their conformity to the environment. Until recently this flaw has been the Achilles heel of orthogenesis. The logic of the concept of the law of multilevel self-organisation (Granovitch 2018) might help eliminate this contradiction and explain the formation of adaptive characters even within the framework of a purely orthogenetic approach.

B. The concept of direct adaptogenesis (DA) (term, Granovitch 2018) of the organism/the morphoprocess of the holobiont under the impact of environmental factors has a venerable history. It is based on two assumptions: (1) conformity of the changes of the morphofunctional features of the morphoprocess to the impact of the environment and (2) direct fixation of such changes of the partial morphoprocess in the general morphoprocess. The first assumption can hardly be challenged: the versatility of phenotypic adaptations of organisms is one of the fundamental properties of living matter. The core of the problem lies in the assessment of the truth of the second assumption. Inheritance of acquired characters has been woven into the evolutionary thought of the last two centuries in the form of numerous "Lamarckian" concepts (see for review Gould 2002; Vorontsov 2004; Nazarov 2005; Bowler 2017; Esposito 2017; Loison and Herring 2017). And though Lamarckism has often been proclaimed defeated and vanquished (Weismann 1883; Luria and Delbrück 1943; Stahl 1988), the descriptions of new potential mechanisms of direct fixation of acquired characters in the general morphoprocess (see below for a brief outline of their range) indicate that this conceptual approach deserves consideration.

Within the framework of the "Lamarckian" approach, the origin of adaptive features is easy to explain. They (adaptive characters, conformity of the organism/morphoprocess to the environment) are a logical extension of the evolutionary mechanism suggesting a direct "fine-tuning" of the morphoprocess by the specific conditions of the environment. Hence, such mechanism can be designated as direct adaptogenesis (DA, Granovitch 2018). It is direct because the acquired morphofunctional features of the partial morphoprocess become the characteristics of the general morphoprocess directly (and not by selection of suitable variants from a group). The environment, in turn, is not merely a selective agent but an active factor forming the adaptive characters of the phenotype (Gilbert et al. 2015).

The question about the mechanisms of DA is associated with the search for the material carriers of transgenerational transfer of information signals of the environment. In other words, the question is how these information signals are transferred from the partial morphoprocess to the general one. Based on the current knowledge, an entire range of these mechanisms may be outlined.

First of all, these are mechanisms that may considerably influence the functioning of the inheritance apparatus and the cell and, at the same time, potentially depend on the impact of the environment. These include epigenetic mechanisms of information transfer such as the composition of hormones, regulatory systems associated with the operation of microRNA, modifications of histones and methylation of the DNA itself (Gilbert et al. 2015; Jablonka 2017; Jablonka and Lamb 2010, 2014; Tikhodeyev 2018; Goldberg et al. 2007; Danchin et al. 2011; Rissman and Adli 2014) and paramutation phenomena associated with the epigenetic regulation of the expression of one allele under the influence of another (Chandler 2010; Suter and Martin 2010; Hollick 2017). Information can also be transferred transgenerationally via self-sustaining metabolic loops and structural templates (Jablonka and Lamb 2010, 2014; Jablonka 2013). Not being associated with DNA structure and protein biosynthesis, these structural and functional features of the morphoprocess may be the target of environmental impacts and consistently translate their results to the next generations.

A special group of mechanisms includes the interactions between the organism and the environment with the help of adaptive physiological and behavioural reactions associated with the care of the offspring. The result is modification of the habitat and altered interactions of the progeny with the environment. These mechanisms, generally known as mechanisms of niche construction, are broadly discussed as potential factors of transformism (Baldwin 1896; Odling-Smee 2010; Pigliucci 2007; Laland et al. 2017).

The most important role in the modification of the general morphoprocess possibly belongs to the mechanisms of genetic accommodation: phenotypic adjustment (accommodation) and genetic assimilation (Waddington 1957; Schmalhausen 1982; West-Eberhard 2003; Jablonka and Lamb 2010; Pocheville and Danchin 2017). Owing to them, phenotypically adaptive modifications of the partial morphoprocess are consistently inherited by the general one.

A vivid, even though a highly specific, example of transgenerational transfer is CRISPR-CAS system of prokaryotes. Using this system, bacteria assimilate in their genome fragments of genetic material of phages interacting with them, so that the next generations of bacteria become resistant to infection (Horvath and Barrangou 2010; Deveau et al. 2010; Koonin 2011; Koonin and Wolf 2009).

Environmental impacts regulate mechanisms of genome reparation and induced mutagenesis and may be responsible not only for the changes in the frequency of genetic reconstructions but even for the reconstructions of certain target genome areas (Lynch 2010; Ponder et al. 2005; Galhardo et al. 2007). Much of the hereditary variation is associated with the transposition of various mobile genetic elements in the genome (McClintock 1984; Gierl 1990; Fedoroff et al. 1995; Asis et al. 2008, Koonin and Wolf 2009; Koonin 2010). In many cases, the migration of mobile genetic elements occurs under the impact of certain environmental factors and may result in quite definite hereditary changes (Rio 1990; Capy et al. 2000; Chuong et al. 2017; Jangam et al. 2017).

In general, the data on the structural limitations of the genome, its dynamic nature and the dependence of both on the impact of external factors (Capy et al. 2000; Shapiro 2009, 2016; Koonin and Wolf 2010; Laos et al. 2014) show quite clearly that the signals carrying information about the changes of environmental conditions may be fixed with various degrees of stability in the systems of transgenerational information transfer. Within the framework of this study, this can be reworded as the transfer of information from the partial morphoprocess to the general one. From the viewpoint of the cellular toolkit of nucleic acids, these views are summarised in the concept postulating a "subdivision" of the genome into two parts (Golubovskii 2000): the obligatory part comprising the functioning genes of the chromosomes in the nucleus and the organelles and the facultative part comprising the fractions of repetitive DNA, mobile elements, pseudogenes and ORFans, cyclic DNA and RNA, plasmids, B-chromosomes and DNA and RNA cytobionts. A direct impact of the environment on the obligatory genome part is difficult (though not impossible). At the same time, the facultative part is fairly dynamic, and the composition of its components may be regulated by the environmental impact. Importantly, the obligatory and the facultative parts interact actively. This suggests that the predominant way of information transfer from the environment is through the facultative components to the obligatory part (Golubovskii 2000), with the corresponding reliable transmission of the signal from the partial morphoprocess to the general one.

The above considerations indicate that DA mechanisms may have ample opportunities of operation. They are associated with various components of the nucleic acid machinery and other structural and functional templates as well as with the general mechanisms of multilevel feedbacks, regulating the stability of the morphoprocess at various levels. Let us note that many of the mechanisms listed above are considered within the eco-evo-devo concept and form a basis of the extended evolutionary synthesis (Pigliucci 2007; Jablonka Lamb 2010; Odling-Smee 2010; Laland et al. 2017; Constant et al. 2018; Clark et al. 2020 and others). And though its champions never tire to emphasise that the discussion of new potential ways of evolutionary transformation does not disrupt the logic of MS, a dispassionate observer would probably arrive at the conclusion that these new data break out of its conceptual framework.

C. The concept of natural selection (tychogenesis-selectogenesis, TS), which has been discussed in detail before (see Sect. 13.2), differs from the others in one feature: its main mechanism (TS) does not imply a direct evolutionary modification of the general morphoprocess. On the contrary, it suggests a rather exotic mechanism—the transfer of information about the environmental factors (adaptogenesis) through a nonspecific signal of the degree of an individual's reproduction/fitness. Hence, it would be logical to call this concept *indirect adaptogenesis (IA)* (term, Granovitch 2018), that is, the formation of evolutionary adaptations not through modification of the general morphoprocess but indirectly, through a selection of partial morphoprocesses.

13.4.2 Combinatory-Based Pathways of Transformism

Our analysis of potential pathways of transformism would have been incomplete, had we confined ourselves to the analysis of the above concepts of transformism. The problem is that all the three focus on the description of the transformation of a genetically homogeneous system, which does not include genetically alien elements. It is becoming clear, however, that not only micro- but also macroevolutionary transitions may be based on the association of genetically diverse components in one holobiont. Let us denote these transformation ways as combinatory-based (CB) modifications of the general morphoprocess. They are associated with the incorporation by the holobiont of genetically heterogeneous components such as separate cells of multicellular organisms, pro- and eukaryotic cells and genetic material of other species. We may identify several aspects of this process, which may apparently result in more or less dramatic evolutionary transformations.

A. Transformations of the general morphoprocess of the holobiont may occur by mutualistic inclusion of new components—genetically alien cells or organoids. This intracellular symbiosis is associated with the loss of individuality by the symbiont and its incorporation into a highly integrated system of the host cell. The result may be the formation of a new holobiont with emergent properties. This pathway is thought to underlie such global evolutionary transitions as the origin of the eukaryotic cell, the photosynthetic apparatus (chloroplasts) and the semi-autonomous system of oxidative phosphorylation (mitochondria) (Margulis and Fester 1991).

   Mutualistic relationships of microbionts with their hosts are also expressed in the cases when the symbionts become an integral part of the holobiont functioning without losing their organismic individuality. Numerous examples of intracellular bacterial symbionts, bacteria and eukaryotes as symbionts of tissues and cavities of multicellular organisms have been given above (see Sect. 13.3.2). Changes in the composition of these organisms may potentially lead to an evolutionary change of the holobiont.

B. Horizontal transfer of genetic material stands apart from other combinatory-based pathways of evolutionary transformations, being one of the mainstream pathways of evolutionary transformations of prokaryotes (Koonin 2011). Newly emerging data indicate the ubiquity of this phenomenon and its importance for the acquisition of key characters of multicellular plants and animals (Gladyshev et al. 2008; Gladyshev and Arkhipova 2010; Boschetti et al. 2011; Husnik et al. 2013; Haegeman et al. 2011).
C. Another variant of combinatory-based evolutionary mechanism is hybridogenesis. In contrast to the previous cases, it is based on the combination of relatively closely related organisms and is implemented through mechanisms of sexual reproduction. The formation of a new holobiont by hybridogenesis implies various pathways of integration of the parent genomes in the common genome of the offspring, as well as the formation of a new complex of symbiotic organisms within a holobiont.

CB mechanisms in general may cause macroevolutionary events, underlying the formation of an entire evolutionary direction (as in the case of formation of constructive types of cells from different evolutionary domains such as Archaeplastida, Ochrophyta, Cryptophyta, Euglenozoa, Dinophyta, etc.) or reticulate evolution of prokaryotes. The role of CB events may be equally important in the formation of adaptations to certain environmental conditions, differentiation of populations and ensuring the possibility of discrete and fast speciation.

13.5 Conclusion

The field of the main evolutionary concepts described in this work should be perceived as a field of possibilities, potential mechanisms of transformation of morphoprocesses. We identified among them a group of mechanisms of transformation of genetically homogeneous systems and a group of combinatory-based mechanisms, yet these groups are not mutually exclusive. The general morphoprocess altered by means of structural transformations of constructional transformism or combinatory based mechanisms may simultaneously undergo transformations of direct adaptogenesis, while in the group of these morphoprocesses selectogenetic processes will undoubtedly be at work.

We should only bear in mind that the mechanism of natural selection (to be more exact, selectogenesis; see Sect. 13.2.1) appears to play a subordinate role in evolution. This role consists in decreasing the morphogenetic potential of a species. Viewed from this angle, selectogenesis should be considered as a conservative force. An intensive selection should prevent the origin of forms and progressive evolution. On the contrary, if the differential mortality is minimal, the genetic and phenotypic diversity of individuals/morphoprocesses should be maximal, promoting fast evolutionary transformations.

Ideas developed in this study are, of course, synthetic. It is an attempt to consider all the mechanisms of evolutionary transformation of biological systems that are potentially possible in the light of modern data. It is interesting that such a generalised approach easily accommodates almost all previously proposed conceptual models, even those that seemed incompatible before. We have intentionally tried to avoid "personification" of evolutionary ideas under labels such as Darwinism, Lamarckism, etc., as more often than not such personification distorts the primary source (see, e.g. reflections on Darwinism and the Modern Synthesis, in Delisle 2017; 2021, and on "orthogenetic" components in B. Rensch and G. Simpson, the "co-architects" of the Modern Synthesis (Levit and Hossfeld 2021; Ochoa 2021). A certain tradition of "adjusting" selectogenesis has formed in the last decades: new data are incorporated and the adjusted concept is branded as a "synthesis"—modern, new evolutionary or extended. This "synthetic" tradition invariably implies that selectogenesis is primary. In the light of the views on the diversity of evolutionary mechanisms propounded in this study, the name "synthesis" should be renounced altogether. What we need is a total *reset* of evolutionary thought or, shall we say, the development of the platform Evolution 2.0.

Acknowledgments I am extremely grateful to Natalia Lentsman for the neat and thoughtful translation of the manuscript. This study was supported by a grant from the Russian Science Foundation (RSF) no. 19-14-00321.

References

Agassiz L (1874) Evolution and permanence of type. Atl Mon 33:92–101

Allis CD, Jenuwein T, Reinberg D (eds) (2007) Epigenetics. Cold Spring Harbour Laboratory Press, New York

Artigiani R (1987) Revolution and evolution: applying Prigogine's dissipative structures model. J Soc Biol Struct 10(3):249–264

Asis RA, Kondrashov EV, Koonin EV, Kondrashov FA (2008) Nested genes and increasing organisational complexity of metazoan genomes. Trends Genet 24:475–478

Attardi G, Schatz G (1988) Biogenesis of mitochondria. Annu Rev Cell Biol 4(1):289–331

Baillie-Johnson P, Van den Brink SC, Balayo T, Turner DA, Arias AM (2015) Generation of aggregates of mouse embryonic stem cells that show symmetry breaking, polarization and emergent collective behavior in vitro. J Vis Exp 105:e53252

Baldwin JM (1896) A new factor in evolution. Am Nat 30(354):441–451

Barros DB (2008) Natural selection as a mechanism. Philos Sci 75:306–322

Bateson W (1894) Materials for the study of variation. Macmillan, London

Beaber JW, Hochhut B, Waldor MK (2004) SOS response promotes horizontal dissemination of antibiotic resistance genes. Nature 427(6969):72–74

Bedzhov I, Bialecka M, Zielinska A et al (2015) Development of the anterior-posterior axis is a self-organizing process in the absence of maternal cues in the mouse embryo. Cell Res 25 (12):1368–1371

Beklemishev VN (1970) On general principles of organization of life. In: Beklemishev VN (ed) Biocenological basics of comparative parasitology. Nauka, Moscow, pp 7–25. (in Russian)

Beklemishev VN (1994) Methodology of systematics. KMK Scientific Press, Moscow. (in Russian)

Berg LS (1977) Works on the theory of evolution. Nauka, Leningrad. (in Russian)

Beverley SM, Coderre JA, Santi DV, Schimke RT (1984) Unstable DNA amplifications in methotrexate-resistant Leishmania consist of extrachromosomal circles which relocalize during stabilization. Cell 38:431–439
Bickford D, Lohman DL, Sodhi NS et al (2006) Cryptic species as a window on diversity and conservation. Trends Ecol Evol 22(3):148–155
Bizzarri M, Masiello MG, Giuliani A, Cucina A (2018) Gravity constraints drive biological systems toward specific organization patterns: commitment of cell specification is constrained by physical cues. BioEssays 40(1):1700138
Bordenstein SR, Theis KR (2015) Host biology in light of the microbiome: ten principles of holobionts and hologenomes. PLoS Biol 13(8):e1002226
Boschetti C, Pouchkina-Stantcheva N, Hoffmann P, Tunnacliffe A (2011) Foreign genes and novel hydrophilic protein genes participate in the desiccation response of the bdelloid rotifer Adineta ricciae. J Exp Biol 214(1):59–68
Bowler PJ (2017) Alternatives to Darwinism in the early twentieth century. In: Delisle RG (ed) The Darwinian tradition in context. Springer, Cham, pp 195–217
Brakefield PM (2010) Radiations of mycalesine butterflies and opening up their exploration of morphospace. Am Nat 176(S1):S77–S87
Brucker RM, Bordenstein SR (2013) The hologenomic basis of speciation: gut bacteria cause hybrid lethality in the genus Nasonia. Science 341(6146):667–669
Capy P, Gasperi G, Biemont C, Bazin C (2000) Stress and transposable elements: co-evolution or useful parasites? Heredity 85(2):101–106
Carazo-Salas RE, Nurse P (2006) Self-organization of interphase microtubule arrays in fission yeast. Nat Cell Biol 8(10):1102–1107
Chandler VL (2010) Paramutation's properties and puzzles. Science 330(6004):628–629
Chuong EB, Elde NC, Feschotte C (2017) Regulatory activities of transposable elements: from conflicts to benefits. Nat Rev Genet 18(2):71–86
Clark AD, Deffner D, Laland K, Odling-Smee J, Endler J (2020) Niche construction affects the variability and strength of natural selection. Am Nat 195(1):16–30
Constant A, Ramstead MJ, Veissiere SP, Campbell JO, Friston KJ (2018) A variational approach to niche construction. J R Soc Interface 15(141):20170685
Cooke J, Zeeman EC (1976) A clock and wavefront model for control of the number of repeated structures during animal morphogenesis. J Theor Biol 58(2):455–476
Cope ED (1887) The origin of the fittest: essays on evolution. D. Appleton, New York
Cordaux R, Michel-Salzat A, Frelon-Raimond M, Rigaud T, Bouchon D (2004) Evidence for a new feminizing Wolbachia strain in the isopod Armadillidium vulgare: evolutionary implications. Heredity 93(1):78–84
Cortes S, Glade N, Chartier I, Tabony J (2006) Microtubule self-organisation by reaction-diffusion processes in miniature cell-sized containers and phospholipid vesicles. Biophys Chem 120 (3):168–177
Cramer F, Prigogine I (1993) Chaos and order: the complex structure of living systems. Weinheim, VCH
Cuvier GJLNFD (1817) Le Règne animal distribué d'aprés son organisation pour servir de base a l'histoire naturelle des animaux et d'introduction a l'anatomie comparé, tome 1. Imprimérie de A, Berlin
Danchin E, Charmantier A, Champagne FA et al (2011) Beyond DNA: integrating inclusive inheritance into an extended theory of evolution. Nat Rev Genet 12:475–486
Darwin CR (1872) The origin of species by means of natural selection, or the preservation of favoured races in the struggle for life. John Murray, London. 6th edition; with additions and corrections. Eleventh thousand
Dawkins R (1976) The selfish gene. Oxford University Press, New York
Dawkins R (1982) The extended phenotype. Oxford University Press, New York
Dayel MJ, Alegado RA, Fairclough SR et al (2011) Cell differentiation and morphogenesis in the colony-forming choanoflagellate Salpingoeca rosetta. Dev Biol 357(1):73–82

de Paoli H, van der Heide T, van den Berg A, Silliman BR, Herman PMJ, van de Koppel J (2017) Behavioral self-organization underlies the resilience of a coastal ecosystem. Proc Nat Acad Sci U S A 114:8035–8040

Delisle RG (2017) From Charles Darwin to the evolutionary synthesis: weak and diffused connections only. In: Delisle RG (ed) The Darwinian tradition in context. Springer, Cham, pp 133–167

Delisle RG (2021) Natural selection as a mere auxiliary hypothesis in Charles Darwin's origin of species. In: Delisle RG (ed) Natural selection - revisiting its explanatory role in evolutionary biology. Springer, Cham, pp 73–104

Denton MJ, Dearden PK, Sowerby SJ (2003) Physical law not natural selection as the major determinant of biological complexity in the subcellular realm: new support for the pre-Darwinian conception of evolution by natural law. Biosystems 71(3):297–303

Deveau H, Garneau JE, Moineau S (2010) CRISPR/Cas system and its role in phage-bacteria interactions. Annu Rev Microbiol 64:475–493

Duarte S, Nunes L, Borges PA, Fossdal CG, Nobre T (2017) Living inside termites: an overview of symbiotic interactions, with emphasis on flagellate protists. Arquipelago Life Mar Sci 34:21–43

Dunbar HE, Wilson AC, Ferguson NR, Moran NA (2007) Aphid thermal tolerance is governed by a point mutation in bacterial symbionts. PLoS Biol 5(5):e96

Dye BR, Hill DR, Ferguson MAH et al (2015) In vitro generation of human pluripotent stem cell derived lung organoids. elife 4:e05098

Echols H (1981) SOS functions, cancer and inducible evolution. Cell 25(1):1–2

Eimer GHT (1890) Organic evolution as the result of the inheritance of acquired characters according to the laws of organic growth. Macmillan, London

Ellison AM, Niklas KJ (1988) Branching patterns of Salicornia europea (Chenopodiaceae) at different succession stages: a comparison of theoretical and real plants. Am J Bot 75(4):501–512

Esposito M (2017) The organismal synthesis: holistic science and developmental evolution in the English-speaking world, 1915–1954. In: Delisle RG (ed) The Darwinian tradition in context. Springer, Cham, pp 219–241

Fedoroff N, Schlappi M, Raina R (1995) Epigenetic regulation of the maize Spm transposon. BioEssays 17(4):291–297

Galhardo RS, Hastings PJ, Rosenberg SM (2007) Mutation as a stress response and the regulation of evolvability. Crit Rev Biochem Mol Biol 42:399–435

Gavrilets S (2004) Fitness landscapes and the origin of species. Princeton University Press, Princeton

Gavrilets S, Gravner J (1997) Percolation on the fitness hypercube and the evolution of reproductive isolation. J Theor Biol 184:51–64

Geoffroy Saint-Hilaire E (1818) Philosophie anatomique: des organes respiratoires sous le rapport de la détermination et de l'identité de leurs pièces osseuses. J.-B. Bailliere, Paris

Gerhart J, Kirschner M (1997) Cells, embryos, and evolution: toward a cellular and developmental understanding of phenotypic variation and evolutionary adaptability. Blackwell science, Malden

Gierl A (1990) How maize transposable elements escape negative selection. Trends Genet 6:155–158

Gilbert SF, Tauber AI (2016) Rethinking individuality: the dialectics of the holobiont. Biol Philos 31(6):839–853

Gilbert SF, Sapp J, Tauber AI (2012) A symbiotic view of life: we have never been individuals. Q Rev Biol 87(4):325–341

Gilbert SF, Bosch TCG, Ledon-Rettig C (2015) Eco-Evo-Devo: developmental symbiosis and developmental plasticity as evolutionary agents. Nat Rev Genet 16:611–622

Gladyshev EA, Arkhipova IR (2010) Genome structure of bdelloid rotifers: shaped by asexuality or desiccation? J Hered 101(S1):S85–S93

Gladyshev EA, Meselson M, Arkhipova IR (2008) Massive horizontal gene transfer in bdelloid rotifers. Science 320:1210–1213

Goldberg AD, Allis CD, Bernstein E (2007) Epigenetics: a landscape takes shape. Cell 128:635–638

Golubovskii MD (2000) The age of genetics: evolution of ideas and concepts. Borei Art, Saint Petersburg. (in Russian)
Gould SJ (2002) The structure of evolutionary theory. The Belknap Press of Harvard University Press, Cambridge
Gould SJ, Eldredge N (1977) Punctuated equilibria: the tempo and mode of evolution reconsidered. Paleobiology 3(2):115–151
Granovitch AI (2018) Is natural selection still have to be regarded a foundation stone of evolutionary process? J Evol Sci 1:14–30
Granovitch AI, Ostrovsky AN, Dobrovolsky AA (2010) Morphoprocess and life cycles of organisms. Zh Obshch Biol 71(6):514–522. (in Russian)
Grant PR, Grant BR (2002) Unpredictable evolution in a 30-year study of Darwin's finches. Science 296:707–711
Grebelnyi SD (2008) Cloning in nature. A part playing by gene recombination stoppage in fauna and flora formation. Gelicon, Saint Petersburg. (in Russian)
Guye P, Ebrahimkhani MR, Kipniss N et al (2016) Genetically engineering self-organization of human pluripotent stem cells into a liver bud-like tissue using Gata6. Nat Commun 7:10243
Haacke W (1893) Gestaltung und verebung: eine entwickelungsmechanik der organismen. Weigel Nachfolger, Leipzig
Haegeman A, Jones JT, Danchin EGJ (2011) Horizontal gene transfer in nematodes: a catalyst for plant parasitism? Mol Plant-Microbe Interact 24:879–887
Hall B (1990) Spontaneous point mutations that occur more often when advantageous than when neutral. Genetics 126:5–16
Hall B (1991) Adaptive evolution that requires multiple spontaneous mutations: mutations involving base substitutions. Proc Natl Acad Sci USA 88:5882–5886
Halley JD, Winkler DA (2008) Consistent concepts of self-organization and self-assembly. Complexity 14(2):10–17
Hastings PJ, Rosenberg SM, Slack A (2004) Antibiotic-induced lateral transfer of antibiotic resistance. Trends Microbiol 12(9):401–404
Hausmann K, Hülsmann N, Radek R (2003) Protistology. E. Schweizerbart'sche Verlagsbuchhandlung (Nägele u. Obermiller). Stuttgart, Berlin
Havstad JC (2011) Problems for natural selection as a mechanism. Philos Sci 78(3):512–523
Hollick JB (2010) Paramutation and development. Annu Rev Cell Dev Biol 26:557–579
Hollick JB (2017) Paramutation and related phenomena in diverse species. Nat Rev Genet 18 (1):5–23
Horenko I, Dittmer E, Lankas F, Maddocks J, Metzner P, Schütte C (2008) Macroscopic dynamics of complex metastable systems: theory, algorithms, and application to B-DNA. SIAM J Appl Dyn Syst 7(2):532–560
Horvath P, Barrangou R (2010) CRISPR/Cas, the immune system of Bacteria and Archaea. Science 327:167–170
Husnik F, Nikoh N, Koga R et al (2013) Horizontal gene transfer from diverse bacteria to an insect genome enables a tripartite nested mealybug symbiosis. Cell 153(7):1567–1578
Hyatt A (1897) Cycle in the life of the individual (ontogeny) and in the evolution of its own group (phylogeny). Science 5(109):161–171
Inagaki N, Katsuno H (2017) Actin waves: origin of cell polarization and migration? Trends Cell Biol 27:515–526
Jablonka E (2013) Epigenetic inheritance and plasticity: the responsive germline. Prog Biophys Mol Biol 111(2–3):99–107
Jablonka E (2017) The evolutionary implications of epigenetic inheritance. Interface Focus 7 (5):20160135
Jablonka E, Lamb MJ (2010) Transgenerational epigenetic inheritance. In: Pigliucci M, Muller GB (eds) Evolution. The extended synthesis. MIT Press, Cambridge, pp 137–174
Jablonka E, Lamb MJ (2014) Evolution in four dimensions, revised edition: genetic, epigenetic, behavioral, and symbolic variation in the history of life. MIT press, Cambridge

Jakab K, Neagu A, Mironov V, Markwald RR, Forgacs G (2004) Engineering biological structures of prescribed shape using self-assembling multicellular systems. Proc Natl Acad Sci U S A 101 (9):2864–2869
Jangam D, Feschotte C, Betran E (2017) Transposable element domestication as an adaptation to evolutionary conflicts. Trends Genet 33(11):817–831
Johnson BR, Lam SK (2010) Self-organization, natural selection, and evolution: cellular hardware and genetic software. BioOne 60(11):879–885
Karen GH, Hawley RS (2007) Epigenetic regulation of chromosome inheritance. In: Allis CD, Jenuwein T, Reinberg D (eds) Epigenetics. Cold Spring Harbour Laboratory Press, New York, pp 263–286
Karsenti E (2008) Self-organization in cell biology: a brief history. Nat Rev Mol Cell Biol 9 (3):255–262
Kellogg VL (1907) Darwinism today. G. Bell & Sons, London
Khesin RB (1984) The inconstancy of genome. Nauka, Moscow. (in Russian)
Kholodenko BN (2006) Cell-signalling dynamics in time and space. Nat Rev Mol Cell Biol 7:165–176
Kim VN, Han J, Siomi MC (2009) Biogenesis of small RNAs in animals. Nat Rev Mol Cell Biol 10:126–139
Koehler KR, Nie J, Longworth-Mills E et al (2017) Generation of inner ear organoids containing functional hair cells from human pluripotent stem cells. Nat Biotechnol 35:583–589
Koonin EV (2010) Taming of the shrewd: novel eukaryotic genes from RNA viruses. BMC Biol 8 (1):2. https://doi.org/10.1186/1741-7007-8-2
Koonin EV (2011) Logic of chance, the nature and origin of biological evolution. FT Press Science, Upper Saddle River, NJ
Koonin EV, Wolf YI (2009) Is evolution Darwinian or/and Lamarckian? Biol Direct 4(1):42. https://doi.org/10.1186/1745-6150-4-42
Koonin EV, Wolf YI (2010) Constraints and plasticity in genome and molecular-phenome evolution. Nat Rev Genet 11:487–498
Kuhn TS (1962) The structure of scientific revolutions. University of Chicago Press, Chicago
Kull K (2014) Adaptive evolution without natural selection. Biol J Linn Soc 112:287–294
Kurakin A (2005) Self-organization versus Watchmaker: stochastic dynamics of cellular organization. Biol Chem 386:247–254
Kurakin A (2007) Self-organization versus Watchmaker: ambiguity of molecular recognition and design charts of cellular circuitry. J Mol Recognit 20:205–214
Laland K, Uller T, Feldman M et al (2014) Does evolutionary theory need a rethink? Nature 514 (7521):161–164
Laland K, Matthews B, Feldman MW (2016) An introduction to niche construction theory. Evol Ecol 30(2):191–202
Laland K, Odling-Smee J, Endler J (2017) Niche construction, sources of selection and trait coevolution. Interface Focus 7(5):20160147
Lamarck JB (1809) Philosophy of zoology. In: Lamarck JB (1955) Selected works.1. Academy of Science Print, Moscow
Landman OE (1991) The inheritance of acquired characteristics. Annu Rev Genet 25:1–20
Laos R, Thomson JM, Benner SA (2014) DNA polymerases engineered by directed evolution to incorporate non-standard nucleotides. Front Microbiol 5:565. https://doi.org/10.3389/fmicb.2014.00565
Levit GS, Hossfeld U (2021) When Panpsychism met Monism: why a philosopher Theodor Ziehen (1862–1950) became a crucial figure for the evolutionary biologist Bernhard Rensch? In: Delisle RG (ed) Natural selection - revisiting its explanatory role in evolutionary biology. Springer, Cham, pp 231–260
Levit GS, Scholz J (2002) The biosphere as a morphoprocess and a new look at the concepts of organism and individuality. Senckenb Lethaea 82:367–372

Lewontin RC (1974) The genetic basis of evolutionary change. Columbia University Press, New York

Liao B-K, Oates AC (2017) Delta-Notch signalling in segmentation. Arthropod Struct Dev 46:429–447

Loison L, Herring E (2017) Lamarckian research programs in French biology (1900–1970). In: Delisle RG (ed) The Darwinian tradition in context. Springer, Cham, pp 243–269

Loose M, Fischer-Friedrich M, Ries J, Kruse K, Schwille P (2008) Spatial regulators for bacterial cell division self-organize into surface waves in vitro. Science 320:789–792

Luria SE, Delbrück M (1943) Mutations of bacteria from virus sensitivity to virus resistance. Genetics 28(6):491–511

Lynch M (2010) Evolution of the mutation rate. Trends Genet 26(8):345–352

Lyubishchev AA (1982) Problems of form, systematics and evolution of organisms. Nauka, Moscow. (in Russian)

Margulis L, Fester R (eds) (1991) Symbiosis as a source of evolutionary innovation: speciation and morphogenesis. MIT Press, Cambridge

Markov A (2010) The birth of complexity. Astrel Publisher; Korpus Publisher, Moscow. (in Russian)

McCauley HA, Wells JM (2017) Pluripotent stem cell-derived organoids: using principles of developmental biology to grow human tissues in a dish. Development 144(6):958–962

McClintock B (1984) The significance of responses of the genome to challenge. Science 26 (4676):792–801

McFall-Ngai M (2008) Are biologists in "future shock"? Symbiosis integrates biology across domains. Nat Rev Microbiol 6(10):789–792

McGhee GR (1980) Shell form in the biconvex articulate Brachiopoda: a geometric analysis. Paleobiology 6(1):57–76

McGhee GR (2007) The geometry of evolution. adaptive landscapes and theoretical morphospaces. Cambridge University Press, Cambridge

McShea DW (2001) The hierarchical structure of organisms: a scale and documentation of a trend in the maximum. Paleobiology 27(2):405–423

Meyen SV (2007) The morphology of Plants in the nomothetic aspect. In: Ignatiev IA (ed) S.V. Meyen: paleobotanist, evolutionist, thinker. Geos, Moscow, pp 162–222. (in Russian)

Millstein RL (2013) Natural selection and causal productivity. In: Chao H-K, Chen S-T, Millstein RL (eds) Mechanism and causality in biology and economics, history, philosophy and theory of life science. Springer Science + Business Media, Dordrecht, pp 147–163

Misteli T (2001) The concept of self-organization in cellular architecture. J Cell Biol 155 (2):181–185

Misteli T (2007) Beyond the sequence: cellular organization of genome function. Cell 128:787–800

Mitchison TJ (1992) Self-organization of polymer-motor systems in the cytoskeleton. Philos Trans R Soc Lond B 336(1276):99–106

Moran NA, Yun Y (2015) Experimental replacement of an obligate insect symbiont. Proc Natl Acad Sci U S A 112(7):2093–2096

Nägeli C (1884) Mechanisch-physiologische theorie der abstammungslehre. R. Oldenbourg, München, Leipzig

Nazarov VI (2005) Evolution is not according to Darwin: a change in the evolutionary model. KomKniga, Moscow. (in Russian)

Nedelec F, Surrey T, Karsenti E (2003) Self-organisation and forces in the microtubule cytoskeleton. Curr Opin Cell Biol 15(1):118–124

Newman SA (2010) Dynamical patterning modules. In: Pigliucci M, Muller GB (eds) Evolution. The extended synthesis. MIT Press, Cambridge, pp 281–306

Ochoa C (2021) G. G. Simpson, an Orthogeneticist?: Simpson's parallelism? In: Delisle RG (ed) Natural selection - revisiting its explanatory role in evolutionary biology. Springer, Cham, pp 261–290

Odling-Smee JF (2010) Niche inheritance. In: Pigliucci M, Muller GB (eds) Evolution – the extended synthesis. MIT press, Cambridge, pp 175–207

Oliver KM, Degnan PH, Hunter MS, Moran NA (2009) Bacteriophages encode factors required for protection in a symbiotic mutualism. Science 325(5943):992–994

Onimaru K, Marcon L, Musy M, Tanaka M, Sharpe J (2016) The fin-to-limb transition as the re-organization of a Turing pattern. Nat Commun 7:11582

Osborn HF (1929) The titanotheres of ancient Wyoming, Dakota, and Nebraska. Dept. of the Interior, U.S. Geological Survey, Washington

Ozone C, Suga H, Eiraku M et al (2016) Functional anterior pituitary generated in self-organizing culture of human embryonic stem cells. Nat Commun 7:10351

Paez-Espino D, Morovic W, Sun CL et al (2013) Strong bias in the bacterial CRISPR elements that confer immunity to phage. Nat Commun 4:1430

Palade GE (1983) Membrane biogenesis: an overview. Methods Enzymol 96:XXIX–LV

Papanikou E, Glick BS (2009) The yeast Golgi apparatus: insights and mysteries. FEBS Lett 583 (23):3746–3751

Pigliucci M (2007) Do we need an extended evolutionary synthesis? Evolution 61(12):2743–2749

Pigliucci M (2010) Phenotypic plasticity. In: Pigliucci M, Muller GB (eds) Evolution. The extended synthesis. MIT Press, Cambridge, pp 281–306

Pocheville A, Danchin E (2017) Genetic assimilation and the paradox of blind variation. In: Walsh DM, Huneman P (eds) Challenging the modern synthesis. Oxford University Press, Oxford, pp 111–136

Ponder RG, Fonville NC, Rosenberg SM (2005) A switch from high-fidelity to error-prone DNA double-strand break repair underlies stress-induced mutation. Mol Cell 19(6):791–804

Ponomarenko AG (2005) Paleontological data on the origin of arthropods. In: Vorobyova EI, Striganova BR (eds) Evolutionary factors of animal diversity formation. KMK Scientific Press, Moscow, pp 146–155. (in Russian)

Ponomarenko AG (2008) Early stages of the arthropods' evolution. In: Mamkaev YV (ed) Evolutional morphology of animals. SPbSU Publishing, Saint Petersburg, pp 43–57. (in Russian)

Popov I (2018) Orthogenesis versus Darwinism. Springer, Cham

Popper K (2009) Darwinism as a metaphysical research programme. In: Ruse M (ed) Philosophy after Darwin: classic and contemporary readings. Princeton University Press, Princeton, pp 167–174

Prigogine I (1978) Time, structure, and fluctuations. Science 201(4358):777–785

Prigogine I, Stengers I (2018) Order out of chaos: man's new dialogue with nature. Verso Books, London

Radman M (1974) Phenomenology of an inducible mutagenic DNA repair pathway in Escherichia coli: SOS repair hypothesis. In: Prakash L, Miller FSM, Lawrence C, Tabor HW (eds) Molecular and environmental aspects of mutagenesis. Charles C. Thomas Publisher, Springfield, pp 128–142

Raspopovic J, Marcon L, Russo L, Sharpe J (2014) Digit patterning is controlled by a Bmp-Sox9-Wnt Turing network modulated by morphogen gradients. Science 345(6196):566–570

Raup D (1966) Geometric analysis of shell coiling: general problems. J Paleontol 40:1178–1190

Raup D (1967) Geometric analysis of shell coiling: coiling in ammonoids. J Paleontol 41:43–65

Raup D, David M, Stanley SM (1978) Principles of paleontology. Macmillan, London

Rio DC (1990) Molecular mechanisms regulating Drosophila P-element transposition. Annu Rev Genet 24:5430–5478

Rissman EF, Adli M (2014) Minireview: transgenerational epigenetic inheritance: focus on endocrine disrupting compounds. Endocrinology 155:2770–2780

Rosenberg E, Zilber-Rosenberg I (2016) Microbes drive evolution of animals and plants: the hologenome concept. mBio 7(2):e01395–e01315

Rozanov AY (1973) Morphological evolution of Archeocyathids and questions of the definition of Lower Cambrian stages. Nauka, Moskow. (in Russian)

Rozhnov SV (2005) Morphological patterns of formation and evolution of higher taxa echinoderms. In: Vorobieva EI, Striganova BR (eds) Evolutionary factors of wildlife diversity. KMK Scientific Press, Moscow, pp 156–170. (in Russian)
Sakaguchi H, Kadoshima T, Soen M et al (2015) Generation of functional hippocampal neurons from self-organizing human embryonic stem cell-derived dorsomedial telencephalic tissue. Nat Commun 6:8896
Sanchez-Canizares C, Jorrin B, Poole PS, Tkacz A (2017) Understanding the holobiont: the interdependence of plants and their microbiome. Curr Opin Microbiol 38:188–196
Sarton G (1953) A history of science: ancient science through the golden age of Greece. Harvard University Press, Cambridge
Sazer S, Schiessel H (2018) The biology and polymer physics underlying large-scale chromosome organization. Traffic 19(2):87–104
Schindewolf OH (1993) Basic questions in paleontology: geologic time, organic evolution, and biological systematics. University of Chicago Press, Chicago
Schmalhausen II (1982) The organism as a whole in its individual and historical development. Nauka, Moscow. (in Russian)
Serzhantov VF (1972) Introduction to the methodology of modern biology. Nauka, Leningrad. (in Russian)
Shapiro JA (2009) Revisiting the central dogma in the 21st century. Ann N Y Acad Sci 1178 (1):6–28
Shapiro JA (2016) The basic concept of the read-write genome: mini-review on cell-mediated DNA modification. Biosystems 140:35–37
Shapiro JA, Sternberg R (2005) Why repetitive DNA is essential to genome function. Biol Rev Camb Philos Soc 80:227–250
Sharon G, Segal D, Ringo JM et al (2010) Commensal bacteria play a role in mating preference of Drosophila melanogaster. Proc Natl Acad Sci U S A 107(46):20051–20056
Sharov AA, Igamberdiev AU (2014) Inferring directions of evolution from patterns of variation: the legacy of Sergei Meyen. Biosystems 123:67–73
Sheth R, Marcon L, Bastida MF et al (2012) Hox genes regulate digit patterning by controlling the wavelength of a Turing-type mechanism. Science 338(6113):1476–1480
Simunovic M, Brivanlou AH (2017) Embryoids, organoids and gastruloids: new approaches to understanding embryogenesis. Development 144(6):976–985
Skippington E, Ragan MA (2011) Lateral genetic transfer and the construction of genetic exchange communities. FEMS Microbiol Rev 35:707–735
Sorek M, Schnytzer Y, Ben-Asher HW et al (2018) Setting the pace: host rhythmic behaviour and gene expression patterns in the facultatively symbiotic cnidarian Aiptasia are determined largely by Symbiodinium. Microbiome 6(1):83
Spencer H (1893) The inadequacy of "natural selection". Contemp Theatr Rev 63:152–166
Stahl FW (1988) A unicorn in the garden. Nature 335(6186):112–113
Suter CM, Martin DIK (2010) Paramutation: the tip of an epigenetic iceberg? Trends Genet 26 (1):9–14
Svetina S (2009) Vesicle budding and the origin of cellular life. ChemPhysChem 10 (16):2769–2776
Taguchi A, Kaku Y, Ohmori T, Sharmin S, Ogawa M, Sasaki H, Nishinakamura R (2014) Redefining the in vivo origin of metanephric nephron progenitors enables generation of complex kidney structures from pluripotent stem cells. Cell Stem Cell 14(1):53–67
Tai V, Carpenter KJ, Weber PK et al (2016) Genome evolution and nitrogen fixation in bacterial ectosymbionts of a protist inhabiting wood-feeding cockroaches. Appl Environ Microbiol 82 (15):4682–4695
Takasato M, Er PX, Chiu HS et al (2015) Kidney organoids from human iPS cells contain multiple lineages and model human nephrogenesis. Nature 526(7574):564–568
Tatarinov LP (1976) Morphological evolution of Theriodonts and general questions of phylogenetics. Nauka, Moscow. (in Russian)

Taylor MJ, Bandi C, Hoerauf A (2005) Wolbachia. Bacterial endosymbionts of filarial nematodes. Adv Parasitol 60:245–284
Thomas RDK, Reif WE (1993) The skeleton space: a finite set of organic designs. Evolution 47:341–360
Thompson D (1961) On growth and form (abridged edition edited by Bonner JT). Cambridge University Press, Cambridge
Tikhodeyev ON (2018) The mechanisms of epigenetic inheritance: how diverse are they? Biol Rev 93:1987–2005
Tsiairis CD, Aulehla A (2016) Self-organization of embryonic genetic oscillators into spatiotemporal wave patterns. Cell 164(4):656–667
Tsuchida T, Koga R, Horikawa M et al (2010) Symbiotic bacterium modifies aphid body color. Science 330(6007):1102–1104
Turing AM (1952) The chemical theory of morphogenesis. Philos T Roy Soc B 237(641):37–72
Turner DA, Baillie-Johnson P, Martinez Arias A (2016) Organoids and the genetically encoded self-assembly of embryonic stem cells. BioEssays 38(2):181–191
Ubukata T (2005) Theoretical morphology of bivalve shell sculptures. Paleobiology 31:643–655
Van de Water JAJM, Allemand D, Ferrier-Pages C (2018) Host-microbe interactions in octocoral holobionts – recent advances and perspectives. Microbiome 6(1):64
Van den Brink SC, Baillie-Johnson P, Balayo T, Hadjantonakis A-K, Nowotschin S, Turner DA, Martinez Arias A (2014) Symmetry breaking, germ layer specification and axial organisation in aggregates of mouse embryonic stem cells. Development 141(22):4231–4242
Vavilov NI (1987) Theoretical basis of selection. Nauka, Moscow. (in Russian)
Vorontsov NN (2004) The development of evolution idea in biology. KMK Scientific Press, Moscow. (in Russian)
Waddington CH (1957) The strategy of the genes: a discussion of some aspects of theoretical biology. Allen & Unwin, London
Waidele L, Korb J, Kuenzel S, Dedeine F, Staubach F (2016) Bacterial but not protist gut microbiota align with ecological specialization in a set of lower termite species. bioRxiv. https://doi.org/10.1101/083683
Wallace AR (1889) Darwinism: an exposition of the theory of natural selection, with some of the applications. Macmillan and Company, London
Walton KD, Freddo AM, Wang S, Gumucio DL (2016) Generation of intestinal surface: an absorbing tale. Development 143(13):2261–2272
Warren G, Wickner W (1996) Organelle inheritance. Cell 84(3):395–400
Weismann A (1883) Ueber die vererbung. Fischer, Jena
Werner S, Vu HT-K, Rink JC (2017) Self-organization in development, regeneration and organoids. Curr Opin Cell Biol 44:102–109
West-Eberhard MJ (2003) Developmental plasticity and evolution. Oxford University Press, Oxford
West-Eberhard MJ (2005) Developmental plasticity and the origin of species differences. Proc Natl Acad Sci U S A 102(1):6543–6549
Williams GC (1966) Adaptation and natural selection. Oxford University Press, Oxford
Wright S (1931) Evolution in Mendelian populations. Genetics 16:97–159
Wright S (1932) The roles of mutation, inbreeding, crossbreeding, and selection in evolution. In: Proceedings of the 6th international congress of genetics, vol 1, pp 356–366
Zilber-Rosenberg I, Rosenberg E (2008) Role of microorganisms in the evolution of animals and plants: the hologenome theory of evolution. FEMS Microbiol Rev 32(5):723–735

Chapter 14
Natural Selection as Agent of Evolutionary Change: A View from Paleoanthropology

Ian Tattersall

Abstract Following the triumph of the Modern Evolutionary Synthesis in the 1940s and 1950s, natural selection became viewed by most anglophone evolutionary biologists as the primary or even the only instrument of evolutionary change. Under the "hardened" version of the Synthesis as it was absorbed into paleoanthropology after 1950, generation-by-generation change under selection's guiding hand became not only a necessary but a sufficient vehicle for explaining the unilinear pattern of human evolution that was perceived under the allied notion that the human "niche" was too broad to admit the existence of more than one hominid species at any one point in time. Subsequent additions to the hominid fossil record, as well as a better understanding of evolutionary process itself, have in contrast revealed a pattern of hominid diversity over time much better explained by sorting propelled by drift and external environmental change than by species-intrinsic natural selection as classically envisaged. Indeed, selection—a mathematical certainty in any population in which more individuals are born than reproduced—very plausibly acted among extinct hominids much more as a crucial homeostatic mechanism ("stabilizing selection"), than as an agent of change.

Keywords Natural selection · Evolution · Modern synthesis · Adaptation · Human fossil record · Paleoanthropology · Phyletic gradualism · Punctuated equilibria

14.1 Introduction

The reductionist human mind loves nothing better than a good story; and a major aspect of Charles Darwin's multifaceted genius was to be an inspired storyteller. Of the many scientific stories Darwin told, none was more effective in selling the idea of evolution to an initially skeptical mid-nineteenth-century audience than his account

I. Tattersall (✉)
Division of Anthropology, American Museum of Natural History, New York, NY, USA
e-mail: iant@amnh.org

R. G. Delisle (ed.), *Natural Selection*, Evolutionary Biology – New Perspectives on Its Development 3, https://doi.org/10.1007/978-3-030-65536-5_14

of natural selection. Of course, Darwin's thumbnail definition of evolution as "descent with modification" captures the two essential attributes of evolution without making reference to any mechanism of change whatever, selection or otherwise; and it still remains incomparably the best capsule formulation of the phenomenon it describes. But nonetheless, in explanatory terms, it is hard to match the sheer "why didn't I think of that?" power of the notion that better adapted individuals leave behind more offspring sharing their hereditary features than less well-favored ones do and that the effect multiplies over the generations. After all, natural selection is a mathematical certainty in any population in which more offspring are born than survive to reproduce; and in the form articulated by Darwin, it is an agency that logically should make gradual change within lineages of organisms virtually inevitable over time.

Accordingly, although the young Darwin had been acutely—and uncomfortably—aware of the many complexities of the natural world that make any reductionist explanation of evolutionary phenomena risky (see Eldredge 1995, 2016), by the time he was goaded by the arch-adaptationist Alfred Russel Wallace (1858) into publishing *On the Origin of Species* in 1859, the mature Darwin had more or less convinced himself that competition among organisms for reproductive success was, and continued to be, the principal driver of evolutionary change, despite, as Ernst Mayr (2001) pointed out, a notable dearth of available and provable examples at the time. Among other things (see Bellon, Chap. 3; Delisle, Chap. 4, Schwartz, Chap. 12), it was this lack of direct evidence for natural selection that prompted ongoing opposition from a wide variety of prominent critics such as the saltationist St George Mivart (1871); but by the end of the nineteenth century, the view of natural selection as *the* agent of evolutionary change had gained considerable public and scientific traction, not only in the anglophone UK and USA but also throughout Europe, not least through the efforts of the conflicted Ernst Haeckel (see Levit and Hossfeld, Chap. 5; Schwartz, Chap. 12).

The Darwinian notion that evolutionary change has usually—or even, if you are an ultra-Darwinian, invariably—been driven by natural selection depends on the phenomenon of heredity, whereby the distinctive biological characteristics of parents are often passed along to their offspring more or less intact. That this was, and is, the case had of course been well-known to animal and plant breeders and genealogists from time immemorial; and such knowledge was enough by itself to allow Wallace and Darwin to formulate their ideas of evolution by natural selection in the absence of an accurate notion of heredity. But a fuller understanding of process in evolution depended on a precise understanding of exactly *how* hereditary characters are transmitted between generations; and the dawning of this key comprehension had to await the turn of the twentieth century and the rediscovery of Mendel's principles by the early geneticists (see the account by Mayr 1982; see also Ochoa, Chap. 6).

By that point, the notion that the living biosphere was in some way the product of evolution had become widely accepted by scientists and educated public alike, partly thanks to the seductive powers of the natural selection concept. But the initial rise of the science of genetics produced some bumpy conditions for selectionists, as the nascent field became polarized between the single-gene "mutationists" in the Mendelian tradition, who fixated on the inheritance of discrete and discontinuous characters (e.g., wrinkled vs. smooth seed pods), and the "biometricians" whose interests lay in continuously varying traits such as body size or stature (for more discussion, see Schwartz, Chap. 12). During the early days, the Mendelians were in the ascendant; and these scientists were in general less receptive than the biometricians to the idea that natural selection was the major propulsive force in evolution. In the end, however, the balance shifted.

What appears in retrospect as the tipping point came in 1918, when the mathematical geneticist R.A. Fisher, confronted with the need to reconcile the results of the Mendelians and the biometricians, came up with what is now known as his "infinitesimal model" of the behavior of genes within populations. Fisher's formulation acknowledged the experimental importance of the single genes of interest to the Mendelians, but it swung the focus back upon the vast majority of traits that were determined by numerous different alleles acting together. In combination, those multiple alleles produced continuous and normally distributed ranges of phenotypes in the populations concerned; and, in its turn, this realization opened the door to quantitative modeling of the action of natural selection (and of other putative influences, such as "mutation pressure") within such populations. The infinitesimal model also led eventually to the recognition of the "norm of reaction," which sees the genotype as specifying a range of potential outcomes among which the environment selects (see Clausen and Hiesey 1958).

In a period of lively debate over evolution, and of natural selection's role in it (see Ceccarelli, Chap. 7), the rise of quantitative population genetics also paved the way for what eventually became known as the "Modern Evolutionary Synthesis" (Huxley 1942; see discussion by Adams, Chap. 8; Schwartz, Chap. 12). This convergence began during the 1920s and 1930s, as the theoretical modeling of Fisher and other luminaries such as Sewall Wright (see account by Provine 1986) and J. B. S. Haldane (see account by Clark 2011) was integrated with empirical data from genetics (Dobzhansky 1937), systematics (Mayr 1942), and paleontology (Simpson 1944) to produce what was later proclaimed (not least by Mayr 1963, 1982) to be a coherent and comprehensive account of the evolutionary process (but see Schwartz, Chap. 12). Most anglophone evolutionary biologists came quite rapidly to subscribe to this account, but while it had been fairly nuanced in its earlier manifestations (for discussions on differences of viewpoint among its founders, see discussions by van der Meer, Chap. 11; Ochoa, Chap. 10; Schwartz, Chap. 12; see also Levit and Hossfeld, Chap. 9; see Granovitch, Chap. 13, for some interesting non-anglophone views), by mid-century, the Synthesis had "hardened," to use Steve Gould's (1983) memorable term, into a simple formula that had shed many of the complexities recognized by earlier naturalists (and summarized by Schwartz, Chap. 12) in favor of reducing the multifaceted evolutionary process (or, rather, processes) to little more than the action of natural selection within gradually transforming lineages of organisms. As Esposito (Chap. 2) implies, this reductionist outcome may well reflect an unduly tidy view of history; and the actual intricacies of the real world are clearly reflected in the myriad popular misunderstandings of evolutionary process elaborated by Watts (Chap. 16). Nonetheless, the ultimate reductive formulation, as energetically promoted by Mayr, was compelling enough that most working systematists were happy to accept that, while identifiable (albeit normally polytypic) at any one point in their history, species actually lost their individuation with time, as they gradually transformed themselves into distinctive new species as selection did its inexorable work.

So persuasive was the reductionist natural selection metaphor as thus deployed (time + natural selection = evolution) that it was many years before the dictates of the Synthesis and the canonical authority of its founding fathers began to be seriously questioned. The inevitable rethinking was eventually set in train by Niles Eldredge and Steve Gould (1972), both invertebrate paleontologists, who pointed out in a now classic but then frequently reviled paper that the Emperor of the Synthesis had been remarkably flimsily clad from the very start (see also Schwartz, Chap. 12). Extrapolating empirically from the invertebrate fossil record, Eldredge and Gould contended that fossil species, instead of gradually evolving into new forms under natural selection, typically showed stasis—non-change—over their frequently quite extended lifetimes. In Eldredge and Gould's view, species characteristically have births (at speciation), histories (as basically stable entities), and deaths (at extinction). During their life spans, they might give rise to descendant species in short-term speciation events, and they might even persist alongside those descendants for long periods of time. But what they do not do is gradually evolve themselves out of existence as envisaged by the Synthesis. In other words, where the Synthesis had taken away the immemorial notion that species were bounded entities, Eldredge and Gould re-established ancient wisdom by arguing on empirical grounds that species really do have real, objective, existences in nature.

In articulating this perspective, Eldredge and Gould also, and inevitably, questioned the role of natural selection as the dominant force guiding evolutionary change. Indeed, they viewed species and populations as basically "homeostatic systems," suggesting that essentially stochastic events, such as invasions of new environments by geographically peripheral populations, were more plausible than selection as drivers of such well-established phenomena of change as adaptive radiation and morphological trends. What's more, in empirically basing their argument on invertebrate fossils, Eldredge and Gould forcefully reinstated the fossil record as the major arbiter of evolutionary mechanism. Under the Synthesis, paleontology had been relegated to the essentially clerical role of documenting the products of the "Phyletic Gradualism" propounded by the geneticists and systematists who produced the basic underpinnings of the Synthesis (the influential mammal paleontologist G. G. Simpson is usually credited as one of the founding triumvirate of the Synthesis, and he certainly achieved a major advance in eliminating orthogenetic echoes from paleontology; but while his *Tempo and Mode in Evolution* (1944) showed that he acutely appreciated the importance of the major discontinuities in the fossil record, the ultimate thrust of the book was to bring paleontology into line with gradualist thinking; see also Ochoa, Chap. 10). Under Eldredge and Gould's "Punctuated Equilibria" model, in contrast, paleontology resumed center stage as the key repository of information about evolutionary process, which was, of course, highly appropriate; paleontology is, after all, our only direct means of accessing, however incompletely, the historical record of how biological history has actually unfolded and on what time scales.

Subsequently, Eldredge (e.g., 1978) took to characterizing the two contrasting views of evolutionary process in slightly modified and more accurately descriptive terms, as "transformational" vs. "taxic," the former depending on long-term natural selection as classically envisaged and the latter relegating selection at best to ephemeral episodes of speciation. Under the taxic paradigm advocated by Eldredge, evolution is principally about the origin of new species and much less about the differential reproductive success of individuals within old ones.

Coincidentally—but not inconsequentially—Eldredge and Gould published their broadside not long after the long-ascendant "expert judgment" paradigm of systematics had begun to be supplanted by the more rigorous demands of the newly surgent cladists. This new breed of systematists insisted that relationships among species should be reconstructed using synapomorphies to link sister taxa (Hennig 1966) and thus that phylogenies should be testable rather than simply declarative and based on little more than informed subjective opinion. And even if Eldredge and Gould's views of evolutionary process had not on their own begun to change minds, the intrusion of cladistics would, by itself, have been bound to throw the emphasis in evolutionary biology back upon taxa as discrete entities, each one defined by characters and with an individuated history. The twin conceptual revolutions in systematics and evolutionary theory of the late 1960s and early 1970s were ontologically unrelated (indeed, some radical cladists even insisted that evolution was not a necessary prerequisite for understanding pattern in nature), but in synergy they radically transformed both the invertebrate and the vertebrate branches of paleontology within a remarkably short space of time. As usual, though, paleoanthropology remained an exception.

14.2 Early Paleoanthropology

Invertebrate paleontology was largely born within the science of geology, while most early vertebrate paleontologists emerged from the very different realm of comparative anatomy. Nonetheless, despite their great disparity in origins, the two disciplines shared from the beginning a profound appreciation of the importance of diversity in nature. After all, the processes of distinguishing among species, and of establishing relationships within larger groups, were of equal importance to the scientists who were, on the one hand, trying to sort out facies and biostratigraphic sequences on varying scales and to those who, on the other hand, were attempting to make evolutionary sense of the bewildering range of structural variations on the basic vertebrate and invertebrate themes. In dramatic contrast, the roots of paleoanthropology lay in the study of human anatomy. This latter field had traditionally been a branch of the medical sciences; and the issues of systematic and functional diversity in nature as a whole had never been perceived as very relevant to its practice. Indeed, to anatomists whose focus was on the minutiae of variation and function within one single species, systematics barely seemed to be relevant at all. As the hominid fossil record enlarged, and it became increasingly evident that the hominids were a pretty diverse clade, this limiting perspective should clearly have altered among paleoanthropologists; but in the event, any fossil that was recognizably human—and that its discoverers had therefore sent to someone with an anatomical background for analysis—was usually seen simply as a variation of some kind on the basic progressive human theme. If, of course, it did not more probably lie somehow directly in the line of descent that culminated in *Homo sapiens*.

From the beginning, the idiosyncratic origin of paleoanthropology promoted a marked degree of insularity in the viewpoint of its practitioners. So marked, indeed, that it would hardly be unfair to characterize that viewpoint as "paleoanthropological exceptionalism." For while paleontologists of other stripes were acutely aware that the very first thing that needed to be done in the case of any newly found fossil was to determine exactly where it had fit into the riotous natural diversity of the world in which it had lived, paleoanthropologists saw no such imperative. To scientists in this blinkered specialty, the subtleties and procedures of systematics and classification, painfully honed through efforts to understand non-hominid organisms, apparently seemed largely irrelevant—if, indeed, they thought about them at all.

The reason for this was simple. For while most other paleontologists saw their job as in one way or another to explain how the vibrant diversity of the biosphere had emerged, paleoanthropologists remained fixated on one single species, *Homo sapiens*. Little else mattered. After all, because it is incontestable that the world today contains only one hominid species, the strong temptation is and was to project that single species back into the past by the simple expedient of fitting fossils into the human lineage according to their chronological positions, rather than by trying to make phylogenetic sense of their morphological characteristics. It was, indeed, no accident that such linear concepts as aristogenesis and orthogenesis found some of their strongest proponents in the ranks of the paleoanthropologists, or at least in paleoanthropological contexts (c.f. Osborn 1915; Weidenreich 1947; see also Ochoa, Chap. 10). A major corollary of this linear mode of thinking was to divert paleoanthropological attention away from issues of morphology and evolutionary mechanism in evolution and toward a minor obsession with stratigraphy. To give one simple but eloquent example, while you will find a vast amount of detailed information about fossil hominid sites and their stratigraphies in the leading anatomist Arthur Keith's influential 1915 book *The Antiquity of Man*, you will search his index in vain for any mention whatever of "natural selection." Indeed, Keith's title says it all: his book is all about human antiquity and the anatomical features of fossils, rather than about evolution and the means by which it might have occurred in the hominid family.

Unsurprisingly, given the resulting profound lack of interest in the proper practice of systematics among students of human evolution (or, perhaps more properly, of human antiquity), the first half of the twentieth century saw a vast proliferation of names in the paleoanthropological literature: a proliferation that was entirely out of scale with the expansion of the hominid fossil record itself, as practically every

Table 14.1 Generic terms commonly in use in the pre-war literature for members of the three genera *Australopithecus*, *Paranthropus*, and *Homo* as generally recognized today

Australopithecus africanus
Plesianthropus transvaalensis
Paranthropus robustus
Pithecanthropus erectus
Meganthropus palaeojavanicus
Sinanthropus pekinensis
Atlanthropus mauritanicus
Palaeoanthropus palestinensis
Palaeanthropus heidelbergensis
Protanthropus neanderthalensis
Cyphanthropus rhodesiensis
Javanthropus soloensis
Telanthropus capensis
Africanthropus helmei
Homo spelaeus

new find was baptized with its own zoological name, regardless of any resemblances it might have borne to fossils already discovered. As quoted by his American Museum of Natural History colleague Ernst Mayr (1950: 109), Franz Weidenreich, the distinguished describer of the Peking Man fossils, puts it this way: "[in anthropology] it always was and still is the custom to give generic and specific names to each new type without much concern for the kind of relationship to other types formerly known." By the mid-twentieth century, there were accordingly at least 15 generic labels in use for members of the three genera *Australopithecus*, *Paranthropus*, and *Homo* as recognized by most paleoanthropologists today (see Table 14.1), not to mention dozens more species names. Paradoxically, this welter of nomina would have given the proverbial observer from Mars the impression of an entirely fictitious diversity of organisms—in a field whose practitioners were signally uninterested in natural diversity as it was understood by other paleontologists! Clearly, some kind of major rethinking was long overdue.

14.3 Paleoanthropology and the Synthesis

The necessary nomenclatural surgery was carried out in 1950 by Ernst Mayr, an ornithologist and evolutionary theorist who unabashedly confessed to "not possessing any first-hand knowledge of paleoanthropology" (Mayr 1950: 109). Instead, Mayr brought the authority of a co-founder of the Synthesis to the self-imposed task of sorting out the "simply bewildering diversity of names" regularly in use by hominid paleontologists, via the expedient of "giv[ing] the categories species and genus a new meaning in the field of anthropology, namely, the same one which in recent years has become the standard in other branches of zoology" (Mayr 1950: 109). By "new meaning," he was referring to the "population thinking" that was a central feature of the Synthesis (and had, indeed, been a major legacy of Darwin's thought). This viewpoint downplayed typology in favor of emphasizing intra-species variability among individuals. In this perspective, according to Mayr, if several species of the fruit fly *Drosophila* were blown up to human size, "it would be apparent even to a lay person that they are probably more different from each other than are the various primates and certainly more than the species of the suborder Anthropoidea." Having thus arbitrarily (and apparently with a straight face) disposed of form as a criterion for recognizing putative (and expectedly polytypic) hominid species and genera, Mayr proceeded to claim that, across the board, genera and even species were much more importantly distinguished by their "adaptive zones" than by their morphologies. His colleague and fellow architect of the Synthesis, Theodosius Dobzhansky, had already surveyed the fossil hominids and concluded that "there is no reason to suppose that more than a single hominid has existed on any time level in the Pleistocene" (Dobzhansky 1944: 264), and Mayr upped the ante by declaring that the hominid adaptive zone/ecological niche is so broad as to obviate *in principle* the existence of more than one hominid lineage at any one point in time. As he put it: "The ecological versatility of man and his slowness in acquiring reproductive isolating mechanisms have prevented the breaking up of *Homo* into several species" (Mayr 1950: 117). See Baravalle (Chap. 15) for an overview of the issues of "cultural evolution" that inevitably intrude in this context.

The systematic conclusion was inevitable: in Mayr's view, only one lineage had ever existed in the entire long history of the hominid family. All members of that consistently evolving polytypic lineage had belonged to the genus *Homo*, within which he perceived a gradually transmuting succession of three species. *Homo transvaalensis* (what we would nowadays call the australopiths) had gradually given rise to *Homo erectus* (as then known from Java and China), which in its turn had insensibly changed over time into *Homo sapiens* (a species that, as Dobzhansky had already proposed, also included the Neanderthals, now differentiated merely at the level of subspecies). Here was the hardened synthesis with a vengeance: a braided stream of hominids flowing and reticulating across time, driven all along by natural selection that somehow contrived simultaneously to respond to local conditions and to maintain the biological coherence of the lineage.

Delivered at an enormously influential conference held on Long Island in 1950, Mayr's diatribe dropped on paleoanthropology like a bombshell. Coming just after the huge disruptions of World War II, and at a time when the old guard of anatomists was ageing out of the profession, the blow had been delivered at what was a propitious moment for the Synthesis. Younger paleoanthropologists, in particular, were vulnerable to the implied—and entirely accurate—accusation that their elders had been operating in what was essentially a theoretical vacuum, and they were traumatized accordingly. From far-away South Africa, John Robinson (1953) did complain that the robust and gracile australopiths formed two demonstrably separate lineages, and Mayr (1953) himself actually conceded this with remarkable rapidity, albeit rather grudgingly and in a footnote. Nonetheless, he still felt obliged to reproach Robinson by tossing in the canard that the latter had not addressed "the serious problem of competition" (Mayr 1953: 281): a "problem" that, of course, existed only if one fixated on Mayr's own unilinear model. Still, aside from Robinson and such odd outliers as Solly Zuckerman (see Ashton and Zuckerman 1951), the capitulation of anglophone paleoanthropology to Mayr's new scenario was both sudden and effectively complete—to such an extent that, for more than a decade, hominid paleontologists hardly dared to use zoological names for the objects of their study, warily avoiding taxonomic designations and referring to individual fossil specimens by their site identifiers (see Tattersall 2018a).

Such was the shock inflicted by Mayr that his minimalist nomenclatural strictures still unhappily reverberate in paleoanthropology some 70 years later, even as the human fossil record, and our perspective on it, has vastly expanded. Still, the fallout from his attack was not limited to the naming and recognition of taxa at the species and genus levels. This was because a major, and much more salutary, effect of Mayr's criticism was to make paleoanthropologists keenly aware of the importance of mechanism in evolutionary studies of all kinds, including their own. Accordingly, as the new generation of hominid paleontologists took over, ideas of natural selection and adaptation entered their discipline to a far greater extent than they had ever done before, and indeed, selection and adaptation (each of them implying and affirming the other) have been the primary foci of interest in modern human evolutionary studies ever since—even as, on the minus side, chronology still rules the day, and it remains distinctly unfashionable to "argue about names" regardless of the self-evident fact that, if you are ever going to properly understand the evolutionary/ecological play, you will first need to have a pretty accurate idea of who the actors are.

Despite this checkered legacy, there can be no doubt that Mayr's intrusion reinvigorated paleoanthropology. There is a distinctly liberated quality to many paleoanthropological publications from the middle 1950s onward, largely due to an overt acknowledgment that the fossils under scrutiny had once been part of living, breathing individuals whose lives could be reconstructed from multiple lines of evidence. Raymond Dart (1925, 1947) and even Franz Weidenreich (1939) had, in their time, exhibited fine fertile imaginations in the domain of ancient hominid behavior, but what was truly new after midcentury was an acceptance that paleoanthropology was a multifaceted science that could substitute empirical fact for imagination and still come up with a dramatic story. Emblematic of this awakening was the early work of F. Clark Howell (1951, 1952), whose studies of European late-middle-to-late Pleistocene hominid fossils possessed a range and a tone that had been missing in pre-war days, integrating morphologies, faunas, environments, and the gradualist evolutionary model to create a picture of a hominid lineage that had flourished through changing and challenging times by diversifying regionally via local adaptation while still retaining its (phylo)genetic coherence.

Howell told a wonderful and erudite story that was both underpinned and animated by a firm Mayrian faith that natural selection could simultaneously act to promote both diversification and coherence within variable and ever-changing lineages. Yet, inevitably, this perspective appears in retrospect as at best incomplete. The reason for this is that when you give primacy in this way to adaptation (usually simply assumed: if it's there, it's an adaptation) in any evolutionary scenario, you run the obvious risk of distracting attention away from the critical taxic aspect of evolution that Niles Eldredge emphasized in 1978. Species are the central actors in the evolutionary play by virtue of being effectively discrete participants in the ecological drama at any one point in time; and their fates depend on the kind of organism they are overall (and with whom they happen to be competing and in what kind of environmental circumstances), rather on the individual excellences of their components, or on what paleontologists might later conclude they became. After all, species consist of populations of organisms that interact with their environments and their peers as integrated wholes, and not as individual characteristics in which we can track change over time.

As a result, while scientists might find it convenient to atomize both taxa and individuals into discrete characters for the purposes of understanding adaptation and other elements of their biology, when they do so, they always run the risk of missing the bigger and more important picture. This is because every individual is an astonishingly complex concatenation of features; and he or she will much more likely succeed or fail in the reproductive stakes as the sum total of them all, rather than because he or she possesses or lacks any one of them. It is, after all, presumably of little survival or reproductive benefit to be the fastest runner in your group if you are also the shortest-sighted or, perhaps even more importantly in the longer term, if your species as a whole is being outcompeted into extinction. What's more, physical characteristics almost invariably turn out to have very complex genomic backgrounds (see discussion in Tattersall and DeSalle 2019), so that there are typically huge constraints on what constitutes a viable change.

Half a century ago, the neurolinguist Eric Lenneberg articulated this key fact with a subsequently unmatched clarity and brevity. He made his remarks in the context of a seminal discussion of the origins of language, but his observations apply much more broadly, and all are as pertinent today as when he wrote:

> We can no longer reconstruct what the selection pressures were or in what order they came, because we know too little that is securely established by hard evidence about the ecological and social conditions of fossil man. Moreover, we do not even know what the targets of actual selection were. This is particularly troublesome because *every genetic alteration brings about several changes at once, some of which must be quite incidental to the selective process*. (Lenneberg 1969: 643; italics added)

Sadly, Lenneberg's sage admonition went largely unheeded by paleoanthropologists who, still in thrall to Mayr, continued diligently dreaming up essentially untestable adaptationist scenarios. For example, it is in the reductionist neo-Darwinian spirit lamented by Lenneberg that the interminable debate over why hominids became bipedal has persisted in churning on, apparently unstoppably. Darwin himself famously thought in terms of a key advantage: he felt that the defining benefit of bipedality was the ability it conferred on individuals to wield clubs for both attack and protection. This attractive proposition has since become empirically indefensible, at least as concerns its timing; but nonetheless, Darwin's successors continue to look for an alternative adaptive advantage that made terrestrial bipedality beneficial for an ancestrally quadrupedal (or perhaps more properly quadrumanous) form living at a time when continuous forest cover was shrinking in Africa, with the consequent expansion of more open woodland and bushland habitats. Raymond Dart (1959) favored the view that standing upright had allowed diminutive hominids to see danger approaching from farther away; Gordon Hewes (1961) thought that the advantage of upright posture lay in the ability it conferred to carry food over long distances; Frank Livingstone (1962) proposed that gorilla-like bipedal display was defensively beneficial at expanding but dangerous forest edges; Cliff Jolly (1970) and others such as Kevin Hunt (1994) have concluded that bipedality in one way or another represented a helpful feeding strategy for a terrestrial primate; Peter Rodman and Henry McHenry (1980) saw improved locomotor efficiency as the fundamental issue; Owen Lovejoy (1981) liked the idea that male provisioning provided the critical advantage in the context of pair bonding; Nancy Tanner (1981) favored phallic display and infant carrying; Pete Wheeler (1984) made a powerful argument for thermoregulation; and the notion (proposed by Alister Hardy in 1960 and later popularized by Elaine Morgan in 1982) that early hominids waded upright in shallow water to exploit aquatic resources still refuses to go away (Kuliukas 2013).

All of these diverse scenarios, and many more, focus on individual key advantages to hominid bipedality, and each one is clearly underpinned, in one way or another, by the notion that adopting this unusual way of getting around was somehow propelled by natural selection acting on a particular feature of the organism. The fallacy is evident, as Lenneberg clearly realized, but from the resulting mindset sprang such extraneous and misleading concepts (rarely mooted outside paleoanthropology but distressingly frequently heard within it) as "mosaic evolution," the nebulous idea that different parts of organisms evolve at different rates. Atomistic notions of this kind are made possible only by transformational views of evolution, and it does not take much reflection to realize just how absurd they are. As already emphasized, morphological (or, for that matter, behavioral) features do not have independent existences. They are packaged into the entire organism. And it is the entire organism, not the feature, that interrelates with the environment and succeeds or fails on the ecological and social stages. When we favor a view of evolution that eliminates the role of whole taxa as interactors, we ignore a crucial dynamic in evolutionary histories, and in the specific context of hominid bipedality, we also lose sight of the fact that once a hominoid has become upright, it enjoys *all* the advantages—and all of the significant *dis*advantages—of the new locomotor strategy. Almost certainly most, if not all, of the posited "key benefits" of bipedality must surely have entered in some way into determining the overall viability of the new locomotor mode; and ultimate evolutionary success or failure (necessarily of the individual and species, not of the feature) must have involved a balance of all of those characteristics that were truly relevant to ecological success.

Adopting the taxic perspective clearly shifts the terms of the debate. The essential question then becomes "Why did the primordial hominid adopt bipedality in the first place?" Clearly a quadrupedal or quadrumanous ape did not come down to the ground to seek new resources and then decide that it would be advantageous to stand tall so that it could see farther, or thermoregulate better, or more effectively attract the ladies. A committed quadruped might facultatively have done any or all of those things; but to escape danger, or to travel any distance, it would have instantly dropped to all fours and scampered away. The only conceivable reason why a hominoid would have stood upright on the ground and would have remained committed to that posture even for traveling was simply that it felt most natural for it to do so (see Tattersall (2016a) for more extensive discussion). And that would only have been the case if the hominoid in question was already highly suspensory and therefore habitually held its body erect for feeding, balancing, and moving around in the trees (see also Tuttle 1981; Thorpe et al. 2007). In turn, this would have made the origins of its putative terrestrial pre-adaptations entirely irrelevant to its later adoption of the new locomotor style. Inevitably, the relevant novel anatomical features had been acquired in another context entirely, only later to be recruited to their new postural and locomotory uses. In this perspective, natural selection cannot have been a driving agent of the change observed. Instead, an adventitious change in environment made an existing structure "adaptive" in an entirely new and unanticipated context. Indeed, it is possible to argue that *everything* we later recognize as an *ad*aptation has to come into existence as an *ex*aptation, that is to say, entirely randomly with respect to any use for which it might be co-opted after becoming fixed in the population. Mutations are, after all, stochastic events.

14.4 Recent Paleoanthropology

After some admittedly fairly shrill initial complaints (e.g., Malmgren et al. 1983; H. Robinson 1986), vertebrate and invertebrate paleontologists quite rapidly and comfortably incorporated both punctuated evolutionary models and cladistics into their world views—not as procedural Holy Grails but simply as additional weapons in their phylogenetic armamentaria alongside emerging technologies such as molecular systematics. As a result, both fields today look very different from their predecessors in the mid-twentieth century. Not so paleoanthropology, which remained generally unreceptive to cladistics and punctuated evolutionary models. Resistance to both remains tenacious, even as such highly selectionist alternatives as the "multiregional" notion of hominid evolution (Wolpoff et al. 1984) have tended to fade or to transmute. True, many paleoanthropologists have adopted at least the terminology of cladistics, along with some of the quantitative phylogenetic modeling methods to which trait-based phylogenetic reconstruction opened the door (e.g., Dembo et al. 2015); and, perhaps even more importantly, the sheer abundance of new morphologies represented in a rapidly expanding human fossil record has by now forced the general recognition of more than two dozen extinct hominid species (see review by Tattersall 2018a). But the overwhelming tendency in paleoanthropology is still to think of the human fossil record in the most minimalist terms possible and to cleave to Mayr's transformationist and linear injunctions to the maximum practical extent. Paleoanthropology is, in other words, still somewhat in thrall to the tired old formula: time + natural selection = evolution.

The ongoing paleoanthropological reluctance to think taxically, and thereby also to bring the systematics of Hominidae (or, if you must, Homininae; in this context the difference is notional) into line with that of other mammalian (sub)families, finds expression in many ways. But perhaps the most egregious example is the apparently inexhaustible willingness of paleoanthropologists to cram ever more diverse morphologies into the already bulging single genus *Homo*, the addition of each new morphology making the genus more "variable" and thus easier to force further ones into. True, Ernst Mayr lost the battle to subsume the australopiths into *Homo* very early on; and a decade and a half after his tirade was published (a period during which anglophone paleoanthropologists hardly dared utter a formal species name), Louis Leakey et al. (1964) had the temerity to apply the new nomen *Homo habilis* to a dental australopith from Olduvai Gorge. Decades later, Meave Leakey and her colleagues (M. Leakey et al. 2001) even dared to propose the new genus *Kenyanthropus*, which was met with a deafening silence where it was not roundly criticized. But otherwise, and very much in the Mayrian spirit, the tendency among paleoanthropologists has increasingly been to dichotomize most hominid fossils into *Australopithecus* vs. *Homo*. The prevailing systematic algorithm seems to be: if it isn't one, it *has* to be the other—regardless of what it actually looks like.

Evidence for the seductive power of this mindset is everywhere. For example, other than that they are plainly not *Australopithecus*, there is precious little morphological or phylogenetic justification for shoehorning the diminutive and tiny-brained hominids from Liang Bua in Flores (Brown et al. 2004) or Rising Star in South Africa (Berger et al. 2015) into the genus *Homo* (Tattersall 2015). The recently announced *Homo luzonensis* (Detroit et al. 2019) also presumably belongs in this category. Perhaps worse yet, there is even less reason for cramming any of the morphologically diverse hominids from Georgia's Dmanisi into the genus that is defined by *Homo sapiens*, let alone into the designated species *Homo erectus*; none of the Dmanisi specimens even remotely resembles the *Homo erectus*-type material from Trinil, in Java (see Schwartz and Tattersall (2005) and Tattersall (2015) for discussion). Indeed, if one were to seek a prime example of systematic exceptionalism in paleoanthropology, one could do no better than to point to the fact that the describers of the most complete and distinctive of the Dmanisi hominid crania (D4500/2600) saw fit to place this remarkable specimen in its own "sub-subspecies"—an otherwise unheard-of taxonomic rank—as *Homo erectus ergaster georgicus* (Lordkipanidze et al. 2013). Even the early anatomists might have scratched their heads at this mind-boggling move; but paleoanthropology itself is left with a systematic disaster, the resolution of which will only be in sight when its practitioners start to admit that Hominidae is a very diverse clade indeed and that more genera than the simple *Australopithecus*/*Homo* dichotomy allows will be needed to express the complex structure that lies within it (Tattersall 2017a). Sadly, the seductive power of the reigning adaptationist/selectionist paradigm helps very effectively to disguise this necessity, and it is clear that in paleoanthropology our received neo-Darwinian assumptions about selection continue to resonate far beyond the immediate fields of evolutionary process and theory.

14.5 Natural Selection and Hominid Phylogeny

A glance at an approximate but fairly mainstream hominid family tree (Fig. 14.1) shows how far, despite our gradualist preferences recent empirical findings in a rapidly expanding fossil record have obliged us to come, in recognizing that ancient hominids really were diverse. Far from reflecting a slow and steady process of refinement over the years, this fairly speciose tree is witness to a story of vigorous

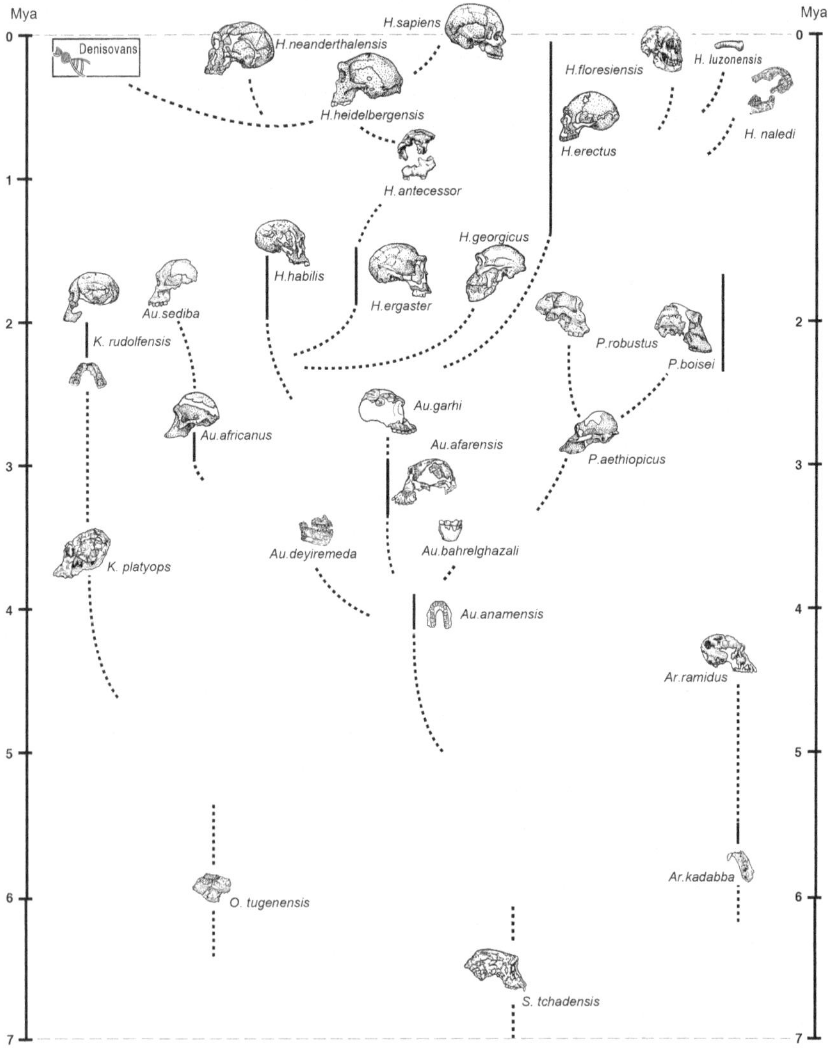

Fig. 14.1 Schematic tree of hominid evolution, showing how, typically, multiple hominid species have shared the planet at any one point in time. It is *Homo sapiens* that is truly unique in being alone in the world. Art by Kayla Younkin

evolutionary experimentation, in which new variations on the hominid theme (new species) were repeatedly generated, and pushed out into the ecological arena to thrive or to perish. Individual taxa in this schema may or may not have been competitively superior to one another, or to unrelated contenders for ecological space; but the overall topography of the tree does little to suggest that the major pattern involved the steady adaptive enhancement of a central evolutionary mainstream.

The belief that natural selection is a major agent of evolutionary change is often as much a matter of faith as of established fact (see also Delisle, Chap. 4). As Mayr pointed out in 1982, there were signally few documented examples of selection-driven change in Darwin's day; and, perhaps oddly, there are not many additional ones now. Such phenomena as industrial melanism and antibiotic resistance are hard to refute, of course, because the cause-effect relationship in such exceptional cases is unusually clear-cut. But it remains true that the best grounds we have for believing in selection-driven change in the ancient past are inferential, rather than observable. That something we can justifiably call Darwinian adaptation is at some level an important part of evolutionary process is pretty strongly suggested by the rampancy of morphological convergence in the natural world. But while the gradualist influence of natural selection might in principle be most powerfully inferred from within-lineage morphological trends observed in the fossil record, the paleoanthropological literature actually yields remarkably few such instances. References to the "mosaic evolution" of fossil hominids abound in the literature, most commonly in the context of functional analyses (e.g., Kivell et al. 2011); but as far as I am aware, there is only one explicit recent test of anagenesis in the human fossil record (Kimbel et al. 2006). Once again, this example was likely compromised by a leaning toward minimalist systematics in the Mayrian tradition (see Schwartz and Tattersall 2005); and it has actually been refuted on the basis of fossil discovery (Haile-Selassie et al. 2019). Without an adequate systematic framework, inferences concerning pattern in the fossil record are essentially worthless.

This having been said, there is one particular line of putative evidence for the long-term transformative action of natural selection that remains, without any question, an intuitively satisfying one. I refer, of course, to the apparent documentation of long-lasting within-lineage morphological trends in the fossil record. But caution is always advisable when interpreting evidence of the kind available. This is because gaps in osteodental morphology between closely related species are typically small, so that in a spottily sampled record, a pattern of continuous change may be closely mimicked by multiple samplings over time of a diversifying clade. Nonetheless, the human fossil record, and specifically that of the genus *Homo* [as rationally defined to exclude anything that is not a reasonably close relative of *Homo sapiens*: see discussion in Collard and Wood (2015) and Tattersall (2016b)] provides us with one of the most striking examples of an evolutionary trend in all of paleontology: the apparently inexorable increase in the volume of hominid brains over the course of the Pleistocene. Australopith cranial volumes in the period before two million years ago already hovered around the 450 ml mark (about a quarter larger than in the very first hominids and the living great apes). A mere couple of hundred thousand years later, early members of *Homo* had brains in the 800 ml range. By a million years ago, the figure was up to around 1000 ml, and by the end of the Pleistocene, the brains of both *Homo sapiens* and *H. neanderthalensis* had reached a mean of about 1500 ml.

In the absence of an adequate systematics of the genus *Homo* (which will only be achieved when the final traces of the linear/selectionist paradigm of human evolution are finally erased), it is impossible to achieve a testable interpretation of the tendency that is somehow buried in these figures. However, within the genus *Homo* as reasonably defined, it is notable that trends toward increasing brain sizes proceeded *independently* in at least three separate lineages: those leading to late *Homo erectus* in eastern Asia to *H. neanderthalensis* in western Eurasia and to *H. sapiens* in Africa. In each lineage, earlier members had smaller brains than later ones. From this we can, I think, legitimately conclude that there is some feature common to members of the genus *Homo* that predisposed all of its constituents to increasing brain size with the passage of time. Exactly what that feature might have been is currently impossible to specify; but it certainly seems reasonable to suggest that natural selection, operating similarly in each lineage because of some shared behavioral or physiological apomorphy, might somehow have been involved in generating the pattern observed. However, while it is also reasonable to surmise that, in some sense, increasing "intelligence" was implicated in the process of brain enlargement in *Homo* (and a stepwise increase in technological complexity is simultaneously documented in the archaeological record), it was certainly not intelligence of the specifically modern human variety. Not only did the very different Neanderthals have brains identical in size (though not in shape) to those of early *H. sapiens*, but the *H. sapiens* brain has shrunk in average volume (by almost 13%) since the archaeological record suggests the modern symbolic cognitive algorithm began to express itself (Tattersall 2018b). And, in any case, only one lineage of the three became fully symbolic.

From a gradualist perspective, the overall trend may seem somehow suggestive; but, given the deficiencies of our existing systematic frameworks, it is difficult to reach any definite conclusions on the action of transformative selection as a guiding force in this aspect of hominid evolution. We can certainly infer that long-lived hominid paleospecies were adequately adapted to their circumstances; but that is very different from claiming that they were in any way optimized for anything (or were even on a trajectory toward optimization) through the action of neo-Darwinian transformative natural selection. And of course, the trick in evolution is evidently not necessarily to be optimized (it is no accident that extinction rates are higher among stenotopes than among eurytopes like hominids), but simply to be good enough to deal successfully with whatever varying conditions present themselves. Evolution is not about engineering; it is about survival and successful reproduction under immediate circumstances, not only at the individual level but at that of the species as well. This means above all maintaining a fit population, rather than merely assuring the success of the "fittest" individuals within it. And it is here that natural selection—which, as we've seen, must inexorably take place in any population in which more individuals are born than survive—almost certainly plays its most vital role in most evolutionary histories.

In obeisance to Ronald Fisher's infinitesimal model, nearly every variable biological feature of every population exhibits a normal distribution (see discussion in Tattersall and DeSalle 2019); and the most critical function of natural selection in any successful species is almost certainly to keep the population fit by trimming off the extremes of those distributions at both ends, in the process maintaining a stable mean (Eldredge and Gould's "homeostasis"). What is more, for the population as a whole (whether locally or as a species), the crucial thing is not how well adapted in their individual features its most outstanding and reproductively successful members are; it is its own unitary success as a whole, something that necessarily involves the entire constellation of its characteristics and members. For while the individual succeeds or fails in the reproductive stakes as the sum of its parts, as far as the species is concerned, the most important role of natural selection is to keep the mean values of relevant characters in the most advantageous positions for the competitiveness of the entire population. This function of selection looms even larger when one considers that, in a hazardous world, the fates of individuals and even of entire populations are often very much at the mercy of chance events.

None of this entirely obviates any potential role for transformational selection in evolution; but given that gradual evolution by natural selection is acknowledged by all to be a very slow process, and that we now know that over hominid history environments have tended to change dramatically and unpredictably on very short time scales, it appears that the external conditions necessary for gradual directional evolution were relatively rare over the tenure of our family. For most of the evolution of the genus *Homo*, certainly, its members seem to have existed in tiny populations that were spread across vast tracts of territory and that were buffeted hither and yon by capricious changes in climate and environment—changes that were entirely random to their existing adaptations or proclivities. In aggregate, those circumstances would have provided ideal conditions not only for genetic drift but for evolutionary change via diversification, reintegration, and competition among and within populations belonging to a clade that was both remarkably flexible behaviorally and highly generalist ecologically. In all likelihood, the form of the hominid family tree in Fig. 14.1 reflects a complex ramifying history of this kind, rather than a basically linear history significantly influenced by gradual and transformational natural selection.

14.6 Conclusion

Independently derived from the study of human anatomy, and isolated from mainstream paleontology and evolutionary theory for the first century of its existence, paleoanthropology has been in thrall to a reductionist version of the Modern Evolutionary Synthesis since 1950. The "hardened" form of the Synthesis involved emphasizes straight-line evolutionary pathways guided by directional natural selection over long periods of time. Despite major developments in evolutionary and systematic theory in the interim, 70 years later, the resulting linear and minimalist mindset continues to dominate human evolutionary studies—which still tend to be largely focused in one way or another on adaptation and the assumed results of transformational selection. This narrow orientation has led to an obsession with the evolution of individual characters and character complexes that are, in reality, inextricably bound up in the whole organisms apart from which they cannot have independent evolutionary histories. And, perhaps predictably enough, it has also led to the chasing of red herrings such as "mosaic evolution," to the effective exclusion both of systematics and of the adequate appreciation of the role of whole individuals and taxa in evolution.

Nonetheless, on a pragmatic level, over time, the pressures of discovery have obliged paleoanthropologists to recognize a certain diversity of species in the hominid fossil record: a diversity that, although still regrettably muted by strong minimalist tendencies, clearly suggests just on its own that the history of the hominids has typically been one of vigorous evolutionary experimentation with the hominid potential, rather than one of the fine-tuning of a major central lineage by classical neo-Darwinian processes. Accordingly, it is hardly surprising that putative long-term evolutionary trends within the human family give equivocal support at best to the idea that transformational natural selection has had the largest effect on human evolution. Instead, it increasingly appears that in an uncertain, unpredictable, and rapidly changing world, natural selection has functioned principally as a force that vitally acts to keep entire populations fit and stable, rather than as one that has favored change over time in individual features we might choose to regard as adaptive. As predicted by Ronald Fisher's century-old infinitesimal model of genotypes and phenotypes, most of the variable characteristics that affect the survival or reproductive success of organisms within a population turn out to be normally distributed. And natural selection—which necessarily operates in any population in which more individuals are born than reproduced—appears to perform the essential function of keeping populations fit by trimming away the maladaptive extremes of those distributions and thereby promoting population homeostasis.

Still, it is undeniable that hominids have come a very long way over the course of the Pleistocene. Indeed, no other species in the world today is phenotypically more dissimilar from its own ancestor of two million years ago than *Homo sapiens* is. And this, of course, leaves us with the issue of explaining this unusually high aggregate rate of change. I have suggested (Tattersall 2017b) that a major cause may have been the unique hominid ingredient of material culture, though not in the gene/culture co-evolutionary context often invoked by evolutionary psychologists. Instead, my suggestion is an essentially demographic one. Namely, that, in a period when tiny hominid populations were thinly spread over vast landscapes, material culture would have permitted wider dispersal than would otherwise have been possible in propitious times. In contrast, it would have provided an incomplete buffer against habitat change in less favorable ones. This effect would have exacerbated the population fragmentation/recoalescence cycle in which genetic novelties could be fixed and sorted, and it would thereby have maximized the probabilities both of evolutionary innovation within hominid populations and of competition among them. In this perspective, hominid evolution was driven to a much greater degree by a combination of demographic and extrinsic factors, than by elements of intrinsic excellence and transformational selection.

Acknowledgments I thank Richard Delisle for the opportunity to express these thoughts here. And my gratitude goes, as always, to the American Museum of Natural History for providing an incomparable ambience in which to record them.

References

Ashton EH, Zuckerman S (1951) Some cranial indices of *Plesianthropus* and other primates. Am J Phys Anthropol 9:283–296

Berger LR, Hawks J, de Ruiter DJ, Churchill SE, Schmid P et al (2015) *Homo naledi*, a new species of the genus *Homo* from the Dinaledi Chamber, South Africa. eLife 4:e09560. https://doi.org/10.7554/eLife.09560

Brown P, Sutikna T, Morwood MJ, Soejono RP, Jatmiko SEW, Due RA (2004) A new small-bodied hominin from the late Pleistocene of Flores, Indonesia. Nature 431:1055–1061

Clark R (2011) J.B.S. The life and work of J. B. S. Haldane. Bloomsbury Books, London

Clausen J, Hiesey WM (1958) Experimental studies on the nature of species, Volume IV: Experimental structure of ecological races. Carnegie Institution of Washington, Publ 615, Washington, DC

Collard M, Wood BA (2015) Defining the genus *Homo*. In: Henke W, Tattersall I (eds) Handbook of paleoanthropology, vol 3. Springer, Heidelberg, pp 2107–2144

Dart RA (1925) *Australopithecus africanus:* the man-ape of South Africa. Nature 115:195–199

Dart RA (1947) The osteodontokeratic culture of *Australopithecus prometheus*. Transvaal Museum Memoir 10:1–105

Dart RA (1959) Adventures with the missing link. Harper, New York

Darwin C (1859) On the origin of species by natural selection. John Murray, London

Dembo M, Matzke NJ, Mooers AO, Collard M (2015) Bayesian analysis of a morphological supermatrix sheds light on controversial fossil hominin relationships. Proc R Soc B 282:20150943. https://doi.org/10.1098/rspb.2015.0943

Detroit F, Mijares AS, Corny J, Daver G, Zanolli C, Dizon E, Robles E, Grun R, Piper PJ (2019) A new species of *Homo* from the late Pleistocene of the Philippines. Nature 568:181–186

Dobzhansky T (1937) Genetics and the origin of species. Columbia University Press, New York

Dobzhansky T (1944) On the species and races of living and fossil man. Am J Phys Anthropol 2:251–265

Eldredge N (1978) Evolution: Taxic or transformational? Syst Zool 27:130–132

Eldredge N (1995) Reinventing Darwin: discovering the tree of life. Norton, New York

Eldredge N (2016) Eternal ephemera: adaptation and the origin of species from the nineteenth century through punctuated equilibria and beyond. Columbia University Press, New York

Eldredge N, Gould SJ (1972) Punctuated equilibria: an alternative to phyletic gradualism. In: Schopf TJM (ed) Models in paleobiology. Freeman Cooper, San Francisco, pp 82–115

Fisher RA (1918) The correlation between relatives on the supposition of Mendelian inheritance. Proc Roy Soc Edinburgh 52:399–433

Gould SJ (1983) The hardening of the evolutionary synthesis. In: Grene M (ed) Dimensions of Darwinism. Cambridge University Press, Cambridge, pp 173–204

Haile-Selassie Y, Melillo SM, Vazzana A, Benazzi S, Ryan TM (2019) A 3.8-million-year-old hominin cranium from Woranso-Mille, Ethiopia. Nature 573:214–221

Hardy AC (1960) Was man more aquatic in the past? New Scientist 174:642–645
Hennig W (1966) Phylogenetic systematics. University of Illinois Press, Urbana, IL
Hewes GW (1961) Food transport and the origins of hominid bipedalism. Am Anthropol 63:687–710
Howell FC (1951) The place of Neanderthal man in human evolution. Am J Phys Anthropol 9:379–416.
Howell FC (1952) Pleistocene glacial ecology and the evolution of "Classic Neanderthal" man. Southwest J Anthropol 8:377–410
Hunt KD (1994) The evolution of human bipedality: ecology and functional morphology. J Hum Evol 26:183–202
Huxley J (1942) Evolution: the new synthesis. Allen and Unwin, London
Jolly CJ (1970) The seed-eaters: a new hypothesis of hominid differentiation based on a baboon analogy. Man 5:1–26
Keith A (1915) The antiquity of man. Williams and Norgate, London
Kimbel WH, Lockwood CA, Ward CV, Leakey MG, Rak Y, Johanson DC (2006) Was *Australopithecus anamensis* ancestral to *A. afarensis*? A case of anagenesis in the hominin fossil record. J Hum Evol 51:134–152
Kivell TL, Kibii JM, Churchill SE, Schmid P, Berger LR (2011) *Australopithecus sediba* hand demonstrates mosaic evolution of locomotor and manipulative abilities. Science 333:1411–1417
Kuliukas AV (2013) Wading hypotheses of the origin of human bipedalism. Hum Evol 28:213–236
Leakey LSB, Tobias PV, Napier JR (1964) A new species of the genus *Homo* from Olduvai Gorge. Nature 202:7–9
Leakey MG, Spoor F, Brown FH, Gathogo PN, Kiarie C, Leakey LN, McDougall I (2001) New hominin genus from eastern Africa shows diverse middle Pliocene lineages. Nature 410:433–440
Lenneberg E (1969) On explaining language. Science 164:635–643
Livingstone FB (1962) Reconstructing man's Pliocene ancestor. Am Anthropol 64:301–305
Lordkipanidze D, Ponce de Leon MS, Margvelashvili A, Rak Y, Rightmire GP, Vekua A, Zollikofer C (2013) A complete skull from Dmanisi, Georgia, and the evolutionary biology of early *Homo*. Science 342:326–331
Lovejoy CO (1981) The origin of man. Science 211:341–350
Malmgren BA, Berggren WA, Lohmann GP (1983) Evidence for punctuated gradualism in the Late Neogene *Globorotalia tumida* lineage of planktonic foraminifera. Paleobiology 9:377–389
Mayr E (1942) Systematics and the origin of species. Columbia University Press, New York
Mayr E (1950) Taxonomic categories in fossil hominids. Cold Spring Harbor Symp Quant Biol XV:109–118
Mayr E (1953) Comments on evolutionary literature. Evolution 7:273–281
Mayr E (1963) Animal species and evolution. Harvard University Press, Cambridge, MA
Mayr E (1982) The growth of biological thought. Belknap Press, Cambridge, MA
Mayr E (2001) What evolution Is. Basic Books, New York
Mivart SG (1871) On the genesis of species. Macmillan, London
Morgan E (1982) The aquatic ape. Stein and Day, New York
Osborn HS (1915) Men of the old stone age: their environment, life and art. Charles Scribner's Sons, New York
Provine WB (1986) Sewall Wright and evolutionary biology. University of Chicago Press, Chicago
Robinson JT (1953) *Meganthropus*, Australopithecines and hominids. Am J Phys Anthropol 11:1–38
Robinson H (1986) A key to the common errors of cladistics. Taxon 35:309–311
Rodman PS, McHenry HM (1980) Bioenergetics and the origin of hominid bipedalism. Am J Phys Anthopol 52:103–106

Schwartz JH, Tattersall I (2005) The human fossil record, Vol 4. Craniodental morphology of early hominids (Genera *Australopithecus, Paranthropus, Orrorin*) and overview. Wiley-Liss, Hoboken N

Simpson GG (1944) Tempo and mode in evolution. Columbia University Press, New York

Tanner NM (1981) On becoming human. Cambridge University Press, Cambridge

Tattersall I (2015) Defining and recognizing the genus *Homo*. Gortania 36:5–22

Tattersall I (2016a) Prerequisites for hominid bipedality. In: Ribot Trafi F (ed) Homenaje al Dr Jose Gibert Clols. Una vida dedicada a la sciencia y al conocimiento de los primeros europeos. Publicaciones Diputacion de Granada, Granada, pp 83–91

Tattersall I (2016b) The genus *Homo*. Inference Int Rev Sci 2(1). http://inference-review.com/article/the-genus-homo

Tattersall I (2017a) Species, genera, and phylogenetic structure in the human fossil record: a modest proposal. Evol Anthropol 26:116–118

Tattersall I (2017b) Why was human evolution so rapid? In: Marom A, Hovers E (eds) Human paleontology and prehistory, essays in honor of Yoel Rak. Springer AG, Heidelberg, pp 1–9

Tattersall I (2018a) History and reality of the genus *Homo*: What is it, and why do we think so? Métode Sci Stud J 9:79–89

Tattersall I (2018b) Brain size and the emergence of modern human cognition. In: Schwartz JH (ed) Rethinking human evolution. MIT Press, Cambridge, MA, pp 319–334

Tattersall I, DeSalle R (2019) The accidental *Homo sapiens*: genetics, behavior and free will. Pegasus Books, New York

Thorpe SKS, Holder RL, Crompton RH (2007) Origin of human bipedalism as an adaptation for locomotion on flexible branches. Science 316:1328–1331

Tuttle RH (1981) Evolution of hominid bipedalism and prehensile capabilities. Philos Trans Roy Soc Lond B 292:89–94

Wallace AR (1858) On the tendency of varieties to depart indefinitely from the original type. Proc Linn Soc (Zool) 3:53–62

Weidenreich F (1939) Six lectures on *Sinanthropus pekinensis* and related problems. Bull Geol Soc China 19:1–110

Weidenreich F (1947) The trend of human evolution. Evolution 1:221–236

Wheeler PE (1984) The evolution of bipedality and loss of functional body hair in hominids. J Hum Evol 13:91–98

Wolpoff MH, Wu XZ, Thorne AG (1984) Modern *Homo sapiens* origins: a general theory of hominid evolution involving the fossil evidence from East Asia. In: Smith FH, Spencer F (eds) The origins of modern humans: a world survey of the fossil evidence. Alan R. Liss, New York, pp 411–484

Chapter 15
Darwinism Without Selection? A Lesson from Cultural Evolutionary Theory

Lorenzo Baravalle

Abstract Wondering about "how Darwinian" cultural change actually is, some authors have recently stressed that there are different degrees to which a process can be considered as evolutionary. Some of them advocate for a central role of selective processes in cultural evolution, while others deny that these are relevant to explain cultural change, if not incidentally. Taking a cue from this debate, in this chapter, I shall discuss a series of theoretical and explanatory commitments usually adopted by those that, like cultural evolutionists, aim to extend evolutionary theory to non-strictly biological domains. My goal is to identify a class of evolutionary factors that, although frequently neglected in the debate, may be actually qualified as Darwinian and, consequently, argue for a more complete picture of evolutionary change. These factors are demographic factors, that is, factors related to the size, density and structure of populations. After having described in some detail in which way they differ from other causes of evolution, I shall relate the discussion in cultural evolutionary theory to a broader debate about the importance of natural selection in Darwinian thinking.

Keywords Cultural evolutionary theory · Cultural change · Natural selection · Population thinking · Demographic factors

15.1 Introduction

Practitioners in cultural evolutionary theory usually consider themselves as full-fledged *Darwinians* (e.g. Richerson and Boyd 2005; Mesoudi 2011; Henrich 2016). Nonetheless, wondering about "how Darwinian" cultural change actually is, Peter Godfrey-Smith (2007, 2009), Tim Lewens (2009, 2013, 2015) and Claidière et al. (2014) have recently stressed that there are different degrees to which a process can

L. Baravalle (✉)
Centro de Filosofia das Ciências, Departamento de História e Filosofia das Ciências, Faculdade de Ciências, Universidade de Lisboa, Lisbon, Portugal
e-mail: lbaravalle@fc.ul.pt

R. G. Delisle (ed.), *Natural Selection*, Evolutionary Biology – New Perspectives on Its Development 3, https://doi.org/10.1007/978-3-030-65536-5_15

be considered Darwinian. Beyond the taxonomic and terminological idiosyncrasies of each author (Brusse 2017), all of them agree that there are three main possible Darwinist interpretations of cultural change. They are—from the strictest (i.e. cultural change is taken to be very similar to biological evolution) to the loosest (i.e. cultural change is only vaguely resembling biological evolution)—the *replicator* interpretation, the *selectionist* interpretation and the *populational* one. According to the replicator interpretation, cultural change is Darwinian insofar as it involves the propagation of discrete units of information analogous to genes, i.e. the memes (Dawkins 1976; Dennett 1995, 2017; Blackmore 1999). More modestly, the selectionist interpretation assumes that cultures are composed by heterogeneous cultural traits or *variants* (Boyd and Richerson 1985; Richerson and Boyd 2005)—that is, traditions, skills, beliefs, techniques, knowledge, etc.—which not necessarily are similar to genes. Yet, given that (I) there is variation between cultural variants, (II) they are relatively faithfully transmitted and (III) they are differentially adopted in virtue of some of their features, then—following a widely accepted conception (Lewontin 1970; Huneman 2015)—it is still possible to say that cultural change instantiates a Darwinian process. Finally, the populational interpretation denies that these conditions are met—if not occasionally—by cultural change and maintains, instead, that cultural evolutionary theory is "evolutionary" just because it relies on a certain extension of *population thinking* to the cultural domain. For the authors supporting the populational view, it is somehow misleading to consider cultural change processes as the effect of selective pressures.

The question is: to what extent a theory that does not acknowledge a central role to selection may be properly considered as "Darwinian"? This is an interesting question not just in the context of the debate over cultural evolution but, more in general, for evolutionary biology as a whole. In fact, throughout the last 50 years or so, many practitioners in the fields of molecular evolution (Kimura 1968; King and Jukes 1969), palaeontology (Gould 1977) and developmental biology (Pigliucci and Muller 2010) have challenged, to a different extent, the prominence of natural selection in genetic evolutionary processes (for detailed discussions of this process of reconsideration of the role of natural selection in evolutionary biology and related fields, see Delisle's, Granovitch's, Tattersall's and Schwartz' contributions to this volume).

In this chapter, I shall interpret some recent developments within cultural evolutionary theory as an example of how an evolutionary theory may be considered Darwinian, even though it does not put selection in the foreground. To be precise, I consider that cultural evolutionary theory may be understood as a *Wrightean* theory. This is because, in addition to explain cultural change by adopting a variational view of populations assimilable to Darwin's one, it also puts emphasis—like the well-known "Shifting-balance theory" by Sewall Wright (1948, 1982)—on structural properties of populations. To this goal, I shall proceed roughly as follows. First, in Sect. 2, I shall discuss, drawing on Lewtonin (1974), the minimal requirements for a theory to be considered as "evolutionary". An evolutionary theory must provide, necessarily, a set of laws of transformation and it must be dynamically sufficient. I shall thus argue that the disagreement between the supporters of the replicator, the selectionist and the populational interpretations is a disagreement about what makes cultural evolutionary theory dynamically sufficient. I shall discuss the differences between these three interpretations in Sect. 3. In Sect. 4, I shall attempt to go beyond these interpretations, by pointing out that both in evolutionary biology and cultural evolutionary theory, practitioners are paying increasingly more attention to the causal import of *demographic* factors of evolution. I shall finally argue, in Sect. 5, that the causal role of these factors can be incorporated in a broad Darwinian/ Wrightian picture of evolution, in which selection, in spite of being an important engine of change, is not necessarily the fundamental one.

15.2 What Is an Evolutionary Theory?

In the first chapter of his classic book, *The Genetic Basis of Evolutionary Change*, Richard Lewontin presents those which, in his opinion, are the basic requirements for a theory to be considered an *evolutionary* theory. They are very general requirements, and they are likely satisfied also by many theories in physics. They should not be intended, thus, as sufficient conditions to identify evolutionary theories *in* biology. Since they are, nonetheless, necessary conditions for any theory aimed to account for evolutionary changes, they provide a valuable starting point for our discussion. Lewontin states that "... the problem of constructing an evolutionary theory is the problem of constructing a state space that will be dynamically sufficient, and a set of laws of transformation in that state space that will transform all the state variables" (1974, p. 8). In other words, the problem of constructing an evolutionary theory is the problem of providing a set of laws—or we may say, more loosely, a set of models—describing the change of the system under study from an instant t to a subsequent instant t', and a suitable *causal* interpretation of the system, such that the change from t to t' can be ascribed to the material factors which are responsible for it.

Thus, for instance, in the *genetic theory of evolution*, we have a set of equations provided by population genetics—which describe different possible transformations in a state space made up of allelic frequencies—that correspond to the laws of transformation of the theory. This theory is, moreover, dynamically sufficient because it identifies, in genetic inheritance and in the differential survival and reproduction of the individuals due to environmental circumstances, the factors that are materially responsible for the transformations in the state space. This is to say that, *at least in principle*, evolutionary biologists may account for the change in allelic frequencies in a specific population by pointing out, with relative precision, the causes of the change.[1]

[1]To be precise, Lewontin introduces a third requirement to be satisfied by evolutionary theories, that is, *empirical sufficiency*. An evolutionary theory is empirically sufficient if scientists are able to *measure* the parameters that they introduce to formulate the laws of transformation. This is tantamount to say that the causal representation offered by an evolutionary theory should not be attainable just *in principle*, but also in practice. Lewontin is sceptical about the capacity of the

Cultural evolutionary theory has received some different mathematical formulations, but the models that are considered by most practitioners as foundational for this disciplinary field are those put forward by Cavalli-Sforza and Feldman (1981) and Boyd and Richerson (1985), within the so-called *dual-inheritance theory*. These models are an intentional extension of population genetics to changing frequencies of cultural traits, or variants. Cavalli-Sforza, Feldman, Boyd and Richerson interpret cultural traits as a class of phenotypical traits transmitted through a channel of inheritance alternative to the genetic one, that is, social learning. The peculiarity of cultural transmission by social learning is that information does not flow mainly vertically (from parents to offspring generation), like in genetic inheritance, but also horizontally (among the members of the same generation), obliquely (from members of the parental generation to members of the offspring generation who are not genetically related to the former) and even from the offspring to the parental generation. Moreover, while the genetic system of inheritance usually determines a fixed number of ancestors for each descendant, social learning allows some individuals to have a variable number of ancestors. Some individuals adopt a cultural trait after having learned it from different and multiple ancestors. Finally, while genetic inheritance is "discontinuous" (in the sense that an allele is either transmitted to the descendant or not), cultural influence comes in degree (that is, a cultural variant can be just partially transmitted).

The change in the distribution of cultural variants in a population from a generation (taken using any arbitrary time interval) to another is, according to this rather standard picture of cultural evolutionary theory, due to both "purely biological" factors (such as selection, migration, drift, etc.—after all, cultural variants are phenotypic traits and, as such, are subject to all these evolutionary forces) and other "properly cultural" factors. These have been conceptualised in a variety of ways. Boyd and Richerson (1985; see also Richerson and Boyd 2005), for instance, talk about distinct kinds of social learning *biases*, which would modify the frequencies of cultural variants depending on the fact that some of them are cognitively more attractive (or more easily memorisable) than others, or are already adopted by prestigious members of the population or by some specific group. Some authors (e.g. Durham 1991; Mesoudi 2011) have suggested to characterise the process resulting from the repeated and large-scale action of social learning biases as *cultural selection*.[2] As we shall see in Sect. 3, other authors consider that this conceptualisation of social learning biases is inaccurate (e.g. Sperber and Claidière 2008; Claidière et al. 2014).

To better understand this criticism we must first, however, consider a preliminary issue. Although all the differences between genetic inheritance and cultural

---

genetic theory of evolution to fully satisfy this requirement. This is certainly a problem also for cultural evolutionary theory, but I shall not discuss it directly here.

[2]Notice that cultural selection does not necessarily favour genetically fittest variants. Since social learning is usually less costly or more effective than individual trial-and-error, partially maladaptive cultural variants (such as certain unhealthy eating habits) may be maintained within a population, evolve and even subvert "genetically-coded" behaviours.

transmission, as well as the effects of the factors producing cultural change, can be, at least in principle, properly formalised (see, for instance, besides the already mentioned Cavalli-Sforza and Feldman 1981; Boyd and Richerson 1985, also Henrich 2001; Henrich and Boyd 2002; El Mouden et al. 2014; Aguilar and Akçay 2018), this is not a guarantee that cultural evolutionary theory is dynamically sufficient. It could be the case, in fact, that we are able to determine the laws of transformation of an evolutionary phenomenon without being able to identify what are, in specific circumstances, the material processes responsible for a certain evolutionary outcome.[3] In particular, in the case of cultural evolution, it has been frequently raised the concern that, differently from what occurs with genetic inheritance, we do not have a satisfactory understanding of the causal processes involved in cultural transmission (e.g. Sperber 1996; Aunger 2001).

15.3 Dynamical Sufficiency in Cultural Evolution

In this section, I shall discuss the three most common interpretations of cultural evolutionary theory as a Darwinian theory. In accordance with the framework presented in the last section, I will argue that they differ mainly because their alternative characterisations of the dynamical features of the processes of cultural change. The replicator interpretation (Sect. 3.1) settles the problem concerning the dynamical sufficiency of cultural evolutionary theory by postulating that the diffusion of cultural traits depends on the cultural transmission of discrete entities analogues to genes, also known as memes. The selectionist interpretation (Sect. 3.2) avoids such a strong analogy between genetic and cultural evolution and grounds the dynamical sufficiency of cultural evolutionary theory on the population-level properties of the processes of cultural change. Finally, the populational interpretation (Sect. 3.3) raises doubts over the correctness of the selectionist interpretation and suggests a deflationary perspective about the dynamical properties of cultural evolutionary processes *per se*. Cultural change, according to this view, would be epiphenomenal with respect to the myriads of social interactions between the members of a population. In the last Sect. 3.4, I shall draw some partial conclusion about the current status of the debate.

15.3.1 The Replicator Interpretation

Originally put forward by Richard Dawkins in his best-seller *The Selfish Gene* (Dawkins 1976), and later popularised by Dennett (1995) and Blackmore (1999), the so-called memetic theory has, at its core, the thesis that the processes of cultural

[3]This is arguably the case of "purely statistical" theories (Matthen and Ariew 2002).

change are carried out by hypothetical entities, i.e. the memes—which would replicate analogously to genes. The problem of the dynamical sufficiency in cultural evolutionary theory is thus easily solved by postulating that cultural evolution works in the same way genetic evolution does. Of course, memeticists are aware of the peculiarities of cultural transmission with respect to genetic inheritance (see Sect. 2 above), as well as of the different paces of cultural evolution and genetic evolution. Nevertheless, these are aspects that empirical research is called to elucidate: they do not necessarily invalidate the analogy (as a matter of fact, even within genetic evolution there are many exceptions to the classical Mendelian and gradualist forms of evolution; see, for instance, Schwartz, this volume).

Naturally, as frequently happens, this apparently easy solution has a high price to pay. Although, in fact, it is true that the analogy between genes and memes may be fruitful even in case it is not perfect (no analogy is!), the burden of proof that the gene is a good model to represent the evolving entities in cultural evolution is on memeticists' side. Memeticists must show that the putative dynamics of memes' replication is consistent with our knowledge about how cultural information is transmitted, and they should ideally provide some evidence that memes indeed exist and are causally efficient. Concerning both issues, however, many difficulties have been raised. I shall summarise here just some of them.

First of all, memeticists are not clear about what entities would exactly count as memes. Memes are arguably not identical to cultural variants (beliefs, skills, traditions, artefacts, etc.), which are commonly considered as a class of *phenotypic* traits (or, in the case of artefacts, objects carrying information potentially influencing phenotypic traits; Odling-Smee et al. 2003). Since memes are by definition analogous to genes, they should be rather considered as the entities which are responsible for the *expression* of cultural variants (just like genes are the entities which are responsible for the expression of non-cultural phenotypic traits). However, this leads to a highly speculative picture of the processes of cultural transmission. Differently from what occurs in genetic inheritance, it would seem that in cultural transmission the phenotypic traits are *directly* transmitted (Sperber 2001; Lewens 2013). In what sense, when we learn a new skill or habit, is there a transmission of *something else* underlying the skill or the habit? And what about the developmental processes supposedly involved in the phenotypic expression of the "memetype"?

The analogy with genes, furthermore, commits memeticists to ascribe to memes a certain fidelity of transmission. Again, it is perhaps not necessary, for the analogy to be tenable, that memetic transmission is as faithful as genetic inheritance. Yet, to the extent that memetic transmission is a process of *replication*, it may well be expected that the socially learned information is approximatively the same as the one that is transmitted. However, as observed by a number of authors (Sperber 2001 is probably the most well-known; see also Sterelny 2017), social learning of skills, knowledge and traditions is not a process of copy-paste, but a far more complex process in which information is reformulated according to the subject's expectations and goals. As evidence of this, notice that it is rarely possible to identify well-defined *lineages* of cultural items or traditions (Wimsatt 1999; Lewens 2013; Morin 2016). Making exception of some rare cases, most interesting cultural traits display a relatively high degree of idiosyncratic variation, which turns the memetic hypothesis quite unlikely.

All these criticisms might perhaps be less pressing if memeticists had been able to provide some empirical evidence of the existence of memes or, at least, of their explanatory import. Neither in this sense, though, memetics seem to have made many steps forward. As some former supporters of memetics themselves noticed (Edmonds 2002, 2005), memetics has failed to produce substantive results. At most, it can be interpreted as a metaphorical (and partial) reformulation of something that can be more fruitfully accounted for within more traditional frameworks of cultural change (Kronfeldner 2011).

15.3.2 The Selectionist Interpretation

According to the "orthodox" view in dual-inheritance theory, the processes of cultural change and accumulation can be considered as analogous to biological evolution even though cultural variants are not analogous to genes (Richerson and Boyd 2005). As already mentioned in the introduction, for the supporters of the selectionist view, all a process needs in order to be properly treated as an evolutionary process is to exhibit *variation*, *heritability* and *variation in fitness*. A process satisfying these requirements is, in addition, not just an evolutionary process, but a *Darwinian* one (Lewontin 1970; Huneman 2015).

Before discussing in some detail the selectionist stance concerning the problem of dynamical sufficiency, it is important to avoid some possible misunderstandings. First of all, we have to distinguish the selectionist view from what we may call a "panadaptationist" view of cultural evolution. According to a supporter of the selectionist view, cultural evolution might be a kind of Darwinian evolution even when selection was not the only factor of change. As a matter of fact, dual-inheritance theorists envisage, as possible causes of cultural change, a broad set of factors, like drift, migration, directed mutation, etc. (Mesoudi 2011; Baravalle 2019), selection thus being just one of them. What characterises the selectionist view against other interpretations of cultural evolution is, rather, a certain emphasis on the population-level properties of the processes of cultural change.

As we have seen in Sect. 2, individuals in a society transmit cultural information through (biased) social learning. Beyond the idiosyncrasies of each interaction between the members of the population, it is possible to find population patterns of distribution of cultural variants. These patterns reflect the overall action of the interactions between individuals in a way that is suitably conceptualised through an evolutionary framework. As a matter of fact, a society may be profitably conceived as a set of different cultural variants (*variation*), changing in composition through time, for reasons due (except when other factors like drift, migration, etc. are in act) to the differential capacity of cultural variants to be adopted—in virtue of their usefulness, attractiveness or memorability (*variation in fitness*).

Of course, this explanation can be interpreted as a merely statistical one. It may be said, for instance, that the population-level description of human societies in terms of changing distributions of cultural variants is nothing more than a useful abstraction (see below, Sect. 3.3). Nonetheless, selectionists seem to have a causally stronger interpretation of the evolutionary framework. In other words, they are committed to the claim that the cultural evolutionary framework is dynamically sufficient. But what does it provide the material support of cultural evolution? In a word, *heritability*. The repeated social interactions between the members of a population constitute a full-fledged *channel* of inheritance, alternative to the genetic one, and it is precisely the persistence of this inheritance channel what guarantees dynamical sufficiency to the cultural evolutionary process.

We have already discussed the characteristics of social learning as channel of inheritance in Sect. 2. The point here is that the selectionist interpretation, differently from the replicator interpretation, dismisses the analogy between cultural variants and genes, but still claims that cultural transmission is faithful enough to be fruitfully compared to genetic inheritance. In spite of the many idiosyncrasies of any single episode of transmission of cultural information, selectionists argue that there is something (i.e. the cultural variant) which is preserved (Henrich and Boyd 2002) and this provides a sufficient basis to consider the processes of cultural change as evolutionary and, more precisely, as Darwinian.

15.3.3 The Populational Interpretation

By analysing a famous model in cultural evolutionary theory on the effects of conformism in the diffusion of a cultural trait (Henrich and Boyd 1998), Lewens (2015) states that it

> aim[s] to show that individuals who learn in a conformist manner are likely to do better than individuals who instead rely on learning from their environment, or who learn by imitating a randomly chosen member of the population. [Henrich and Boyd's] model takes account of the effects of learning, but not in a way that relies on a notion of Darwinian struggle, or selection, going on among cultural traits (Lewens 2015, p. 38).

After having extended similar considerations to another set of models (Henrich and Boyd 2002; on the characteristics of cultural transmission), Lewens concludes that "the explanatory pay-off of the evolutionary stance derives from the surprising nature of aggregation, made visible by the use of mathematical models. In neither case does the pay-off derive from a notion of cultural selection" (Lewens 2015, p. 38).

If the notion of cultural selection is not crucial to cultural evolutionary explanations, it is natural to wonder: in what sense are they Darwinian? In order to answer this question, Lewens invokes bibliographic evidence, mainly drawn from Boyd and Richerson' writings and, in particular, Richerson and Boyd (2005). He observes that:

> in the opening pages of that book, where their basic methodological and explanatory assumptions are laid out, they [Richerson and Boyd] affiliate their approach to a Darwinian tradition in a manner that does not place selection in the foreground, but which instead stresses the importance of population thinking (Lewens 2015, p. 16; emphasis in the original).

Now, in a frequently cited passage, Mayr states that population thinking is the claim that

> All organisms and organic phenomena are composed by unique features and can be described collectively only in statistical terms. Individuals, or any kind of organic entities, form populations of which we can determine the arithmetic mean and the statistics of variation. Averages are mere statistical abstractions; only the individuals of which the populations are composed have reality (1959 [2006], p. 326).

If we adapt this conception to cultural evolution, we may reasonably think—Lewens suggests—that, according to cultural evolutionary theory, only the individual organisms, endowed by specific psychological biases, are "real", while the properties of the populations, like the distribution of cultural variants, are mere statistical abstractions (Lewens 2009, p. 248–9). Cultural change, accordingly, would be—like a change in temperature or pressure in the kinetic theory of gases—just the aggregate of myriads of individual-level interactions: no population-level factors would play a causal role in addition to these interactions. Population-level descriptions certainly have an invaluable explanatory import, because they allow to bookkeep the otherwise epistemically inaccessible contingent circumstances in which individual organisms learn and transmit information. Nevertheless, the population-level properties that occur in such descriptions are causally inert, mere epiphenomena, which are in need to be further explained in terms of individual-level interactions.

This interpretation is also shared by a group of cultural evolutionists frequently labelled as *cultural epidemiologists* (Sperber 1996; Claidière et al. 2014; Morin 2016; Scott-Phillips et al. 2018). Besides being critics of memetics, they are sceptical about the selectionist interpretation for similar reasons: in their opinion, cultural transmission is not faithful enough to be properly considered as a channel. During the many interactions between individuals in a population, cultural information is continuously reformulated. While, *a posteriori*, it is frequently possible to reconstruct a unitary narrative of certain cultural traditions, there is nothing like *a* cultural trait persisting through all the interactions. This may be interpreted as a claim that, *as evolutionary theory*, cultural evolutionary theory is not dynamically sufficient. Of course, there is "something underlying" the process of cultural change, but cultural change in itself is not a phenomenon admitting an autonomous causal characterisation (contrariwise genetic evolutionary processes).

15.3.4 Provisional Remarks

As already mentioned in the introduction, my goal in this chapter is to show that there is at least another alternative way to interpret a process as evolutionary and, more specifically, as Darwinian—besides the ones usually considered in the literature. Before proceeding to that, however, I would like to pause our discussion and highlight some, in my opinion, important points.

First of all, I would like to stress that the interpretation of population thinking advocated by Lewens and cultural epidemiologists is just one among others possible. I argued more extensively for this claim elsewhere (Baravalle 2019), but I think it is useful here to shortly return on it. According to two champions of dual-inheritance theory, McElreath and Henrich (2007), for instance, population thinking does not entail a reduction of population-level dynamics to individual-level interactions (this would be more characteristic of methodological individualism in social sciences) nor a reduction of individual-level interactions to population-level dynamics (this would be methodologically equivalent to adopting a sort of holism). Instead, it entails an interplay between the two levels. Given a certain demic structure, individuals interact with neighbours and with the environment, thus determining the composition of the population in terms of cultural variants. However, the other way around, cultural variants influence, in virtue of their distribution and functional properties, the demic structure and, therefore, individuals' behaviours and interactions. A similar conception was indeed already defended by Boyd and Richerson (1985, pp. 23–4).

According to this alternative interpretation of population thinking—which is in contrast with the one supported by Lewens and cultural epidemiologists—in order to explain cultural change it is not enough to decompose the phenomenon into its basic elements, but it is also necessary to take into account the characteristics of the population as a whole. To provide a complete picture of cultural change, populations *must* be represented as "changing distributions of cultural variants". This is not simply a useful description of something that can also be expressed otherwise (i.e. in terms of additive individual interactions); it denotes a standalone feature of the processes being modelled.

Even admitting this alternative view of population thinking, which is certainly more friendly towards a selectionist interpretation of cultural evolutionary processes, the problem of the fidelity of cultural transmission is still there. On the one hand, and independently from any specific conception of population thinking, a purely populational interpretation of cultural change, although perhaps sufficient to characterise cultural evolutionary theory as "evolutionary" (in the sense that it provides information about the entities carrying the causal weight of the processes under study), is probably too weak to make it a "Darwinian" theory. Not just Darwin's theory, but many other theories which we would not consider as properly Darwinian—for instance, in epidemiology, sociology or economics—adopt some populational view akin to the ones presented in this and the last subsection (see Illari and Russo 2014, Chap. 5).

On the other hand, the issue is that—if we consider the selectionist view as the one stating the minimal requirements for a theory to be Darwinian—it is not clear that cultural transmission is faithful enough to count as an inheritance channel. I think that this problem is prominently empirical. In this sense, further research on the nature of the transmission of information in social interactions and the social learning processes is certainly needed (and it is already ongoing; see, for instance, Kendal et al. 2018). In the remainder of this chapter, though, I shall suggest that all this debate is somehow vitiated by an unwarranted assumption, that is, that the dynamical sufficiency of cultural evolutionary theory entirely (or, at least, mostly) depends on the fidelity of cultural transmission.

15.4 Demographic Factors in Genetic and Cultural Evolution

In Sect. 2, we have seen that the theory of genetic evolution is a full-fledged evolutionary theory, in conformity with Lewontin's characterisation, because it supposedly provides a set of transformation laws and a dynamical characterisation of the processes under study. This theory is dynamically sufficient because it identifies the factors that are materially responsible for the transformations in the state space of allelic frequencies: these factors are genetic inheritance *and* the differential survival and reproduction of the individuals due to environmental circumstances. The debate over the evolutionary interpretation of cultural evolutionary theory discussed in Sect. 3 has been almost entirely focused on the putative similarities between cultural transmission and genetic inheritance, largely neglecting the importance of environmental factors in the dynamical characterisation of evolutionary processes. In this section, I shall focus on a type of environmental factors that, in my opinion, are especially relevant in the causal characterisation of cultural evolution, that is, *demographic* factors.

Interestingly, demographic factors have been somewhat neglected also in evolutionary biology. For this reason, I shall first, in Sect. 4.1, briefly discuss how they have been recently incorporated within traditional models of genetic evolution. Then, in Sect. 4.2, I shall argue that they have always played a prominent role also in cultural evolutionary models. In Sect. 5, I shall finally show how the consideration of demographic factors helps to provide a more articulated *Darwinian* picture of cultural evolutionary theory.

15.4.1 Demographic Factors in Genetic Evolution

Although, as we shall discuss in some detail in Sect. 5, the relevance of demographic factors in evolution was already stressed by an architect of the Modern Synthesis such as Sewall Wright, it is just in the last 30 years or so that their importance in bridging ecological and evolutionary knowledge has started to be fully appreciated. As Lowe et al. (2017) notice:

> The rift between longstanding population genetics theory and current eco-evolutionary research underscores the challenge of fully addressing the forces that drive evolution at the population level. This challenge is not new ([Darwin 1859], p. 64), and is embodied in a historical debate over the power of selection to drive evolutionary change in the face of other, non-adaptive forces—a debate that is largely settled in the field of evolutionary biology, but the root of a narrow view of evolution in many current eco-evolutionary studies (Lowe et al. 2017, p. 142).

In Lowe and his colleagues' opinion, the study of the demic structure of a population is the key to understand the conditions under which selective pressures and other evolutionary forces can act, and with which intensity. Lowe and colleagues identify three main properties of a population that can bias evolutionary dynamics: population size, population density and population connectivity.

Population size is, perhaps, the less surprising one. As a matter of fact, the effects of population size on population dynamics has long been noticed, both theoretically (Wright 1932) and experimentally (Dobzhansky and Pavlovsky 1957). Typically, a reduced population size prevents natural selection to act sorting the fittest traits, increases stochastic fluctuations of allelic frequencies (i.e. genetic drift) and, ultimately, results in a reduction of variation. The other way around, to the extent population size (or, more correctly, *effective* population size; Charlesworth 2009) increases, drift effects diminish and natural selection can act undisturbed.

The effects of population density are more interesting. Population density favours processes of *soft selection* (Wallace 1975). In Lowe et al.'s words:

> Under soft selection, the availability of different niches in the environment and competition among genotypes results in adaptive change. There are no optimal genotypes, but instead a range of successful genotypes *determined by the combination of extrinsic forces*, density and the frequency of other genotypes in the population (Lowe et al. 2017, pp. 142–143; emphasis added).

In other words, soft selection—which, contrarily to "hard" selection, is not an eliminative force, but permits the maintenance of a broad range of different phenotypes—is a force directly related to demographic factors. We may say that population density favours *specialisation*. The high density of a population is generally the effect of the absence of ecological factors causing hard selection (like, for instance, predators or the extreme scarcity of resources, which yield a high mortality). In such conditions, as Darwin had already noted, the "struggle for survival" may acquire less dramatic tones; nonetheless, since resources are usually not unlimited, selection can act more subtly, by promoting survival strategies that resort to resources not exploited by other members of the population.[4]

[4]This is indeed, according to some authors, the most common scenario in evolution (see, for a discussion of ideas related to this claim, Tattersall's chapter in this volume).

Finally, population connectivity is surely the most complex and heterogenous demographic cause of evolution. As a matter of fact, it would be more correct to conceive population connectivity as a class of demographic factors, rather than a single one. What all these factors have in common is that they depend on migration between demes in a structured population. The migration of individuals from a deme to another may generate a series of evolutionary dynamics, from the promotion or the prevention of adaptive evolution, to stochastic fluctuations. Notably, one of the contexts in which the effects of migration in a connected population have been most extensively studied is in the diffusion of behavioural traits. Animal personality, for instance, has been linked to different eco-evo dynamics (Sih et al. 2012). Animals with "exploratory" personalities, who are more propense to migrate than "shy" ones, can be heavily influential in the behaviour of the individuals belonging to the deme which receives them. Populations with a high level of connectivity are more subject to this kind of "invasions" than less connected ones, and different demic structures may favour or prevent a broad range of possible dynamics.

A common feature of all these demographic factors is that they can act very quickly, from an evolutionary point of view, reflecting in almost real-time the effects of the ecological circumstances in which the population happens to find itself. The demographic factors accelerate complex evolutionary dynamics affecting the composition of the population, even when the traits undergoing the evolutionary dynamics have not a well-defined genetic basis (or it is unknown).

15.4.2 Demographic Factors in Cultural Evolution

All the demographic factors discussed in the previous sub-section have also been studied in the context of cultural evolution.

Concerning the effects of population size, the possibly most well-known case is the loss of technology in Tasmania, originally described in cultural evolutionary terms by Henrich (2004). About 18,000 years ago, New Guinea and Tasmania were a unique land with Australia. When the process of insulation started, Tasmanian populations found themselves separated from their Australian cousins. This resulted in a drastic reduction of the population size of Tasmanian populations. One of the effects of this geological transformation was a quick loss of technologies, especially related to fishing, in Tasmanian populations. Like in genetic evolution, also in cultural evolution population size reduction entails a reduction of variation in the population (in the Tasmanian case, the death of fishermen who never taught fishing techniques irremediably resulted in the disappearance of such techniques).

Population density and connectivity are both decisive factors, according to Lycett and Norton (2010), for the emergence of Palaeolithic technological evolution. In their opinion, we can reconstruct three ideal stages of such evolution (occurred during different epochs depending on the geographical region), roughly corresponding to three distinct "demographic levels", or modes. In the first level, human populations are relatively small, with low density and weak interconnectedness. Human populations at this stage exhibit little technological variation. The second and the third modes correspond to a progressive enlargement of human populations, a greater density and higher levels of social interconnectedness. All this contributes to a more diverse cultural variation (similar conclusions are attained by Powell et al. 2009).

These studies do not reveal what kind of cultural variants are favoured by population density and connectivity, but just that these are crucial conditions for having cultural variation. Nonetheless, other works developed in the context of *cliodynamics* (Turchin 2003, 2008) seem to provide more information about frequency-dependent dynamics between specific kinds of cultural variants throughout the history of human societies. Cliodynamics is a dynamical systems approach applied to human history, which accounts for the evolution of a society as the effect of the interaction between many subsystems. In order to formulate reliable hypotheses about historical phenomena, Turchin and his collaborators elaborate data from *Seshat*, an impressive databank which collects information about the social and political organisation of virtually any human group along the history of our species, from Neolithic to modern societies (http://seshatdatabank.info).

Turchin et al. (2018) provide surprising evidence about the predictability of many characteristics of complex societies across different world regions. These researchers captured information on 51 variables reflecting nine characteristics of human societies, such as social scale, economy, features of governance and information systems. Even though Cliodynamics adopts an approach which is different from the one adopted by dual-inheritance theory, we may reasonably consider such variables as classes of cultural variants. Turchin et al. found out that some aspects of social organisation are functionally connected and, for this reason, the classes of behaviour related to them are expected to coevolve in a predictable way. For instance,

> scale variables are likely to be tightly linked, since increases or decreases in size may require changes in the degree of hierarchy (both too few and too many decision-making levels create organizational problems). A similar argument has been put forward for size and governance. The production of public goods, such as infrastructure, may require solutions to collective action problems, and these can be provided by governance institutions and professional officials (Turchin et al. 2018, p. E147).

The spread of religious, philosophical or scientific beliefs are, as it may be expected, correlated with the diffusion of information systems, such as alphabet and writing. On the contrary, other classes of variables, like those related to money, are just loosely related to the evolution of most other variants. However, all these variables are somehow dependent on the demographic and organisational features of the societies.

15.5 Is this Darwinism?

All the evidence discussed in the previous section supports, in my view, the claim that, even if we eventually discovered that cultural transmission is significantly different from genetic inheritance, cultural evolutionary theory would keep being to a certain extent dynamically sufficient in virtue of demographic factors. Demographic factors seemingly play a crucial role both in the emergence and in the evolution of culture. They are what Wimsatt (2019) would call an "external scaffold" of cultural evolution: a set of structural features that are, at the same time, the condition of possibility of cultural dynamics, and determine (although, of course, with a certain flexibility) the paths that societies are likely to follow once certain cultural variants spread.

We are now ready to reconsider our original question: is a, so to speak, "evo-demo" theory of cultural evolution Darwinian? The question is interesting because it has long been thought that the notions of population involved, respectively, in demographic and evolutionary studies were different and, to a large extent, incompatible (Kreager 2009). While Darwinian populations would be characterised by the variation between the members that make them up (in accordance with the interpretation of population thinking supported by Lewens and cultural epidemiologists), the notion of population at stake in demography—especially after the work of Alfred Lotka (1925)—would stress "typological" features of populations (such as fertility, mortality, etc.), shared by all of them. The two notions of population also differ because they suggest distinct conceptualisations of the *environment* in which a population is found. On the one hand, Darwinian populations are typically conceived as "open" populations, in the sense that they are sensitive to ecological influences, which can constitute selective pressure and, thus, change the composition of the population. Lotkean populations, on the other hand, would be "closed", in the sense that the demographic factors are intended as sufficient for determining the changes that the population undergo through time.

These differences between Darwinian and Lotkean populations—although perhaps methodologically relevant in some practices of, respectively, evolutionary biologists and demographers—are not, as far as I can see, an impediment to a fruitful synthesis between the two approaches. As a matter of fact, the basis for this synthesis was already laid, long time ago, by a fervent Darwinist, that is, Sewall Wright.

The work of Wright is best understood in opposition to that of another Darwinian, Ronald Aylmer Fisher. Fisher first exposed his theoretical proposal about evolutionary processes in a series of papers between the 1910s and 1920s and later, more in detail, in *The Genetical Theory of Natural Selection* (1930). Fisher's theory of evolution has been traditionally considered as a synthesis between the hitherto incompatible views of biometricians and Mendelians. In Fisher's view, Biometricians were right in believing that phenotypic variation is continuous and that this fact requires an explanation, but they erroneously explained the transgenerational change in the frequencies of phenotypic variants by postulating a mechanism of blending inheritance and massive mutations. On the other hand, Mendelians correctly believed that the units of inheritance—that is, of course, genes—were discrete entities and that Mendelian inheritance was sufficient to guarantee the large amount of variability necessary to undertake the evolutionary processes, without postulating massive mutations. Nevertheless, they were wrong in underestimating the creative power of natural selection.[5]

The challenge was thus, for Fisher, to explain how discrete entities like genes could generate continuous phenotypic variation and gradual evolutionary novelties. To this goal, he conceived allelic genes as individually responsible for very small phenotypic differences and selectable for any, even weak, adaptive benefit. Fisher conceived natural populations, at a certain generation, as arrays of alleles distributed along a multidimensional Gaussian curve, with the most common combinations in the middle. Ideally, an indefinitely large panmictic population where no evolutionary force is acting persists in this equilibrium. However, this is not what happens in the real world: in real populations, this equilibrium is constantly perturbed. Fisher considered that natural selection is the single most relevant cause of adaptive change in genetic frequencies. It constantly operates, in Fisher's view, on every single locus—with completely addictive effect—by eliminating less fit genotypes, thus gradually increasing the average fitness of the population. This claim is expressed by Fisher's *Fundamental Theorem of Natural Selection*, according to which "the rate of increase in the mean fitness of a population ascribable to gene-frequency changes is exactly equal to the additive genetic variance in fitness" (Edwards 2002, p. 335).

Wright regarded at Fisher's model as overly simplistic (Wright 1931). Following a typical Darwinian *modus operandi* (that is, the comparison between natural and artificial populations), he observed that when breeders try to select a single advantageous character in a large population (this technique is usually called "mass selection"), the consequences are frequently negative: "mass selection has a tendency to lower fitness in the population by turning up all sorts of unwanted gene combinations, normally hidden in heterozygotes, eventually inducing infertility" (Depew and Weber 1996, p. 280). Even when this is not the case, the fixation of the selected trait is extremely slow. In Wright's opinion, the effectiveness of mass selection is severely limited by the fact that genes usually do not phenotypically express themselves separately, but through complex networks of interactions, which are not necessarily genetically inheritable. A more effective way to select a trait would be to breed only those specimens whose networks of interactions are genetically transmissible as a whole.

Unfortunately, it is difficult to identify such specimens. In order to overcome this difficulty, "skilled breeders take the best specimens that can be produced by inbreeding and then outbreeding them with the best specimens of a separate population" (Depew and Weber 1996, p. 280). The inbreeding phase favours the expression of a variety of inheritable interaction networks, thus allowing selection to act

[5]For some more accurate remarks on the controversy between Mendelians and biometricians, and the origins of population genetics, see Adams's chapter and Ochoa's first contribution in this volume.

more efficiently. Wright guessed that something similar had to happen in nature as well. Consequently, he conceived the three-step model known as *shifting-balance* model. According to it, species in nature are subdivided in relatively small inbreeding sub-populations, called demes. During a first phase, every deme, due to its reduced size, undergoes genetic drift. This process favours genetic recombination and increases selectable variation. Once inheritable interaction networks are made phenotypically visible, mass selection may act on the fittest variants, eventually raising the mean fitness of some demes (second step). Finally, interdemic selection, favoured by the interdemic migration of the most adapted individuals, raises global fitness (third step). In order to represent his model graphically, Wright conceived the famous diagram known as *adaptive landscape* (Wright 1932), depicting the evolutionary space as a complex array of valleys and hills, instead of a linear pathway to adaptation, as in Fisher's conception.

Through his shifting-balance model, Wright overcame a simplistic distinction between open and closed populations. Demes within a metapopulation, as well as the metapopulation itself, are open and closed at the same time, and both "openness" and "closedness" play a crucial role in population dynamics. It is precisely because demes are closed that they can undertake those processes (i.e. drift and recombination) that eventually permit intrademic selection to act. At the same time, demes need to be open to allow migration and interdemic selection to act. At a greater scale, the metapopulation is open insofar as, as a whole, it is subject to environmental conditions that influence its composition. Yet, in another sense, it is closed because its demic structure is a "typological" condition, which is necessary in order to produce the characteristic outcome of the shifting-balance process.

Nowadays, most biologists consider that the shifting-balance model is, strictly speaking, flawed (Coyne et al. 1997). This is because the three phases of the shifting-balance process predicted by Wright have been rarely (if ever) observed. Nonetheless, this does not diminish the importance of Wright's approach. Even though natural populations do not behave exactly like Wright thought, the consideration of the structural and demographic features of populations is, in many scenarios, crucial for correctly depicting evolutionary processes (Wade and Goodnight 1998; Wade 2016; see also Sect. 4.1 above).

All this reinforces the impression that, in the context of cultural evolution, the debate about "how Darwinian" is cultural evolutionary theory has been artificially narrowed. First of all, it is not necessary that natural selection is ubiquitous for considering an evolutionary dynamic as Darwinian (see Bellon, Delisle; this volume). In Wrightian dynamics, selection is always accompanied by many other processes modulating the population change. This is indeed something that the supporters of the selectionist approach acknowledge, but it has not been fully appreciated by Lewens and cultural epidemiologists in their criticisms. Both the selectionist and the populational interpretations, nonetheless, are limited to the extent that they focus almost exclusively on the putative resemblances (or differences) between genetic and cultural inheritance. The persistence of a relatively faithful channel of inheritance is, of course, a fundamental condition for the implementation of a Darwinian process. Nevertheless, there are other factors that can play an important role in directing evolutionary dynamics.

As seen in Sect. 4, demographic factors are powerful causes of evolution. They sometimes simply favour some specific kinds of selective dynamics. However, in other cases, they create the conditions for more complex dynamics to occur. The demographic features of a population can drive the population towards certain specific compositions, in terms of cultural variants, or even to large-scale transitions (like in the case of Palaeolithic technological evolution).

15.6 Conclusion

In this chapter, I have assessed some explanatory and methodological issues related to the possibility of considering the theory of cultural evolution as a Darwinian theory. First of all, in Sect. 2, I have presented a minimal characterisation of an evolutionary theory as a theory containing a set of transition laws and a dynamical characterisation of those laws (i.e. a causal description of the evolving system). The genetic theory of evolution is an evolutionary theory because it provides a set of transition laws describing the change of the allelic frequencies in a population, and it dynamically characterises this process by spelling out the mechanisms governing genetic inheritance and the possible ecological factors instantiating evolutionary forces.

In the context of cultural evolutionary theory, there is a disagreement concerning whether it is possible to provide a causal characterisation of cultural change that is, to some extent, analogous to that of the genetic theory of evolution. In this debate, it is more or less implicitly assumed that it is depending on the similarities between genetic inheritance and cultural transmission that it is possible to qualify cultural evolutionary theory as a Darwinian theory (and cultural change as a Darwinian process). In Sect. 3, I have reviewed the three main positions in this debate, and I have concluded that the plausibility of two of them—that is, the selectionist and the populational interpretations—has to be assessed in light of empirical evidence. Still, I have also argued that the emphasis on the similarities or the differences between genetic inheritance and cultural transmission has obscured other possible similarities between biological and cultural evolution.

In Sect. 4, I have stressed that demographic factors—understood as a set of factors related to the structure of populations—are calling attention of both evolutionary biologists and cultural evolutionists as an important cause of evolution. The point relevant for the debate concerning cultural evolution is that, even if we found out that cultural transmission is dissimilar to genetic inheritance in crucial aspects, this would not imply that we need to give up the idea that cultural evolutionary theory is dynamically sufficient in a way that resembles evolutionary theory in biology. To address the concern that such a theory, although evolutionary, would not be Darwinian, I have argued in Sect. 5, that there is a long tradition in evolutionary biology that considers demographic factors as influential in evolutionary dynamics. This way of conceptualising evolutionary change, albeit perhaps not strictly speaking Darwinian, would be, at least, *Wrigthean*.

In conclusion, regarding the role of selection in cultural evolution and the appropriateness of considering cultural change as a selective process, I would say that this should not be the focus of the discussion. Indirectly, the same may be said concerning the broader debate about the centrality of natural selection in biological evolutionary theory. Natural selection is certainly a fundamental process in evolution, and a fundamental process in the original Darwinian formulation of his theory. Nonetheless, theories change through time. In the case of biological evolutionary theory, the Modern Synthesis and, especially, the work of Sewall Wright showed that Darwinian theory is able to incorporate elements that were not included in the Darwinian formulation of the theory (as also emphasised by Esposito, this volume). Rather than discussing whether cultural evolutionary theory is a Darwinian theory because it properly mimics certain aspects of the genetic theory of evolution, we should be open to extend our analysis to other—perhaps sometimes considered as peripheral—aspects of evolutionary biology.

In this sense, the parallel drawn in Sects. 4 and 5 between genetic and cultural evolution concerning the impact of demographic factors in evolution should not be just considered as an invitation to cultural evolutionists to seek the Darwinian origins of cultural evolutionary theory outside the debate over the similarities between genetic inheritance and cultural transmission. Also, it has the goal of stressing the ability of Darwinian theory of including topics and descriptions of evolutionary phenomena elaborated within other disciplinary fields, even when these descriptions do not put selection in the foreground.

Acknowledgements I would like to thank the Brazilian National Council for Scientific and Technological Development (CNPq Grant N°402619/2016-1) and the Fundação para a Ciência e a Tecnologia de Portugal (FCT Contract N° DL57/2016/CP1479/CT0064) for the financial support.

References

Aguilar E, Akçay E (2018) Gene-culture co-inheritance of a behavioral trait. Am Nat 23:311–320

Aunger R (2001) Introduction. In: Aunger R (ed) Darwinizing culture: the status of Memetics as a science. Oxford University Press, Oxford, pp 1–23

Baravalle L (2019) Cultural evolutionary theory as a theory of forces. Synthese. https://doi.org/10.1007/s11229-019-02247-0

Blackmore S (1999) The meme machine. Oxford University Press, Oxford

Boyd R, Richerson PJ (1985) Culture and the evolutionary process. The University of Chicago Press, Chicago

Brusse C (2017) Making do without selection. Biol Philos 32:307–319

Cavalli-Sforza L, Feldman M (1981) Cultural transmission and evolution: a quantitative approach. Princeton University Press, Princeton

Charlesworth B (2009) Effective population size and patterns of molecular evolution and variation. Nat Rev Genet 10:195–205

Claidière N, Scott-Phillips TC, Sperber D (2014) How Darwinian is cultural evolution? Philos Trans R Soc Lond Ser B Biol Sci 369
Coyne JA, Barton NH, Turelli M (1997) Perspective: a critique of Sewall Wright's shifting balance theory of evolution. Evolution 51:643–671
Darwin C (1859) On the origin of species by means of natural selection. Murray, London
Dawkins R (1976) The selfish gene. Oxford University Press, Oxford
Dennett DC (1995) Darwin's dangerous idea: evolution and the meaning of life. Simon & Schuster, New York
Dennett DC (2017) From bacteria to Bach and back: the evolution of minds. W. W. Norton, New York
Depew DJ, Weber BH (1996) Darwinism evolving: system dynamics and the genealogy of natural selection. The MIT Press, Cambridge (MA)
Dobzhansky T, Pavlovsky O (1957) An experimental study of interaction between genetic drift and natural selection. Evolution 11:311–319
Durham WH (1991) Coevolution: genes, culture, and human diversity. Stanford University Press, Stanford
Edmonds B (2002) Three challenges for the survival of memetics. JOM 6
Edmonds B (2005) The revealed poverty of the gene-meme analogy – why memetics per se has failed to produce substantive results. JOM 9
Edwards AWF (2002) The fundamental theorem of natural selection. Theor Popul Biol 61:335–337
El Mouden C, André J-B, Morin O, Nettle D (2014) Cultural transmission and the evolution of human behaviour: a general approach based on the Price equation. J Evol Biol 27:231–241
Fisher RA (1930) The genetical theory of natural selection. Clarendon, Oxford
Godfrey-Smith P (2007) Conditions for evolution by natural selection. J Philos 104:489–516
Godfrey-Smith P (2009) Darwinian populations and natural selection. Oxford University Press, Oxford
Gould SJ (1977) Ontogeny and phylogeny. Harvard University Press, Cambridge (MA)
Henrich J (2001) Cultural transmission and the diffusion of innovations: adoption dynamics indicate that biased cultural transmission is the predominate force in behavioural change. Am Anthropol 103:992–1013
Henrich J (2004) Demography and cultural evolution: how adaptive cultural processes can produce maladaptive losses: the Tasmanian case. Am Antiq 68:197–214
Henrich J (2016) The secret of our success: how culture is driving human evolution domesticating our species and making us smarter. Princeton University Press, Princeton (NJ)
Henrich J, Boyd R (1998) The evolution of conformist transmission and the emergence of between-group differences. Evol Hum Behav 19:215–242
Henrich J, Boyd R (2002) On modeling cultural evolution: why replicators are not necessary for cultural evolution. J Cogn Cult 2:87–112
Huneman P (2015) Selection. In: Heams T, Huneman P, Lecointre G, Silberstein M (eds) Handbook of evolutionary thinking in the sciences. Springer, Dordrecht, pp 37–76
Illari P, Russo F (2014) Causality. Philosophical theory meets scientific practice. Oxford University Press, Oxford
Kendal RL et al (2018) Social learning strategies: Bridge-building between fields. Trends Cogn Sci 22:P651–P665
Kimura M (1968) Evolutionary rate at the molecular level. Nature 217:624–626
King JL, Jukes TH (1969) Non-Darwinian evolution. Science 3881:788–798
Kreager P (2009) Darwin and Lotka: two concepts of population. Demogr Res 21:469–502
Kronfeldner M (2011) Darwinian creativity and memetics. Routledge, New York
Lewens T (2009) Innovation and population. In: Krohs U, Kroes P (eds) Functions in biological and artificial worlds: comparative philosophical perspectives. The MIT Press, Cambridge (MA), pp 243–258
Lewens T (2013) Cultural evolution. In: ZaltaEN (ed), The Stanford encyclopedia of philosophy, Spring 2013 edition. http://plato.stanford.edu/archives/spr2013/entries/evolution-cultural/

Lewens T (2015) Cultural evolution. Oxford University Press, Oxford
Lewontin RC (1970) The units of selection. Annu Rev Ecol Syst 1:1–18
Lewtonin RC (1974) The genetic basis of evolutionary change. Columbia University Press, New York
Lotka AJ (1925) Elements of physical biology. Williams & Wilkins, Baltimore
Lowe WH, Kovach RP, Allendorf FW (2017) Population genetics and demography unite ecology and evolution. Trends Ecol Evol 32:141–152
Lycett SJ, Norton CJ (2010) A demographic model for Palaeolithic technological evolution: the case of East Asia and the Movius Line. Quatern Int 211:55–65
Matthen M, Ariew A (2002) Two ways of thinking about fitness and natural selection. J Philos 99:55–83
Mayr E (1959 [2006]) Typological versus population thinking. In: Sober E (ed) Conceptual issues in evolutionary biology. The MIT Press, Cambridge (MA), pp 325–328
McElreath J, Henrich G (2007) Modelling cultural evolution. In: Dunbar R, Barrett L (eds) Oxford handbook of evolutionary psychology. Oxford University Press, Oxford, pp 571–585
Mesoudi A (2011) Cultural evolution: how Darwinian theory can explain human culture and synthesize the social sciences. The University of Chicago Press, Chicago
Morin O (2016) How traditions live and die. Oxford University Press, Oxford
Odling-Smee FJ, Laland KN, Feldman MW (2003) Niche construction: the neglected process in evolution. Princeton University Press, Princeton
Pigliucci M, Muller G (2010) Evolution – the extended synthesis. MIT Press, Cambridge
Powell A, Shennan S, Thomas MG (2009) Late Pleistocene demography and the appearance of modern human behavior. Science 324:1298–1301
Richerson PJ, Boyd R (2005) Not by genes alone: how culture transformed human evolution. The University of Chicago Press, Chicago
Scott-Phillips T, Blancke S, Heintz C (2018) Four misunderstandings about cultural attraction. EvolAnthropol 27:162–173
Sih A et al (2012) Ecological implications of behavioural syndromes. Ecol Lett 15:278–289
Sperber D (1996) Explaining culture. A naturalistic approach. Blackwell, Oxford
Sperber D (2001) An Objection to the memetic approach to culture. In: Aunger R (ed) Darwinizing culture: the status of Memetics as a science. Oxford University Press, Oxford, pp 163–174
Sperber D, Claidière N (2008) Defining and explaining culture (comments on Richerson and Boyd, Not by genes alone). Biol Philos 23:283–292
Sterelny K (2017) Cultural evolution in California and Paris. Stud Hist Phil Biol Biomed Sci 62:42–50
Turchin P (2003) Complex population dynamics: a theoretical/empirical synthesis. Princeton University Press, Princeton
Turchin P (2008) Arise "Cliodynamics". Nature 454:34–35
Turchin P et al (2018) Quantitative historical analysis uncovers a single dimension of complexity that structures global variation in human social organization. PNAS 115:E144–E151
Wade MJ (2016) Adaptation in metapopulations. University of Chicago Press, Chicago
Wade MJ, Goodnight CJ (1998) Genetics and adaptation in metapopulations: when nature does many small experiments. Evolution 52:1537–1553
Wallace B (1975) Hard and soft selection revisited. Evolution 29:465–473
Wimsatt WC (1999) Genes, memes, and cultural heredity. Biol Philos 14:279–310
Wimsatt WC (2019) Articulating Babel: a conceptual geography for cultural evolution. In: Love AC, Wimsatt WC (eds) Beyond the meme. University of Minnesota Press, Minneapolis, pp 1–41
Wright SW (1931) Evolution in Mendelian populations. Genetics 16:97–159
Wright SW (1932) The roles of mutation, inbreeding, crossbreeding, and selection in evolution. Proceedings of sixth international congress on genetics 1:356–366
Wright SW (1948) On the roles of directed and random changes in gene frequency in the genetics of populations. Evolution 2:279–294
Wright SW (1982) The shifting balance theory and macroevolution. Annu Rev Genet 16:1–19

Part VI
Teaching Evolution

Chapter 16
Beyond Survival of the Fittest: A Look at Students' Misconceptions About Natural Selection and Evolutionary Theory

Elizabeth Marie Watts

Abstract The emphasis on natural selection when teaching about evolutionary mechanisms has led to a number of misconceptions about the validity of the theory of evolution and the societal dangers of the theory of evolution. Natural selection is often falsely and categorically translated as a "struggle for survival" or "survival of the fittest" allowing many antievolution groups to grab ahold of the concept of natural selection like the perceived Achilles' heel of evolution. General antievolution arguments claim that evolution—seen synonymously with the concept of natural selection—teaches us to be selfish and that accepting the theory of evolution causes people to abandon their morals and develop callous, narcissistic or even inhuman behavioral patterns. It is clear that none of these claims reflect a true understanding of evolution, nor do they reflect the intent of educational institutions. It is thus imperative to rethink the means by which we teach students about the concept of natural selection—particularly with respect to human evolution. This paper will look at concrete misconceptions about the theory of evolution with a particular focus on the misinterpretation of natural selection and how these might be amended.

Keywords Charles Darwin · Natural selection · Evolution · Adaptation · Science education · Misconceptions · Conceptual change

16.1 Introduction

Evolutionary theory is a foundational concept in the biological sciences and research has shown that students' understanding of evolution greatly impacts their overall learning potentials in the field of biology (Dolter 2016). Unfortunately multiple studies have affirmed that the theory of evolution is one of the most poorly accepted and understood scientific theories (Barnes et al. 2017a). When examining the

E. M. Watts (✉)
Biology Education Research Group, Institute for Zoology and Evolutionary Biology, Jena, Germany
e-mail: elizabeth.watts@uni-jena.de

R. G. Delisle (ed.), *Natural Selection*, Evolutionary Biology – New Perspectives on Its Development 3, https://doi.org/10.1007/978-3-030-65536-5_16

concept of "evolution", most associate the concept with Charles Darwin, but there were many ideas about biological evolution that predated the publication of *On the Origin of Species* (1859) anywhere from decades to millennia whereby Darwin's most important contribution to the conversation was the idea of natural selection (Delisle 2017).

The basic concept of natural selection is not excessively complex as Timothy Shanahan explains:

> Living things tend to differ slightly from one another in ways that confer on some a small advantage in the struggle for survival and reproduction. Some of these characteristics are heritable and are passed on to offspring, who in turn exhibit differential fitness with respect to their own (often slightly different) environments. Over time, kinds of living things become better adapted to their diverse environments and tend to further diverge from one another. Adaptation and diversification are thereby explained by appeal to natural causes alone (Shanahan 2017).

Understanding natural selection is not only important in a theoretical sense but is also becoming increasingly important in policy decisions within fields of medicine, agriculture, and resource management (Gregory 2009).

Due to the central importance of natural selection as the mechanism that Darwin used to describe the evolutionary process, it is not surprising that high school instruction on evolution focuses primarily on the process of natural selection (Catley 2006). Yet despite the apparent simplicity of the idea, studies have shown that high school students (Dolter 2016; Ferrari and Chi 1998; Anderson et al. 2002; Hokayem and BouJaoude 2008; Nehm and Schonfeld 2008; Nehm et al. 2009; Lawson and Worsnop 1992) and teachers (Nehm et al. 2009) struggle with understanding the concept and in its application to evolutionary theory. Many of these studies regarding students' difficulty with natural selection have further shown that these difficulties are not limited to school students but also re-emerge in college level biology courses (Dolter 2016; Anderson et al. 2002; Dagher and Boujaoude 2005; Hokayem and BouJaoude 2008; Nehm and Schonfeld 2008; Novick et al. 2014).

A number of hypotheses exist regarding the underlying cause for the difficulty in understanding natural selection, which can be broadly broken down into two categories: choice and education (Gregory 2009). In other words, understanding the mechanism of evolution necessitates that a person first accepts that evolution as a valid theory. When a person generally rejects the concept of evolution, it is not surprising that they do not have a strong grasp of evolutionary mechanisms. The second group of hypotheses involves the manner in which evolution is taught or mistaught. In this chapter, we will examine the misunderstanding of natural selection from both of these separate standpoints and attempt to develop educational interventions that could lead to an increased general understanding of the mechanism of natural selection and its application to evolutionary theory.

16.2 Misconceptions and Misunderstandings: Rejection Based on Historical, Cultural, and Societal Misunderstandings of Evolutionary Theory

16.2.1 Historical Battle Between Religion and Science

The question as whether there is an organic conflict between religion and science has already been addressed and answered by many experts. Michael Ruse, for example, has explicitly stated that this struggle is more legend than truth (2001). Stephen Jay Gould with his concept of "Nonoverlapping Magisteria" has also vehemently proclaimed that there is an absolute lack of conflict due to the two very different realms of religion and science: "The lack of conflict between science and religion arises from a lack of overlap between their respective domains of professional expertise—science in the empirical constitution of the universe, and religion in the search for proper ethical values and the spiritual meaning of our lives" (1997). Even Pope Benedict XVI and his predecessor Pope John-Paul II have both praised the role of science in the evolution of humanity and acknowledged the strength of the theory of evolution allowing Catholics to avoid any conflict between their belief system and scientific progress (Numbers 1998). The statement made by Pope Benedict XVI is quite similar in fact to the basis of Gould's argument as he explained: "The story of the dust of the earth and the breath of God, which we just heard, does not in fact explain how human persons come to be but rather what they are. . . . And vice versa, the theory of evolution seeks to understand and describe biological developments. . . . To that extent we are faced here with two complementary—rather than mutually exclusive—realities" (Ratzinger 1995).

Yet, despite these proclamations from both scientists and religious leaders, creationists continue to fight against the teaching of evolution claiming that it contradicts the biblical account of special creation and thus leads to a loss of faith (Ham 2012, 2013; Morris 1961, 1972, 1974, 1985, 1989, 2001). The reasoning behind this fear lies therein, that, if evolution tells a different story than what is in the Bible and if evolution was true then the Bible would be false or allegorical at best. If the Bible is no longer seen as the literal word of God, then doubt arises to whether or not there is a God, which according to creationists' claims could cause moral demise through the loss of faith or Christian values (Watts 2019; Numbers 1992, 2006). As founder of the Creation Museum, Ken Ham has explained—the Genesis story forms the foundation of Christianity and if Genesis were to be lost, Christianity would certainly tumble (2012).

Yet, according to current and historical biblical scholars, the Bible was never intended to be read literally (Hyers 1984; Ehrman 2005, 2009). This has been central to Christian teachings since St. Augustine (354–430) who actively argued against the literal interpretation of biblical texts, explaining that the Bible was written in a language that should be understood by relatively uneducated people since this was the characteristic of the mass population at the time that the Bible was revealed to human kind (Dixon 2008).

This is known as the principle of accommodation and according to this principle; Genesis does not need to be read as a literal account of the creation of the Earth for it to provide a foundation of the Judeo-Christian belief system that revolves around the concept of a single, almighty, omniscient God. When Genesis is read in this manner, it poses no problems with evolutionary biology, as can be seen by the theistic evolution individuals who are able to maintain their faith while simultaneously embracing science (Scott 2009).

What does however appear to be true is that there is an organic conflict between fundamentalism and science whereby fundamentalism is defined as: "A form of a religion, especially Islam or Protestant Christianity, that upholds belief in the strict, literal interpretation of scripture" (Oxford dictionary). Clearly the objection to evolutionary theory that is raised by Christians is not caused by a natural conflict between science and religion but is instead caused by a specific belief in the strict and literal interpretation of the Bible which is directly contradicted by evolution (Ruse 2001; Watts et al. 2016b).

According to Eugenie Scott, this form of Christian fundamentalism formed the basis in the United States for the antievolutionism of the 1920s Scopes era as well as the present day whereby the emphasis has always been placed on five main points of literal interpretation: (1) the inerrancy of Scripture, (2) the Virgin Birth of Christ, (3) Christ's atonement for our sins on the cross, (4) His bodily resurrection, and (5) the objective reality of His miracles (Scott 2009).

16.2.2 Survival of the Fittest and Social Darwinism

It is important to look at how Darwin's theory of biological evolution transformed into other concepts such as "Social Darwinism" since it is this aspect that fuels so many of the fears that creationist harbor, in other words: teaching evolution will cause people to adopt "Darwinistic" belief-systems that guide their behavior (Watts et al. 2016a). Social Darwinism can be understood as the transfer of the Darwinian concept of struggle for existence in nature to the realm of human existence at the individual level, races or nations (Bowler 2003). While this concept is associated with Darwin, Darwin was not responsible for this transformation of his theory; as early as the 1860s philosophers began to apply his biological concepts to sociological structures and politics (Claeys 2000). While Darwin's theory did not give rise to social Darwinistic ideologies, by referring to these biological arguments regarding nature's laws, philosophers and political leaders were able to substantiate their own opinions by lending it a waft of scientific validity (La Vergata 1994; Junker and Hossfeld 2009). While there is no uniform ideology of Social Darwinism, it can be understood as a complex web of ideologies that exploit the idea of "survival of the fittest" in different ways (Bowler 2003; Paul 1988).

Although, the concept of "survival of the fittest" is associated with Darwin and his theory of natural selection, Darwin did not coin the phrase or even use it in the original 1859 publication of *On the Origin of Species*. The term was actually first used by Herbert Spencer in his book *Principles of Biology* in 1864 (Bowler 2003). Spencer was a polymath who was interested in Darwin's work and was the first to use extend Darwin's ideas into the realms of sociology and ethics and thus he is often seen as the father of Social Darwinism (Hodgson 2004). Darwin did not make any public statement in opposition to or support of Spencer's usage of his theory and did decide to incorporate the term "survival of the fittest" into the 1869 version, the *Origin of Species* at Alfred Russel Wallace's insistence (Hodgson 2004; Claeys 2000).

Karl Marx (1818–1883) also attempted to substantiate his own theories on modern communism by applying Darwinian principles and made direct references to Darwin in *Das Kapital* (1873). Marx's admiration for Darwin's theory can also be seen in his personal communications (Gould 1977, 1992). In a letter to his friend Ferdinand Lasalle in January 1861, for example, Marx wrote: "Darwin's work is most important and suits my purpose in that it provides a basis in natural science for the historical class struggle" (Marx 1861).

Friedrich Nietzsche (1844–1900) also saw evolution as an accurate explanation of biological history and quickly realized that it could have far-reaching effects on other philosophical realms. If evolution is true, then there is no longer a need for God as He is no longer a requirement to explain the existence of humankind. This meant to Nietzsche that evolution would lead to the collapse of all traditional values and moral sediment of society (Birx 2000). This caused Nietzsche to equate morality with herd instinct as he stated in his 1882 publication *The Gay Science*:

> *Herd-Instinct.* Wherever we meet with a morality we find a valuation and order of rank of the human impulses and activities. These valuations and orders of rank are always the expression of the needs of a community or herd: that which is in the first place to its advantage—and in the second place and third place—is also the authoritative standard for the worth of every individual. By morality, the individual is taught to become a function of the herd, and to ascribe to himself value only as a function. As the conditions for the maintenance of one community have been very different from those of another community, there have been very different moralities; and in respect to the future essential transformations of herds and communities, states and societies, one can prophesy that there will still be very divergent moralities. Morality is the herd-instinct in the individual (p. 116).

He also made his most well-known statement "God is dead" in this same publication. Nietzsche also formulated the concept of the "Übermensch" which was later used by Adolf Hitler (1889–1945) to describe the superiority of the "Aryan" or Germanic master race (Alexander et al. 2011).

The idea of "survival of the fittest" has been blamed for causing both world wars and Nazism (Hodgson 2004). In 1917, Vernon Kellogg (1867–1937) drew a connection between German war atrocities and Darwin's concept of natural selection or survival of the fittest in his book *Headquarters Night*; a record of conversations and experiences at the headquarters of the German army in France and Belgium. As Kellogg wrote:

> Well, I say it dispassionately but with conviction: if I understand theirs, it is a point of view that will never allow any land or people controlled by it to exist peacefully by the side of a people governed by our point of view. For their point of view does not permit of a live-and-

> let-live kind of carrying on. It is a point of view that justifies itself by a whole-hearted acceptance of the worst of Neo-Darwinism, the Allmacht of natural selection applied rigorously to human life and society and Kultur (p. 22).

Strong ties have also been drawn between Hitler and the concept of survival of the fittest. Although Hitler does not ever mention Darwin by name, he does use biological concepts to argue his opinion about the necessity of maintaining the purity of the superior Aryan race as he stated in *Mein Kampf* (1925):

> Nature supplies this by establishing rigorous conditions of life to which the weaker will have to submit and will thereby be numerically restricted; but even that portion which survives cannot indiscriminately multiply, for here a new and rigorous selection takes place, according to strength and health. If Nature does not wish that weaker individuals should mate with the stronger, she wishes even less that a superior race should intermingle with an inferior one; because in such a case all her efforts, throughout hundreds of thousands of years, to establish an evolutionary higher stage of being, may thus be rendered futile.
>
> History furnishes us with innumerable instances that prove this law. It shows, with a startling clarity, that whenever Aryans have mingled their blood with that of an inferior race, the result has been the downfall of the people who were the standard-bearers of a higher culture (p. 223).

Through the use of Darwin's concept of natural selection to substantiate particular philosophies and ideologies, it became associated with a worldview that showed little or no sympathy to those individuals who could not support themselves (Bowler 2003). Those who oppose evolution often claim that the teaching of evolution leads to ideologies such as communism, atheism, fascism. Moreover, there is often an attempt made to hold Darwin's theory morally responsible for the crimes committed by Hitler in order to undermine the theory of evolution (Richards 2013).

Misunderstanding Evolutionary Theory from an Educational Perspective

Confusion Despite Apparent Simplicity

While the intricate details involved in evolutionary theory can be complex, particularly when considering the new synthesis, the basic concept of natural selection is remarkably simple (Gould 1996; Mayr 1997). As Ferrari and Chi point out, natural selection makes three basic claims followed by an syllogistic conclusion, i.e., (1) organisms produce more offspring than can survive, yet populations remain stable, (2) offsprings are related to their parents but are not immutable, (3) potential variations are passed down from generation to generation (1998). This is followed by the inference that survivors will tend to be those whose variations are best suited to meet the changing conditions of the local environment (Ferrari and Chi 1998). Yet, misconceptions[1] about the basic principles of evolution abound and are robust

[1] In the literature, a differentiation is made between misconceptions and false information where the latter are much easier to correct (Chi 1997).

despite educational interventions (Bishop and Anderson 1990; Demastes et al. 1995, 1996; Jensen and Findley 1996; Settlage 1994; Zuzovsky 1994).

Two general hypotheses abound for why this misunderstanding of natural selection is so widespread, namely (1) due to the fact that the theory of evolution is still entirely rejected by a large portion of the population and (2) due to lack of formal biology education (Gregory 2009). These hypotheses appear to be supported by the popularity of so-called "alternative theories" in the United States such as intelligent design, creationism, and creation science. Studies have shown that despite the relative simplicity of Darwin's theory, it remains a source of controversy and confusion among the general public with only 35% of Americans believing that the theory of evolution is well supported by evidence (Newport 2004). Interestingly, these surveys have also showed that respondents do not state that they are unaware of the evidence but rather that they strongly believe such evidence does not exist, i.e., that the theory is *not* well supported by evidence. In fact, it has been found that one in three American adults firmly rejects the concept of evolution (Miller et al. 2006), while instead convinced of alternative "theories" such as creationism, creation science, or intelligent design. The authors of the seminal 2006 piece hypothesize that the popularity of such alternative explanations is based on the high prevalence of Christian fundamentalism in the United States and because evolution education has been politicized, it becomes part of the partisanship divides there.

Yet, the degree of controversy that surrounds evolutionary theory does not appear to be adequate in explaining the degree of confusion that arises amongst the general public regarding evolutionary mechanisms, as there appears to be both complex cognitive and emotional factors that also contribute to this widespread confusion. From a cognitive level, many explanations exist to explain the existence of misconceptions despite formal education, namely: (1) students' difficulty of understanding underlying concepts such as population, adaptation or frequencies, (2) students' difficulty in reconciling different levels of organization, i.e., differentiating between genes, individuals, populations, and species, and (3) students' difficulty in understanding natural dynamics or the time frame of evolution (Ferrari and Chi 1998). In addition to the multiple causes for misconceptions, there are also a number of different varieties of misconceptions that can be broken down into three main categories: maintenance of naïve explanations, e.g., Lamarckian or teleological explanations, category mistakes or preference for so-called "alternative theories", e.g., intelligent design or creationism.

Naïve Explanations (Shtulman)

Darwin was not the first to contemplate how different species arose. In fact ideas about species' ability to adapt to their environment were put forth by Greek philosophers in the seventh century BCE (Mayr 1982). According to Mayr, the major distinction between Darwin's theory of evolution and the theories developed by his predecessors is that Darwin developed a "variational" theory of evolution, as opposed to the "transformational" theories of his predecessors and contemporaries

(Mayr 2001). In other words, while other theories explained adaptation as the transformation of a species or a species' "essence", Darwin explained these adaptions as the selective propagation of randomly occurring mutations within a population (Shtulman 2006).

While variationist explanations see adaptation as a two-step process involving (1) mutations leading to individual variation and (2) the elimination or reproduction of these differences in accordance with their utility to survival and reproduction, transformationists combine these two concepts into one (Shtulman 2006). There is a prevalence of transformational theories in the history of biology such as Lamarck's acquired traits (1809), Haeckel's recapitulation of ontogeny (1876), and Berg's chemical structure of protoplasm (1926). The prevalence of such theories appears to be based on the fact that individuals are predisposed to essentialize species, which explains why early evolutionary theorists tended to interpret evolution as the transformation of a species essence and according to recent studies conducted by Shtulman (2006), this also predisposes modern students to likewise rely on transformationist explanations.

Shtulman's hypothesis in his study was that the tendency for early evolutionary theorists to essentialize species is inherently human and would thus also cause modern students to use the same type of explanations. His hypothesis was based on multiple other studies which have already established that the individual cognitive development often parallels theory development in the history of science (2006), whereby students' naïve explanations resemble earlier theories in fields such as mechanics (McCloskey 1983), thermodynamics (Wiser 1988), acoustics (Mazens and Lautrey 2003), physical chemistry (Smith et al. 1985), and cosmology (Vosniadou and Brewer 1992).

In his study, participants were tested on their understanding of six major evolutionary phenomena: variation, inheritance, adaptation, domestication, speciation, and extinction. Specifically, Shtulman wanted to determine whether the students would interpret all phenomena within a *variational* framework or a *transformational* framework. His study showed that although transformational theories have not existed within the scientific community for almost 100 years (Bowler 1983), they continue to be used within modern populations as many students continue to rely on transformationist phrases such as "need to", "have to", "in order to", or "must" in describing evolutionary processes; students of all ages prefer teleological explanation of biological change to mechanic ones (Shtulman 2006). As Werth writes, "students tend naturally to see all evolutionary change as adaptive, progressive, optimal, and teleological, with improvement achieved as needed or desired by organism (if not as planned as advanced)" (2012). According to Barnes et al., it is this preference for teleological explanations that presents the largest hurdle with regards to understanding natural selection and ultimately the acceptance of evolution according to a recent study (2017a).

Mistaken Categorization (Ferrari & Chi)

Category mistakes occur when students assign a concept to the wrong category. This occurs most commonly due to perceptual attributes, e.g., a young child mistakenly believes that dolphins are fish rather than mammals based on the characteristic attributes. Chi states that such category mistakes hinder students from learning and understanding certain scientific concepts (Chi 1992, 1997), particularly when students assign concepts to an ontologically distinct category from the true category (Ferrari and Chi 1998).

As Ferrari and Chi explain, one can best understand what is meant by ontologically distinct categories by examining the difference between artifacts and animate objects. Here it is clear that an animate object such as a dog could not inherit the properties of artifacts, i.e., one could not refer to the dog as being "finely crafted" since animate objects support predicates such as "grow" whereas artifacts support predicates such as "hand-made". During learning processes, ontological shifts may be necessary as in the case of recategorizing the dolphin as a mammal, while other learning only requires a refinement of one's knowledge without an ontological shift. This lower level cognitive operation has been called belief revision (Chi 1992, 1997).

With regard to science education, Chi (1997) proposes that natural processes can be divided into two basic categories: events and equilibration, whereby six features differentiate events from equilibration. Events (1) consist of distinct actions, (2) whereby actions are bounded with a clear beginning and end, (3) actions of an event occur in sequential order, (4) where the sequence of actions are contingent or causal, (5) events are contingent also goal-directed, and (6) the event is completed when the goal is achieved. Equilibration processes on the other hand have properties that are diametrically opposed to these event-like features. Equilibration processes (1) have uniform actions, (2) are continuous or ongoing without a beginning or end, (3) occur simultaneously, (4) involve actions that occur randomly and independently, (5) are the net effect of the random actions, and (6) are continuous and dynamic (Ferrari and Chi 1998). According to these attributes, Ferrari and Chi propose that the modern theory of evolution and specifically the mechanism of natural selection is an equilibration concept and have developed an educational theory which proposes that students' failure to understand natural selection is due to the attribution of event-like properties to the concept of evolution.

To test this theory, Ferrari and Chi conducted interviews with forty college students with no prior college courses in biology. The students were asked to solve five prediction-explanation problems which were used to assess their understanding of the five basic principles of the Darwinian explanatory pattern (intraspecies variability, heritability, differential survival rate, differential reproduction rate, and accumulation of change over time). The students' answers were then analyzed using two different coding schemas.

The results of the study showed that when students provide false explanations of evolution, they tend to use Lamarckian arguments, whereby organisms realize what features they require, develop these features, and then pass these features on to their offspring. As Ferrari and Chi explain "Lamarckian notions are prevalent and are consistent with a causal, intentional, event-like process" (1998, p. 24). They posit that this type of explanation is more intuitive as humans have a predisposition to perceive all processes as events and that this predisposition explains why it is so difficult to undo already existing misconceptions when they are based on this type of category mistakes.

Hinderances and Obstacles in Science Education

The Necessity of Conceptual Change

The term conceptual change describes the process by which one's perceptions or thought schemas transform over the course of a person's lifetime or throughout human development. While the term is often attributed to Thomas Kühn's explanation of conceptual change as a paradigm shift within the history of science, in education and psychology, the idea of conceptual change is often traced back to Jean Piaget (Sinatra et al. 2008). Piaget (1896–1980) was a Swiss psychologist who became the first psychologist to make a systematic study of cognitive development. He is most well-known for his work on child development and his theory of cognitive development and epistemological view are together called "genetic epistemology".

Within the framework of science education, the term "conceptual change" has received a number of various definitions. According to Reinders Duit and David Treagust, the term "conceptual change" is often misunderstood as "an exchange of pre-instructional conceptions for the science concepts", yet they propose that "conceptual change" should be understood as the fundamental restructuring of preinstructional conceptual structures in order to allow for the acquisition of science concepts (2003). In this sense, conceptual change is much more a description of the pathway from the preconception to the acquisition of the intended knowledge (Duit 1999). The mechanisms of conceptual change in education are often described using Piaget's notions of *assimilation* and *accommodation*, which were part of his larger cognitive development theory.

According to Piaget, cognitive development is a progressive restructuring of the brain which results through the combination of biological maturation and environmental stimulus. Piaget believed that as children mature, they do not just learn about their environment in a passive way but instead develop certain schemas or patterns of knowledge in order to help them remember, organize, and process information. According to Piaget's theory, when children encounter new information, they attempt to reconcile this new information with existing thought patterns. He proposed that the human brain was programmed to seek a state of equilibrium, which it achieves educationally through the processes of assimilation and accommodation, whereby Piaget defines assimilation as the process by which new information is integrated into pre-existing cognitive schemes, and accommodation as the means by which humans alter pre-existing cognitive schemes so that they become aligned with newly acquired information (Berger 2008). In other words, assimilation processes use prior knowledge structures to understand new information while accommodation focuses more attention on the new information and requires a restructuring of prior knowledge (Linnenbrink and Pintrich 2004).

As understanding natural selection as a mechanism for evolution is not an inherent observation that children make during their natural development, it requires that they either assimilate or accommodate this knowledge into existing cognitive structures during formal education. As Shtulman states, "the fact that evolution is simply not evident from one's interaction with a seemingly static biological world suggests that few students would have contemplated the problem of biological adaptation prior to taking a biology class, and even fewer students would have devised their own solution to this problem" (2006).

Thus, conceptual change is almost always necessary for students to acquire a correct understanding of mechanisms of natural selection. However, there are multiple obstacles to conceptual change according to Sinatra, Brem, and Evans, who point out three main barriers to conceptual change: developmental constraints, prior knowledge, as well as emotions and motivation (2008).

Cognitive Obstacles to Accepting Evolution by Natural Selection

As multiple authors have stated, natural selection in itself is not particularly difficult to understand. Thagard and Findlay attribute its simplicity to the fact that "concepts of evolutionary change, genetic variation, struggle for existence, and natural selection can all be described qualitatively, without the mathematical complexities that impede understanding of major theories in physics such as relativity and quantum mechanics" (2009). Yet, students' understanding of natural selection is often plagued by misconceptions and misunderstandings based on intuitive ideas about intentionality, teleology, and essentialism (Cooper 2016). In other words, the conceptual difficulties that accompany natural selection are due to the fact that the statistical and emergent characteristic of the biological processes involved in evolutionary theory do not fit well with innate "common sense explanations" as humans are most familiar with mechanical and intentional processes, whereby intentional processes are driven by decisions, beliefs, and desires, while mechanical processes depend on causal relationships that can be broken down into the interaction of various parts (Thagard and Findlay 2009).

The so-called *population thinking* (Mayr 1982), which is required to understand evolution is inherently difficult for human cognition since it requires an understanding of probability theory and the theory of statistical inference, which has only developed within the past few centuries of human thought (Hacking 1975). Additionally, the true understanding of natural selection requires an internalization of the emergent processes, whereby large effects result from microscopic operations. Emergent processes produce emergent properties, which are defined as a property belonging to a system but not to any of the system's parts (Bunge 2003). As Thagard and Findlay explain, these types of properties abound in the natural world and are seen in simple examples such as how the effect that salt has in stimulating our taste buds is difficult to predict based on the individual properties of sodium and chloride.

Research on students' understanding of emergent properties has already shown that while students can readily understand the state and properties of a given entity, they have a particularly difficult time understanding processes that occur over time. In these instances, students appear to more easily understand those processes that are direct and have an identifiable causal agent (Chi 2005, 2008). Clearly neither of these is true in the case of evolution.

Ironically, one of the largest cognitive obstacles to understanding natural selection is when students are asked to contemplate how natural selection could have influenced the development of such cognitive processes, i.e., how natural selection could have led to the development of human thought. Students are not the only ones who have difficulty with this concept as apparently even Alfred Wallace was unable to accept that the process of natural selection applied to the human mind since he believed it to have an irreducible spirituality (Thagard and Findlay 2009). Recent studies have shown that students find Darwinian explanation of the development of human thought to be implausible (Ramney and Thanukos 2009). According to Thagard and Findlay, the major obstacle herein lies in the understanding of emergent processes because a person's comprehension of the evolution of human thought requires a two-fold understanding of emergent properties: (1) the understanding of the evolution of the human brain through natural selection as well as (2) the production of human thought through the interactions of neurons within the brain (2009).

Emotional Barriers to Conceptual Change

Traditionally, the focus placed on understanding and describing the cognitive mechanisms involved in the learning process means those complex affective characteristics such as attitudes, beliefs, motives, and emotions are often overlooked. Yet, the seminal paper by Pintrich et al. in 1993 highlighted the need to contemplate these affective characteristics in order to include motivational constructs into conceptual change models. According to Sinatra et al., Pintrich's publication led to new trends in educational psychology whereby conceptual change has been increasingly characterized as social, contextual, motivational, and affective in nature (Dole and Sinatra 1998; Sinatra et al. 2008). This has also led to new lines of investigation that examine how learning science affects students emotionally, culturally, and personally.

Part of the emotionally difficult aspect of learning about evolution is that it can be both emotionally and personally uncomfortable for individuals to learn about (Watts 2019). Sinatra et al. offer anecdotal evidence presented by Richard Dawkins who has received numerous remarks about how his publications have left readers feeling as if "life was empty and purposeless" (Dawkins 1998). Yet, they also present studies which have shown that many people's beliefs about evolution are remarkably similar as many surveyed undergraduates reported believing that the acceptance of evolution causes people to become more racist and selfish while simultaneously diminishing their sense of purpose, self-determination, and spirituality (Brem et al. 2003).

According to Thagard and Findlay, the problem is that accepting Darwin's theory is not only in direct conflict with many people's religious and psychological beliefs but that it also conflicts with their deepest personal motivations (2009). It is not that individuals *think* that natural selection is false—but rather that they *want* it to be false (ibid). In fact, studies have even shown that those who do accept and understand evolutionary theory would prefer for it to be false (Brem et al. 2003).

With regards to emotional barriers, the foundation of these emotional issues appears to lie therein that natural selection leaves little room for divinity or a personal relationship with God. In other words, students' learning about natural selection is not a cognitive exercise about how scientists have been able to illuminate the naturalistic origins of our species but is instead an emotional choice between two competing value systems. As Thagard and Findlay state, "On the one hand, there is the familiar, reassuring religious picture that includes a caring God, immortality, free will, moral responsibility, and meaningful lives. On the other hand, there is the gloomy scientific picture of humans as specks in the vast universe, irrevocably doomed to die after a brief life devoid of freedom, morality, and purpose" (2009).

In this way, the emotional barriers to learning about natural selection are closely related to the (perceived) historical, social, and cultural conflicts discussed at the beginning of this chapter, whereby natural selection is closely linked to "survival of the fittest" and projected as a value system onto human society.

Conclusion: New Educational Approaches

While many educators continue to believe that students' lack of knowledge or understanding about evolution can be simply amended through the introduction of factual information, research has shown that students face multiple roadblocks including deeply entrenched misconceptions and emotional discomfort with topics that challenge pre-existing worldviews (Werth 2012). Due to the cognitive and emotional hinderances that present themselves within the context of formal science education, we must address the manner in which evolution and natural selection are communicated in these contexts. Barnes et al. in fact argue that it is *necessary* to address both the cognitive and cultural factors (such as family attitudes and religiosity), if instructors want to increase both understanding of natural selection and acceptance of evolution (2017a).

At a cognitive level it is important, for example, to address category mistakes and an individual's propensity for teleological thinking. Here Shtulman suggests that it is important in science education and communication to become more aware of the vocabulary we are using, as the word "adaptation" can be interpreted as adaptation of individuals rather than populations, and phylogenetic trees can be seen either as depictions of common ancestry or as depictions of linear evolution (2006). Here he points out that while such "ambiguous modes of discourse are unlikely to be the cause of evolutionary misconceptions, they facilitate the persistence of such misconceptions" (Shtulman 2006). It is therefore important that to address this cognitive hurdle that students are informed about their own intuitive cognitive biases, particularly teleological reasoning (Barnes et al. 2017a).

We must however recognize that the cognitive propensity to recognize order and assume purpose and causality (Carroll 2001; Zeigler 2008) is also engrained in many students' culturally as notions of progress are often linked to students' religious worldviews (Werth 2012). At a cultural and emotional level, it is important to realize that while the general concept of natural selection appears to be simple, research continues to show that facts are not automatically persuasive, particularly when they conflict with beliefs that are central to an individual's identity (Barnes et al. 2017b; Evans 2013).

While instructors must be careful not to attack students' beliefs which fall outside the boundaries of science (Werth 2012; Scott and Branch 2009; Lovely and Kondrick 2008; Meadows et al. 2000), it is important that the teachers are cognizant of the discomfort that this subject causes students and try to adjust their delivery of course material accordingly (Barnes et al. 2017a, b).

Lastly, as Thagard and Findlay point out, the fear that Darwinian explanations of the origin of humans predisposes us to be greedy, violent, and selfish, is based on a deep misunderstanding of evolutionary theory (2009), and educators should strive to teach natural selection in a manner that allows for an understanding of how this mechanism can also explain the origin of desirable human characteristics such as empathy and altruism (Sober and Wilson 1998; Thagard 2010).

References

Alexander J, Thompson K, Edles LD (2011) A contemporary introduction to sociology: culture and society in transition. Routledge, New York

Anderson DL, Fisher KM, Norman GJ (2002) Development and evaluation of the conceptual inventory of natural selection. J Res Sci Teach 39(10):952–978. https://doi.org/10.1002/tea.10053

Barnes ME, Evans EM, Hazel A, Brownell SE, Nesse RM (2017a) Teleological reasoning, not acceptance of evolution, impacts students' ability to learn natural selection. Evol Educ Outreach 10(1). https://doi.org/10.1186/s12052-017-0070-6

Barnes ME, Truong JM, Brownell SE (2017b) Experiences of Judeo-Christian students in undergraduate biology. CBE Life Sci Educ 16(1). https://doi.org/10.1187/cbe.16-04-0153

Berger KS (2008) The developing person through the life span. Worth

Birx HJ (2000) Nietzsche & evolution. Philosophy Now 29:24–25

Bishop BA, Anderson CW (1990) Student conceptions of natural selection and its role in evolution. J Res Sci Teach 27:415–427

Bowler PJ (1983) The eclipse of Darwinism: anti-Darwinian theories in the decades around 1900. John Hopkins University Press, Baltimore

Bowler PJ (2003) Evolution: the history of an idea, 3rd edn. University of California Press, Berkley, CA

Brem SK, Ranney M, Schindel J (2003) Perceived consequences of evolution: college students perceive negative personal and social impact in evolutionary theory. Sci Educ 87(2):181–206

Bunge M (2003) Emergence and convergence: qualitative novelty and the unity of knowledge. University of Toronto Press, Toronto
Carroll S (2001) Chance and necessity: the evolution of morphological complexity and diversity. Nature 409:1102–1109
Catley KM (2006) Darwin's missing link—a novel paradigm for evolution education. Sci Educ 90(5):767–783. https://doi.org/10.1002/sce.20152
Chi MTH (1992) Conceptual change within and across ontological categories: examples from learning and discovery in science. In: Giere R (ed) Cognitive models of science: Minnesota studies in the philosophy of science. University of Minnesota Press, Minneapolis, MN
Chi MTH (1997) Creativity: shifting across ontological categories flexibly. In: Ward TB, Smith SM, Vaid J (eds) Creative thought: an investigation of conceptual structures and processes. American Psychological Association, Washington, DC, pp 209–234
Chi MTH (2005) Commonsense conceptions of emergent processes: why some misconceptions are robust. J Learn Sci 14:161–199
Chi MTH (2008) Three types of conceptual change: belief revision, mental mode transformation, and categorical shift. In: Vosniadou S (ed) International handbook of research in conceptual change. Routledge, New York, pp 61–82
Claeys G (2000) The 'survival of the fittest' and the origins of social Darwinism. J Hist Ideas 61 (2):223–240
Cooper RA (2016) Natural selection as an emergent process: instructional implications. J Biol Educ 51(3):247–260. https://doi.org/10.1080/00219266.2016.1217905
Dagher ZR, Boujaoude S (2005) Students' perceptions of the nature of evolutionary theory. Sci Educ 89(3):378–391. https://doi.org/10.1002/sce.20054
Dawkins R (1998) Unweaving the rainbow: science, delusion, and the appetite for wonder. Houghton Mifflin, New York
Delisle RG (ed) (2017) The Darwinian tradition in context – research programs in evolutionary biology. Springer, Cham
Demastes SS, Settlage J, Good R (1995) Students' conceptions of natural selection and its role in evolution: cases of replication and comparison. J Res Sci Teach 32:535–550
Demastes SS, Good RG, Peebles P (1996) Patterns of conceptual change in evolution. J Res Sci Teach 33:407–431
Dixon T (2008) Science and religion: a very short introduction. Oxford University Press, Oxford
Dole JA, Sinatra GM (1998) Reconceptualizing change in the cognitive construction of knowledge. Educ Psychol 33(2/3):109–128
Dolter DM (2016) Effects of discourse on high school students' conceptual understanding of natural selection. Master's thesis, Point Loma Nazarene University, San Diego, CA
Duit R (1999) Conceptual change approaches in science education. In: Schnotz W, Vosniadou S, Carretero M (eds) New perspectives on conceptual change. Elsevier Science, Kidlington, Oxford, pp 263–282
Duit R, Treagust D (2003) Conceptual change: a powerful framework for improving science teaching and learning. Int J Sci Educ 25:671–688
Ehrman B (2005) Misquoting Jesus: the story behind who changed the Bible and why. HarperCollin, New York
Ehrman B (2009) Jesus, interrupted: revealing the hidden contradictions in the Bible. HaperCollins, New York
Evans EM (2013) Conceptual change and evolutionary biology: taking a developmental perspective. In: Vosniadou S (ed) International handbook of research on conceptual change. Routledge, New York, pp 220–239
Ferrari M, Chi MTH (1998) The nature of naive explanations of natural selection. Int J Sci Educ 20 (10):1231–1256. https://doi.org/10.1080/0950069980201005
Gould SJ (1977) Ontogeny and phylogeny. Harvard University Press, Cambridge, MA
Gould SJ (1992) Ever since Darwin: reflections in natural history. W. W. Norton, New York

Gould SJ (1996) Full House: the spread of excellence from Plato to Darwin. Harmony Books, New York
Gould SJ (1997) Nonoverlapping magisteria. Nat Hist 106(March):16–22
Gregory TR (2009) Understanding natural selection: essential concepts and common misconceptions. Evol Educ Outreach 2(2):156–175. https://doi.org/10.1007/s12052-009-0128-1
Hacking I (1975) The emergence of probability. Cambridge University Press, Cambridge
Ham K (2012) The lie: evolution (revised and expanded). Master Books, Green Forest, AR
Ham K (2013) Six days: the age of the earth and the decline of the Church. Master Books, Green Forest, AR
Hitler A (1925) Mein Kampf. München Zentralverlag der NSDAP., Frz. Eher Nachf., G.m.b.H
Hodgson J (2004) Social Darwinism in anglophone academic journals: a contribution to the history of the term. J Hist Sociol 17(4):428–463
Hokayem HFA, BouJaoude S (2008) College students perceptions of the theory of evolution. J Res Sci Teach 45:395–419
Hyers C (1984) The meaning of creation: genesis and modern science. Westminster John Knox Press
Jensen MS, Findley FN (1996) Changes in students' understanding of evolution resulting from different curricular and instructional strategies. J Res Sci Teach 33:879–900
Junker T, Hossfeld U (2009) Die Entdeckung der Evolution. Eine revolutionäre Theorie und ihre Geschichte, 2nd edn. WBG, Darmstadt
La Vergata A (1994) Evolution and Wars, 1871–1918. Nuncius 9(1):143–163
Lawson AE, Worsnop WA (1992) Learning about evolution and rejecting a belief in special creation: effects of reflective reasoning skill, prior knowledge, prior belief and religious commitment. J Res Sci Teach 29(2):143–166
Linnenbrink EA, Pintrich PR (2004) Role of affect in cognitive processing in academic contexts. In: Dai DY (ed) Motivation, emotion, and cognition: integrative perspectives on intellectual functioning and development. Lawrence Erlbaum Associates, Mahwah, NJ, pp 57–87
Lovely E, Kondrick L (2008) Teaching evolution: challenging religious preconceptions. Integr Comp Biol 48(2):164–174
Marx K (1861) Letter to F. Lassalle in Berlin. London, 16 January 1861. In: Lassalle F (ed) (pp 246–247). Marx & Engels Collected Works
Mayr E (1982) The growth of biological thought. Harvard University Press, Cambridge, MA
Mayr E (1997) This is biology. Harvard University Press, Cambridge, MA
Mayr E (2001) What evolution is. Basic Books, New York
Mazens K, Lautrey J (2003) Conceptual change in physics: children's naïve representations of sound. Cogn Dev 18:159–176
McCloskey M (1983) Naïve theories of motion. In: Gentner D, Stevens A (eds) Mental models. Lawrence Erlbaum, Hillsdale, NJ
Meadows L, Doster E, Jackson D (2000) Managing the conflict between evolution and religion. Am Biol Teach 62(2):102–107
Miller JD, Scott EC, Okamoto S (2006) Science communication. Public acceptance of evolution. Science 313(5788):765–766. https://doi.org/10.1126/science.1126746
Morris HM (1961) The genesis flood. Presbyterian and Reformed Publishing Company
Morris HM (1972) The remarkable birth of planet earth. Bethany House Publications
Morris HM (1974) Many infallible truths. Master Books, Green Forest, AR
Morris HM (1985) Scientific creationism. Master Books, Green Forest, AR
Morris HM (1989) The Long War Against God: the history and impact of the creation/evolution conflict. Master Books, Green Forest, AR
Morris HM (2001) Biblical creationism: what each book of the Bible teaches about creation and the flood. Master Books, Green Forest, AR
Nehm RH, Schonfeld IS (2008) Measuring knowledge of natural selection: a comparison of the CINS, an open-response instrument, and an oral interview. J Res Sci Teach 45(10):1131–1160. https://doi.org/10.1002/tea.20251

Nehm RH, Kim SY, Sheppard K (2009) Academic preparation in biology and advocacy for teaching evolution: biology versus non-biology teachers. Sci Educ 93(6):1122–1146

Newport F (2004) Third of Americans say evidence has supported Darwin's evolution theory. The Gallup Organization, Princeton, NJ

Nietzsche F (1882) Die fröhliche Wissenschaft. Chemnitz

Novick LR, Schreiber EG, Catley KM (2014) Deconstructing evolution education: the relationship between micro- and macroevolution. J Res Sci Teach 51(6):759–788. https://doi.org/10.1002/tea.21161

Numbers R (1992) The creationists. University of California Press, Berkeley and LA

Numbers R (1998) Darwinism comes to America. Harvard University Press, Cambridge

Numbers R (2006) The creationists: from scientific creationism to intelligent design. Harvard University Press, Expanded Edition

Paul D (1988) The selection of 'survival of the fittest'. J Hist Biol 21:411–424

Pintrich PR, Marx RW, Boyle RA (1993) Beyond cold conceptual change: the role of motivational beliefs and classroom contextual factors in the process of conceptual change. Rev Educ Res 63 (2):167–199. https://doi.org/10.3102/00346543063002167

Ramney M, Thanukos A (2009) Accepting evolution or creation in people, critters, plants, and classrooms: the maelstrom of American cognition about biological change. In: Taylor R, Ferrari M (eds) Evolution, epistemology, and science education. Routledge, Milton Park

Ratzinger J (1995) In the beginning: a Catholic understanding of the story of creation and the fall. Eerdmans

Richards RJ (2013) Was hitler a Darwinian? Disputed questions in the history of evolutionary theory. University of Chicago Press, Chicago

Ruse M (2001) The evolution wars: a guide to the debates. Rutgers University Press, New Brunswick, NJ

Scott EC (2009) Evolution vs. creationism: an introduction. Greenwood Press, Westport, CT

Scott EC, Branch G (2009) Don't call it "Darwinism". Evol Educ Outreach 2:90–94

Settlage J (1994) Conceptions of natural selection: a snapshot of the sense-making process. J Res Sci Teach 31:449–457

Shanahan T (2017) Selfish genes and lucky breaks: Richard Dawkins' and Stephen Jay Gould's divergent Darwinian agendas. In: Delisle RG (ed) The Darwinian tradition in context. Springer, Cham, pp 11–36

Shtulman A (2006) Qualitative differences between naive and scientific theories of evolution. Cogn Psychol 52(2):170–194. https://doi.org/10.1016/j.cogpsych.2005.10.001

Sinatra GM, Brem SK, Evans EM (2008) Changing minds? Implications of conceptual change for teaching and learning about biological evolution. Evol Educ Outreach 1(2):189–195. https://doi.org/10.1007/s12052-008-0037-8

Smith C, Carey S, Wiser M (1985) On differentiation: a case study of the development of the concepts of size, weight, and density. Cognition 21:177–237

Sober E, Wilson DS (1998) Unto others: the evolution and psychology of unselfish behavior. Harvard University Press, Cambridge, MA

Thagard P (2010) The brain and the meaning of life. Princeton University Press, Princeton, NJ

Thagard P, Findlay S (2009) Getting to Darwin: obstacles to accepting evolution by natural selection. Sci & Educ 19(6–8):625–636. https://doi.org/10.1007/s11191-009-9204-8

Vosniadou S, Brewer W (1992) Mental models of the earth: a study of conceptual change in childhood. Cogn Psychol 24:535–585

Watts E (2019) Neanderthals in the classroom. Taylor & Francis, Boca Raton, FL

Watts E, Hossfeld U, Tolstikova II, Levit GS (2016a) Beyond borders: on the influence of creationist movement on the educational landscape in the USA and Russia. Theory Biosci 136(1–2):31–48

Watts E, Levit GS, Hossfeld U (2016b) Science standards: the foundation of evolution education in the United States. Persp Sci 10:59–65

Werth A (2012) Avoiding the pitfall of progress and associated perils of evolutionary education. Evol Educ Outreach 5(2):249–265. https://doi.org/10.1007/s12052-012-0417-y
Wiser M (1988) The differentiation of heat and temperature: history of science and novice-expert shift. In: Strauss S (ed) Ontogeny, phylogeny, and historical development. Ablex, Norwood, NJ
Zeigler D (2008) The question of purpose. Evol Educ Outreach 1:44–45
Zuzovsky R (1994) Conceptualizing a teaching experience on the development of the idea of evolution: an epistemological approach to the education of science teachers. J Res Sci Teach 31:557–574

The manufacturer's authorised representative in the EU is Springer
Nature Customer Service Centre GmbH, Europaplatz 3, 69115 Heidelberg,
Germany. If you have any concerns regarding our products, please
contact ProductSafety@springernature.com

Printed and bound by CPI Group (UK) Ltd, Croydon, CR0 4YY
30/09/2026
02225508-0018